Quantum [Un]speakables

AF708243

Springer
*Berlin*
*Heidelberg*
*New York*
*Barcelona*
*Hong Kong*
*London*
*Milan*
*Paris*
*Tokyo*

**Physics and Astronomy** ONLINE LIBRARY

http://www.springer.de/phys/

R.A. Bertlmann A. Zeilinger

# Quantum [Un]speakables

## From Bell to Quantum Information

With 141 Figures, 4 in Color

Springer

Professor Dr. Reinhold A. Bertlmann
University of Vienna, Institute for Theoretical Physics
Boltzmanngasse 5, 1090 Vienna, Austria
e-mail: reinhold.bertlmann@univie.ac.at

Professor Dr. Anton Zeilinger
University of Vienna, Institute for Experimental Physics
Boltzmanngasse 5, 1090 Vienna, Austria
e-mail: anton.zeilinger@univie.ac.at

Library of Congress Cataloging-in-Publication Data
Quantum [un]speakables : from Bell to quantum information / [edited by] R.A. Bertlmann, A. Zeilinger. p. cm. Includes biblographical references and index.
1. Bell's theorem--Congresses. 3. Bell, J.S.--Congress. I. Bell, J.S. II. Bertlmann, Reinhold A. III. Zeilinger, Anton. QC174.17.B45 Q36 2002 530.12--dc21 2002021641

ISBN 978-3-642-07664-0

This work is subject to copyright. All rights are reserved, whether the whole or part of the material is concerned, specifically the rights of translation, reprinting, reuse of illustrations, recitation, broadcasting, reproduction on microfilm or in any other way, and storage in data banks. Duplication of this publication or parts thereof is permitted only under the provisions of the German Copyright Law of September 9, 1965, in its current version, and permission for use must always be obtained from Springer-Verlag. Violations are liable for prosecution under the German Copyright Law.

Springer-Verlag Berlin Heidelberg New York
a member of BertelsmannSpringer Science+Business Media GmbH

http://www.springer.de

© Springer-Verlag Berlin Heidelberg 2010
Printed in Germany

The use of general descriptive names, registered names, trademarks, etc. in this publication does not imply, even in the absence of a specific statement, that such names are exempt from the relevant protective laws and regulations and therefore free for general use.

Cover design: *design & production*, Heidelberg

Printed on acid-free paper

John Bell

©Renate Bertlmann, 1980

# Preface – From Bell to Quantum Information

John Stewart Bell was certainly one of the really outstanding scientists of the Twentieth Century. The theorem named after him was most aptly called by Henry Stapp the most profound discovery since Copernicus. Born in Belfast in 1928, his scientific career up to his untimely death in 1990 covered not only the foundations of quantum mechanics, but also the areas of quantum field theory and the physics of accelerators, where he made outstanding contributions. So, the conference organized in his honor at the University of Vienna, 10–14 November 2000, to commemorate the Tenth Anniversary of his death brought together friends and colleagues working in all three areas.

John Bell's scientific career began in Harwell and Malvern, where he made calculations of particle trajectories in accelerators and became an expert in high demand for the focusing of particle beams. During this time, he met his wife, Mary, who was also an accelerator physicist. We were very happy and grateful to have Mary Bell as the Guest of Honor at the Vienna Conference.

Later, while working on his thesis with Rudolf Peierls at Birmingham, he discovered independently from G. Lüders and W. Pauli the CPT theorem, one of the most basic theorems in physics. This is probably the most fundamental symmetry principle which states that the joint action of charge conjugation, inversion of parity and time reversal leaves any physical system unchanged.

Of his contributions to quantum field theory, two others could be mentioned. One is the discovery, together with Roman Jackiw, of what is today called the Bell–Jackiw–Adler anomaly. It explains some hitherto-not-understood pion decays by adding divergences of the quantum current to the description of the system. The other most important contribution of John Bell is his suggestion in 1967 that there should exist a gauge theory describing weak interactions. This idea had an immense influence on Martinus Veltman and Gerard 't Hooft and resulted in a full understanding of the weak force. Ultimately, gauge theories became the standard tool of modern particle physics.

While it is difficult to underestimate the importance of John Bell's contributions to accelerator physics and to field theory, his most important contributions were certainly in the foundations of quantum mechanics. Evidently, John Bell had been interested from a very early stage in the philosophical issues raised by quantum theory, a topic not very popular during his student days at Queen's University, Belfast. Apparently, John Bell, who had been interested in the Bohr–Einstein dialogue, always took the position of Albert Einstein on philosophical issues. He also felt that a completion of quantum mechanics using so-called "hidden variables" would be highly desired, as it would help to regain a realistic and objective picture of the world. That way, Bell hoped one would be able to arrive at a physics where "measurement" would not play such a central role as in the Copenhagen interpretation of quantum mechanics.

Then, a most interesting sequence of events set in. In 1952, David Bohm had achieved something which had earlier been proclaimed impossible. It had been proved by John von Neumann that no hidden variable theory could agree with quantum mechanics. Bohm actually formulated such a theory, where each particle at any time has both a well-defined position and a well-defined momentum. The conflict raised between von Neumann and Bohm was elegantly resolved by Bell, who showed that von Neumann's proof contained a physically unjustifiable assumption. So while John Bell had flung open the door widely for hidden variable theories, he immediately dealt them a major blow. In 1964, in his celebrated paper "On the Einstein–Podolsky–Rosen Paradox", he showed that any hidden variable theory, which obeys Einstein's requirement of locality, i.e. no influence travelling faster than the speed of light, would automatically be in conflict with quantum mechanics. This is the celebrated Bell's theorem, which even found entrance into comics.

Surprisingly, while quantum mechanics was already well established at the time of the publication of Bell's theorem, no experiments existed which definitely allowed one to rule out a local realistic interpretation. Which means that no experiment had measured for those specific two-particle correlations which are so necessary to demonstrate the violation of Bell's inequality, the quantitative measure of the border of the validity of local realism. Since then, experimentation has become better and better, and by now an impressive body of evidence has been collected, supporting quantum mechanics and being in conflict with local realism. While a very tiny loophole in principle remains for local realism, it is a very safe position to assume that quantum mechanics has definitely been shown to be the right theory. Thus, a very deep philosophical question, namely, whether or not events observed in the quantum world can be described by an underlying deterministic theory, has been answered by experiment, thanks to the momentous achievement of John Bell. This achievement is even more remarkable as he was able to rule out a gigantic class of theories, without having to know any details of the theories.

Today, experimental development has gone far beyond this. Bell's inequality and the underlying physics of entangled states have become cornerstones of the newly evolving technology of quantum information. There, information is encoded, transmitted and processed in completely novel ways based on quantum laws. A bit of information can be encoded in a quantum superposition. Two quantum bits or q-bits can be entangled over long distances, and entanglement can be used to encode information in a novel way unprecedented in classical physics. Experimentally, it has been shown that, using entanglement, one can encode more information than one bit in a two-state system, one can teleport a quantum state over large distances and one can use quantum entanglement to provide a cryptographic method which is secure against eavesdropping by the laws of physics and not by a trick of the experimentalist. Finally, quantum computation promises exponential speed-up for certain problems, and first steps have been taken in the direction of quantum computation in various laboratories all over the world.

It is most interesting to note that, despite his own theorem, John Bell continued to be an advocate of realistic hidden variable theories, which now, according to his own findings, have to be non-local. Therefore, he became an advocate of theories proposed by Ghirardi, Rimini, Weber, Peierls and Gisin, theories that, no matter how little, deviate from quantum mechanics and thus might be ruled out by experiment someday. Thus, even if John Bell in the end might not have turned out to be on the right side when it comes to the fundamentals of the interpretation of quantum mechanics, this should never be held against him. In contrast, he was one of the few, like Albert Einstein, who realized how extremely strange the consequences of quantum mechanics, if it finally should turn out to be the ultimate correct theory, are for our view of the world, and it is a sign of his high moral distinction that he was not at all willing to give in easily.

In conclusion, it might very well turn out in the future that Bell's theorem paved the way to a momentous change of our conception of the world.

Nobody has expressed his appreciation of John Bell, which we fully share, better than Jack Steinberger:
"One of the greatest privileges of my rewarding life in physics has been the contact with (...) John Bell. (...) Trying to learn (...) the behavior of neutral kaons in the light of CP violation, I had the pleasure of benefiting from John´s penetrating understanding and insight, and of his readiness to share this. Bell was among the most brilliant physicists I have known; in addition he had the very important human qualities of being accessible, unpretentious and kind to his less gifted colleagues."

The organization of the conference in Vienna would have been impossible without the financial support of the Austrian Bundesministerium für Bildung, Wissenschaft und Kultur, the Stadt Wien, Wissenschafts- und Forschungsförderung, the International Erwin Schrödinger Institut (ESI), the TMR-Network "The Physics of Quantum Information" and the University of Vienna. Many thanks also to Larissa Cox and Andrea Aglibut for handling the correspondence with the speakers and the editing of the present book, to Gabriele Zobl-Kratschmann and numerous other helpers for organizing the conference, to the Austrian Central Library for Physics for the video documentation and to Renate Bertlmann for taking a complete photographic record of the event.

Vienna,
January 2002

Reinhold Bertlmann
Anton Zeilinger

# Contents

**Part I John S. Bell – The Man**

**1 Some Reminiscences**
Mary Bell . . . 3

**2 John Bell in Belfast: Early Years and Education**
Andrew Whitaker . . . 7

2.1 Introduction . . . 7
2.2 Family and Early Life . . . 7
2.3 Queen's University as a Technician . . . 8
2.4 A Student at Queen's . . . 10
2.5 Early Ideas on Quantum Theory . . . 14
2.6 John Bell and Michael Faraday . . . 18

**3 My Interaction with John Bell**
Bernard d'Espagnat . . . 21

**4 Magic Moments: A Collaboration with John Bell**
Reinhold A. Bertlmann . . . 29

4.1 Prologue . . . 29
4.2 Duality in Hadronic Reactions . . . 30
4.2.1 Hadron Production in $e^+e^-$ Collisions . . . 30
4.2.2 Duality . . . 32
4.3 Nonrelativistic Potential Theory . . . 32
4.3.1 Résumé . . . 34
4.4 Magic Moments . . . 35
4.4.1 Moments . . . 37
4.4.2 Ground State . . . 38
4.4.3 Balance . . . 39
4.5 Equivalent Potential . . . 40
4.6 Epilogue . . . 42

## Part II Tests of Bell's Inequalities

**5 John S. Bell: Some Reminiscences and Reflections**
Abner Shimony ... 51

5.1 Why was Bell the Discoverer of Bell's Theorem? ... 51
5.2 Some Reminiscences ... 55
5.3 In What Direction Does Bell's Theorem Point? ... 57

**6 Early History of Bell's Theorem**
John F. Clauser ... 61

6.1 Introduction ... 61
6.2 An Unresolved Issue Left by the Founding Fathers ... 63
6.3 An Untidy Legacy Left by the Founders of Quantum Mechanics .. 66
6.4 Evangelical Theoreticians Dominate ... 69
6.5 The Development of a "Stigma" ... 71
6.6 Challenging the Common Wisdom ... 74
6.7 John Bell Unravels the Confusion ... 75
6.8 Theoretical Truth Versus Experimental Truth ... 77
6.9 Beware of the Experimentalists Lurking About ... 81
6.10 Generalization of the Bell and CHSH Results to Constrain Local Realism and Space-Time ... 82
6.11 Common Confusion About Count-Rate Normalization ... 87
6.12 Bell's Response to CH and "Local Beables" ... 88
6.13 The Quantum-Optics Community Encounters Related Problems .. 89
6.14 Splitting Photons? ... 91
6.15 Remaining Locality Loopholes ... 94
6.16 Conclusions ... 96

**7 On Four Decades of Interaction with John Bell**
Michael Horne ... 99

**8 Atom Based Tests of the Bell Inequalities – the Legacy of John Bell Continues ...**
Edward S. Fry and Thomas Walther ... 103

8.1 Historical Overview ... 103
8.2 The Bell–Clauser–Horne Inequality ... 106
8.3 Loopholes ... 107
8.3.1 The First Loophole ... 107
8.3.2 The Second Loophole ... 108
8.3.3 The Third Loophole ... 108
8.4 Atom Based Experiments ... 109
8.4.1 The Paris Experiments ... 109

8.4.2 The Boulder Experiments ... 110
8.4.3 The Texas A&M Experiment ... 111
8.5 Summary ... 116

**9 Bell's Theorem: The Naive View of an Experimentalist**
Alain Aspect ... 119

9.1 Introduction ... 119
9.2 Why Have Supplementary Parameters?
The Einstein–Podolsky–Rosen–Bohm Gedankenexperiment ... 121
9.2.1 Experimental Scheme ... 121
9.2.2 Correlations ... 122
9.2.3 Difficulty of an Image
Derived from the Formalism of Quantum Mechanics ... 123
9.2.4 Supplementary Parameters ... 124
9.3 Bell's Inequalities ... 125
9.3.1 Formalism ... 125
9.3.2 A (Naive) Example
of the Supplementary-Parameter Theory ... 126
9.3.3 The Inequalities ... 127
9.4 Conflict with Quantum Mechanics ... 128
9.4.1 Evidence ... 128
9.4.2 Maximum Conflict ... 129
9.5 Discussion: The Locality Condition ... 130
9.6 Gedankenexperiment with Variable Analyzers:
The Locality Condition as a Consequence of Einstein's Causality . 131
9.7 From Bell's Theorem to a Realistic Experiment ... 133
9.7.1 Experimentally Testing Bell's Inequalities ... 133
9.7.2 Sensitive Situations Are Rare ... 133
9.7.3 Production of Pairs of Photons in an EPR State ... 134
9.7.4 Realistic Experiment ... 135
9.7.5 Timing Conditions ... 136
9.8 First-Generation Experiments ... 136
9.8.1 Experiments with a One-Channel Polarizer ... 136
9.8.2 Results ... 137
9.9 Orsay Experiments (1980–1982) ... 138
9.9.1 The Source ... 138
9.9.2 Detection–Coincidence Counting ... 139
9.9.3 Experiment with One-Channel Polarizers ... 140
9.9.4 Experiment with Two-Channel Analyzers ... 141
9.9.5 Timing Experiment ... 143
9.10 Third Generation: Experiments with Pairs of Photons
Produced in Parametric Down Conversion ... 146
9.11 Conclusion ... 149

**10 Bell's Theorem for Space-Like Separation**
Gregor Weihs ... 155

10.1 From Gedanken to Real Experiments ... 155
10.2 Efficiency ... 156
10.2.1 Reducing the Required Efficiency ... 157
10.2.2 State of the Art and Development of Optical Experiments 158
10.2.3 A Proposal Using Mercury Atoms ... 158
10.2.4 Closure of the Detection Efficiency Loophole with an Ion Trap Experiment ... 159
10.3 Locality ... 159
10.3.1 Experiments with Large Separation Between Measurement Stations ... 160
10.3.2 Analyzer Switching ... 161

**11 The EPR Paradox in Massive Systems or about Strange Particles**
Reinhold A. Bertlmann, Walter Grimus and Beatrix C. Hiesmayr ... 163

11.1 Introduction ... 164
11.2 The Bell–CHSH Inequality for Photons and for Kaons ... 164
11.3 A Neutral Kaon Introduces Itself ... 167
11.4 The Experiment at CERN and Possible Decoherence ... 169
11.5 The Generalized Bell Inequality and Unitary Time Evolution ... 175
11.5.1 The Choice of the Strangeness Eigenstate ... 178
11.5.2 The Choice Sensitive to the $CP$ Violating Parameter, $\varepsilon$ .. 178
11.6 Connection of the Bell Inequality and the Decoherence Approach . 179
11.7 Final Remark ... 180

**Part III Quantum Information**

**12 Are There Measurements?**
Stig Stenholm ... 185

12.1 An Encounter with John Bell ... 185
12.2 Measurements or Not? ... 187
12.3 Describing a Measurement ... 189
12.4 How to Map Reality ... 193
12.5 Conclusion ... 196

**13 Sundays in a Quantum Engineer's Life**
Nicolas Gisin ... 199

13.1 I am a Quantum Engineer, but on Sundays I Have Principles ... 199
13.2 Quantum Cryptography on Sundays ... 199
13.3 Let's Assume That the Collapse Is Real ... 201

13.4 ... and Relativity? ..... 204
13.5 Conclusion ..... 206

**14 Secret Sides of Bell's Theorem**
Artur Ekert ..... 209

14.1 Is the Bell Theorem of any Practical Use? ..... 209
14.2 Is There a Perfect Cipher? ..... 210
14.3 Quantum Key Distribution ..... 213
14.4 Eavesdropping Revisited ..... 216
14.5 Quantum Privacy Amplification ..... 218
14.6 Concluding Remarks ..... 219

**15 An Impossible Necklace**
Lev Vaidman ..... 221

**16 Multi-Photon Entanglement and Quantum Non-Locality**
Jian-Wei Pan and Anton Zeilinger ..... 225

16.1 Introduction ..... 225
16.2 The GHZ Theorem ..... 226
16.3 Experimental Multi-Photon GHZ Entanglement ..... 229
16.4 Experimental Test of Quantum Non-Locality ..... 233
16.5 Discussions and Prospects ..... 237

**17 Bell's Theorem, Information and Quantum Physics**
Anton Zeilinger ..... 241

17.1 Introduction ..... 241
17.2 Information and Interference ..... 242
17.3 Information and Entanglement ..... 244
17.4 Bell's Theorem, Quantum Communication and Quantum Information ..... 246
17.4.1 Quantum Dense Coding ..... 246
17.4.2 Quantum Teleportation ..... 248
17.4.3 Teleportation of Entanglement ..... 249
17.4.4 Quantum Cryptography ..... 251
17.5 John Bell's Desiderata and the Interpretation of Quantum Mechanics ..... 252

**Part IV Quantum Ideas**

**18 The Geometry of the Quantum Paradoxes**
John Conway and Simon Kochen ..... 257

18.1 Introduction ..... 257
18.2 Old Paradoxes ..... 258

18.3 New Paradoxes and Their Geometry ... 262
18.4 Reconstructing Quantum Mechanics ... 267

**19 Whose Knowledge?**
N. David Mermin ... 271

**20 The History of the GHZ Paper**
Daniel M. Greenberger ... 281

**21 John Stewart Bell and the Dynamical Reduction Program**
GianCarlo Ghirardi ... 287

21.1 John Stewart Bell and His Role ... 288
21.2 His Role Before 1985 ... 289
21.3 The Birth of Dynamical Reduction Theories ... 291
21.3.1 Stochastic Equations for the Description of Decay Processes ... 291
21.3.2 Stochastic Differential Equations ... 291
21.3.3 The Breakthrough – Quantum Mechanics with Spontaneous Localizations ... 292
21.4 His Role Immediately After the Publication of Our Paper ... 293
21.4.1 The Letter ... 294
21.4.2 The Paper: Are There Quantum Jumps? ... 296
21.4.3 The Relativistic Issue ... 297
21.5 From QMSL to Continuous Spontaneous Localization (CSL) ... 298
21.6 The Subsequent Role of J.S. Bell in the Dynamical Reduction Program ... 301
21.6.1 Problems of Interpretation ... 301
21.6.2 The Relativistic Program ... 302
21.7 Concluding Remarks ... 303

**22 How Does God Play Dice? (Pre-)Determinism at the Planck Scale**
Gerard 't Hooft ... 307

22.1 Dice ... 307
22.2 Quantum Interference ... 314
22.3 Quantum Coherence and Entanglement ... 314
22.4 The Einstein–Rosen–Podolsky Paradox and the Violation of the Bell Inequalities ... 315

## Part V Quantum Specials

**23 John Bell, State Reduction, and Quanglement**
Roger Penrose ... 319

23.1 Encounters with John Bell ... 319
23.2 Gravitationally Induced State Reduction: A Minimalist View ... 321
23.3 Quanglement ... 326

**24 Interferometry with Macromolecules: Quantum Paradigms Tested in the Mesoscopic World**
Markus Arndt, Olaf Nairz, Anton Zeilinger ... 333

24.1 A Prototype Quantum Experiment ... 333
24.2 Interference of Fullerenes ... 336
24.3 Decoherence ... 340
24.3.1 Vibrational Transitions ... 342
24.3.2 Emission of Blackbody Radiation ... 343
24.3.3 Absorption of Blackbody Radiation ... 344
24.3.4 Rayleigh Scattering of Thermal Radiation ... 344
24.3.5 Fragmentation and Ionization ... 345
24.3.6 Influence of Collisions with the Residual Gas ... 345
24.3.7 Quasi-Static Interactions ... 346
24.3.8 Magnetic Interactions ... 346
24.3.9 Electric Interactions ... 347
24.3.10 Inertial Forces ... 347
24.3.11 Which-Way Information in Internal Clocks ... 348
24.4 Conclusion ... 349

**25 Towards More Quantum Complete Neutron Experiments**
Helmut Rauch ... 351

25.1 Introduction – Basic Relations ... 351
25.2 Classic Neutron Interference Experiments ... 355
25.2.1 Gravity Experiments ... 355
25.2.2 Neutron Fizeau Effect ... 356
25.2.3 $4\pi$ Spinor Symmetry ... 356
25.2.4 Spin Superposition ... 357
25.2.5 Neutron Josephson Effect ... 359
25.2.6 Stochastic Versus Deterministic Beam-Path Detection ... 360
25.3 Postselection Experiments ... 362
25.3.1 Postselection of Momentum States ... 362
25.3.2 Contrast Retrieval by Phase Echo ... 366
25.4 Phase-Space Coupling ... 366
25.5 Topological Effects ... 368
25.6 Discussion ... 370

## Part VI Particles and Fields

26 John Bell's Observations on the Chiral Anomaly and Some Properties of Its Descendants
Roman Jackiw ........ 377

26.1 John Bell and the Chiral Anomaly ........ 377
26.2 Descendants of the Anomaly ........ 379

27 Fractional Charge
R. Rajaraman ........ 383

27.1 Introduction ........ 383
27.2 Fractional Charge in Field Theory ........ 384
27.2.1 Vacuum Sector ........ 385
27.2.2 Soliton Sector ........ 387
27.3 Polyacetylene ........ 389
27.4 Eigenvalue or Expectation Value? ........ 391
27.4.1 Polyacetylene Re-Visited ........ 396
27.5 Conclusion ........ 397

28 Thermal Excitations of Accelerated Electrons
Jon Magne Leinaas ........ 401

28.1 Electrons as a Thermometer ........ 401
28.2 Linear Acceleration and the Unruh Effect ........ 403
28.3 Stationary World Lines ........ 406
28.4 Electrons in a Storage Ring ........ 407
28.5 Concluding Remarks ........ 411

29 Bell's Spaceships and Special Relativity
Franco Selleri ........ 413

29.1 The Generalised Transformations ........ 413
29.2 Michelson-Type Experiments ........ 416
29.3 Jupiter's Satellites Occultations ........ 417
29.4 Aberration ........ 419
29.5 Radar Ranging of Planets ........ 420
29.6 Proofs of Absolute Simultaneity ........ 421
29.7 The Inertial Transformations ........ 425

30 John Bell and the Ten Challenges of Subnuclear Physics
Antonino Zichichi ........ 429

30.1 Introduction ........ 429
30.2 The Remarkable Value of John Bell's Support to My Physics ........ 431
30.3 The LAA Project in a Few Words ........ 447
30.4 Facilities and the Basic Steps ........ 448

30.5 The Ten Challenges ........................................ 451
30.5.1 The Physics of Imaginary Masses (SSB) ................ 451
30.5.2 Matter–Antimatter Symmetry ........................... 454
30.5.3 Supersymmetry ........................................ 455
30.5.4 Non-perturbative QCD ................................. 457
30.5.5 Anomalies and Instantons ............................. 458
30.5.6 Flavour Mixing in the Quark Sector ................... 462
30.5.7 Flavour Mixing in the Leptonic Sector ................ 464
30.5.8 The Problem of the Missing Mass in the Universe ...... 466
30.5.9 The Problem of Hierarchy ............................. 468
30.5.10 The Physics at the Planck Scale: The Gap and the Number of Expanded Dimensions ............ 469
30.6 The ELN Project in a Few Words ............................ 471
30.7 Conclusions ................................................ 473

**Subject Index** ................................................ 479

# List of Contributors

**Markus Arndt**
Institute of Experimental Physics
University of Vienna
Vienna, Austria

**Alain Aspect**
Laboratoire Charles Fabry
Institut d'Optique Théorique et Appliquée
Orsay Cedex, France

**Mary Bell**
Geneva, Switzerland

**Reinhold A. Bertlmann**
Institut of Theoretical Physics
University of Vienna
Vienna, Austria

**John F. Clauser**
J.F. Clauser & Assoc.
Walnut Creek, CA, USA

**John Conway**
Department of Mathematics
Princeton University
Princeton, NS, USA

**Artur Ekert**
Centre for Quantum Computation
Clarendon Laboratory
Oxford University
Oxford, UK

**Bernard D'Espagnat**
Professeur émérite de l'Université de Paris-Sud, Membre de l'Institut
Paris, France

**Edward S. Fry**
Department of Physics
Texas A&M University
College Station, TX, USA

**GianCarlo Ghirardi**
Dipartimento di Fisica Teorica
Dell' Universita' degli Studi di
Trieste, Italy

**Nicolas Gisin**
Group of Applied Physics
University of Geneva
Geneva, Switzerland

**Daniel Greenberger**
Department of Physics
CCNY
New York, NY, USA

**Walter Grimus**
Institut of Theoretical Physics
University of Vienna
Vienna, Austria

**Beatrix C. Hiesmayr**
Institut of Theoretical Physics
University of Vienna
Vienna, Austria

**Gerard 't Hooft**
Institute for Theoretical Physics
University of Utrecht
Utrecht, The Netherlands

**Michael Horne**
Physics Department
Stonehill College
Easton, MA, USA

**Roman Jackiw**
Center for Theoretical Physics
Massachusetts Institute of Technology
Cambridge, MA, USA

**Simon Kochen**
Princeton University
Department of Mathematics
Princeton, NJ, USA

**Jon Magne Leinaas**
Department of Physics
University of Oslo
Blindern, Oslo, Norway

**N. David Mermin**
Laboratory of Atomic and Solid State Physics
Cornell University
Ithaca, NY, USA

**Olaf Nairz**
Institute of Experimental Physics
University of Vienna
Vienna, Austria

**Jian-Wei Pan**
Institute of Experimental Physics
University of Vienna
Vienna, Austria

**Roger Penrose**
The Mathematical Institute
University of Oxford
Oxford, UK

**R. Rajaraman**
School of Physical Sciences
Jawaharlal Nehru University
New Delhi, India

**Helmut Rauch**
Atominstitut der Österreichischen Universitäten
Vienna, Austria

**Franco Selleri**
Università di Bari
Dipartimento di Fisica
Bari, Italy

**Abner Shimony**
Cupertino, CA, USA

**Jack Steinberger**
CERN
Geneva, Switzerland

**Stig Stenholm**
Royal Institute of Technology Kansli
Stockholm, Sweden

**Lev Vaidman**
Faculty of Exact Sciences, Physics
Tel Aviv University
Tel Aviv, Israel

**Thomas Walther**
Department of Physics
Texas A&M University
College Station, TX, USA

**Gregor Weihs**
Institute of Experimental Physics
University of Vienna
Vienna, Austria

**Andrew Whitaker**
Department of Physics
Queen's University Belfast
Belfast, Northern Ireland

**Anton Zeilinger**
Institute of Experimental Physics
University of Vienna
Vienna, Austria

**Antonino Zichichi**
CERN
EP Division
Geneva, Switzerland

Part I

# John S. Bell – The Man

# 1 Some Reminiscences

Mary Bell

I have been asked to talk about things that other people do not know. So much has been written about John that this is rather difficult. So I will restrict myself to a few personal recollections. I hope that it is not too much of a hotchpotch.

I first saw John, without beard, in 1949 when he came to AERE (Atomic Energy Research Establishment) at Harwell, Berks., where I had worked on reactors since 1947. After a very short time he was sent to Malvern, Worcs., to join the small group of W. Walkinshaw. This group was devoted to the theory of particle accelerators. During World War II the radar establishment TRE (Telecommunications Research Establishment) was based there after a stay at Swanage in the South of England. At the end of the war a natural development for some members of the staff was to move into the field of accelerators. (J.B. Adams, a future director general of CERN, was one of these). The work at that time was on disc-loaded linear electron accelerators, some for medical use. I was also sent to Malvern in 1950, and I spent a year there. Most of John's accelerator papers were on these electron linacs. They included a theoretical study of dielectric-loaded waveguides. As they were all internal reports they were not considered as published papers. The World Scientific volume lists all papers, but, as the book would have been too large if all had been included, only one is there. I selected it as I knew that John was well pleased with it. Indeed, he did submit it to a magazine. One referee said it was too short, the other that it was too long. He gave up then as he was leaving the accelerator field and was short of time. It is amusing that many years later people were still making the same mistake which his paper had pointed out. The errors always arose because approximations were made which did not obey Hamiltonian dynamics.

At Malvern John lived in Geraldine Road hostel and was one of the group of young men all with motor cycles which they took to pieces regularly.

Most of TRE moved gradually to Harwell during 1951, and the Walkinshaw group was one of those forming the Theory Division there. With the setting-up of CERN, interest turned to proton accelerators. John contributed there also, and attended one of the early CERN meetings in Geneva.

When the David Bohm papers appeared in 1952, I can remember how excited he was. In his own words, "the papers were for me a revelation". When he had digested them, he gave a talk about them to the Theory Division.

There were interruptions, of course, from Franz Mandl, with whom he had had many fierce arguments.

In 1953, while still paid by Harwell, he was selected to spend some time at a University. He was the second person to have that privilege – the first was Brian Flowers, now Lord Flowers. I have no idea if the scheme continued. His aim was to study field theory. He considered Glasgow where J. Gunn was professor, but eventually chose Birmingham where Rudolf Peierls was chief of Theoretical Physics. There he saw little of Peierls but the late P.T. Matthews was the person of whom he saw most. In May 1954 we married. It was convenient that we were both vegetarians. In the autumn he returned to Harwell where he joined the group of the late Tony Skyrme. Many years later, as most of you will know, the word 'Skyrmions' became popular. At Harwell, he completed his thesis for Birmingham. Most of it was typed by me in the evenings. It had two sections – *Time Reversal in Field Theory* and *Some Functional Methods in Field Theory.* The first section, which was written while he was at Birmingham, contained the CPT theorem.

For a number of years he worked on nuclear physics in the Skyrme group. Other members were John Perring, Roger Phillips, Euan Squires, Tony Lane, and, of course, Franz Mandl.

In November 1960, we came to CERN, John to the Theory Division and I to the Accelerator Research group. John's aim was to move into High Energy Physics. Among his collaborators, who are present at this meeting, were J. Steinberger, D. Sutherland, R. Jackiw, E. de Rafael, R. Bertlmann, J. Leinaas, and R. Rajaraman.

At the instigation of Sam Berman, who had been a visitor at CERN, we went to SLAC at the end of 1963 (we left Geneva on the morning when the news of John F. Kennedy's assassination had come through) and remained there until the end of 1964. There John had time to work on his "hobby", and the well-known paper with the inequality was written. The preprint appeared quickly. For a number of reasons, among them the inefficiency of the SLAC post office, it was quite a time before the published paper appeared in a short-lived magazine. At first, there was very little reaction; if any, it was negative, but gradually more and more references were made to it. It was even mentioned in the Tin-Tin comic strips, and, after the experiments of Clauser et al., Fry et al. and later of A. Aspect, there was a fair amount of interest.

In the early eighties, although normally in the PS division, I joined temporarily the small group of Frank Krienen to work on electron cooling. John took an interest and we collaborated on a few theoretical papers. When the project was finished, I was asked by Frank to prepare the big paper for publication. This involved a lot of condensing to make the paper a reasonable size, and John kindly helped me. He particularly liked to be acknowledged at the end for "helping with the typing". I should like here to emphasise that he was just as pleased with his accelerator and high energy papers as with the quantum mechanics papers.

I will now add a few personal details.

All his life he suffered from migraine attacks. Sometimes they lasted a week. Although they nearly disappeared for a number of years, towards the end of his life he had many short attacks.

As some of you know, he liked things expressed very exactly. On one occasion, a visiting theorist, a good friend, had sent him a paper for comments. He told me, fortunately while laughing, that John, when he questioned him, had replied, with apologies, that he had found the paper "too vague to comment on".

He always liked the theory of anything to be well understood. As a boy he studied "Every Boy and Girl a Swimmer", and then followed the instructions. Similarly, he learned ballroom dancing from the Victor Sylvester book, and, on arrival in Geneva, liked very much the French skiing book, which was full of dynamics. When at Belfast Technical College he had a theoretical course in bricklaying, but, as far as I know, no practical experience.

Although an atheist for most of his life, while at Queen's University he had many discussions with a Catholic friend, Denis McConalogue, about the devil, and even attended some meetings of the Student Christian Movement for the sake of argument.

I will finish by quoting a passage which he had marked of special interest in some writings of Schrödinger:

"Whence come I and whither go I? That is the great unfathomable question, the same for every one of us. Science has no answer to it".[1]

[1] Taken from 'Quantum Questions' ed. by Ken Wilber, New Science Library (Shambhala Publications, London 1984).

# 2 John Bell in Belfast: Early Years and Education

Andrew Whitaker

## 2.1 Introduction

In this paper I shall sketch an account of John Bell's upbringing in Belfast, his family influences and his education [1]. As well as giving a general description, I shall attempt to see if there may be anything in his upbringing that can explain in any way the direction his career took. Of course there can and should be no attempt to explain genius, but we may wonder why his particular genius took him towards analysis of the structure and interpretation of the quantum theory, an area of physics almost completely disregarded at the time he took it up.

## 2.2 Family and Early Life

John Stewart Bell was born on 28th July 1928 in Tate's Avenue, Belfast. His parents Annie and John already had a daughter Ruby, and later two more sons were born – David and Robert. In 1936, the family took the opportunity to move along Tate's Avenue, from the city end, not too far from Queen's University, but rather dominated by the railway and railway bridge, to the more rural end, which in the 1920s and 30s was surrounded by fields. By the year 2000 it is surrounded instead by a link-road, a motorway and out-of-town shopping, but before the war, and with the hills in the distance, it must have been a pleasant area to grow up in.

The area was staunchly Protestant. Though the families of both Bell's parents had lived in the area for several generations, Annie's had originally come from Scotland; her actual name is Brownlee, but the 'Stewart' in Bell's name was her original family name, and, of course, it resonates entirely of Scotland. John Bell was, in fact, called 'Stewart' until he left home for University. The Bell family were good Protestants themselves, but it was possible for them to have close friends across the religious divide, and Annie knitted First Communion garments for several Catholic friends.

In the 1930s, of course, times were tough. Tate's Avenue was not an affluent area, most of the people surviving on jobs like small-scale trading, labouring and factory work. In this period, John Bell Senior, again like most people in the area, was in and out of work. When the war came, it brought mixed fortunes to the city and in particular to the Bell family. There was substantial bombing, and Bell's father, like many others, spent most of the war away in the army. But at least soldiering was a regular job with a regular income, and the war brought considerable prosperity to the city. When he returned at the end of the war, he was appointed to a responsible job managing the British Legion car parks in the city.

However it was probably Annie Bell who had much the greater influence on the children. She very clearly saw education as the route to prosperity; as she told them, she wanted them to be able to wear their Sunday suits every day of the week. Even so, in these days before the post-war education reforms, clearly the difficulties were enormous. John Bell was the only one of the children to move on a secondary school at the age of 11, or indeed to stay at school much beyond the age of 14. That lack of ability was certainly not the problem was clear when David, who declined to take the required exam to progress further at 11, subsequently became an electrical engineer by evening study, and concluded his career lecturing at Lambton College in Canada and publishing several successful textbooks. Robert developed a sound retail business, while Ruby has also had a successful life, chiefly as wife and mother.

John Bell himself needed no encouragement towards education. From his earliest school days at Ulsterville Avenue and Fane Street Schools he was highly successful. Outside school he read voraciously, and indeed at home he was known as "The Prof" because he had amassed huge quantities of information on which he would freely expound. Even at the age of 11 he announced that he wanted to be a scientist. He passed his qualifying exam for grammar school with ease, but in those days the costs of attending one of the more prestigious Belfast schools were still prohibitive for children from not exceptionally privileged backgrounds. Fortunately, enough money was found to enable him to attend the Belfast Technical High School, which was less prestigious, but certainly suited Bell exceptionally well. Practical courses, which he enjoyed, were mixed with a full academic curriculum, which allowed the brightest students to progress to University. In fact he had the required qualifications while still only 16, a year younger than the minimum age for admission to Queen's University.

I will briefly comment on one facet of his life at this time, which certainly shows a very developed moral sensibility – his vegetarianism. From early teenage years he wished to be a vegetarian – at Christmas time, for example, when the turkey was being cooked, he would remark that he smelled a corpse burning. However, when you are living in somebody else's house, and they are non-vegetarian, it is difficult to be consistent, and it is probably fair to say that it was not till he left Belfast that he was able to embrace the life-style fully.

## 2.3 Queen's University as a Technician

Bell spent the year after he left school as a technician in the Physics laboratories at Queen's, the very laboratories he would return to the following year as a student. After 55 years or so, the precise sequence of events is difficult to unravel. There is, in fact, a long-standing tradition, perhaps myth, in the Department of Physics at Queen's that finance prohibited Bell becoming a student, and that the job as a technician was intended to be permanent; further that the Physics staff at the time, Professor Emeleus and Dr. Sloane, recognised Bell's abilities, obtained financial support for him, and thus set him on the way to his great career achievements.

Other accounts imply that entry as a student once Bell was 17 was already arranged, and that the year was always intended to be just a filling-in year (though certainly also providing some money which Bell saved for his student years). Probably the truth lies in the middle, that Bell certainly hoped and even perhaps expected to change from technician to student, but financial details required clarification. In particular a relatively small grant obtained from the Co-operative Society appears to have made a crucial contribution to the decision that Bell should enroll as a student. What is certainly true is that Emeleus and Sloane recognised Bell's gifts and helped him in every way possible, both in his year as a technician and subsequently. In particular, they allowed him, even when he was a technician, to attend the first-year lectures, and lent him books.

There may be more to this than meets the eye, and a plausible reconstruction may be as follows. At Queen's at that time there were two levels of entrance. The bulk of students entered at what I will call Level 0, and took three years to obtain a pass degree, four to obtain Honours. However a few students who had particularly high qualifications might enter at Level 1, and proceed to an Honours degree in three years. The type of qualification Bell would have been able to obtain at the Technical High School would only have allowed him to enter at Level 0 (despite the fact that, of course, his performance in the examinations would have been exceptional). His lecture attendance while he was a technician may then have excused him from Level 0 Physics, and towards the end of that year he was able to take a higher level mathematics examination to qualify him for Level 1 in that discipline. Actually when he enrolled as a student, he was entered for the lower Level 0 in his third subject, Mathematical Physics (strangely for somebody who would be one of the best mathematical physicists of the century), but at the end of the year he was able to jump to Level 2 in this topic as well. Thus Emeleus and Sloane may have helped him to save a year of his studies, very important, of course when finance was crucial. To put it another way, they may have enabled him to undertake an extra year of studies at the end of his course, as will shortly be explained.

It may be interesting to wonder whether this year as a technician, following on from his education in the technical school, had any effect on Bell's approach to physics later in his career. It might be suggested (a Marxist analysis might certainly point in this direction) that the understanding of and concentration on experimental aspects of physics in this formative period, played a crucial part in stimulating his later awareness of the experimental consequences of his theoretical ideas. Again it could be argued that, both in quantum mechanics and elsewhere, this awareness was very much to his eventual benefit. Clearly his famous quantum papers were very close to suggesting actual performable experiments; towards both the beginning and the end of his career he did excellent work on very practical problems of accelerator design; and even his particle physics remained closer to experimental fact and possibility than that of at least some other workers.

## 2.4 A Student at Queen's

John Bell now entered Queen's University as a physics student. It was an intimate department that he joined, the number of students graduating in Physics with Honours each year being around five, with two or three others in Mathematical Physics. A similar number would obtain Pass degrees specialising mainly in Physics. There would, though, be large classes at lower levels, where scientists, engineers and medical students would all do some elementary physics including physics laboratory work.

On the staffing side, while there were a couple of assistant lecturers and demonstrators, temporary posts usually occupied by graduates of Queen's working for PhDs, certainly the powerful personalities were Emeleus and Sloane. Professor Karl George Emeleus was born in 1900. His family was originally Huguenot, but his father was a Swedish-speaking Finn, who emigrated to the United States, and later to Sussex in England, where he worked as a pharmacist and where Emeleus was brought up. Karl Emeleus studied physics at Cambridge and was working at the Cavendish Laboratory in Cambridge in the early 1920s, coming to Belfast as a Lecturer in 1927, and being promoted to Professor in 1933. He was originally an atomic physicist, though becoming increasingly involved with conduction of electricity in gases, applied physics and electronics. He was an excellent scientist, publishing widely and becoming an MRIA – Member of the Royal Irish Academy, though he was disappointed not to become a Fellow of the Royal Society, unlike his brother, Harry Julius Emeleus, who was Professor of Inorganic Chemistry at Cambridge University from 1945. Karl Emeleus retired in 1966 but continued publishing around three papers a year on the physics of electronic devices right up to his death in 1989.

Richard Sloane was a protégé of Emeleus and became a Lecturer in 1939. He was also an atomic physicist, and had a successful career, being promoted to Reader (Associate Professor as it might be called elsewhere, a grade of which there were only a handful of holders at Queen's) in 1948, and in the same year becoming an MRIA, a striking achievement for one who was not a professor. The point I wish to stress is that while both men were extremely good physicists, and Emeleus in particular had an exceptionally broad knowledge of his subject, the situation was very far removed from one where each area of physics was taught by an expert in that area. In particular, obviously neither man could have been regarded as extremely knowledgeable in modern quantum theory; Emeleus would have been self-taught, while in turn he would have taught Sloane.

To Bell there may have appeared a number of lessons. Most important may have been the realisation that it was possible and desirable to obtain a thorough knowledge across all the areas of physics. This certainly stood Bell in good stead through his career. His deep understanding of electromagnetism helped him in his early Harwell work on accelerator beams and strong focussing; it is interesting that only three years after leaving Queen's with just a BSc, he was a consultant on the Proton Synchrotron project at CERN. A broader knowledge of classical physics was essential for the later accelerator work, particularly thermodynamics for his discussion of the Unruh effect. Of course, for much of the rest of his work he needed quantum theory, quantum field theory, nuclear physics and what there was at the time of elementary particle physics. He also showed a knowledge and understanding in depth of relativity and its history in his discussion on how it should be taught; he suggested that following and understanding the Lorentz–Poincaré route was a very useful supplement to the usual Einstein approach.

Just as important for Bell must have been the understanding of fallibility. However able Emeleus and Sloane were, there must have been areas of physics where they were vulnerable to the alert questioner. And the area where this must have been most true was, of course, quantum theory, which I shall discuss in detail later.

Bell entered Queen's in 1945, and for a little background on his life at that time I turn to a memoir of *Dr. Lesley Kerr* [2], who was one year behind Bell academically, but a good friend at the time and for a few years afterwards during the Malvern/Harwell years, though they met very infrequently once the Bells had moved to Geneva. Kerr remembers going to "hops" at the Students' Union with a small group including Bell, and also, on one occasion, walking with the same group in the hills surrounding Belfast. On the latter occasion he remembers coming across a number of bowl-shaped depressions perhaps 10 feet deep and 20 feet in diameter. In one of them there were several boulders, mostly at the bottom but a few higher up. Bell announced that they were in a Boltzmann distribution and in fact became somewhat heated when Kerr gave his own opinion – that there was no reason why they should be in a Boltzmann distribution, even in an ensemble of such craters.

On the same walk Kerr remembers discussion of quantum theory, and the problems of wave/particle duality and two-slit interference – certainly not, as Kerr points out, the EPR work, which had been produced ten years before, but had not received much general attention away from the specialists. Kerr recognised that Bell had already thought deeply about the problems of quantum theory, indeed seemingly much along the lines he would follow in his crucial papers fifteen years later. Bell became somewhat irritated when the conversation went round in a circle returning to points on which he had already expounded. But again I will come back to quantum theory later.

In *Jeremy Bernstein*'s article in the Bibliography [1], Bell talks about a rather brief interest in philosophy, mainly Greek philosophy, during his teenage years, and Kerr found him interested in, and well able to argue on important philosophical matters. This was the heyday of logical positivism, at least in Britain with the predominance of A.J. Ayer. Bell was at first a great upholder of this philosophy, which he felt had made all the great problems of philosophy irrelevant, but he then changed his mind, as he became convinced that logical positivism must lead to solipsism. For Kerr, Bell's explanation was the first refutation of logical positivism he had heard, and he says that, like most of Bell's ideas, it was beautifully simple and convincing.

On the academic front, Bell found few if any problems, and graduated on July 9th 1948 with First-Class Honours in Experimental Physics. Of the seven Honours graduates in Experimental Physics and Mathematical Physics, actually six obtained Firsts. One is reminded that it was a small institution; the total number graduating at all levels and in all disciplines was 357, with a further 100 or so graduating in December. Incidentally well over a third were medical students – Queen's had been described in the 1930s as a medical college with a University attached.

Because he had already taken an advanced Applied Maths course, Bell was in the position of being able to complete a second degree – in Mathematical Physics – by one year's further study, and this he did, graduating, again with First-Class Honours, in 1949. His main lecturer here could hardly have been less like the conscientious and helpful, but rather dignified Emeleus and Sloane. Peter Paul Ewald, who had been born in Berlin in 1888, was an exceptionally famous figure from the very start of X-ray crystallography. It was his dynamical theory of X-ray interference that had inspired von Laue's original ideas of 1912, and, being unimpressed by the lack of rigour of Bragg's useful but rather naive theory, he responded by producing his own more satisfactory version, inventing, on the way, the reciprocal lattice, central in all subsequent work in the theory of solids, and the sphere of reflection or "Ewald sphere". He had to leave Germany in 1937, being, as Bell put it, washed up on the shores of Ireland as a refugee from the Nazis.

He came to Queen's as a lecturer in 1939. It seems that the appointment committee told him that although the advertised position was as a Lecturer, there was no doubt that such a distinguished scientist as himself would immediately gain promotion to a chair. Unfortunately the University then brought in a regulation that, for the duration of the war, there would be no promotions, so he did not, in fact, become a Professor until 1945. He left Queen's in 1949 to become Chairman of the Physics Department at the Brooklyn Polytechnic University (now the New York Polytechnic University), so Bell was one of the last of his students in Belfast. Ewald spent the remainder of his life in the United States, becoming a Fellow of the Royal Society in 1958 at the age of 70, and dying in 1985 at the age of 97.

Ewald was notably un-stuffy, and would talk to Bell about anything, including, as is mentioned by *Bernstein* [1], saying that his assistant was mad. On the academic front, during this year, under Ewald's supervision, Bell wrote an essay on the quantum mechanics of long chain molecules. Ewald was extremely impressed by this, and there was talk of it being published, either under Bell's name only or as a joint publication with Ewald, though it seems that nothing came of this.

At the conclusion of his undergraduate studies, Bell would certainly have liked to continue studying for a PhD, Ewald suggesting that the natural choice of supervisor was Rudolf Peierls in Birmingham. At the time this was certainly the leading department in the UK for what might be called (relatively) down-to-earth theoretical physics, as distinct from the more mathematical physics that might be studied at Cambridge. Another possibility that was mentioned was University College London, with which physics at Queen's had good links. However Bell already felt a little guilty about what he felt of as living off his family. After his first year, he had received, in fact, a Queen's scholarship of £60 per year, and a Belfast City scholarship of £20 per year, a total of £80, perhaps around £2500 in today's terms. This was probably enough to live on, but not enough to make the contribution at home that he would have liked. Conceivably if Ewald had stayed in Belfast there might have been a possibility for him to have stayed at home and made his contribution, but of course that was ruled out as well by Ewald's departure.

So he started his career without a PhD in the Scientific Civil Service. He was clearly a little concerned about whether his progress might be limited, as he asked Ewald's opinion on the matter, but he was very cheered by the answer – that it did not matter where he started 'as long as he had elbows'. As everybody knows, not, or at least hardly at all, by the use of elbows but by ability and diligence, John Bell's progress was certainly not limited. He was to get the chance to work with Peierls, he was to obtain a PhD, and of course he achieved far more as well.

It seems fair to say that Bell had profited from his time in Belfast. He had been supported both emotionally and, to some extent, financially by his family, and he had been taught well at various schools. He had received a sound general education in physics from Emeleus in particular, and from Sloane and Ewald, and as I shall shortly mention, almost serendipitously Sloane's attempts to defend the orthodox in quantum theory may have been very useful for Bell as well. Through the rest of his life, he visited his family in Belfast regularly of course and gave several invited talks at the University, and in 1988 he was awarded an Honorary DSc. (In a period of days, he got honorary degrees from Queen's and from Trinity College Dublin.) It is particularly interesting, though again very typical of Bell, that when in the 1980s, Lesley Kerr and others wrote round to former students of Emeleus requesting contributions to an Emeleus Prize for the best physics student each year, the highest contribution came from John Bell.

## 2.5 Early Ideas on Quantum Theory

I now want to discuss Bell's views on quantum theory during his time at Queen's, using as evidence, apart from the rather general remarks of Lesley Kerr, the interviews of Bell by *Jeremy Bernstein* [1]. These took place, of course, around 1990, over forty years after the events described, but there seems no reason not to consider them, at the very least, as reasonably accurate accounts of his ideas at the time.

The most memorable passage in Bernstein's account is of the clash between Bell and Sloane. Earlier in the account Bell had already indicated his distaste for Sloane's course on atomic spectra. It may be worth clarifying the content of the various courses. Emeleus gave what Lesley Kerr calls an excellent course on theoretical physics – quantum theory, but also statistical mechanics, thermodynamics and so on. Sloane then gave a course of which half was on atomic spectra, the other half on radio-frequency spectroscopy one year and accelerators the other year. The atomic spectra part of the course started from the vector model and considered all the various types of atom in great detail. It related to a major interest in atomic physics at Queen's, but Bell was not alone in finding it somewhat tedious. He, of course, would have liked to learn more of the philosophy of quantum theory. (It may be fair to Sloane to mention that in the late 1940s, the topics of radio-frequency spectroscopy and accelerators were very new, and it is interesting to see them appearing at the undergraduate level.)

The actual struggle with Sloane, culminating in Bell's charge of dishonesty, may be discussed in various ways. Obviously it shows Bell's passionate concern, even at that time, with quantum theory. Obviously it also shows that Bell could, at that time, be excitable in discussion of physical principle. It is ironic that it was Sloane who bore the brunt of Bell's tongue. For one thing, it had been Sloane in particular who had supported and encouraged Bell, but also, in addition to being a good physicist, and almost more important to Sloane, he was a man of great rectitude, and to be accused of dishonesty must have been rather unpalatable.

But one can obtain much more detailed information from the account of the disagreement. It centered round the Heisenberg principle, and it is very clear that the whole discussion was in the context of the Bohr approach, and did not intimately concern the idea of measurement, at least of a preliminary measurement or state preparation. This would be very typical at least for UK physics courses at the time, and probably for twenty years or so afterwards, and I suspect it would not be restricted to the UK. In this approach, the Born postulate tended to mean primarily probability density, and even then not taken very definitely as referring to a *measurement* of position. While the accuracy of simultaneous measurements was certainly mentioned, it was not made very clear how that related to the state of the measured system either before or after the measurement. Indeed measurement was often treated as a semi-qualitative problem (qualitative apart from the use of the Heisenberg principle) not closely connected with the mathematical formalism of the rest of the quantum theory. (These fairly general comments may be broadly substantiated by reference to books such as that of *Schiff* [3], first published in the late 1940s, and a book used at Queen's.)

Of course what is often called the von Neumann approach with its attention to measurement, repeated measurement, collapse of wave-function or projection postulate was available, appearing, for example, in Dirac's book, the first edition of which was published in 1930. But in general I suspect that most general discussions did not follow that line; it was felt that everything could be explained by following Bohr in an entirely verbal sense, provided one could follow him closely enough, though that, it would have been admitted, was not easy.

In particular, though the Heisenberg principle clearly related the values of $\Delta x$ and $\Delta p$ obtained in a measurement, and Bohr's "parables" as Bell called them, could maybe be used to justify this point, it was left totally unclear what each of $\Delta x$ and $\Delta p$ should be. And to go one step further back, if there were no measurement, and one was just considering the values of $\Delta x$ and $\Delta x$ for a particular system, one could easily get the impression that it was almost purely a subjective matter. One could *choose*, for example to think of one or the other as well defined, and the other would then be totally undefined.

It was roughly at this (rather incoherent) point that the discussion with Sloane must have started. Sloane, it must be remembered, would have been far from an expert in quantum theory himself, but would obviously have learned the standard mathematical developments and the usual Bohrian type of arguments. Doubtless he was perfectly capable of presenting the formalism nicely, and solving the simple conceptual difficulties of the average student. But equally doubtless he was completely incapable of answering the testing questions of John Bell, or even of responding to his worries or complaints in a logical way, even, one might say, in an intellectually respectable way. Hence the charge of "dishonesty". Whatever the merits of Bohr's approach, it did not really facilitate answering awkward questions; it was better at giving verbally dexterous accounts of why they could not be answered.

Sloane was clearly not verbally dexterous enough, and this may have turned out to be a massive great fortune for Bell. Not only did he have a chance to produce and try out some of his arguments about Bohr's Copenhagen interpretation, he could see that these arguments were not answerable by an intelligent and reasonably knowledgeable man claiming to understand the orthodox point of view. This certainly would have encouraged Bell to continue to think out his own position, and to dream of one day clarifying the whole area.

Others may also have been sceptical of Copenhagen, but may not perhaps have generated the self-confidence that they might be able to put things right. One might contrast Bell's position with that of another highly intelligent physicist who also had considerable doubts concerning what he was told to believe about quantum theory. The late Euan Squires was to do important work on nuclear physics with Bell at Harwell in the late 1950s, and in the late 1980s he was to change his field with good effect from particle physics to the foundations of quantum theory. In the earlier 1950s he has told how he was a research student at Manchester University, where Leon Rosenfeld, Bohr's acolyte and later successor as standard-bearer of Copenhagen, was Professor. At Christmas, *Squires* [4] says, the students sang (to the tune of "*The boar's head in hand bear I*")

*At Bohr's feet I lay me down*
*For I have no theories of my own.*
*His principles perplex my mind*
*But he is so very kind.*
*Correspondence is my cry; I don't know why; I don't know why.*

But as Squires commented, they were too afraid to ask. And one might add that, if they had asked, Rosenfeld or, behind him, Bohr would certainly have been able to produce an answer, not necessarily specific enough to enlighten them, but certainly erudite enough to suggest to them that the limitation was purely their own, not anything lacking in the Bohr's complementarity framework itself. It is arguable that it may have been to Bell's great long-term advantage that Sloane, his opponent in this type of argument, was less able to win by sheer force of wordpower.

The Sloane argument had convinced Bell of the central importance of the concept of measurement in any meaningful discussion of quantum theory as it was practiced at the time. It was necessary first to identify this as a central element, in order to be able to move on to criticise it, by pointing out the illegitimacy of using "measurement" as a primary term; the process of "measurement", Bell would argue, should surely be reducible to perfectly normal physical processes, rather than being treated as a special type of process with its own rules. This argument was to become, of course, a major component of the notion of "unspeakability", culminating in the Against "measurement" paper of 1990 [5].

At the time the aspect of this which most concerned Bell was the divide or "Heisenberg cut" between quantum and classical. He fully accepted Bohr's acknowledgement that the apparatus must be treated classically, but definitely did not accept his arguments about the relation between quantum and classical. Actually it would be true to say that he did not recognise where Bohr's actual argument was supposed to be, though he knew that Bohr *claimed* to have solved the problem. If possible, Bell wished to eliminate the distinction between quantum and classical altogether, and the most obvious way to attempt to do that was with hidden variables. As he said, once you give definite properties to the elementary quantum particles, you don't need to worry that the classical apparatus has definite properties, because there is no discontinuity between classical and quantum. It may be just that the properties are more under our control for big things than for little things.

However, as is well-known, at this time Bell read *Max Born*'s very readable and interesting book, *Natural Philosophy of Cause and Chance* [6]. Born argues in general terms towards the difficulty of constructing hidden variable theories which would duplicate the quantum predictions, but then states that, in any case, opinions on the matter were unnecessary because von Neumann had actually *proved* that it could not be done. Even had he felt inclined to do so, Bell could not have consulted von Neumann's book, because this was still only available in German, which Bell did not know. But it seems that he was inclined, in any case, to take Born's word for it, though still remaining deeply unhappy about the Copenhagen approach to quantum theory. In any case he recognised that he could not at that stage develop his career by thinking about quantum theory, so deliberately walked away from that subject and towards accelerators.

It would seem, though, that, by the time he left Belfast, most of the ingredients of Bell's subsequent work in quantum theory were present, though it is not clear to what extent he had been exposed to or had thought about the EPR argument. He had identified the role measurement had to play in a consistent theory, but also recognised that it was a role that, as a non-primary term, it was not actually able to play. He had thought a great deal about the cut between quantum and classical, and in that context had also given a great deal of thought to hidden variables, though he was temporarily moved offside by Born's remarks.

Apart possibly from more study of EPR, the only further ingredient required was the Bohm hidden variable theory of 1952 when Bell "saw the impossible done". When he went to Birmingham on his year's leave from Harwell in 1953, Peierls asked him to choose a topic for a talk. Bell suggested either accelerators or the foundations of quantum theory, but Peierls was convinced that the foundations of quantum theory required no elucidation, and chose accelerators. It is fascinating to contemplate what Bell might have achieved in the early 1950s if the choice had gone the other way.[1]

[1] At the conference in Vienna, John Clauser commented that what he would have achieved was making himself unemployable! Worrying about the foundations of quantum theory was definitely not regarded as respectable at the time.

## 2.6 John Bell and Michael Faraday

Having discussed Bell's origins and education it may be natural or at least allowable, to inquire how he may be compared in this way with other great physicists. I might mention that Lesley Kerr, basing his remarks mainly on the early years, makes the following comment: 'John was easily the most brilliant physicist I have ever known. Even on a short acquaintance one was impressed by the simplicity and brilliance of his insight and his easy intellectual ability to find a solution. He was something quite different from most able men and in fact the only person who ever reminded me of John, both in apparent personality and known intellect, was Richard Feynman'.

Bell and Feynman may certainly be comparable in intellect, but I would be less convinced by the mention of personality, especially if one takes into account the whole of Bell's life rather than just the earlier parts, and I would bring another name into the arena: Michael Faraday.

It is very easy, on the one hand to recognise Faraday as one of the great physicists, and then, on the other, to start subtly patronising him. Certainly he performed extremely important experiments, and of course he made very nice models, but, we may suggest, he was very limited by having practically no mathematics at all, indeed, for all his honesty and endeavour, not really educated to any reasonable standard. His ideas, we may suggest, actually had to wait for Maxwell to come along and treat them mathematically before they really took their place in physics.

I would suggest that this is highly misleading. An alternative point of view is that Faraday was a great mathematician in the truest sense of the word, that of relating concepts, just not very experienced in manipulating symbols. Most people have heard of the Science Wars, the clashes between scientists and practitioners of science studies, I'll use the term sociologists, over the true meaning and significance of science. The sociologist will argue that all scientific knowledge is social in nature. One aspect of this centres on the kind of claim made by Stephen Weinberg that if there exists intelligent life on another planet, the science the people there would discover would be the same as ours, in particular the Periodic Table there would have the same form as here. The sociologist may reply: if it's true for the Periodic Table, it's going to be just as true for the Eiffel Tower.

We will probably agree with Weinberg that the science will be the same, but there could well be some surprises for us as to the ideas that the scientists on the other planet might use to build their theories. In particular, is it just possible that these people might never have met the idea of the field? After all, until Faraday, much good work could be and was carried out with the idea of action at a distance between charges, magnetic dipoles and so on. Is it just conceivable that we might have continued on through strong and weak nuclear interactions, through quantum field theory or an equivalent, gauge theories and so on, without this fundamental unifying concept. Perhaps it seems very unlikely, but even to ask the question emphasises the massive and crucial step taken by Faraday, which really made the rest of theoretical physics, as we see it, possible.

Was there anything special about Faraday's background that made it possible for him to take this step. The standard answer is that it was precisely his lack of mathematical background, as compared to the other people studying electricity and magnetism at the same time. While they were prepared to look for complicated mathematical expressions for the various interactions, it may be said that Faraday sought the genuine source of the relevant physics, sought to understand what was going on in physical terms.

While he was certainly possessed of genius and great physical intuition, it may be said that it was his luck that he, in a sense, came from behind in comparison with his more overtly mathematical competitors. It was his reward to take subjects that were already reasonably well established, and to see how one could gain from them something infinitely richer in conceptual possibilities. For a few more general comments on Faraday, it may be said that, though never backward in stating his case, he would always do so with dignity and good nature. He was a great communicator, regarding it as an important part of the job of scientists to tell others about their work, and giving beautiful coherent accounts of this work. He was a man of the highest principle, being, of course, extremely religious.

Let us now consider a comparison with John Bell. If one was to say that Bell came from behind, one might mean that he came from rather an ordinary background compared to some of his peers, compared, for example, to von Neumann; that he studied at a comparatively humble University rather than Cambridge or Harvard or Göttingen; that he started his career without a PhD and without quite the freedom he might have had in an academic post. But none of this should be exaggerated – he received a good education, and his job was interesting and quickly brought him prestige.

In the subject of quantum theory, though, it could be said that he came from behind merely by not being Bohr or Einstein, Heisenberg or Pauli, Schrödinger or Born. While all these had had their say before a listening public long before the 1960s, Bell was not guaranteed an audience. Nor could he gain one by the use of clever or double-edged words, or by lofty aspiration (references to Bohr and Einstein respectively) but only by a re-thinking of what the issues actually were, and a recognition that progress – genuine scientific progress – was possible.

As with Faraday the achievement was to take rather well-worn ideas, and to show that they could be re-interpreted in a completely new way, bringing immense increase in genuine understanding. As with Faraday, the progress can be described in comfortable retrospect as a return to basics, as applying a novel slant to down-to-earth ideas. Unusually in a scientific context, one feels that, had Faraday or Bell not done the work, it might never have been done; science would just be lacking the major insights their work provided.

In character also we may see parallels between Faraday and Bell. Bell also, while, of course, prepared to argue his case with vigour, was honourable and good-hearted even in such argument. Like Faraday, he was an excellent communicator, having a totally clear view of his subject, and being able to present his opinions with the utmost clarity and enthusiasm. And also like Faraday he was a man of the highest principle, scrupulously honest, and willing to give his time and efforts for others. He did not follow Faraday in being religious, but it could be said that his standards of behaviour were a precise mapping for the twentieth century from those which would automatically have followed from genuine religious belief in the nineteenth.

Like Faraday, his ideas may be described as simple, not in the sense of being easily arrived at, but in the sense of being clearly fundamental once arrived at, and being the foundation for a range of important concepts. Like those of Faraday, the ideas of John Bell are seen, year by year, decade by decade, of being of steadily increasing importance in our approach to understand the Universe.

## Acknowledgements

I would like to thank Mrs. Annie Bell for many interesting discussions, Lesley Kerr for providing his memoir, and Reggie Scott for some useful information. I would particularly like to thank Mary Bell for reading through an earlier version of this paper, and making some helpful comments. Appreciation is also due to Reinhold Bertlmann and Anton Zeilinger for general encouragement.

## References

1. For general biographical information on John Bell see, for example, J. Bernstein: *Quantum Profiles* (Princeton Univ. Press, Princeton 1991)
   M. Bell: in *Quantum Mechanics, High Energy Physics and Accelerators: Selected Papers of John S. Bell*, ed. by M. Bell, K. Gottfried, M. Veltman (World Scientific, Singapore 1995) pp. 1–2
   A. Whitaker: Phys. World **11**(12), 29 (1998)
   P.G. Burke, I.C. Percival: Biographical Memoirs of Fellows of the Royal Society of London **45**, 1 (1999)
   M. Bell: in *Quantum Reflections*, ed. by J. Ellis, D. Amati (Cambridge Univ. Press, Cambridge 2000) pp. $xi - xiv$
2. L.W. Kerr: unpublished memoir
3. L.I. Schiff: *Quantum Mechanics*, 1st edn. (McGraw-Hill, New York 1949)
4. E.J. Squires: *The Mystery of the Quantum World*, 2nd edn. (Institute of Physics, Bristol 1994)
5. J.S. Bell: Physics World **3**(8), 33 (1990)
6. M. Born: *Natural Philosophy of Cause and Chance* (Oxford Univ. Press, Oxford 1949)

# 3 My Interaction with John Bell

Bernard d'Espagnat

*Personal recollections are given of the author's longstanding relationship with John Bell. Glimpses are offered on the nature of their agreement and disagreement concerning such philosophical issues as the notion of physical reality and what physics is actually about.*

Heidegger used to say that physicists do not think. This, of course, is vastly true but not completely so. Sometimes *some* physicists *do* think. The trouble is that their thoughts remain almost totally unnoticed, because they mostly do it privately and "on the sly" so to speak. At least, this was the case at CERN in the 1960s, when we both worked there, John Bell and I (he full time, I part time only).

At that time, I had problems with quantum mechanics and *he* had problems with quantum mechanics. But he did not know I had and I did not know he had. I was once told by a friend that, for some reason or other, he was sort of suspected of having some. And indeed I had a confirmation of this when I spotted a heretical book on John's bookshelf. A most heretical book indeed. A book in French by Louis de Broglie! Can you imagine such a thing? I was not particularly attracted at that time by the pilot-wave theory – I still am not – but nevertheless this evidence that, in CERN, somebody other than myself could take interest in essentially conceptual problems was for me a revelation.

It is a long time now since these events took place, and I do not remember their exact chronology. But I remember that we then began discussions and that, some time later, in the mid-sixties, when each one of us was offered to publish a paper in a little book in honor of Victor Weisskopf, we both produced one that had to do with the interpretation of quantum mechanics. A few years later the Italian Physical Society proposed that I should organize a Varenna Summer School on the foundations of quantum mechanics. This took place in 1970, and of course I invited John Bell and also Abner Shimony, who, together with Clauser, Horne and Holt, had just issued their famous generalization of the Bell inequalities, which made experiments possible. So, as you see, this gathering came up at just the right time. It certainly fostered discussions concerning the best ways to test these inequalities, and in the following years we had a few workshops on that subject. I remember one I organized in Paris with John Bell, Shimony, Fry and a few others and also one in Erice, in April 1976, of which John Bell and I were co-organizers (I still have its announcement poster). In the meantime I had made Aspect's acquaintance and he had shown great interest for this question, so, of course, he came to Erice, and ... well you know the rest.

Now that you know the facts, let me come back to a moment in the early sixties and have a look at the ideas. At that time I was already interested, as I just said, in the foundations of quantum mechanics. But I had not yet fully realized how many apparently obvious "common-sense" views are incompatible with them. For example, without having thought the matter over seriously, I tended to imagine that when you produce spin 1/2 particle pairs in a singlet state, each pair is somehow composed of two spins pointing in definite (and opposite) directions, these directions being isotropically distributed within the ensemble thus created. I was impressed when I rediscovered for myself what, in fact, Schrödinger had – in broad terms – pointed out a long time before, namely that, irrespective of the distance between the particles in one given pair, descriptions of such a kind are just simply false, since they yield observational predictions incompatible with those yielded by the said ensemble (the singlet state). In other words, I had found out something that I had not read about in any book, namely nonseparability. When I spoke of this with John he agreed of course. He also agreed that redefining states by means of the density-matrix concept is not an answer since, in the considered phenomenon, correlations are thereby lost. But at the same time he pointed out that arguments of such a type are not sufficient to discard local realism, since they rely on the implicit assumption that supplementary variables do not exist. I then had to confess that, when this possibility is considered, his 1964 argument in *Physics* is the only convincing one. And from then on I became an enthusiastic supporter of the argument in question. Sometimes I even think that I was one of its very first supporters. Anyhow, I had become convinced that indeed Bell's derivation is a major clue in "our efforts at grasping what quantum mechanics is really trying to tell us", as John Wheeler nicely puts it. Nay, I soon became convinced that the clue is even deeper, since the experimental violation of the Bell inequalities proves nonlocality quite independently of whether quantum mechanics is right or wrong. This of course is most important. We all know of the reasons why Dirac was not worried by the so-called paradoxical aspects of quantum mechanics. He simply pointed at the historical fact that up to now great theories all had but a finite lifetime and that the theories that replaced them were, as a rule, based on quite different concepts. So, he said, at one time or other the same must happen to quantum mechanics, and since the new theory will rest on different concepts the quantum-mechanical conceptual riddles will vanish. This, I think, is a most reasonable standpoint. But it does not apply in the present case since, if we forget about the remaining, so-called experimental loopholes in the Aspect-like experiments, nonlocality is proved by the latter independently from quantum mechanics.

All this of course clearly shows how important it was, in the seventies, that experimental tests of the Bell inequalities should be performed. But a question then arises: Why is it that it was so difficult to convince the physics community? One reason, of course, is that the very nature of the question puts it apart from the usual ones such as "Is there a boson with such and such a mass?" and so on. So, it ran against the routine. But I think there is a deeper reason. I am not, by far, a fan of Heidegger and of philosophers in general, but I still think that there is a grain of truth in the famous saying that I ventured to put up as a joke at the beginning of this talk: that physicists do not think. It is true, I believe, that on the whole most contemporary physicists are much too busy to really think, and that consequently they tend to consider genuine thinking as quite an obsolete activity. And please note that this is more or less what Bell implied in a footnote to his paper with Nauenberg (the one I spoke of, that appeared in Weisskopf's festschrift). They mentioned there the burning question of the foundations of quantum mechanics and wrote "The typical physicist feels that these questions have long been answered and that he will fully understand just how if ever he can spare twenty minutes to think about it."

This brings me back to my main and deep "point of agreement" with John Bell. You see: in most papers and books written by theoretical physicists, you find a host of brilliant ideas. But practically all of them are ideas about methods. Recently I read a few textbooks on superstring theory. I learned a lot of things. But frankly I could not find out what the basic ideas of this theory really are. Apart from, at the beginning and in the conclusions, a few simplistic allegories about strings floating in a peculiar kind of space, everything was about computing methods.

Well, John was good at elaborating computing methods, at finding out "rules" leading to correct predictions and, more generally, at exerting the normal activity of a theoretical physicist. During his "working hours" he actively and brillantly contributed to developments in such fields. Also, he liked to discuss with experimentalists about conceivable experiments in high-energy physics. He saw this, he told me, as his normal job: what he was paid for. But still, his main interest clearly was in ideas, not in methods. And this is why I felt so close to him. I was older than he was but I actually felt a great respect for him because of his interest and his success in the area of ideas.

Now the trouble with ideas is that the structure they constitute does not rest on a firm and unquestionable basis. When you ponder on some basic idea you soon feel prompted to derive it from some even more basic one, and so on. But this more basic idea, since it is more basic, will also be more abstract and more philosophical. When philosophers say we do not think I sometimes wonder whether they fully realize how difficult it is to seriously think. At any rate, what seems quite clear is that it is more difficult to agree on abstract, philosophical ideas than on just computing methods. When we take a look at the lifelong disagreement between Einstein and Bohr, for example, we cannot but observe that the reason why they could never satisfactorily resolve it is that it was most deeply philosophical. Perhaps even more so than they thought. Fortunately, it sometimes happen that one, particular, deep question that seemed to be of an essentially philosophical nature and on which conflicting views were entertained is suddenly shown, through some momentous discovery, to be answerable by scientific means. Then it is resolved and conflict on it disappears. This is precisely what happened, as we all know, concerning the question "Is a realistic, local theory of the universe acceptable?" Thanks to John Bell, we know that the answer is no. But such advances are extremely rare. And they are only partial ones. Bell's Theorem deeply changed the nature of the Einstein–Bohr controversy, but it did not resolve it. I remember that – in the late eighties, I think – a private organization in Geneva invited John Bell to a day-long discussion with them. He thought that the burden would be a little too heavy for one person, and he asked me to share it. So I came, I spoke of course, but I could also observe how John answered various questions. The point the people on the floor insisted on was that, through his proof of nonlocality, he had very much weakened the original Einstein position, and that he had thereby reinforced Bohr's. As far as I can remember, John's answer was that yes, of course, Einstein's standpoint had been very much centered on locality, that in this respect it could no longer be maintained, and that consequently the structure of the Einstein–Podolsky–Rosen attack on the Copenhagen interpretation had to be viewed as obsolete. He granted that an external observer could well consider this as a point scored by Bohr and the Copenhagen school. But he insisted that this did not touch on the utmost core of Einstein's – and his own! – requirement, which concerned reality, not locality; that theories do exist that are nonlocal but ontologically interpretable (such as, for example, the Broglie–Bohm model); and that this condition – of being ontologically interpretable – was, in his opinion, the only truly crucial one, for a physical theory to make sense.

Now, to repeat what I just said, when you decide to ground your research work not on mere methods but on *the* truly basic idea – on the idea that *must* underlie our representation of the World and our relationship to it – you may reach the troublesome finding that this truly basic idea does not exist. You may find that the idea that, for a physicist, is the obvious candidate, namely the idea of a scientifically knowable Reality, *may* not be the absolutely stable rock on which all our knowledge can safely be constructed. Does this "finding out" make sense, or is it just a manifestation of some facile Humean skepticism? It is a fact that realism cannot be proved. And the very successes of quantum mechanics do show that it is in no way a precondition to scientific research either, as people once used to think. Nevertheless, I shall leave the question open. Let me simply note that it is one on which John and I always entertained opposite views. I, for one, am a realist but quite an unconventional one. I am convinced that the notion of a Mind-Independent Reality – a kind of an absolute – does make sense. But I take, at the start, a quite agnostic standpoint as to whether it is scientifically describable or not.

And in view of the turn physics has actually taken I am inclined to consider that what science describes is merely an "empirical reality", that is, nothing else than the set of the phenomena in a Kantian sense of the word. In other terms: a set of collective appearances. To John Bell this idea was absolutely repellent. Indeed I remember that, beginning in 1990, I sent John a preprint of mine, the title of which contained the expression "empirical reality". And he sent me back something – either my own preprint or a paper by him, I don't remember – on which he had diagonally written in large letters: "I do not believe in empirical reality!".

Unfortunately, John Bell and I never seriously discussed the question together. This was mainly my fault. I was not incisive enough. Sometimes I think that I did not bring this point in because I was afraid that, by undermining his conviction, I should at the same time undermine his creative power. For it is, I believe, a well-established fact – think of Kepler, for instance – that, for scientists just as for others, creative power is often linked with extremely strong convictions. But, quite frankly, I am not sure that I was moved by such lofty views. More likely I was just afraid that, with a gentle smile, he would strike me down with one of the crushing arguments he had the secret of, and that were so clearly true (once he had formulated them!) that his opponents were left speechless. If this is really the explanation, then I have reasons not to feel oversatisfied. But to my defence I can say that I was genuinely convinced that it was best to wait for a good opportunity and that we would, at one time or other, finally tackle the question. At times we came near to doing so. I remember a nice restaurant in Paris (vegetarian of course!) where John and Mary had kindly invited us (he was there, I guess, for a seminar). The conversation happened to come to the preposterous statements people sometimes utter. John pointed out that even first-class physicists are liable to make some. Even Feynman, he said. I think he was referring to the place in Feynman's lectures where, after having discussed the interference-fringe phenomenon, Feynman bluntly states that "nobody can explain more than we have explained... We have no idea whatsoever of a more basic mechanism from which the foregoing results [the interference fringes] could be deduced." John insisted that, taken literally, such statements are just simply false. The Broglie–Bohm model *does* indeed describe a basic mechanism from which the fringe phenomenon can be deduced. Even if we do not believe in the model, we cannot say that we have no idea of how such a mechanism could be conceived.

Well, once again, I could not but agree. And it is true that, seen under that light, the Broglie–Bohm model, which offers a nice explanation in terms of Man-Independent Reality, is more satisfactory than just the quantum-mechanical formalism, which yields but observational predictive rules. But I felt that the question had better be transferred to a more general framework. After all, John himself was far from being totally satisfied with the Broglie–Bohm model, as shown by the fact that, as soon as the Ghirardi–Rimini–

Weber theory appeared, he transferred, so to speak, his hopes to it. Clearly his position was not "I know the answer" but "I crave for an answer that will be both ontologically interpretable and scientifically convincing." This, anyway, is how I interpret what he told me on some other occasion, when – referring to analyses of contextuality and so on – I asked him how *his theory* dealt with this or that point. For he then retorted: "I have no theory, *I wish I had*!"

Now – to come back to the question of the explanatory power of theories as it had been raised at that dinner I told you of – what I then would have liked to point out to John is just what I was telling you a moment ago, namely, that physical realism is not proved. That it is a fact that when you take a serious, critical look at the elementary arguments that seem to obviously prove its truth you discover that this obviousness is, to a great extent, a lure. You find that the thesis of a *fully knowable* Man-Independent Reality is, in a way, but a working hypothesis. That, as with other theses, it should be compared with the facts, and rated according to the extent to which it helps us, or prevents us, synthesize and understand the said facts. But then, alas, the waiter came with the food, and we spoke about something else.

In view of such differences, which are, admittedly, of a basic nature, the question might again be raised of the reason why my interaction with John Bell was so important for me. I think there are essentially three such reasons.

The first one is, I think, common to all of us, and I have already mentioned it. It is the fact that, through his theorem, John brought a basic question, that of locality, down from the cloudy heavens of philosophy to the more solid realm of scientific research. Such performances are extremely rare. I can only think of a few of them: the discovery by Eratosthenius that the Earth is not flat, Copernicus' finding that the Earth is not the center of the Universe, Einstein's proof that time is not universal... . It is interesting, by the way, that the most salient aspects of these discoveries are best stated by negative sentences. As if Nature were definitely more inclined to tell us what she is *not* than what she is. In the case of nonlocality this is quite especially clear. Maybe the Broglie–Bohm model is right, maybe collapse is real, as Ghirardi and Penrose and some others would have it, maybe the modal theory is the correct one, etc., but anyhow, whatever realistic model is the right one – if any! – we know for sure that the one that is is not local.

This momentous finding of nonlocality has, in common with that of Einstein concerning time, the additional feature that it disproves the validity, not of a "view of the World", but of a deeply ingrained *concept*. And this brings me to the second reason. It is that this disproof of a deeply ingrained concept pointed in fact in a direction quite consonant with my own line of thought. In a way, it brings us back to Descartes, for, as we all know, Descartes was the first scientist who dared to question our common views, including even all the notions that had always seemed so primitive and obvious that thinkers, scientists and so on never hesitated in making use of them. He found out that, at the start, he could doubt of everything but his own thinking, and in this, according to Hegel, he was a hero. Unfortunately he then constructed a grand metaphysical argument that led him to the view that, after all, since God is not a liar, the "obvious" realistic concepts must apply. He thereby founded mechanicism, which is the theory that, apart from thought, everything has to be described by the exclusive means of familiar concepts. We know, of course, how deficient such a view is. But I think Descartes' really significant contribution on these matters is *not* mechanicism. It is what I just said. In other words, it is the realization that a sharp distinction has to be made between rationality on the one hand and the use of seemingly obvious concepts on the other hand. And that therefore, if you are a rational person you cannot demand that science should be based exclusively on seemingly obvious concepts without first logically justifying this demand. This Descartes tried to do but, since his argument based on God not being a liar is now considered as not convincing enough, we are not bound to his conclusion. Indeed, I consider that we are not even bound to the idea that physics should be expressed in an ontological language, which was more or less Einstein's view. The older Einstein seems to have considered that Reality, and even physical objects in the plural, can be described as they really are, if not by familiar concepts, at least by unfamiliar ones, such as those borrowed from mathematics. I do not think this is necessarily true. I consider that mere predictive rules, such as the Born rule in quantum mechanics, count as fully fledged explanations, or, more precisely, as constituting, when all taken together, a first step in an explanation, the second step being the philosophical idea that these predictive rules dimly reflect some existing, largely hidden structures of Mind-Independent Reality. Of course, these speculations of mine go much further than nonlocality. But you understand that they receive some support from it.

Explaining the third reason will bring me back to what I said at the beginning. I think that scientific research is very much driven by contingent facts. This is true for science in general, but it is also true for each one of us in particular. We become interested in such-and-such field because people around us show interest in that field. We engage in such-and-such investigations because it happens that we are in a position in which we can carry them out with some hope of success. We have ideas without distinctly knowing where they come from. But still: we do not totally float dowstream, so to speak. To some extent at least, we can, and often do, organize our quest. A quest, however, is something like a theorem. You cannot prove a theorem out of nothing. You have to assume some premises, and you then derive your theorem starting from them. Similarly, significant scientific quests are necessarily grounded on some explicit or tacit ideas or principles. I think it would be reasonable to establish some coherence between these principles and what we actually do while referring to these principles. In other words, in theoretical physics just as in ethics we should consider two questions: "What do we set as principles?" and "Is what we actually do consistent with these principles?" I do not mean, of course, that we should decide on the principles once and for all. If they definitely do not work we had better change them. But we should try to preserve consistency. I think that Bohr, for example, followed this rule. His principle was that the purpose of science is to account for communicable human experience, and, on the whole, his interpretation of quantum mechanics was consistent with this principle.

The trouble is that a great number of the physicists who came after Bohr clung to the formalism, procedures, predictive rules, etc., developed in Copenhagen and resisted any attempt at changing anything in them, but at the same time they adopted quite a different motivation principle, namely the one of scientific realism. In so doing they gave up consistency. As a student and in the first part of my career as a physicist I was deeply worried by this. At some places in the books and papers I read, the argument seemed to go through only by virtue of an axiom that correct prediction of what is observed is all we may expect from of science (all the rest just being "empty words"). But at other places (in introductions, summaries, applications to solid-state physics, thermodynamics or cosmology) the same authors would unabashedly express themselves as if the probabilities and other items they juggled with did refer to events "as they really happen." There they expected me to take such a meaning for granted. This was irritating. So, you can imagine what a relief it was for me to discover the existence of a physicist both actively taking part in the most advanced parts of present-day theoretical research and just as worried as I was about this inconsistency.

Finally, let me mention some aspects of John Bell that we all appreciated, I think. One of them we can still enjoy now, by reading his papers and his book. It is a fact that he was a writer, and a quite remarkable one. I think that even a non-English-speaker is entitled to testify to this. His prose is simple, alert and to the point, and his texts are sprinkled with turns of phrase that could serve as paradigmatic examples of fine English (or should I say Irish?) humour. In discussions this was just the same. Thank God he was not witty. He seriously discussed serious things. But he was one of those people who instinctively feel when an amusing remark fits, and he immediately knew how to concoct it. We discussed languages, of course. I remember he teased me about such things as the existence, in French, of two words to say the same thing, such as *Avenir* and *Futur*. My answers were biting, of course, but I'll spare you. I prefer mentioning my delight when, in the last paragraph of one of John's preprints – a very serious and enlightening paper on the Everett theory – I read a poetic quotation, in French, of one of my favorite XIXth century authors, François-René, vicomte de Chateaubriand. So, you see, to sum up, John was a real European, fully percipient of the very subtle unity that underlies our creative diversity. And, as a result, he was for me one of the best friends I ever had.

# 4 Magic Moments: A Collaboration with John Bell

Reinhold A. Bertlmann

I want to give an impression of the time I spent together with John S. Bell, of the atmosphere of our collaboration and friendship. I briefly review our work, the methods of nonrelativistic approximations to quantum field theory for calculating the properties of heavy quark–antiquark bound states.

## 4.1 Prologue

The working place of John Bell was CERN. Near the entrance of this huge laboratory stands the building where the Theory Division is located. There John Stewart Bell – often abbreviated just by JSB – had his office on the first floor.

It was in 1978 when I came to CERN for the first time as a young Austrian with a Fellowship there. CERN was very impressive for its experimental facilities, the big colliders, and as a place where one could meet all the distinguished physicists of the field.

One of the great physicists in the Theory Division was John Bell. He was highly respected and often consulted by his colleagues. He could judge if a theory was right or wrong, or as he phrased it *sound* or *rotten.* JSB was called the Oracle of CERN; there was a certain aura around him and his office. My first impression of his office was a room full of boxes where he filed letters and collected works from the several fields. He himself was sitting dignified in an armchair which sometimes dangerously bent backwards and only he could use. In the middle of the room was standing a double-desk; on the walls were attached two blackboards opposite each other. In Fig. 4.1 you can see a little bit of his desk.

You can imagine how exciting it was for me to come into contact with this man. I remember very well having tea in the common room after a seminar on theory division. There he approached me, the newcomer, and introduced himself: "I am John Bell. Where are you from? What are you working on here?" My interest was calculating the properties of heavy quark–antiquark systems – quarkonium – specifically the wavefunctions, a hot subject at that time since charmonium (the $J/\psi$ family) had been found in 1974 and the even heavier system bottonium (the $\Upsilon$ family) more recently in 1977.

Bell was very interested in my explanations, he even liked my results, since they were in accordance with his Thomas–Fermi model calculations.

**Fig. 4.1.** Desk of J.S. Bell at CERN

So our discussions in this field began and became rather quickly a closer investigation into what is called *duality*.

## 4.2 Duality in Hadronic Reactions

The concept of duality is often used in physics. What I refer to is when two seemingly different phenomena are strongly correlated with each other; they appear as the *dual* aspects of one and the same reality [1].

### 4.2.1 Hadron Production in $e^+e^-$ Collisions

Let us consider the production of hadrons (strongly interacting particles) in electron–positron collisions. Among the hadrons there are also resonances produced, the $\rho, \phi, J/\psi, \Upsilon$ families of particles:

$$
\begin{array}{llllll}
e^+\,e^- \longrightarrow \text{hadrons} & & & & & \\
 & \rho & \phi & J/\psi & \Upsilon \ldots & \text{resonances} \\
 & \Downarrow & \Downarrow & \Downarrow & \Downarrow & \\
 & (u\bar{u}+d\bar{d}) & (s\bar{s}) & (c\bar{c}) & (b\bar{b}) & \text{bound states}
\end{array}
\tag{4.1}
$$

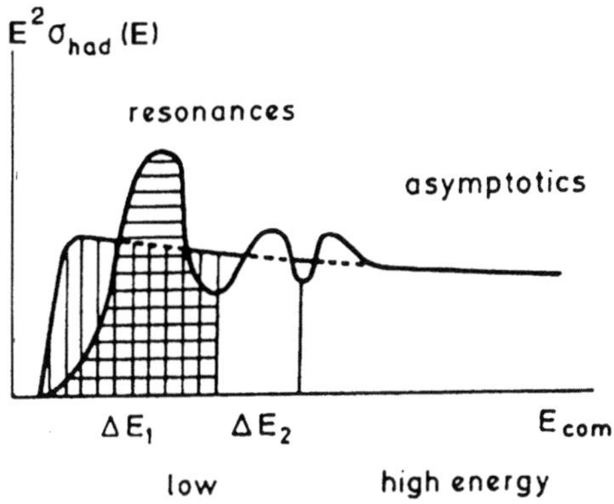


**Fig. 4.2.** The hadronic cross-section in $e^+e^-$ collisions shown qualitatively

The cross-section for these processes are plotted qualitatively in Fig. 4.2; it exhibits the following feature: At low energies some pumps appear, the resonances, but at high energies the curve becomes quite flat or asymptotically smooth. These are two different phenomena, the resonance production and the asymptotic production of hadrons.

*Question: Are they related to each other in a dual sense?*
*Answer: Yes, they are.*
*Question: How can we understand that?*

Hadrons consist of quarks which are confined and interact strongly via gluons (the colour force). Now, at high energies, which corresponds to short distances, the quarks behave as quasi-free particles. The corresponding field theory – quantum chromodynamics (QCD) – is asymptotically free. The process

$$e^+ \, e^- \quad \longrightarrow \quad q \, \bar{q} \qquad \text{quasi-free (anti-)quark} \tag{4.2}$$

is a good approximation. Then the cross-section approaches a constant,

$$E^2 \sigma_{q\bar{q}}(E) \quad \longrightarrow \quad \frac{4\pi\alpha^2}{3} 3 \sum_q e_q^2 \; = \; \text{constant} \; , \tag{4.3}$$

which is proportional to the sum of the quark charges squared, $e_q^2$, times the colour factor 3. So we find the asymptotics correctly.

However, at low energies, where the quarks can penetrate into larger distances (the typical distance is of the order of a fermi), they are confined and generate bound states which show up as resonances. These states are named quarkonium in analogy to positronium.

### 4.2.2 Duality

It was *Sakurai* [2] who formulated duality quantitatively. He found that if you average the cross-section suitably over *all* resonances then it agrees with the averaged asymptotics

$$\int_{\Delta E} E^2 \sigma_{\text{res}}(E)\,\mathrm{d}E = \int_{\Delta E} E^2 \sigma_{q\bar{q}}(E)\,\mathrm{d}E\,. \tag{4.4}$$

Relation (4.4) is called *global duality.*

Here is the point where *Bell* and *I* began our investigations [3]. We simply asked: How far can we push duality?

*How many resonances do we need in order to reproduce the asymptotics?*

The answer is quite simple:

*One individual resonance is enough!*

This we called *local duality* [3]:

$$\Delta E \quad \longrightarrow \quad \Delta E_{\text{res}}\,. \tag{4.5}$$

The explanation for why it works we found in nonrelativistic potential theory.

## 4.3 Nonrelativistic Potential Theory

Although the bound states of quark–antiquark pairs are not ideal nonrelativistic systems, they decay, and the relative velocity of the quarks is rather high ($v \approx 0.3c$). Nonrelativistic potential theory is an amazingly powerful tool to describe their properties.

So for our purpose of energy averaging the cross-section we can approximate the resonances very well by a sum of delta functions:

$$E^2 \sigma_{\text{res}}(E) = 6\pi^2 \sum_n \Gamma_n^{ee} \delta(E - E_n)\,. \tag{4.6}$$

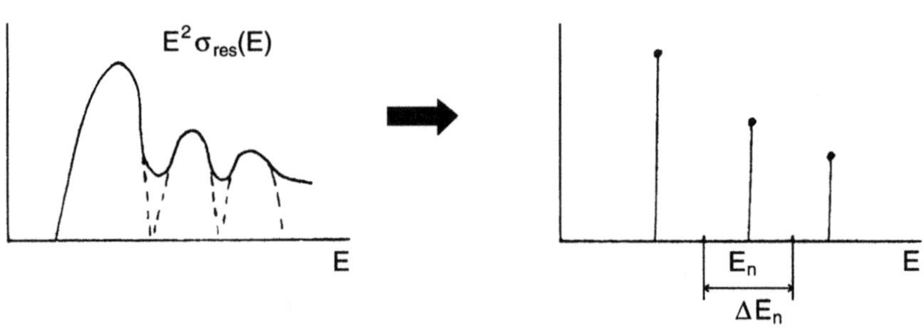


**Fig. 4.3.** Approximation of resonances by delta functions

This replacement of the cross-section is illustrated in Fig. 4.3. The leptonic width of the resonance is related to the wavefunction at the origin by

$$\Gamma_n^{ee} = N\frac{|\psi_n(0)|^2}{m^2}\,, \quad N = \frac{4\pi\alpha^2}{3}3e_q^2\,. \tag{4.7}$$

On the other hand, in a nonrelativistic approximation the cross-section for quasi-free quarks only depends on their relative velocity, $v = \sqrt{E/m}$:

$$E^2\sigma_{q\bar{q}}(E) = N\frac{3v}{2}K(v)\,.$$

$\Downarrow$

Coulomb enhancement factor (4.8)

The function $K(v)$ arises from the short-distance part of the potential and is calculable. For example, in the case of the Coulomb potential it denotes the well-known Coulomb enhancement factor.

At this stage I must tell a story which happened when I was calculating some specific short distance functions, $K(v)$. One day John Bell said "Show me your calculations at home!" So in the evening I went to his apartment, in a big new building in Geneva, took the lift to the second floor, and was very excited when I rang the bell. John Bell opened the door, he was alone and offered me a drink, an orange juice. But I was so excited that I forgot to close my fingers on the glass, and all the orange juice spilled over his beautiful carpet. I was shocked but John smiled and said "You sit down and think – I'll clean the floor". I certainly couldn't think; I nearly fainted.

Coming back to expressions (4.6) and (4.8), if we compare both we can check quantitatively whether duality (4.4) in its local form (4.5) is true. The result for *local duality* is

$$4\pi|\psi_n(0)|^2 = \frac{m^2}{\pi}\int_{\Delta E_n} \mathrm{d}E\, v(E)K(E)\,,$$

$\Uparrow$ $\Uparrow$

OK to within a few % (4.9)

where we have chosen for each resonance the energy interval

$$\Delta E \quad \longrightarrow \quad \Delta E_n = \frac{E_n - E_{n-1}}{2} \,. \tag{4.10}$$

Calculating now the wavefunctions and the integrals in (4.9) explicitly, it turns out that this local duality relation holds surprisingly well. For the ground state the duality integral in (4.9) approximates the wavefunction within a few percent, for the higher levels it practically coincides, and this is rather independent of the potentials used (for details see [3]).

*What's the reason for it?*

Let us further apply the mean value theorem to the duality integral in (4.9). We find an expression which is quite familiar [4–6]. It's the *WKB relation*

$$\begin{aligned} 4\pi|\psi_n(0)|^2 \quad &= \quad \frac{m^{\frac{3}{2}}}{\pi}\sqrt{E_n}\, K(E_n)\frac{dE_n}{dn} \,, \\ \text{wavefunction} \Longleftrightarrow &\qquad \text{energy spectrum} \end{aligned} \tag{4.11}$$

which relates the wavefunction at the origin to the energy spectrum. So we can trace back duality to a well-known result in quantum mechanics.

When we had our results and also thought we understood them, John proposed: "You begin to write the paper!", I felt very proud, worked all night, and showed him the paper the next day. John smiled and said, "Oh, that looks nice, I will have a closer look at it." The day afterwards he returned the paper. I was shocked to find that there was not one word left in the place where I had put it – it was a completely different paper! But, thank God, I slowly improved.

### 4.3.1 Résumé

By investigating local duality, the energy smearing of a resonance cross-section by a quasi-free quark cross-section, we find that (4.9) is very accurate independent of the considered potential.

This feature can be understood quite naively [1]. In the duality relation we allow for an energy spread, which means – via the uncertainty relation –

Nuclear Physics B177 (1981) 218–236
© North-Holland Publishing Company

CERN
SERVICE D'INFORMATION
SCIENTIFIQUE

MAGIC MOMENTS

J.S. BELL and R.A. BERTLMANN[1]

*CERN, Geneva, Switzerland*

Received 30 June 1980

The moment method of Shifman, Vainshtein and Zakharov, for calculating bound-state energies in QCD, is tested in the context of potential models. For simple power-law potentials of low degree a refined version of the method works surprisingly well. The cruder version actually used by Shifman, Vainshtein and Zakharov for charmonium works less well, and the composite potentials usually envisaged for charmonium are less accurately dealt with than simple power potentials. We conjecture then that the magnitude of their confinement parameter $\phi$ has been substantially underestimated by those authors.

**Fig. 4.4.** Paper of *magic moments*

that we focus on small values of the conjugate variable time. But for short times the corresponding wave cannot spread far enough to feel the details of the long-distance part, the confining potential. So this part can be neglected, but certainly not the short-distance part.

For this reason we can predict from a quasi-free $q\bar{q}$ pair the wavefunction at the origin of the bound state, the leptonic width or the area of resonance:

$$\text{quasifree } q\bar{q} \quad \longrightarrow \quad |\psi_n(0)|^2 \ \sim \ \Gamma_{\text{res}}^{ee} \ \sim \ \text{area of resonance}\,. \tag{4.12}$$

However, if we want to push the idea of duality even further in order for it to become sensitive to the position of the state, the mass of the resonance, then we penetrate into larger distances and we must have some information about confinement.

That's the subject Bell and I got interested in next and it resulted in a wonderful collaboration which we called *magic moments* (see Fig. 4.4).

## 4.4 Magic Moments

One of John's habits was to have a *4 o'clock tea*. So in our afternoon discussions in John's office we always had a break. It was like a ritual, at two minutes to four we left the office and went down to the CERN cafeteria.

There John ordered in his typical British accent, “deux infusions verveine, s’il vous plâit”, John’s favorite tea. There, in a relaxed atmosphere, we talked not only about physics but also about politics, philosophy, and when Renata Bertlmann was with us we also had heated discussions about modern art.

When we worked at home a similar tea ritual took place in his flat, and as you can see from Fig. 4.5, it took us quite a time to choose the right sort of verveine.

**Fig. 4.5.** John Bell and Reinhold Bertlmann choosing the right sort of tea

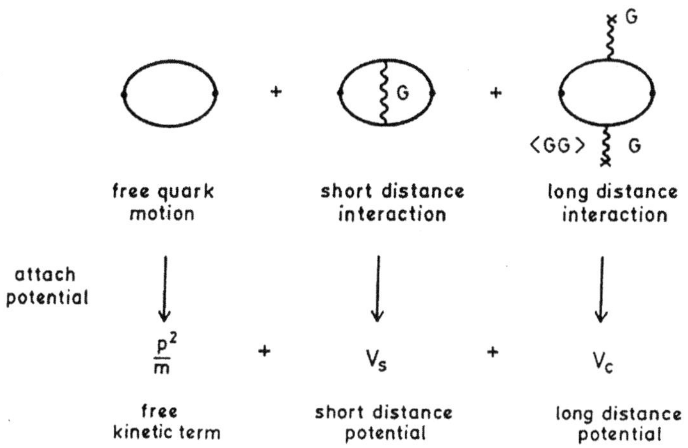


**Fig. 4.6.** Feynman diagrams for the vacuum polarization tensor

So it was in this *atmosphere verveine* that John and I discussed how to include confinement in our duality idea in order to determine a bound state position.

Our starting point was the vacuum polarization tensor in quantum field theory. This is the Fourier transform of the vacuum expectation value of the time-ordered product of two currents

$$i \int dx\ e^{iqx} \langle \Omega | T\ j_\mu(x) j_\nu(0) | \Omega \rangle = \Pi(q^2)(q_\mu q_\nu - q^2 g_{\mu\nu}) \, . \tag{4.13}$$

It can be calculated within field theory, QCD, with help of Feynman diagrams, loop diagrams which are depicted in Fig. 4.6. The novel thing is that the usual perturbation series, representing the short-distance interaction, is modified by the addition of a small part (third diagram in Fig. 4.6), the so-called gluon condensate, $\langle GG \rangle$, the vacuum expectation value of two gluon field strength tensors. This part had been introduced at that time by a Russian group [7] in order to account for the longer distances, the influence of confinement.

### 4.4.1 Moments

Again, our investigation was within potential theory, where we could calculate both the perturbative *and* the exact result. For the energy smearing we chose exponentials (they worked best) and we defined the following *nonrelativistic moment* [8]:

$$M(\tau) = \int dE\ e^{-E\tau}\ \mathrm{Im}\, \Pi(E) \, , \tag{4.14}$$

where $\mathrm{Im}\, \Pi(E)$ is the imaginary part of the vacuum polarization function calculable via the Feynman diagrams of Fig. 4.6.

When we do the calculation, which is actually a perturbation theory calculation with respect to an imaginary time, the result is the following:

$$M(\tau) = \frac{3}{8m^2} 4\pi \left(\frac{m}{4\pi\tau}\right)^{\frac{3}{2}} \left\{ 1 + \frac{4}{3}\alpha_S \sqrt{\pi m}\, \tau^{\frac{1}{2}} - \frac{4\pi^2}{288m} \left\langle \frac{\alpha_S}{\pi} GG \right\rangle \tau^3 \right\} \quad (4.15)$$

The leading term corresponds to the free motion of the quarks (first diagram in Fig. 4.6); it is perturbed by the $\alpha_S$ term, representing the short-distance interaction (second diagram), and by the gluon condensate $\langle GG \rangle$ term, responsible for the longer distances (third diagram).

On the other hand, the exact moment – which is our "experimental" moment containing the resonances, the bound states – is given by the optical theorem

$$\mathrm{Im}\, \Pi(E) \sim \sigma(E) \,, \quad (4.16)$$

which relates the imaginary part of the vacuum polarization function (the forward scattering amplitude) to the total cross-section. Inserting this cross-section within our potential theory, (4.6) together with (4.7), we find

$$\mathrm{Im}\, \Pi(E) = \frac{3}{8m} \sum_n 4\pi |\psi_n(0)|^2 \delta(E - E_n) \,. \quad (4.17)$$

This is the expression we have to insert into (4.14) to find the exact nonrelativistic moment.

### 4.4.2 Ground State

*How do we obtain the ground state level?*

To find the ground state level, $E_1$, the mass of the resonance $M = 2m + E_1$, we finally work with a logarithmic derivative, the *ratio of moments*

$$R(\tau) = -\frac{d}{d\tau} \log M(\tau) = \frac{\int \mathrm{d}E\, E\, \mathrm{e}^{-E\tau}\, \mathrm{Im}\, \Pi(E)}{\int \mathrm{d}E\, \mathrm{e}^{-E\tau}\, \mathrm{Im}\, \Pi(E)} \overset{\tau \to \infty}{\longrightarrow} E_1 \,. \quad (4.18)$$

For large (imaginary) times, $\tau$, the ratio of moments cuts off the contributions from the higher states and projects the ground state energy, $E_1$.

In the corresponding theoretical – perturbative – expression we regard the minimum value, even though not occurring at infinity, as an approximation to $E_1$,

$$\min \mathcal{R}(\tau) = E_1 \,,$$

$\Uparrow$ $\Uparrow$

OK to within a few % (4.19)

where the theoretical ratio of moments is given by a very simple formula:

$$R(\tau) = \frac{3}{2\tau} - \frac{2}{3}\alpha_S \sqrt{\pi m}\, \tau^{-\frac{1}{2}} + \frac{4\pi^2}{96m} \left\langle \frac{\alpha_S}{\pi} GG \right\rangle \tau^2 \,. \quad (4.20)$$

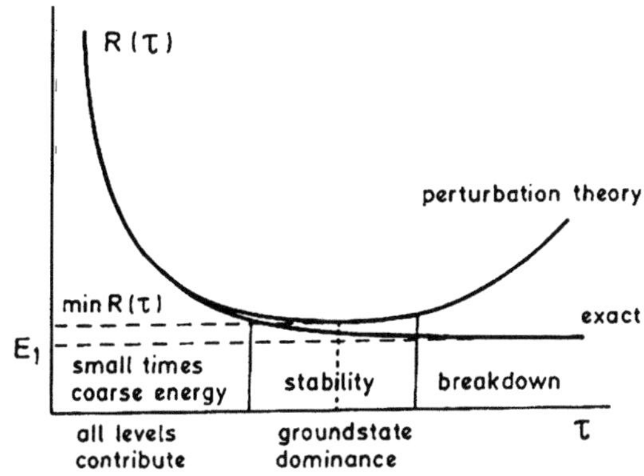


**Fig. 4.7.** ratio of moments plotted qualitatively

The mass of the corresponding resonance is then $M = 2m + \min R(\tau)$.

*What's the result now?*

Let me discuss a typical example, charmonium, with values $m = 1.4\,\mathrm{GeV}$, $\alpha_\mathrm{S} = 0.3$ and $\langle \frac{\alpha_\mathrm{S}}{\pi} GG \rangle = 0.02\,\mathrm{GeV}^4$. I have plotted the ratio $R(\tau)$ in Fig. 4.7. It shows the following typical features: The exact ratio approaches rather quickly its limit, $E_1$. The theoretical ratio agrees perfectly for small times and stabilizes for large times. This stability, the minimum, happens to already be close to the ground state, quantitatively within 12%. So we obtain a good prediction for the position of the ground state.

### 4.4.3 Balance

We could show within potential theory that there exists a balance which we can phrase in the following way [8]:

$$\begin{array}{lcl} \text{coarse enough} & & \text{fine enough} \\ \text{pertubation theory OK} & \Longleftrightarrow & E_1 \text{ accurate} \end{array} \qquad (4.21)$$

*The energy average can be made coarse enough – involving small times – for the modified perturbation theory to work, but on the other hand fine enough for the individual levels to emerge clearly.*

Surprising? Yes! Intuitively we had expected that for a clearly emerging level the confinement force must be dominant and not just a small additional perturbation. The moments, however, forced us to re-educate our intuition; when modifying the perturbation, levels do appear for *magical* reasons.

## 4.5 Equivalent Potential

Since John and I worked within potential theory it was quite natural for us to ask whether one can attach a potential to this gluon condensate effect. This led us to our next collaboration [9, 10].

One of the nice things about being a physicist is that you travel around the world and meet your collaborators and friends everywhere. So it happened that in 1983 I stayed at the University of Marseille together with Eduardo de Rafael. He also invited John and Mary Bell to visit the institute. There we had interesting discussions and we also enjoyed the beautiful surroundings. One of our walks to the Calanque of Port Alon you can see in Fig. 4.8. We even arranged some nice British weather for John, which was not so easy to get in the South of France.

When starting from quantum field theory we faced the essential difficulty of representing the gluon condensate insertion, in the gluon propagator, by a potential. Whereas from the short-distance part of the gluon propagator a potential can be extracted in the usual way, the familiar Coulomb potential, the long-distance part, the gluon condensate contribution, diverges in this procedure. So we had to look for regularized quantities. But those we had already close at hand – they were our *magic moments.*

We know already the gluon condensate effect of quantum field theory in a nonrelativistic approximation, it is expression (4.15). So what we have to do is to calculate the nonrelativistic moment (4.14) within quantum mechanics determined by the Hamiltonian

$$H = \frac{p^2}{m} + V \,, \tag{4.22}$$

where $V$ represents now the potential.

**Fig. 4.8.** Reinhold Bertlmann, John Bell and Mary Bell on a walk in the Calanque of Port Alon, South of France, 1983

The moment defined by (4.14) we can rewrite, by virtue of (4.17), as

$$
\begin{aligned}
M(\tau) = \int \mathrm{d}E \; \mathrm{e}^{-E\tau}\, \mathrm{Im}\, \Pi(E) &= \frac{3}{8m^2} \sum_n 4\pi |\psi_n(0)|^2 \; \mathrm{e}^{-E_n \tau} \\
&= \frac{3}{8m^2} \langle \boldsymbol{x} = 0 |\, \mathrm{e}^{-H\tau} | \boldsymbol{x} = 0 \rangle \,, \qquad (4.23)
\end{aligned}
$$

and we recover the (imaginary) time-dependent Green's function at $\boldsymbol{x} = 0$.

According to our previous procedure, we perturb the kinetic term by the potential with respect to the time, $\tau$:

$$
\mathrm{e}^{-H\tau} = \exp\left[-\frac{p^2}{m}\tau\right] - \int_0^\tau \mathrm{d}\tau' \, \exp\left[-\frac{p^2}{m}(\tau - \tau')\right] V \exp\left(-\frac{p^2}{m}\tau'\right) . \quad (4.24)
$$

The perturbation integral is determined by the potential and the familiar free Green's function. For power potentials such as

$$
V = \sum_s \lambda_s r^s \,, \qquad (4.25)
$$

the result for the moment is quite simple [8]:

$$
M(\tau) = \frac{3}{8m^2} 4\pi \left(\frac{m}{4\pi\tau}\right)^{\frac{3}{2}} \left[1 - \sum_s \lambda_s \Gamma\left(\frac{s}{2} + 1\right) m \left(\frac{\tau}{m}\right)^{\frac{s}{2}+1}\right] . \qquad (4.26)
$$

This perturbation formula we have to compare with the corresponding field-theoretic expression (4.15). Identifying term by term ($s = -1$ and $s = 4$) leads to the *equivalent potential of Bell and Bertlmann* [9]:

$$V_{\mathrm{BB}} = -\frac{4\alpha_{\mathrm{S}}}{3r} + \frac{\pi^2}{144}\left\langle\frac{\alpha_{\mathrm{S}}}{\pi}GG\right\rangle mr^4 \,. \tag{4.27}$$

It is a superposition of Coulomb and quartic potentials, it is quite steep and mass (or flavour) dependent. So it is rather different to those favoured by the potential modellers [11].

Finally, John and I also considered very heavy quarkonium systems which are very compact and dominated by the Coulomb interaction [12, 13]. There we also found a static potential reproducing precisely the gluon condensate contribution; however, it is also mass dependent [14].

In conclusion, no adequate bridge is found between field theory with a gluon condensate contribution, on the one hand, and popular potential models on the other. For an overview of this field I refer the reader to [15].

## 4.6 Epilogue

John Bell was a deep and sharp thinker, a philosopher of Nature but with the tools of a theoretical physicist. And it was his great critical intellect that led him to his profound discoveries. So he found together with Roman Jackiw [16] the celebrated *anomalies of quantum field theory*,[1] which opened the door to a deeper understanding of Nature (see Jackiw's contribution to this book [17]) and which I could enjoy discussing with John [18].

Bell's famous dictum,

> "*Speakable and unspeakable in quantum mechanics,*"

which collects his quantum papers to a milestone book [19], or its variation *Quantum* [Un]*speakables*, the title of the conference commemorating John Bell [20], expresses truly his character. The use of words, their meaning, must be very precise. I experienced it in all our works.

> *Concern about terminology!*

This was Bell's demand. Like a moralist he attacked *bad habits*, the imprecise use of words. His attacks culminated in his brilliant article *Against 'measurement'* [21], where he branded the words that should be forbidden in any serious discussion:

> "*system, apparatus, environment, microscopic, macroscopic, reversible, irreversible, observable, information, measurement.*"

[1] This is by far Bell's most quoted paper!

For instance, *observables* should be replaced by his favorite concept, the *beables* [22], or *measurements* by *experiments.* But it is his great humor (for me his typical Irish wit) which makes his attacks so delightful. For example,

"*ordinary quantum mechanics is just fine FAPP*"

(FAPP = for all practical purposes).

Bell's wit you can find in all his actions, in every day life, and also when he communicated by drawing little sketches. In one sketch he characterized himself, and indeed, John was a gifted cartoonist. But before I show it I have to explain it a little bit.

In winter time, when it was cold, John liked to wear a Norwegian cap. On the cap one could see the letters: APPI LAPPI embroidered. Fig. 4.9 shows the *real* cap. With this APPI LAPPI cap on his head, John has sketched himself – as you can see from Fig. 4.10, the cap, the hairs, the glasses, the beard, his image is just perfect.

**Fig. 4.9.** The *real* Norwegian APPI LAPPI cap of John Bell

Now I come to the story some of you might know. I also became a victim of his splendid wit. His article *Bertlmann's socks and the nature of reality* [23], where he described the Einstein–Podolsky–Rosen (EPR) correlations with Bertlmann's socks, came out of the blue for me. In our first years of collaboration he never mentioned his quantum works to me. And I had not the slightest idea that he had noticed my habits of wearing socks of different colours (a habit I had had since my student days). Completely unexpected, this article appeared and pushed me *instantaneously* into the debate of quantum mechanics, and actually it was the cartoon (see Fig. 4.11) showing me with my odd socks in an EPR-like situation, which *really* changed my life. Everybody wants to see my socks now.

**Fig. 4.10.** John Bell sketched by himself

JOURNAL DE PHYSIQUE

*Colloque C2, supplément au n°3, Tome 42, mars 1981* *page* C2-41

# BERTLMANN'S SOCKS AND THE NATURE OF REALITY

J.S. Bell

*CERN, CH-1211, Geneve 23, Suisse*

Fig. 1.

**Fig. 4.11.** "Bertlmann's socks", sketch by John Bell

**Fig. 4.12.** Conclusion of the preprint dedicated to John Bell

My answer to Bell's article was a paper which I named *Bell's theorem and the nature of reality* and which I dedicated to him on the occasion of his 60th birthday. So I wrote a preprint in July 1988 [24] and sent it to all the universities, which was the customary procedure at that time. It appeared later on in *Foundation of Physics* [25]. In the conclusion of the paper I took a little revenge by also drawing a cartoon, emphasizing the *spooky action at a distance* (see Fig. 4.12). It amused John, who was rather unaccustomed to alcohol, very much that the *spooky action* escaped from a Bell's Whisky bottle which *really* does exist.

When I recall now my collaboration with John Bell and the time we spent together, I really can say it was a great and wonderful time; they were *magic moments* indeed, and I hope I have been able to give some impression of this in these pages.

## References


1. R.A. Bertlmann: Acta Phys. Austr. **53**, 305 (1981)
2. J.J. Sakurai: Phys. Lett. B **46**, 207 (1973)
3. J.S. Bell, R.A. Bertlmann: Z. Phys. C **4**, 11 (1980)
4. M. Krammer, P. Leal-Ferreira: Rev. Bras. Fis. **6**, 7 (1976)
5. C. Quigg, J.L. Rosner: Phys. Rev. D **17**, 2364 (1978)
6. J.S. Bell, J. Pasupathy: Phys. Lett. B **83**, 389 (1979)
7. M.A. Shifman, V.I. Vainshtein, V.I. Zakharov: Nucl. Phys. B **147**, 385, 448, 519 (1979)
8. J.S. Bell, R.A. Bertlmann: Nucl. Phys. B **177**, 218 (1981)
9. J.S. Bell, R.A. Bertlmann: Nucl. Phys. B **187**, 285 (1981)
10. J.S. Bell, R.A. Bertlmann: Phys. Lett. B **137**, 107 (1984)
11. W. Lucha, F.F. Schöberl, D. Gromes: Phys. Rep. **200**, 127 (1991)
12. H. Leutwyler: Phys. Lett. B **98**, 447 (1981)
13. M.B. Voloshin: Nucl. Phys. B **154**, 365 (1979)
14. R.A. Bertlmann, J.S. Bell: Nucl. Phys. B **227**, 435 (1983)
15. R.A. Bertlmann: Nucl. Phys. B (Proc. Suppl.) B **23**, 307 (1991)
16. J.S. Bell, R. Jackiw: Nuovo Cim. A **60**, 47 (1969)
17. R. Jackiw: John Bell's observation on the chiral anomaly and some properties of its descendants, hep-th/0011274
18. R.A. Bertlmann: *Anomalies in Quantum Field Theory*, Int. Ser. Monogr. Phys. **91** (Clarendon, Oxford 1996)
19. J.S. Bell: *Speakable and Unspeakable in Quantum Mechanics* (Cambridge Univ. Press, Cambridge 1987)
20. *Quantum* [Un]*speakables*, Conference in commemoration of John S. Bell, 10–14 November 2000, University of Vienna
21. J.S. Bell: Phys. World **3**, 33 (1990)
22. J.S. Bell: Beables for quantum field theory, Article 19 of [19]
23. J.S. Bell: Bertlmann's socks and the nature of reality, Article 16 of [19]
24. R.A. Bertlmann: Bell's Theorem and the nature of reality, preprint University of Vienna, UWTh-Ph-1988-35 (1988)
25. R.A. Bertlmann: Found. Phys. **20**, 1191 (1990)

# Part II

# Tests of Bell's Inequalities

# 5 John S. Bell: Some Reminiscences and Reflections

Abner Shimony

Section 5.1 considers the question, "Why Was Bell the Discoverer of Bell's Theorem?", and proposes his moral character as an answer. Section 5.2 gives three reminiscences of encounters with Bell. Section 5.3 proposes that the nonlocality exhibited by Bell's Theorem and the Bell-inspired experiments is best understood in the framework of the "pregeometry" of Michael Heller and collaborators, which is based upon the noncommutative geometry of Alain Connes.

## 5.1 Why was Bell the Discoverer of Bell's Theorem?

Before presenting a few reminiscences of John Bell, which are factually accurate but not very important historically, I wish to consider a question of psycho-history – notoriously a domain of uncertainties. The question, stated as the title of Sect. 5.1, is extraordinarily interesting and merits a conjecture regardless of the intrinsic difficulty of any exercise in psycho-history.

The question can be sharpened by pointing out that Bell's proofs of his theorem [1] – *no hidden variables theory satisfying his locality condition (which under sufficiently ideal experimental conditions appears to be a consequence of relativistic locality) agrees with all of the observable predictions of certain quantum states* – are elegant but not very difficult.

Not only is the reasoning of his theorem easily within the powers of the mighty quantum mechanicians of the period from the mid-twenties to the mid-sixties of the twentieth century, but it doesn't even look too difficult for an intelligent student. Well, one might comment, Bell's genius lay not in proving his theorem but in formulating the theorem as a hypothesis to be investigated. No doubt there is some truth in this statement, but I don't think it completely answers the question posed. There was a non-negligible number of thoughtful and independent-minded people during the period in question who reflected upon the completeness or incompleteness of the quantum mechanical description of physical systems and upon the "spookiness" of correlations implied by quantum mechanically entangled states, and these reflections could have supplied the heuristics for formulating Bell's Theorem as a hypothesis.

My answer to the question that has been posed is that *Bell's moral character is primarily responsible for his discovery of Bell's Theorem.* Even his passionate interest in the foundations of quantum mechanics, which of course was a necessary condition for his discovery, was in large part a consequence of character. To establish my thesis I shall review half a dozen steps in Bell's career as a physicist, culminating in the formulation and proof of his theorem.

1. In Bell's last two years (1948–1949) at Queen's University, Belfast, he argued strenuously and passionately about quantum mechanics with Dr. Richard Sloane, from whom he took a course on atomic spectra. According to *Andrew Whitaker* [2] Bell questioned Sloane about fundamentals of quantum mechanics, especially Heisenberg's Uncertainty Principle.

"*In particular, though the Heisenberg principle clearly related the values of* $\Delta x$ *and* $\Delta p$ *obtained in a measurement, and Bohr's 'parables', as Bell called them, could maybe be used to justify this point, it was left totally unclear what each of* $\Delta x$ *and* $\Delta p$ *should be. And to go one step further back, if there were no measurement, and one was just considering* $\Delta x$ *and* $\Delta p$ *for a particular system, one could easily get the impression that it was a subjective matter.*"

Bell's dissatisfaction with Sloane's presentation of the Copenhagen interpretation culminated in accusing his teacher – unfairly, in the light of Sloane's rectitude – of dishonesty. But Whitaker conjectures persuasively that Bell's frustration had the beneficial consequences of leading him to take hidden variables theories seriously, even though they were very much out of favor in most of the physics community, and of stimulating his own meditations. Bell's independence on this matter was an indication of his strength of character. *A propos* of hidden variables Whitaker notes that during Bell's Belfast days "it is not clear to what extent he had been exposed to or had thought about the EPR argument". But Bell did read *Max Born*'s *Natural Philosophy of Cause and Chance* [3], which argued in general terms against hidden variables and in addition cited with approbation von Neumann's proof of their impossibility in *Mathematische Grundlagen der Quantenmechanik* [4] (not directly accessible to Bell because it had not yet been translated from German).

2. Bell took seriously the two papers by *Bohm* of 1952 [5], in spite of their dismissal by most of the physics community on various grounds (e.g. that Bohm's model is a reactionary reversion to classical mechanics, that Bohm favors the position representation and thereby disregards the democracy of bases characteristic of quantum mechanics, that Bohm's model does not make predictions in the atomic domain different from those of quantum mechanics and therefore gives only unnecessary innovations). As Bell viewed the matter, Bohm's model is a counter-example to von Neumann's claim that a hidden variables interpretation is incompatible with quantum mechanics, thereby showing that there is a field for further investigation. Again, Bell's resistance to prevalent opinion and to indecisive argumentation were manifestations of character.

3. When Bell learned the details of the proof given in von Neumann's book (probably from discussions with *Franz Mandl* [6] at Harwell in 1952) he noticed the implausibility of von Neumann's assumption that in a hidden variables model the expectation value of the sum of any two physical quan-

tities equals the sum of their expectation values singly. This assumption is correct in quantum mechanics even when the quantities in question are incompatible, i.e. represented by non-commuting self-adjoint operators; but that is a peculiarity of quantum mechanics, not to be anticipated in a hidden variables model, since the procedure for measuring the sum is incompatible with the procedures for measuring the summands separately. Thus, although the proof of von Neumann's theorem is mathematically correct, its application to physics is undermined by the physical unreasonableness of one of its premisses. This argument is presented in Sect. 5.3 of his 1966 paper "On the problem of hidden variables in quantum mechanics" [7]. Furthermore, in Sect. 5.2 of this paper Bell constructed, for a spin-1/2 particle without translational motion (associated by quantum mechanics with a Hilbert space of dimension two), a counter-example different from Bohm's to von Neumann's claim that a hidden variables model of quantum mechanics is mathematically impossible. Bell's tenacity in pursuing this investiagation in spite of von Neumann's prestige is another indication of his character.

4. In Geneva Bell heard from *Jauch* [8] about *Gleason*'s powerful theorem [9] that the only probability measures on the lattice of subspaces of a Hilbert space of dimension greater than two are those provided in the standard way by statistical operators (quantum mechanical pure states being special cases). A condition for the probability measure assumed by Gleason is that the probability of the sum of finitely or denumerably many *commuting* projections equals the sum of the probabilities of those projections singly. The restriction to *commuting* projections frees Gleason's theorem from the taint of physical inapplicability that affected von Neumann's theorem. The impossibility of a hidden variables model for a quantum mechanical system of dimension greater than two, in the sense of a definite simultaneous assignment of values to every self-adjoint operator of the system, is a simple corollary of Gleason's theorem, and this corollary was of greater intrest to Bell than the main theorem itself. The model that Bell had constructed as a counter-example to von Neumann is consistent with Gleason's theorem, since the Hilbert space involved has dimension two. Although Bell was inclined to believe that Gleason's theorem is correct, the proof is very hard to follow, and Bell remarked (at a conference, perhaps Varenna 1970) that he either had to understand Gleason's proof or invent a proof of his own of the corollary which concerned him, and he knew the latter would be easier. And that is what he did in [7]. (A proof of the same corollary was independently published by *Kochen* and *Specker* [10] a year later). Bell's tenacity in obtaining his own demonstration of the impossibility of a program of hidden variables that had long been congenial to him is another manifestation of character.

5. But there is a remarkable turnabout at the end of section 5 of [7]. He points out that his theorem earlier in that section does not, after all, completely demolish the hidden variables program.

"*That so much follows from such apparently innocent assumptions leads us to question their innocence ... It was tacitly assumed that measurement of an observable must yield the same value independently of what other measurements are made simultaneously. Thus as well as $P(\phi_3)$ say, one might measure either $P(\phi_2)$ or $P(\psi_2)$, where $\phi_2$ and $\psi_2$ are orthogonal to $\phi_3$ but not to one another. These different possibilities require different experimental arrangements; there is no reason to believe that the results for $P(\phi_3)$ should be the same. The result of an observation may reasonably depend not only on the state of the system (including hidden variables) but also on the complete disposition of the apparatus; see again the quotation from Bohr at the end of section 1.*"

Bell has thus kept the program of hidden variables alive by suggesting an entirely new class of hidden variables models, which have come to be called "contextual hidden variables models". Character is operative here. Bell had proved an important theorem, which seemed to accomplish what von Neumann had failed to do, that is, to exhibit the impossibility of the class of hidden variables theories which had been the center of attention; but instead of trumpeting the decisiveness of his theorem, Bell exhibited a physically plausible weakening of his own premisses, thereby giving a new lease of life to the hidden variables program as a whole. I remark tangentially that there is delicate humor both in "leads us to question their innocence" and in the citation of Bohr, a strong opponent of hidden variables theories, as support for the proposal of a new class of such theories. Self-confidence and a freely ranging mind, both of which are aspects of strong moral character, are the underpinnings of Bell's humorous sallies.

6. The foregoing steps bring us to the threshold of Bell's Theorem. As noted in step 2, Bell took Bohm's hidden variables model of 1952 seriously. In particular, he paid attention to a feature of the model that Bohm candidly discussed – namely, its manifest nonlocality.

If the systems under consideration consists of two particles, the measurement of the position of one of them effectively reduces the wave packet, thereby instantaneously changing the values of observables of the other particle. Bell then asked whether the nonlocality of Bohm's model could be removed by a more clever construction. This question stimulated the investigation of his paper [11] of 1964, in which the first version of Bell's Theorem was formulated and proved: no hidden variables model satisfying a certain condition of locality and assigning definite values to all observables can agree with all the predictions of quantum mechanics. Also heuristic for Bell's Theorem was reflection on the contextual hidden variables theories proposed in [7]. The contexts of measurement in a model studied in section 6 of that paper are grossly nonlocal, leading Bell to comment:

"*It would therefore be interesting to pursue some further 'impossibility proofs', replacing the arbitrary axioms objected to above by some condition of locality, or of separability of distant systems.*"

I do not know the exact chronology of the heuristics of Bell's Theorem and would very much welcome further information on this fascinating matter; the historical accident that [11] was published before [7], even though [11] presents a later discovery, is one cause of obscurity concerning the chronology. I can only say that two currents of thought in his previous investigations on the foundations of quantum mechanics seem to have flowed together to suggest the crucial question that his theorem answers, and that Bell's moral character was exhibited in following them tenaciously to a wonderful conclusion.

The evidence assembled to support the thesis of Sect. 5.1 has been drawn from Bell's written works, supplemented by a few remarks at conferences. It is important to add, however, that the character manifested in his writings appeared also in his speech and demeanor, in his behavior towards the people he encountered, and in his social and political beliefs. John Bell was a dignified and somewhat reserved man, who must have felt a need to protect his privacy; but this reserve was combined with courtesy, consideration, and generosity. He was passionate in his pursuit of clarity, understanding, and social justice, with the consequence that his argumentation was sharp; but a benevolent humor almost always mollified his critical remarks. I had the impression that Bell's presence in a gathering raised the collective level of thinking, speaking, and listening. Others who attended meetings with him may testify to the same impression.

## 5.2 Some Reminiscences

(*i*) In 1969 I had a brief exchange of letters with Bell, but met him for the first time at one of the sessions of the International School of Physics "Enrico Fermi" in Varenna on Lake Como in the midsummer of 1970. Almost immediately after we met, he asked me whether I had proved a generalization of his theorem without using the assumption that the hidden variables fix the values of all the quantum mechanical observables. He asked this, evidently, because he did not want a much less well known person than himself, who might have obtained this generalization, to lose credit for the discovery. I told him that I had indeed thought about the generalization but had not succeeded in proving it, and then he felt free to present it in his lecture (see paper 4 in *Speakable and Unspeakable in Quantum Mechanics*, [1]). Here is an instance of his consideration and generosity, that I mentioned above in general terms.

(*ii*) During the academic year 1977–1978 I was a visitor at the Department of Physics of the University of Geneva and occasionally visited Bell at CERN. My most memorable encounter with him, however, was at Seelisberg, Switzerland, near Lucerne, the location of MERU – Maharishi European Research University. This institution was founded by Maharishi Mahesh Yogi, who had developed the techniques of Transcendental Meditation. I was told that Maharishi had a bachelor's degree in physics, and he certainly was interested in physics and in its possible relation to the mind. On his staff of advisers there were several physicists, and several famous physicists used to visit MERU in order to meditate and to study with him. In late winter or early spring of 1978 Maharishi organized a two-day conference entitled "Physics and Consciousness", to which a number of physicists and philosophers in western Europe were invited, among them Bell and I. I decided to go out of curiosity, if for no other reason than to make a report to my wife Annemarie, who was an anthropologist. I was not disappointed in the least, for it was one of the most exotic experiences of my life. MERU's site was very beautiful, the vegetarian cuisine was excellent, and the disciples in saffron robes created a serene atmosphere. There was an EKG laboratory, where we were shown tapes of people in several stages – before meditating, during meditating, and then coming out of the meditating state – and it was remarkable how much more regular, in almost all cases, the alpha rhythms were during meditation than before or after. We were told, I suspect without exaggeration, that this laboratory had the largest bank in the world of tapes of people in the meditating state. Various banners and placards in the main rotunda of MERU, with slogans like "Fifth Year of World Government" and "Invincibility to All Nations", made it difficult for my skeptical mind to suspend disbelief, although I did acknowledge that meditation was psychologically helpful to some people.

I had the impression that Bell was initially more favorably inclined than I towards the discipline, if not the doctrine, of Transcendental Meditation. He told me that he had tried it himself and it had not worked for him, but that he personally knew people who had benefited from it.

We gathered in an auditorium to discuss the topic of the conference. Maharishi sat on a dais, holding a flower, and looking at the screen of a small closed circuit television set, where he could see any one who spoke. The proceedings began with a talk by one of Maharishi's assistants, a physicist who was able to summarize certain physical phenomena and theories that he considered relevant to the doctrine of Transcendental Meditation. Maharishi then spoke and wove together some of what the assistant said with his own psychological theses. As I recall, two propositions were emphasized. One was the peculiarity of the vacuum in quantum field theory, which is not a Democritean void, but is rather an entity rich in potentialities and correlations. This vacuum he compared to the mind in the meditating state, when it is emptied of mundane content, and then is not void but rich and fertile. The other proposition was that amplitudes in quantum mechanics are added prior to taking an absolute square in order to obtain a probability, so that doubling the amplitude implies quadrupling the probability. The corollary, Maharishi claimed, is that ten thousand people meditating coherently would have a net psychological effect equal to one hundred million people meditating incoherently. This corollary was known at MERU as "the Maharishi Effect", and it was the basis for the *prima facie* strange thesis that the relatively small number of devotees of Transcendental Meditation could change the world. The explanation given was that coherence guarantees an effect proportional to the square of the number of coherent meditators.

After Maharishi's lecture the audience was invited to participate. My memory is that Bell did not comment at once, but eventually he did. He was courteous and respectful to Maharishi, and I had the feeling that there was some tension between his intention to be polite and his intellectual conscience. In his culminating statement the latter predominated. He said that the Maharishi Effect is only "the shadow of an analogy" unless there are controlled experimentation and analysis. I regret that I do not have a verbatim report of his entire statement. If there are unpublished notes by Bell about the episode, their publication would be valuable as a corrective to undisciplined writing on the relation between physics and psychology.

(*iii*) In the summer of 1990, only a few months before Bell's death, we both attended a workshop on the foundations of quantum mechanics at Amherst College. The lecture "Against 'Measurement'" [12] which he gave at this workshop is an important addition to his famous collection of papers *Speakable and Unspeakable in Quantum Mechanics* [1], published three years earlier. We discussed, took our meals, and slept in Marsh House, a very pleasant institution at the College. Menus were handed to the participants at the beginning of each meal, and at the first breakfast Bell was puzzled by an item on the menu and asked me, "What is a *bagel*?" I was delighted to know something that Bell didn't know and answered him, "A bagel is a toroid." He then asked, "Does it matter what it is made of?" I said, "Yes, it is made of dough", but I regret omitting the prescription that the dough should be boiled in order to achieve the proper texture. Although Bell was a consummate theoretical physicist, he was also – witness his expertise in accelerator physics – a practical man.

## 5.3 In What Direction Does Bell's Theorem Point?

The conclusion of Bell's famous paper "Bertlmann's Socks and the Nature of Reality" [13], states four possibilities, with no pretense at exhaustiveness, concerning the interpretation of his theorem and of the experiments inspired by it. He expresses reservations about all of them, but seems guardedly to prefer the third:

"*it may be that causal influences <u>do</u> go faster than light. The role of Lorentz invariance in the completed theory would then be very problematic. An 'aether' would be the cheapest solution .... But the unobservability of this aether would be disturbing. So would the impossibility of 'messages' faster than light, which follows from ordinary relativistic quantum mechanics in so far as it is unambiguous and adequate for procedures we can actually perform. The exact elucidation of concepts like 'message' and 'we', would be a formidable challenge.*"

Similar reservations are expressed in "Beables for Quantum Field Theory" [14], where elements of reality, "beables", are used to construct an alternative to ordinary quantum field theory, which centers on "observables". Near the conclusion he says,

"*So I am unable to prove, or even formulate clearly, the proposition that a sharp formulation of quantum field theory, such as that set out here, must disrespect serious Lorentz invariance. But it seems to me that this is probably so.*"

Bell was evidently willing to entertain radical innovations in physics, and to devote some effort to investigating them mathematically, in order to do justice to the apparent nonlocality revealed by correlation experiments and at the same time to hold on to his conception of a physical reality underlying observations.

I wish now to say why I think Bell's Theorem points to a yet more radical proposal than that of the foregoing quotations. Bell's Theorem shows that there is some tension between quantum mechanics and the space-time structure of special relativity, even though the impossibility of using quantum mechanical entanglement to send superluminal messages prevents outright inconsistency between the two theories. There are, however, two other areas of tension between quantum mechanics and contemporary space-time theory, concerning general rather than special relativity. First, in general relativity the metric field is a dynamical entity rather than a fixed structure, and therefore it has to be quantized if quantum mechanics applies to physical reality in full generality. But the difficulties encountered in attempting to quantize general relativity have been so great that one suspects them to be not be merely technical and mathematical in character but conceptual. Second, at the Planck level – around $10^{-33}$ cm – the quantum fluctuations of the space-time metric become as large as or larger than the expectation values of lengths under consideration, so that the metric structure may no longer be well defined. Both of these difficulties suggest that the union of quantum mechanics with general relativity may require the modification of one or both. It is not unreasonable to hope that if these difficulties are resolved, the tension that Bell's Theorem exhibits between quantum mechanics and special relativistic locality may find a deep and natural resolution.

For these reasons, the radical program of *Michael Heller* and his collaborators [15], based on the noncommutative geometry of *Alain Connes* [16], is very attractive. This program aims at generalization of space-time theory that is appropriate below the Planck threshold. The "pregeometry" of the theater of physics is postulated not to be a differentiable manifold, and is fundamentally nonlocal. Indeed, the concepts of a point and of a neighborhood are not well defined in this pregeometry, and hence there are no locally defined tangent spaces with space-like and time-like vectors. What takes the place of the point as a basic entity of the theater of physics is something analogous to a quantum state, which is an intrinsically global entity even in ordinary quantum mechanics. One may wonder, of course, how a theater in which concepts like point, neighborhood, and continuity are meaningless could accommodate anything like dynamics. But the pioneers of the new program have made use of the algebraic definition of a derivation – an operation that is linear and obeys Leibniz's rule – to generalize differential geometry without depending upon these familiar concepts. Among the virtues ascribed to the pregeometry is elimination of the dichotomy of singular and regular space-time points – since there are no points!

To be physically relevant, the program based on noncommutative geometry must be able to recover ordinary general relativity and ordinary quantum mechanics above the Planck threshold. They propose – but seem only partially to have mathematically described – certain phase transitions from the pervasively global sub-Planck domain to a domain where one can meaningfully speak of locality, neighborhoods, number of particles, and other familiar mathematical and physical concepts. Ordinary quantum mechanics and ordinary general relativity are attained in appropriate correspondence limits. The nonlocality implied by Bell's Theorem and Bell-inspired experiments holds even in the correspondence limits, as a residue of the pervasive nonlocality of the sub-Planck domain. Instead of being a mystery or an anomaly, Bell's nonlocality is a phenomonon comprehensible in terms of a deep and systematic theory, which simultaneously aims to clear up other mysteries of contemporary physics.

After this effusive summary of some ideas of Connes and Heller et al. I have to admit that I am not a well qualified advocate. I know very little of their mathematics and am particularly baffled by the phase transitions postulated in their program, which do not seem to be illuminated by analogies with the familiar phase transitions of condensed matter physics. Nevertheless, I am so impressed by what I do understand of their program that I intend to devote much effort to it in the years to come.

What would Bell's reaction be? That is a question that goes beyond respectable psycho-history. I should like to think that he would find the program attractive. But another scenario is also possible. After listening courteously Bell would say, "The mathematics of this program is very difficult; it is masochistic mathematics. I prefer something simpler." To which I would reply, "So do I, but remember Einstein's maxim, 'Make things as simple as possible, but not simpler.'" At that point he would smile skeptically, but with good-nature and tolerance, and say, "Good luck to you."

## References


1. J.S. Bell: On the Einstein–Podolsky–Rosen paradox, Phys. **1**, 195 (1964), and Introduction to the hidden-variable question, in *Foundations of Quantum Mechanics*, ed. by B. d'Espagnat (Academic, New York 1971), pp. 171–181
   The second paper proves a stronger version of Bell's theorem than the first, for it assumes that the hidden variables ascribe only probability distributions rather than definite values to the outcomes of measurements. Both of these pioneering papers are reprinted in
   J.S. Bell: *Speakable and Unspeakable in Quantum Mechanics* (Cambridge Univ. Press, Cambridge 1987). Papers 7, 10, 13, and 16 of this collection also present expositions of and comments on the Theorem
2. A. Whitaker: John Bell in Belfast: Early Years and Education, this volume
3. M. Born: *Natural Philosophy of Cause and Chance* (Oxford Univ. Press, London 1949)
4. J. von Neumann: *Mathematische Grundlagen der Quantenmechanik* (Springer, Berlin 1932). English translation by
   R.T. Beyer: *Mathematical Foundations of Quantum Mechanics* (Princeton Univ. Press, Princeton 1955). In the translation the proof of the impossibility of hidden variables is on pp. 305–324
5. D. Bohm: A suggested interpretation of the quantum theory in terms of 'hidden variables', Parts I, II, Phys. Rev. **85**, 166 (1952); ibid p. 180 (1952)
6. J. Bernstein: *Quantum Profiles* (Princeton Univ. Press, Princeton 1991)
7. J.S. Bell: On the problem of hidden variables in quantum mechanics, Rev. Mod. Phys. **38**, 447 (1966)
8. See [7], footnote 4
9. A. Gleason: Measures on the closed subspaces of a Hilbert space, J. Math. Mech. **6**, 885 (1957)
10. S. Kochen, E. Specker: The problem of hidden variables in quantum mechanics, J. Math. Mech. **17**, 59 (1967)
11. J.S. Bell: On the Einstein–Podolsky–Rosen paradox, Phys. **1**, 195 (1964)
12. J.S. Bell: Against 'Measurement', Phys. World **3**, 33 (1990)
13. J.S. Bell: *Speakable and Unspeakable in Quantum Mechanics* (Cambridge Univ. Press, Cambridge 1987) pp. 154–155
14. J.S. Bell: *Speakable and Unspeakable in Quantum Mechanics* (Cambridge Univ. Press, Cambridge 1987) p. 180
15. M. Heller, W. Sasin, D. Lambert: Groupoid approach to noncommutative quantization of gravity, J. Math. Phys. **38**, 5840 (1997)
    M. Heller, W. Sasin: Emergence of time, Phys. Lett. A **250**, 48 (1998)
    M. Heller, W. Sasin: Einstein–Podolski–Rosen experiment from noncommutative quantum gravity, in *Particles, Fields, and Gravitation*, ed. by J. Rembielinski (Am. Inst. Phys., Woodbury, NY 1998) pp. 234–241
    M. Heller, W. Sasin: Noncommutative unification of general relativity and quantum mechanics, Int. J. Theor. Phys. **38**, 1619 (1999)
16. A. Connes: *Noncommutative Geometry* (Academic, New York 1994)

# 6 Early History of Bell's Theorem

John F. Clauser

This article is dedicated to the memory of John Bell, whose work exerted a profound influence on my own life and professional career as an experimental physicist. In this article I attempt to recount as accurately as possible the important events in the development of one of the most profound results in physics of the twentieth century – Bell's Theorem.

## 6.1 Introduction

Bell's Theorem provides us with the somewhat rare opportunity for experimental physics to answer questions in natural philosophy. Correspondingly, *Abner Shimony* and I open our 1978 review article, *Bell's Theorem: experimental tests and implications* [1], with the sentence,

"*Realism is a philosophical view in which external reality is assumed to exist and have definite properties, whether or not they are observed by someone*".

In that review we describe how the experimental evidence in hand in 1978 allows us reasonably to assert that

"*The conclusions* (from Bell's Theorem) *are philosophically startling; either one must totally abandon the realistic philosophy of most working scientists, or dramatically revise our concept of space-time.*"

Despite numerous repetitions and extensions of the reviewed experiments and theory, along with considerable scrutiny of both, our 1978 conclusion appears to stand.

The present-day heuristic impact of Bell's Theorem is immense. It clearly and explicitly reveals highly peculiar and remarkable properties of quantum-mechanically "entangled" states that were previously not appreciated. Its elucidation of these properties, in turn, has inspired various efforts, both to quantify and to exploit entanglement on a wide variety of physical systems. As a result, it has directly or indirectly fostered many spin-off practical applications and created whole new areas of study, including quantum cryptography, quantum communication, quantum computing, entanglement sharing, quantum cloning, etc. The spin-offs from Bell's Theorem are now so widely ranging that it is difficult to open any recent issue of Physical Review Letters or Nature without finding at least one article that is a legitimate progeny. Its concepts are now common in undergraduate physics textbooks, as well as in a large body of scientific literature for the layman. One colossal recent spin-off proposal is by the US National Security Agency, who now envisions a quantum Internet that is absolutely secure against eavesdropping, and that is linked together by an international network of quantum computers, wherein quantum entanglement is then shared world-wide via a network of satellites. Thus, not only has Bell's Theorem deepened our understanding of quantum mechanics and the foundations of physics and natural philosophy, its full impact far exceeds the realm of natural philosophy, and has become an important conceptual tool for promoting revolutionary useful technology.

Bell's Theorem also did one more very important thing. It cleaned up a very important "untidiness" left by the founding fathers of quantum mechanics. The word "untidy" is used here to describe a state of affairs in which the physics community had become bifurcated into two camps, with each camp confidently holding beliefs that were just the opposite of that of the other camp, with neither camp holding any experimental justification for its beliefs, with opposite appropriate future directions for physics research implied by these opposing beliefs, and with any rational discussion by these two groups about their differences prohibited by a ludicrous and debilitating social convention. How did this absurd situation arise, and how did Bell's Theorem end it?

This note provides an overview of the history (through the early 1980s) of the development of Bell's Theorem. As a parallel narrative, I offer some first-hand personal recollections about the conduct of physics during this development period. Young physicists may find it difficult to believe that thirty years ago, most of the above ideas and subject matter represented forbidden thinking for practicing physicists. Indeed, any open inquiry into the wonders and peculiarities of quantum mechanics and quantum entanglement that went outside of a rigorous "party line" was then virtually prohibited by the existence of various religious stigmas and social pressures, that taken together, amounted to an evangelical crusade against such thinking. As a result of this evangelism, much of the early important work on Bell's Theorem was published only in an "underground" newspaper[1], whose circulation was limited to members of a "quantum-subculture", and that probably cannot be found in most physics libraries. The achievement of the results that led to our review article may then be seen to be all the more remarkable, given then existing sociological barriers to normal scientific inquiry.

The reader may soon feel that much of my portrayal of this history resembles a strange mixture of serious physics with a cheap soap opera. While my inclusion of the latter aspect of this history may detract from my portrayal of the former, I beg the indulgence of reader. I include some of

[1] *Epistemological Letters* (Association Ferdinand Gonseth, Institut de la Methode, Case Postale 1081, CH-2501, Bienne.) This newspaper was somewhat unique for its time, in that it openly proclaimed that the usual stigma against hidden-variable theories, and the like, was to be absent for publications within it.

the sleazy aspects of this history (but still attempt to portray it as accurately as possible), because it had a significant impact on both the workers and on the works of this field, because it may explain why some important ideas were not published in a timely fashion (including much of John Bell's work), because it explains why many important published ideas languished unread for prolonged periods (including John Bell's most important 1964 paper), and finally, because, to my knowledge, it is not described elsewhere. I apologize ahead of time to anyone who is either offended or bored by my sacrilegious approach. Such "sensitive" readers are advised to bypass the two Sections entitled, "Evangelical theoreticians dominate", and "The development of a 'stigma'".

Also, what the heck! How often is a simple experimental physicist (who spends much of his time mucking about in cutting oil and chasing down vacuum-system leaks and electronics and software bugs) given poetic license to offer his own editorial opinions about the sociology of physics? I sincerely hope that John Bell would have approved.

## 6.2 An Unresolved Issue Left by the Founding Fathers

While quantum mechanics was being initially formulated, there were many discussions about its meaning, its self-consistency, and about whether or not it provides a complete description of physical reality. Albert Einstein's famous debate with Niels Bohr at the 1927 Solvay Conference was primarily focused on the self-consistency of quantum mechanics *vis-à-vis* Heisenberg's uncertainty principle and the Copenhagen interpretation. At the end of the debate, Einstein conceded that quantum mechanics is indeed self-consistent. However, he remained adamantly undeterred in his belief that quantum mechanics, as a framework for physics, is incomplete. In his view, additional "hidden parameters" are needed to supplement quantum mechanics, and thereby to provide a full and complete description of physical reality.

If quantum mechanics is an incomplete theory, then the probabilities that one calculates from it are similar to those defined by Bernoulli and Pascal, i.e. they are quantitative measures of ignorance. Via classical probability theory, probabilities are modeled classically as sampling frequencies, in a conceptual parallel to the sampling of colored balls with varying weights from an urn. That is, given a blind choice of a ball from an urn, one estimates, for example, the probability (i.e. the sampling frequency) for choosing a ball with say a prescribed color. Alternatively, since the balls have weight as a second distinguishing characteristic among them, one may also estimate a probability for obtaining a prescribed weight. If the ball's weights are somehow correlated with their colors, one may also estimate a joint probability distribution of weight and color for these balls. In classical probability theory, however, no one ever doubts the existence of some actual set of balls within the urn, immediately prior to a sampling. That set correspondingly still has some preexisting color distribution, weight distribution, and joint color and weight distribution, although these distributions' properties may or may not be known. Thus, even when, "weight" and "color" are not quantum-mechanically "commuting observables", Einstein still expected that probabilities arise within quantum mechanics in exactly the same way that they do in classical probability theory, i.e. through ignorance of details about the sampling process, and/or ignorance of details of the parent distribution.

The measurement of some characteristic feature of a quantum-mechanical system is, however, not quite as simple as simply weighing a ball and/or observing its color. During Einstein's initial discussions with Bohr, Bohr talked about a "physical disturbance" that occurs during a measuring operation that is caused by the measuring apparatus, itself. Given only a limited set of then-known apparatus choices, Bohr argued that this disturbance is unavoidable in a quantum-mechanical measurement. In effect, Bohr was saying that one can measure a ball's weight, or its color, but not both, and somehow a measurement of a ball's weight can change its color. Nonetheless, the existence of such a physical disturbance mechanism is fully consistent with Einstein's view, and does not contradict any classical probability concepts. In Einstein's view, random results occur here because one is ignorant about both the details of the physical disturbance mechanism and the details of the *à priori* probability distribution of any physical variable or set of variables being measured. If one somehow were given a hypothetical "improved" apparatus that produces only a negligible disturbance, then the preexisting value of a variable could be measured in a single trial, and the *à priori* probability distribution could be determined with a large set of trials. In effect, such a hypothetical apparatus can still weigh the ball without thereby changing its color. Equivalently, given such a hypothetical apparatus, one can then measure simultaneously the preexisting values of two non-commuting variables, along with their *à priori* joint probability distribution. Bohr's initial arguments thus offered nothing to Einstein to counter his notion about the origins of randomness within quantum theory, and Einstein continued to believe that the uncertainty principle sets only practical limits to measurement precision, and that it is not an inherent fundamental limitation to the precision of quantum measurements.

Bohr, however, also maintained that quantum mechanics is a complete theory. If so, then the randomness observed in experimental outcomes is something very different from the probabilistic behavior of events envisaged by Bernoulli and Pascal, and the non-existence of said hypothetical apparatus is highly relevant. In Bohr's view, probabilities arise within quantum mechanics, not through ignorance of the internal workings of nature, but in a fundamentally new and different way. In a complete theory, there is no preexisting *à priori* joint distribution of actual values that one may attempt to measure, since that complete distribution is not prescribed by quantum mechanics.[2] The prohibition against peeking inside the urn prior to sampling is now one of principle, and is not due to a limitation of the availability of suitable measuring devices. Bohr thus insisted that there is no such distribution, nor are there any details available in principle about the disturbance mechanism. In such case, one may not ask questions about details of said distributions, nor may one ask questions about details of the measurement process for any given trial.

A moment's reflection will convince the reader that Bohr's insistence on the nonexistence of an *à priori* joint distribution of actual values, is also a denial of the preexistence of any one value in any one trial. In essence, Bohr is denying the whole realistic philosophy, as defined above in the opening paragraph.[3] The random results observed in a quantum-mechanical measurement of a specified parameter are somehow a manifestation of the lack of a preexistence of a value for that parameter. Einstein evidently recognized that Bohr was denying realism. At one point in their discussions, Einstein reportedly asked Bohr "Do you really believe that the moon doesn't exist when no one is looking at it?" A recognition that the basic difference between their opposite views is simply acceptance versus denial of realism, then clarifies both of their logically following conclusions. Einstein is often quoted as saying that "God does not play at dice!" A more cogent metaphoric response that describes Bohr's view is then "Of course not; God has no dice to play with!" The apparent nonexistence of the abovementioned hypothetical "improved" apparatus is a *necessary* condition for Bohr's view to hold. However, from Einstein's view, issues about the existence or nonexistence of such an apparatus are all red herrings, since the apparent nonexistence of said hypothetical apparatus is, by no means, a *sufficient* condition for a denial of realism. Neither man, however, expected that the whole realistic philosophy would become amenable to experimental testing. It is thus probable that if Bell's Theorem had been available to them, their discussions and debates would have been remarkably different, and that both men would have taken keen interest in Bell's Theorem!

Whether or not quantum mechanics is a complete theory is a nontrivial question, and its answer significantly determines the appropriate direction of future research in physics. Clearly, if quantum mechanics is incomplete, then a complete theory presumably exists that should be sought. On the other

---

[2] The Wigner distribution is obviously not such a distribution, since it is not always non-negative, and thus it cannot be viewed as a sampling-frequency distribution.

[3] Max Jammer, in Sect. 4.2 of his book, *The Conceptual Development of Quantum Mechanics* (McGraw-Hill, New York 1966) argues that Bohr's denial of realism was in response to his being strongly influenced by the contemporary philosophers, Kierkegaard and Høffding. Chapter 4 of this book is noteworthy in that it summarizes the important experimental results available when quantum mechanics was originally being formulated, while Chap. 8 provides an abbreviated list of experimental confirmations of the theory that were then promoted universal belief in the "standard religion".

hand, if it is complete, then one may ask Einstein's questions, but one can never hope to find their answers, nor can one even hope to understand why it is vain to ask these questions.

## 6.3 An Untidy Legacy Left by the Founders of Quantum Mechanics

Soon after the Solvay conference debates had ended, two heuristically important works were published. Both were written for the purpose of settling the completeness issue, once and for all. Unfortunately, the opposite result occurred, since these two works reached exactly opposite conclusions. In turn, these were immediately followed by additional works that further sharpened their content.

One of these two important works is by *John von Neumann.* In his book, *Mathematische Grundlagen der Quantenmechanik* [2], von Neumann offers a "proof" of the non-existence of "hidden variables" within quantum mechanics. As a basis for this mathematical proof, he assumes that a linear operator (on a Hilbert space) is associated with each measurement apparatus, and that any measured result from this apparatus is an eigenvalues of this operator. He then proves that, within the mathematical structure of quantum mechanics, there cannot exist any additional (hidden) variables that then may be used to distinguish any further differences between two particles in the same "pure" state. Thereby, he shows that quantum mechanics is a complete theory. First, he concludes that

"... *The only formal theory existing at the present time which orders and summarizes our experiences in this area in a half-way satisfactory manner, i.e. quantum mechanics, is in compelling logical contradiction with causality. Of course it would be an exaggeration to maintain that causality has thereby been done away with: quantum mechanics has, in its present form, several serious lacunae, and it may be even that it is false, although this latter possibility is highly unlikely, in the face of its startling capacity in the qualitative explanation of general problems, and in the quantitative calculation of special ones.* ..."

Further on, he strengthens this conclusion by stating that

"... *we may say that there is at present no occasion and no reason to speak of causality in nature – because no experiment indicates its presence, since the macroscopic are unsuitable in principle, and the only known theory which is compatible with our experiences relative to elementary processes, quantum mechanics, contradicts it.* ... *Under such circumstances, is it sensible to sacrifice a reasonable physical theory for its sake?*"

Given the careful wording of these passages it should be noted, in all fairness to von Neumann, that he does not prove, nor does he claim to prove the nonexistence of some other alternative theory that does not use his linear operator calculus, and whose predictions are equivalent to those of quantum mechanics. He just suggests that he believes that the existence of such an alternative theory is "highly unlikely". Unfortunately, many subsequent workers have misinterpreted his work as a "proof" of the non-existence of any such alternative theory.

The other one or the two important works is by *Einstein*, *Podolsky* and *Rosen* [3] (EPR). In their paper, they construct a hypothetical configuration that involves a pair of spatially separated but quantum-mechanically correlated particles. They further specify that these two particles are sufficiently far apart that it is impossible for the particles to physically interact with each other in any way. In this manner, EPR evade Bohr's "physical-disturbance" argument. EPR contend that a measurement at one location on one particle of the pair cannot "physically disturb" the other particle, nor can it disturb the other particle's measurement outcome. In the above ball and urn analogy, there are now two urns. At one urn, the ball's weight is measured, and at the other, the ball's color is measured. Knowing say the color correlation between the two balls, one can then simultaneously determine each ball's weight and color. Quantum mechanics, however, does not simultaneously specify these two parameters, and therefore it must be considered incomplete. Correspondingly, it then must be supplemented with "hidden parameters" that provide such a specification for it to become complete.

*Bohr* then offered an important rebutal [4] to EPR that further complicated the saga. It significantly diminished the elegance of EPR's careful reasoning by its providing an important counter-argument. Bohr therein contends that, depending on how one words the "completeness" question, EPR's configuration can be viewed in a manner that escapes their conclusion. While many physicists view Bohr's arguments as a "refutation" of EPR's arguments, in fact, Bohr simply shows that his own view may be held consistently, even for EPR's configuration. First he concedes a lack of full generality to his own earlier arguments regarding a (presumably local) physical disturbance of either particle by its associated apparatus. Thus, he generalizes his earlier idea of a local physical disturbance now to include a possible global disturbance to the two-particle composite system by the two-component composite apparatus. If either of the component apparatuses that comprise the global apparatus is changed, then the net global disturbance by the composite apparatus is also changed. Since this global disturbance interacts with the composite two-particle system, then the predicted outcome for the joint measurement is also changed.

To anyone intent on obtaining a microscopic view of details of this interaction process, then action-at-a-distance clearly seems to be necessary for Bohr's global disturbance to occur. Bohr's argument is more readily understood, once that one recognizes that it is based on his denial of realism. That is, it is impossible to physically disturb something that doesn't exist! Thus, if there is no physical-space description of either the quantum mechanically entangled individual systems or of the actual disturbance process, then the required associated existence of non-local action-at-a-distance is a non-issue. The objectionable aspect of Bohr's description (action-at-a-distance) is then, consistently with Bohr's assertion of completeness, simply nonexistent, by definition!

To an avowed realist (such as Einstein), Bohr's "refutation" of the EPR argument seems like simply playing with words. Thus, Einstein remained adamantly unconvinced by Bohr's argument, saying,

"*To believe this is logically possible without contradiction, but it is so very contrary to my scientific instinct that I cannot forego the search for a more complete conception.*"

Reportedly, Einstein was also unconvinced by von Neumann's argument, but never published a critique of it.[4] Thus, despite the aims of von Neumann (and Bohr) and of EPR, their antipodal conclusions left the completeness issue in a very untidy state.

Additional related comments about EPR were then offered by *Schrödinger* [5] and by *Furry* [6]. They both considered what would happen if the quantum-mechanical correlation between the particles somehow ceased to exist, once the two particles became sufficiently separated. Both men arrive at exactly the same conclusion. If this were to happen, then the specific experimental predictions by quantum mechanics would be altered significantly. Schrödinger's paper is also noteworthy in that it introduces the term "entanglement" into our vocabulary for describing the two-particle correlation. Interestingly, these two authors also offer exactly opposite personal impressions of entanglement. Schrödinger calls entanglement "sinister", and views its presencc as indicating a fatal flaw in quantum mechanics. Furry, on the other hand, views these result as indicating that any modified theory that

[4] Unfortunately, there does not appear to be any evidence that von Neumann and Einstein (whose offices at Princeton were just down the hall from each other) ever directly discussed the validity of von Neumann's "completeness proof". However, Abner Shimony relates a story, told to him by Peter Bergmann, about a time when he (Bergmann) asked Einstein for an opinion of this proof. Reportedly, Einstein was quite familiar with it. In response to Bergmann, he fetched and opened von Neumann's book, and then pointed to one of von Neumann's assumptions, upon which the proof is based. He then said that he saw absolutely no reason to believe that this assumption should hold in general for all alternative theories. Sadly, Einstein never published his criticism of von Neumann's proof. Evidently, the assumption that he then fingered is the same one (linearity) that is later criticized by Bell in his 1965 paper, and found there to be untenable. Given the careful wording of von Neumann's above-quoted passages, perhaps Einstein saw no compelling reason to publish such criticism, since von Neumann, himself, never actually claimed in his book that hidden variables are impossible.

omits entanglement must be immediately viewed as wrong, since quantum mechanics is "obviously" a correct theory.

## 6.4 Evangelical Theoreticians Dominate

The untidy legacy left by the above authors was particularly acute, since neither side attempted to justify its position with hard experimental data. It was automatically assumed by them that, without any doubt, quantum mechanics is "obviously" *the* correct theory.[5] Indeed, none of the above authors ever cite any experimental evidence at all to back up their arguments. Instead, hard experimental evidence is replaced in the above arguments by a simple reliance on the self-consistency of the theory, whereupon only *Gedankenexperiments* are then needed for these arguments to proceed. Thus, it seems that all of these authors believed in what I will call here the "standard religion", that may be summarized as follows:

*If a theory is self-consistent and elegant, and if it explains a significant body of experimental data that are inconsistent with previously held theories, then the theory* must *be accepted as gospel. Correspondingly, the theory needs no further testing, even in areas where its predictions may seem to be surprising and/or paradoxical.*

In later years, the phrase "it needs no further testing" became reinterpreted to mean, *it is sacrilege to even suggest that further tests should be performed.*

Said "religion" follows, in turn, directly from the above-quoted passages by von Neumann. Unfortunately, von Neumann's work evidences a total lack of any direct contact with any actual experiments. His "proof" is based only on the self-consistency of the operator calculus used by quantum mechanics. Throughout his book, he cites only one experiment, that by *Compton* and *Simon* [7],[6] and then only to refute the short-lived *Bohr–Kramers–Slater* [8] predecessor theory. But after all, von Neumann was a mathematician. Unfortunately and correspondingly, his totally mathematical approach leaves an

---

[5] Whenever I asked anyone during the 1960s about the experimental justification for using quantum mechanics to describe entangled-state systems, I was told that its accurate prediction of the spectrum of helium proved it to be universally correct for all such systems. However, not much sophistication is required (especially now in hindsight) to recognize that the energy levels exhibited by two electrons bound in a helium atom do not provide, by themselves, a very unconvincing example of the full range of remarkable phenomenology exhibited by an EPR system.

[6] This experiment is a more precise repetition of an earlier experiment by W. Bothe, H. Geiger: Z. Phys., **26**, 44 (1924).
Schrödinger, however, subsequently found a semi-classical explanation of this experiment, and went on to propose an important relevant experiment. See "Splitting photons?".

experimental physicist in the very awkward position of having *no* specifically defined experiment (or set of experiments) on which quantum mechanics is purportedly based! Needless to say, von Neumann also does not specify any allowed error bounds for these unspecified experiments, which, if violated, might overturn or help confirm the far reaching conclusions that he draws.

Von Neumann is not alone in this regard. Bohr, EPR, Furry and Schrödinger also cite no actual experiments. Perhaps these authors all tacitly accepted and promoted the aforementioned religious belief because they were all theorists and felt that correspondingly they should act as clergymen. At the very least, they clearly show negligible respect for experimental physics as *the* final arbiter in physics. To their discredit, none of these authors even suggests in the above-cited works that experimental testing of the predictions for their *Gedankenexperiments* is in order. Perhaps their tacit reliance on and promotion of this religion was due to the fact that in the thirties, experimental physics was still in a rather crude state of development, and theorists then had no choice but to base their arguments on *Gedankenexperiments*, and then to stop at that point, because realizations of these *Gedankenexperiments* were well beyond the technology of the day.

Whatever justification these men may have had in this regard is perhaps unimportant now. Given their ecumenical leadership, and especially given Bohr's strong leadership, the net legacy of their arguments is that the overwhelming majority of the physics community accepted Bohr's "Copenhagen" interpretation as gospel, and totally rejected Einstein's viewpoint. Also, given von Neumann's daunting intellect, his proof was held to be sacrosanct, even to those who had never even perused its details. Moreover, since its details were rarely perused, over time, von Neumann's proof became commonly (and incorrectly) misinterpreted to imply that

*No theory based on hidden variables is possible that gives the same experimental predictions as quantum mechanics!*

Religious dogmatism then quickly promoted a nearly universal acceptance of quantum theory and its Copenhagen interpretation as gospel, along with a total unwillingness to even mildly question the theory's foundations. Such dogmatism persisted well into the 1970s. A standard phrase that was (and frequently still is) offered in quantum theory's defense, as a logical missal to the "standard religion" (as a nearly direct paraphrasing of von Neumann), is then,

*Quantum theory is so successful, it is simply inconceivable that its foundations and/or predictions can be unsettled, misunderstood, or, God forbid, in error.*

Given the omnipresent religious zeal, it would seem to be sheer folly for anyone to propose new experimental tests of quantum theory. What if their results disagree with quantum theory's prediction? We certainly will all perish! I offer the two following anecdotes as typical examples of this kind of dogmatism.

While I was a physics graduate student at Columbia University, I went to Prof. Bob Serber (a former student of Oppenheimer) to ask for his opinion of the 2-spin EPR configuration, and also to describe Bell's Theorem to him, along with my opinion of the importance of experimental tests of this configuration. To him, it was inconceivable that there might be any possible difference between the quantum mechanical prediction for this configuration and the experimental results. He dismissed me from his office, saying "Well *that* (the Bell's Theorem prediction) might be worth pointing out in a letter to the editor, but no decent experimentalist would ever go to the effort of actually trying to measure it with *that* in mind!"

As a second example, while I was actually performing the first experimental test of the CHSH-Bell predictions as a postdoc at UC-Berkeley (see below), I visited my undergraduate *alma-mater*, Caltech, and made an appointment with Prof. Richard Feynmann to discuss these same questions. Feynmann was very impatient with me. As soon as I told him that I was performing an experiment to test the predictions of quantum mechanics for the EPR configuration, he immediately threw me out off his office saying "Well, when you have found an error in quantum-theory's experimental predictions, come back then, and we can discuss *your* problem with it."

Given the religious impact of Bohr's teaching, little progress was made during the subsequent two to three decades towards any further development of our fundamental understanding of the meaning of quantum mechanics, or especially towards a resolution the completeness problem. To a large extent, further work in this field got put on hold by the intervention of World War II. The atomic bomb, radar, rocketry, and flight from Nazi Germany by many then necessarily occupied much of physicists' attention, and little time was then available to ponder the philosophical implications of quantum mechanics. Following the war, new technology that had seen unprecedented rapid development as a result of the war effort (e.g. sophisticated electronics, particle accelerators, microwaves, high-vacuum technology, etc.) brought the capabilities of experimental physicists to a dramatically higher level. Previously impossible experiments became straightforward. Maybe now would be the time to tidy up the foundations of quantum mechanics, perhaps now with some help from experimentalists? Not yet!

## 6.5 The Development of a "Stigma"

During the post-war years, the United States quickly became embroiled both in the cold-war and in an internal anti-communist frenzy. Driven by Sen. Joe McCarthy, stigmas then came into vogue. Indeed, McCarthy's stigmatizing attacks on physicist, Robert Oppenheimer, are well known and well documented. Unfortunately, acceptance of stigmas by the populace appears to foster the malignant creation of additional stigmas, especially in an environment that is dominated by religious fanaticism. Thus, in keeping with the times, a very powerful secondary stigma began to develop within the physics community towards anyone who sacrilegiously was critical of quantum theory's fundamentals. This stigma long outlived the McCarthy era and persisted well into the 1970s and 1980s. Sadly, it effectively kept buried most of the untidiness left behind by quantum theory's founders, and physicists went on about their business in other areas. The net impact of this stigma was that any physicist who openly criticized or even seriously questioned these foundations (or predictions) was immediately branded as a "quack". Quacks naturally found it difficult to find decent jobs within the profession.

To be sure, there, remained alive a minority of the theory's founders (notably Einstein, Schrödinger, and de Broglie) who were still critical of the theory's foundations. These men were obviously not quacks. Indeed, they all had Nobel Prizes! Instead, gossip among physicists branded these men "senile". This is not a joke. On many occasions, I was personally told as a student that these men had become senile, and that clearly their opinions could no longer be trusted in this regard. This gossip was repeated to me by a large number of well-known physicists from many different prestigious institutions. Given this branding, their leadership role in charting the course of progress in physics was thereby severely undermined. Under the stigma's unspoken "rules", the worst sin that one might commit was to follow Einstein's teaching and to search for an explanation of quantum mechanics in terms of hidden variables, as Bohm and de Broglie did.

This stigma was very real and had more than just a mild social impact on the conduct of physics. For example, even as late as the early and mid seventies, whenever a manuscript discussing the foundations of quantum mechanics (and especially one discussing hidden variables) was submitted to either the Physical Review or Physical Review Letters, editor Samuel Goudsmit would, in turn, enclose a one-page APS policy statement along with it to the manuscript's referee. That policy, in essence, urged the referee summarily to reject any paper on this subject, unless the paper was both mathematically based *and* gave new quantitative experimental predictions. Bohr's response to EPR certainly could not have been published under Goudsmit's stated criteria.

Religious zeal among physicists prompted an associated powerful proselytism of students. As part of the "common wisdom" taught in typical undergraduate and graduate physics curricula, students were told simply that Bohr was right and Einstein was wrong. That was the end of the story, and the end of the discussion. Of course it was the end, because the concluding chapters of the story were not yet written. Bohm's and de Broglie's alternative theories (see below) were neither taught, nor even cited.[7] Any student who questioned the theory's foundations, or, God forbid, considered studying the associated problems as a legitimate pursuit in physics was sternly advised that he would ruin his career by doing so. I was given this advice as a student

[7] John Bell laments this fact in his book.

on many occasions by many famous physicists on my faculty at Columbia, and Dick Holt's faculty at Harvard gave him similar advice.

Fortunately, within the community, there were a few physicists who actually enjoyed challenging this common wisdom. These are the iconoclasts of our profession, who, unfortunately, took the major brunt of the stigma. One important player was David Bohm. Given the era, no one dared trust his opinion since he was openly communist. At one point in his career, the best job Bohm could find was a professorship in far away Brazil, until Oppenheimer brought him back to Princeton. Louis de Broglie, held a prominent seat on the French National Academy, and was thus comparatively immune from the stigma's curse. However, his publications went largely unread, since, of course, everyone already "knew" that he was senile.

An obviously central figure in the development of Bell's Theorem is the hero of our story, John Bell, himself, who appeared to be keenly aware of the power and potential danger available within this stigma. Generally, he appeared to exercise great caution to avoid its wrath. Perhaps, for this reason, not much of his work on Bell's Theorem was formally published (see below) until it much of it was later collected into his book, *Speakable and unspeakable in quantum mechanics* [9]. John Bell clearly evidences his awareness of the stigma and its associated limitation to open discussion, via the double-entendre manifest in this book's title. His article, *Against Measurement* [10], is somewhat more poignant and outspoken on this issue. I offer here three anecdotes further supporting my opinion of his extremely cautious attitude towards the stigma.

In 1982, John Bell and I were announced to be joint winners of "The Reality Foundation Prize", and to share its $6000 award between us. Before accepting this award, however, he wrote to me and asked whether or not this was a "quack" group that was offering it, and whether if not accepting it might hurt his own reputation as a legitimate scientist. I assured him that such was not the case, and that the money was coming from one of the founders of Federal Express (Charles Brandon), who personally had a great interest in promoting further legitimate study of the foundations of physics. Nonetheless, Bell did not attend the awards ceremony to accept the Prize. Instead, Bernard d'Espagnat, who did attend, accepted it on his behalf.

Another example of John Bell's caution relative to the stigma's wrath is that he never openly discussed his study of the foundations of quantum mechanics with any of his co-workers in the high-energy physics theory group at CERN. The first time that any of these coworkers found that this man of mystery had been leading a double life, is when Bell, Zichichi and d'Espagnat put together an International "Ettore Majoranna" Conference in Erice, Sicily on "Experimental Quantum Mechanics". Until then, his coworkers had never even heard the term "Bell's Theorem". I speculate that this double-life was chosen primarily to avoid the stigma's curse!

A final example is from Alain Aspect. When Alain was preparing to do his own experimental tests of the CHSH-Bell predictions in the early 1980s to confirm and extend our earlier work, he visited Bell at CERN. Fearing that Aspect, himself, might fall afoul of the stigma, Bell's first question of him was "Do you have a permanent position?"

## 6.6 Challenging the Common Wisdom

The next important step in the saga occurred in 1950, when *David Bohm* (a former student of Oppenheimer) published a new textbook, *Quantum Theory* [11]. The book's treatment of the underlying conceptual foundations of quantum mechanics is quite similar to other textbooks of the era. In it, Bohm expresses strong doubts about the existence of hidden parameters within quantum mechanics. However, he stops short of asserting that their existence is impossible. While he does cite von Neumann's book as a reference, he does not mention its "impossibility proof". Instead, he discusses the completeness of quantum mechanics via reference to the EPR configuration, and argues for completeness by defending Bohr's criticism of EPR's argument. However, in the process, Bohm adds one very important new idea to the discussion that significantly sharpens both the EPR configuration and its associated argument.

Originally, EPR had discussed entanglement in terms of the continuous variables, momentum and position. In this book, Bohm reformulates the configuration in terms of entangled discrete-state systems, i.e. in terms of the spin-component correlations of two spin-1/2 atoms in a spin-singlet entangled state. In a historical context, this sharpening is crucial for the work that follows.

One problem with writing a textbook is that it makes you think carefully about the book's subject matter. Following publication of this book, *Bohm* broke with tradition in two important ways. First, with his new student *Yakir Aharonov*, he published a paper [12] where, for the first time, they sought real experimental evidence for the peculiar properties of entanglement that are exhibited by the EPR configuration. Using a two-photon analogy of the two-spin configuration outlined in Bohm's book, they noted that the polarization correlation of gamma-ray pairs emitted during positronium annihilation is a real-life example of EPR's configuration. That correlation's magnitude had, in fact, recently been measured by *Wu* and *Shaknov* [13] at Columbia University. Bohm and Aharonov then point out that the correlation's measured magnitude agrees with quantum theory's prediction (with entanglement present), and significantly disagrees with the prediction by the Schrödinger–Furry hypothesis (with entanglement absent, once the particles become significantly separated). By publishing this paper, Bohm and Aharonov transcend the "standard religion", and actually consult an experiment! They also prove here what Furry could only assume but could not prove, i.e. that the Schrödinger–Furry hypothesis cannot be held as a correct description for nature.

Bohm, however, was just beginning his iconoclastic endeavors. Despite the fact that in his book he expresses strong doubts that hidden variables might exist, he now adds extreme untidiness to the existing situation by formulating a new "causal" theory that actually incorporates hidden parameters within it [14].[8] Given the stigma against such work, few people took Bohm's paper seriously. One who did take it seriously was Louis de Broglie. Inspired by this work, *de Broglie* reentered the arena and published a whole series of books [15] and papers promoting a similar theory. (Bohm and de Broglie's theories are both examples of what are commonly referred to as "pilot-wave theories".) The state of untidiness surrounding the issue of quantum theory's completeness had now blossomed to become a real mess, let alone a now seriously important skeleton in quantum theory's closet. Von Neumann's "proof" purportedly asserts that these theories cannot exist. But they obviously now do! Something clearly must be wrong, somewhere. Aware of von Neumann's "impossibility proof" Bohm attempts to clean things up a little, and offers an argument as to how von Neumann's proof may not apply to his theory.

## 6.7 John Bell Unravels the Confusion

The stage was finally set for the brilliant iconoclast John Bell to enter the picture to begin unraveling much of the accrued (but largely hushed-up) mess. While both von Neumann's and EPR's arguments are directed towards settling the issue about the possibility (or impossibility) of hidden variables within quantum mechanics, ironically Bell now shows, in a sense, that both arguments are effectively "wrong", at least in terms of their effective conclusions.

Bell first critically looked at von Neumann's argument. Given the existence of Bohm and de Broglie's theories, von Neumann's conclusion (or at least his perceived conclusion) obviously must be wrong! Bohm had given one argument as to how that might be so. However, Bell is neither very complimentary towards nor impressed by Bohm's counter to von Neumann's argument. In a masterfully written article that was published in a 1965 [16], Bell scrutinizes von Neumann's requirement for a linear superposability of eigenvalues along with the linear superposability their associated operators. He then points out that this assumed requirement has no *à priori* reason to hold, in general, for a hidden-variables theory, especially when different operators correspond to physically different measuring apparatuses, and when these eigenvalues are also to be taken as the possible measurement outcomes for the associated apparatus. Furthermore, without this linearity

[8] Fortunately, this work was published before the APS policy had been formulated. Presumably, it would have not passed muster under that policy.

assumption, von Neumann's argument neither obtains, nor constrains hidden-variable theories. Correspondingly, hidden variables are then no longer found to be "impossible" for alternative theories (to quantum mechanics) by von Neumann's argument.

At that time, many physicists who had looked carefully at these issues generally put much more reliance on von Neumann's rigorous mathematical argument than on Bohr's somewhat vague philosophical argument in casting their votes against hidden variables. However, given that von Neumann's argument can no longer be relied upon, Bell next wondered if EPR's precisely worded argument might be relied upon instead. Given that Bohm and de Broglie's theories had provided good test cases for von Neumann's argument, Bell (in his 1965 paper) offers a generalization of these theories to allow them to include the two-particle case needed by the EPR configuration. Bell's generalization does not follow exactly along the lines of what Bohm and de Broglie had wanted.[9] In fact, Bell found that no matter how hard he tried to formulate a two-particle generalization of these theories while simultaneously maintaining agreement with quantum theory's predictions, he could only succeed by incorporating non-local interactions between the separated particles into the generalization. But it is just these non-local interactions that EPR reasonably had insisted cannot and do not exist, given the arbitrarily wide separation of the two particles. Given Bell's difficulties in generalizing these theories in a purely local way, he then began to speculate that this difficulty might be generic to any possible hidden-variables generalization operating in the EPR configuration. Following this line of thinking further, Bell discovered an astonishing new result. This result (along with its many subsequent generalizations by others) is (are collectively) now referred to as Bell's Theorem. *Bell* evidently considered his new result so important that he published it in 1964 [17], before publishing his 1965 review of von Neumann's impossibility proof.

Recall that EPR had argued earlier that considerations of their configuration show that hidden variables must exist. With supreme irony, Bell shows in his 1964 paper that further considerations of EPR's configuration indicate that exactly the opposite is true! Therein he proves that for Bohm's two-spin configuration, the prediction by any deterministic local hidden variable theory for the polarization correlation coincidence rate's normalized angular dependence will necessarily be different from that rate's angular dependence predicted by quantum theory. Quantum mechanics predicts that the dependence will be precisely sinusoidal with unit amplitude. Bell's paper shows that for *any* deterministic local theory based on hidden variables, this dependence

[9] de Broglie had earlier admitted to having encountered serious difficulties in generalizing his theory to describe more than one particle, and in affecting a transition from the configuration space in which quantum theory is formulated, to physical space, in which his and Bohm's theories are formulated.

must be strongly non-sinusoidal if it has at least one point with the maximal correlation value that is required by quantum theory.

## 6.8 Theoretical Truth Versus Experimental Truth

Initially, Bell expected that his new result would be adopted as a straightforward replacement for von Neumann's argument. With this expectation, however, he initially fell into the trap of accepting the "standard religion". His 1964 paper cites no experimental data as to what the observed dependence actually is. Of course, he couldn't cite any such data, because none had yet been measured. Still, he appears to assume tacitly in that paper that experimental evidence for this rather dramatic prediction is already in hand. After concluding that quantum mechanics and a theory with local hidden variables give different statistical predictions, he further states

"*Of course, the situation is different if the quantum mechanical predictions are of limited validity. Conceivably they might apply only to experiments in which the settings of the instruments are made sufficiently in advance to allow them to reach some mutual rapport by exchanging signals with velocity less than the speed of light. In this connection, experiments of the type proposed by Bohm and Aharonov, in which the settings are changed during the flight of the particles, are crucial.*"

Bell's wording of this quoted passage has been very carefully chosen. Its converse conclusion is that it is correspondingly inconceivable that the quantum-mechanical predictions do not correctly predict experiments with static instrument settings, i.e. that the quantum-mechanical predictions for static instrument settings are indeed correct, and have been verified experimentally to be correct. While the passage does not actually assert that this converse conclusion holds true, it strongly suggests that it does. Thus, the passage tacitly implies and gives the misimpression that the existence of hidden variables (at least in situations with static analyzers) is, *ipso-facto*, refuted by his proof.

Nonetheless, Bell hedges his bets. Given that his arguments only consider the predictions for non-relativistic quantum theory, he appears to suggest that new keys to the hidden-variables mystery will be revealed only via considerations of relativistic corrections to quantum theory. Correspondingly, he goes on to promote Bohm and Aharonov's second-generation experiment with rapidly changed analyzers. However, such a proposal is clearly silly, unless one acknowledges the existence of first-generation experimental results with static analyzers. He was clearly bluffing, and did not really know, at this point, what the experimental status of his result really was. I personally will claim much of the credit for calling his bluff.

In the late sixties I was a graduate student at Columbia University, and was struggling, trying to understand quantum mechanics. I had read EPR's paper, and also had read Bohm's and de Broglie's work. While I had difficulty understanding the Copenhagen interpretation, the arguments by its critics seemed far more reasonable to me at that time. Since quantum mechanics is formulated in a configuration space, it then provides no physical-space model to consult, whereupon it is then difficult to visualize what might actually be happening, especially in a two- (or more-) particle system. I found Bohm's and de Broglie's works refreshing, since they do give real physical-space models of what is happening. I also found EPR's arguments much more persuasive than Bohr's, and was only vaguely aware of von Neumann's "proof" at this time. Hidden variables thus seemed (to me then) to be a perfectly logical solution to the problem.[10] By holding that opinion (given the above-mentioned stigma), I was then certainly branded as a heretic by many, and undoubtedly as a quack by others.

Then, I read Bell's 1965 and 1964 papers. Given my familiarity with Bohm and de Broglie's work, I was not particularly surprised by Bell's 1965 result. On the other hand, once I comprehended what the 1964 paper actually said, I was astounded by its result. However, I was not yet willing to accept the paper's far-reaching implications until I finally saw some experimental evidence that decided between its two significantly different predictions. Since Bell's paper is conspicuously vague on the experimental status of its prediction (but crystal clear on everything else), I suspected that Bell was bluffing and started to search through the literature for experimental results. I found none, other than that by Wu and Shaknov.

With regard to experimental testing, I saw at least one serious problem with Bell's 1964 result. That problem is that Bell's form for his result applies *only* to his idealized configuration, and thus cannot be directly tested experimentally. Bell (being a theorist) assumes an ideal apparatus and an ideal two-particle entangled-state preparation. However, his proof then relies on this idealization, and further uses it to show that a perfect correlation between the two particle's polarizations occurs for at least one setting of the polarization analyzers. Unfortunately, even an infinitesimal departure by the considered system from this idealized behavior then invalidates the applicability of this proof to it, since then, the required single point with a maximal correlation no longer exists. Correspondingly, his result then applies only to ideal systems and not to real ones! (Damn theorists!) Obviously, in any real experiment, Bell's required perfect correlation cannot and will not obtain.

Does my observation here then imply that Bell's exciting new result is no more applicable to the completeness problem in real-life situations than is von Neumann's result? I thus asked myself whether or not Bell's assumption of an idealized apparatus and system is, in fact, necessary for an alternative proof of his Theorem to obtain. Looking further, I found that this assumption

---

[10] John Bell confesses in the preface to his book, *Speakable and Unspeakale in Quantum Mechanics*, to being similarly enamored with these features of pilot-wave theories.

can indeed be avoided. Thus, I succeeded in deriving a new more general form for his result that then *can* be tested experimentally. Thus holding an experimentally testable prediction, I then returned to the question about the result's experimental status. It became immediately clear that no definitive conclusion then obtains when this testable prediction is compared with the data available from the Wu–Shaknov experiment. Moreover, I found that it is straightforward to build a simple *ad-hoc* hidden-variables counter-example that yields predictions that are exactly the same as those from the quantum-mechanical prediction for the Wu–Shaknov experiment. Thus, I concluded that something beyond a simple extension of the Wu–Shaknov experiment is needed for a valid experimental test. What can it be?

I recalled a paper by *Peres* and *Singer* [18], who (following Bohm and Aharonov) had proposed additional tests of the Schrödinger–Furry hypothesis.[11] They had suggested using 90° scattering as a method for producing Bohm's two-particle spin-singlet entangled state. Hearing that Dan Kleppner and Dave Pritchard at MIT were doing crossed-beam scattering experiments with alkali metal atoms, I went to MIT and gave Kleppner's group a seminar on Bell's Theorem, and suggested that their experiments might be reconfigured so as to provide an experimental test. A newly arrived postdoc in the group, Carl Kocher, attended the seminar. For his PhD thesis at UC Berkeley, he had just finished performing an experiment with Gene Commins, wherein they had measured the polarization correlation of photon pairs emitted in an atomic cascade. At the end of my talk, Dave Pritchard asked "Carl wouldn't your experiment test this?" Kocher replied, "Of course, that's why we did it!" *Kocher* had no reprints on hand, so I returned to New York and eagerly located his published results in the library [19]. However, contrary to Kocher's claim, I discovered that their experiment was totally inadequate as a test. Kocher and Commins had only measured coincidences at 0° and 90° relative analyzer orientation. As per Bell's assumption, these data points are located exactly where the two theories can give the same prediction. Also, as per my own generalization of Bell's result to provide an actual experimental prediction, I noted that their polarizer efficiencies were nowhere high enough for the experiment to be conclusive, in any case. Thus, their experiment provides no data at all for a test of hidden-variable theories. What is clearly needed at that point are data taken at angles intermediate between their measured points with an experiment that is similar to theirs, but that uses much better polarizers.

Finally, I decided to call Bell's bluff. I wrote letters to Bell, Bohm and de Broglie, asking all three of them (a) did they know of any experiments

[11] The Schrödinger–Furry hypothesis represents probably the simplest form of a hidden-variable theory. It predicts a reduced-amplitude sinusoidal correlation. Strictly speaking, Bell's 1964 proof does not apply to it, since Bell's required perfect correlation (for at least one point) never obtains for it. This hypothesis, however, is constrained by the CHSH and CH inequalities.

testing the result, and (b) did they agree that a repeat of the Kocher–Commins experiment with improved polarizers at intermediate angles would be convincing, and (c) how did they view the importance of such tests [20].[12] All three courteously responded and replied NO to (a) and YES to (b). Bell was particularly enthusiastic about the idea, and simultaneously also revealed his own agnostic religious beliefs about quantum theory. In response to (c), he said

"*In view of the general success of quantum mechanics, it is very hard for me to doubt the outcome of such experiments. However, I would prefer these experiments, in which the crucial concepts are very directly tested, to have been done and the results on record. Moreover, there is always the slim chance of an unexpected result, which would shake the world!*"

The McCarthy era was now in the distant past. Instead, the Vietnam War dominated the political thoughts of my generation. Being a young student living in this era of revolutionary thinking, I naturally wanted to "shake the world". Since I already believed that hidden variables may indeed exist, I figured that this was obviously the crucial experiment for finally revealing their existence. But if they do exist, then quantum mechanics must be verifiably wrong here, with its error having gone undiscovered heretofore. However, since there was virtually no experimental evidence then available for this configuration (plus or minus), that "slim" possibility could not be discounted. To me, the possibility of actually experimentally discovering a flaw in quantum mechanics was mind-boggling. Thus, I drafted an abstract for the Washington DC Spring Meeting of the American Physical Society proposing this experiment [21]. Immediately as it appeared in print in the APS Bulletin, I received a phone call from Abner Shimony (a former student of Eugene Wigner). He said that he and his student Mike Horne had come to the same conclusions.[13] Their pursuit of experimental evidence had led them to Frank Pipkin at Harvard, who had just performed an experiment that was similar to the Kocher–Commins experiment, in that it observed two-photon coincidences from an atomic cascade, but that did not include any polarization analysis. Pipkin's PhD student, Dick Holt, was now setting up to perform the experiment that I had just proposed. After comparing notes, we all agreed to coauthor a Phys. Rev. Letter formally presenting our conclusions [22].[14] Our collaborative effort is now commonly referred to by

[12] According to David Wick, my 1969 letter was the first response to his 1964 paper that Bell had received.

[13] Shimony's name was already familiar to me when he called, because I had earlier read some of his work on the "measurement problem" in quantum mechanics. This problem then leads to the Schrödinger's-Cat and Wigner's-Friend paradoxes, that are another untidy aspect of the foundations of quantum mechanics. Both paradoxes result from the peculiar properties of quantum entanglements.

[14] This paper first coined the term "Bell's Theorem".

the initials, CHSH, with the paper's experimental prediction contained in what is now correspondingly known as the CHSH inequality. In the process of writing this paper, Abner, Mike and I forged a long lasting friendship that was to spawn many subsequent collaborations.

## 6.9 Beware of the Experimentalists Lurking About

Upon receiving my PhD from Columbia, I moved to UC Berkeley to work in radio astronomy as a postdoc for Charlie Townes. Since Gene Commins was also at UC Berkeley, I suggested to both Townes and Commins that they allow me to resurrect the Kocher–Commins experimental apparatus, and to use it to test the CHSH-Bell prediction. Townes was intrigued by the idea, took it seriously, and offered half of my time to this project, while the other half was to be spent on radio astronomy. Commins offered Stuart Freedman, his new graduate student, to work with me on the project. Fortunately, Townes was very tolerant as my radio astronomy projects languished. Stu and I published our results in 1972 [23], and Stu got his PhD with this experiment.

Our results agree exactly with the quantum-mechanical prediction, but exclude theories based on local hidden variables by about five standard deviations. This experiment is noteworthy in that it is the first to actually be able to draw such a conclusion! Its count rate was so low that the results are effectively unchanged, even if "accidental" (background) coincidences are not subtracted from the totals used in a comparison with the CHSH inequality.

Meanwhile, in competition with our experiment at Berkeley, Dick Holt and Frank Pipkin at Harvard had been pursuing their parallel experiment. It was quite similar to ours, except that it used a different atomic species and a different kind of polarizers. While our experiment was still in progress, they actually got the first results. Unfortunately, their results disagreed with the quantum-mechanical prediction, but did agree with the local hidden-variables prediction by the CHSH inequality. They decided to be cautious, however, and not to publish their results until they saw what our experiment at Berkeley revealed. Once we announced our results, they then opted not to publish theirs formally. Instead, they distributed an informal preprint that outlined what they had done and observed. In the end, they concluded that their result was due to an underlying systematic error.

By declining publication of their experimental results, they left ours as the only published results available. Given the importance of the issues at hand, our own results obviously needed further confirmation. Obviously, I was no longer actively working in radio astronomy at this point. Fortunately, Howard Shugart allowed me to continue my work at UC Berkeley as a member of his atomic-beams group. Following my photon-splitting experiment (described below) I proceeded to repeat Holt and Pipkin's experiment with only a few minor modifications. Again, my new result [24] agrees with quantum theory's prediction, but significantly disagrees with the CHSH-Bell prediction. I also went further and tested the predictions of quantum theory for the circular polarization correlation [25]. These results also agree with quantum theory's prediction.

It should be noted that when we began our experiments at Berkeley, tunable lasers were not yet commercially available. Thus, all of the Berkeley experiments were done without their benefit. As a result, the count rates were very low, and the experiments were very slow and tedious. Finally, *Fry* and his student *Thompson* [26] performed a test of the CHSH prediction using one of the first commercially available dye lasers. Given the short counting periods afforded by their use of this laser, they could more readily search for possible systematic errors. Their experiment confirmed our "Berkeley" results, but disagreed with Harvard's results. The score was now three to one, in favor of quantum mechanics. That score was sufficiently convincing to allow *Abner Shimony* and me to draw the above-described conclusions in our review paper [1].

## 6.10 Generalization of the Bell and CHSH Results to Constrain Local Realism and Space-Time

While I was actively pursing my experimental efforts at UC-Berkeley, in parallel with that work, I also pondered some remaining details that still had not yet been addressed by CHSH and Bell. There were then at least four important unanswered questions:

(1) What are the fundamental assumptions underlying Bell's Theorem?
(2) What is the affect of the large number of unobserved particles (especially when the particles are optical photons) associated with a finite detector solid-angles and finite detector efficiencies? Also, given unobserved particles, what are the requirements for an experiment's design that then still allow its results to provide conclusive arguments to be made regarding the truth or falsity of the fundamental assumptions that, in fact, underlie Bell's Theorem?
(3) If these experimental requirements cannot readily be met in practice, then what is the least objectionable supplementary assumption that one can make to allow reasonably convincing inferences to be drawn from experiments that, for example, can now be performed with optical photons?
(4) Also, if these requirements are not met in practice, is there at least one counterexample consistent with the fundamental assumptions underlying Bell's Theorem for experiments performed with say optical photons? Unfortunately, if such a counterexample exists, then only an experiment performed within these limits can be considered to be fully conclusive as a test of the fundamental assumptions underlying Bell's Theorem.

Mike Horne and I addressed and answered all four of these questions in our 1974 paper [27]. That paper is now commonly referred to by the initials, CH.

Regarding question (1), it may seem surprising and strange, at this point in the development, that the fundamental assumptions underlying Bell's Theorem were still not yet well defined. As a parallel problem to the gap between Bell's 1964 result and the CHSH result, this lack of definition again arises from differences between the idealized configuration considered by Bell, and the realizable configuration considered by CHSH. Here and elsewhere, it represents an example of crucial details falling through a gap that exists between a theorist's view of an experiment and an experimentalist's view of the same experiment.

Following EPR's lead, both Bell and CHSH initially had assumed that determinism is the basic underlying assumption for Bell's Theorem. This notion follows from EPR's *definition* of an "element of reality", as per

*"If, without in any way disturbing a system, we can predict with certainty (i.e. with probability equal to unity) the value of a physical quantity, then there exists an element of physical reality corresponding to this physical quantity."*

While predictions with certainty are possible for idealized systems, unfortunately, no such predictions can be made for realizable systems (except, of course, for death and taxes). Following EPR, Bell effectively derives determinism in his 1964 paper for the EPR particles, via a use of the at least one data point with a perfect correlation. Unfortunately and conversely, determinism cannot be derived via this reasoning for any real system, which, of course, has no such point with Bell's perfect correlation, and thus no predictions with probability equal to unity. Given this fact but still following EPR's ideas, CHSH simply and explicitly assumed determinism to hold for the purposes of their derivation [28].[15] However, Mike Horne and I noticed the fact that the CHSH prediction also appears to hold for models in which determinism is neither assumed, nor in fact, even holds. This fact suggested to us that an assumption of determinism is not really necessary for Bell's Theorem to apply. Independently, *John Bell* had also noticed this fact and mentions it in a book review [29], wherein he comments that a hidden-variables model cannot give the probabilistic quantum-mechanical prediction for the two-spin system, even if a random-noise generator at every point in space-time is assumed to influence the space-time evolution of the hidden variables. Such a system, by definition, does not have a deterministic evolution.

Mike and I thus asked ourselves the following questions: If determinism is not required of a system for Bell's Theorem's constraints to hold for it, what then is the fundamental characteristic of a physical system, such that it is constrained by Bell's Theorem. Also, how does one go about specifying said

[15] Shortly following CHSH, Bell did extend the range of applicability of the CHSH result further to include the effects of additional hidden variables in each apparatus. However, in doing so he adds no new premises to Bell's Theorem in this new derivation, and similarly explicitly assumes determinism to hold.

system's assumed nature within the Theorem's derivation? Our 1974 paper was the first publication that specifically answers these questions.

Historically, the CHSH result can be seen as a logical continuation of a sequence of quandaries faced by Einstein during his development of special relativity. In formulating special relativity, Einstein noted that one cannot ask the universal questions "When" and "Where" in a precise fashion, without first defining the associated primitive entities, "time" and "distance". He found that, at best, he could only provide purely operational definitions for these entities. That is, he defines "time" to be the stuff you measure with a clock, and "distance" to be the stuff you measure with a ruler. Similarly, to allow Einstein to ask the universal question "What" in a similarly precise fashion, EPR's definition of an "element of reality" effectively provides his initial attempt to help one to define the associated primitive entity, "object". But, as noted above, that definition does not apply to realizable systems. CH thus found it necessary to offer an improved and explicit operational definition for the notion of an "object". The historical development of Bell's Theorem is then seen to have a direct parallel to the development of special relativity. Once that a suitably precise definition is given for some fundamental "stuff" of nature, then and only then can new testable physics emerge.

In their 1974 paper, CH operationally define an "object" (within a local realistic theory) as stuff with measurable properties that also can be put in a box, i.e. stuff (along with an associated measuring apparatus) that can be spatially surrounded by a space-time Gaussian surface. CH supplement this definition with an associated definition of an "objective local theory" (subsequently also called a "local realistic theory" and/or a "theory of local realism" by *Clauser* and *Shimony* in [1]), as a theory that describes such objects, wherein action-at-a-distance is precluded, and wherein a measurement reveals local properties of an object (i.e. of said stuff within said box).[16] Moreover, which properties of an object are to be measured may be arbitrarily chosen at the free will of the experimenter who operates the apparatus. Thus, for objects within such a theory, if the correlated properties of two objects in two disjoint boxes are measured at space-like separated measurement events, then absence of action-at-a-distance prevents the experimenter's choice of property measured in the first box (e.g. his choice of color vs. weight, in the earlier examples) from affecting the results of a measurement made on/in the second arbitrarily distant box. CH found that these very simple and naïve premises and definitions for *any* such local realistic theory of nature, then lead to a very general formulation of Bell's Theorem, that, in turn, then constrains the experimental predictions for said theory. The CH specific experimental predictions are contained in a new inequality that then must be satisfied by any such theory within local realism. That inequality is now commonly referred to as the CH inequality.

[16] Under the CH definition, a "deterministic local hidden variables theory" then becomes a special case of the more general class of "local realistic theories".

Under CH's definition of an object, Bell's Theorem then not only provides hard new experimental predictions for a local realistic theory, but it also provides surprising and profound implications for quantum mechanics as well. One first surprise is that these two different theories (local realism and quantum mechanics) give measurably different predictions for realizable experiments. A second surprise is that Bell's Theorem shows that quantum mechanics does not describe objects that fall within the scope of CH's naïve definition. Thus, it appears that within quantum mechanics, a "quantum" object (however else one may choose to define it) is *not* something that can be put in a box! Unfortunately, it does not appear that Bohr (or anyone else) has ever provided a clear concise definition for an "object" within quantum mechanics. On the contrary (as is often noted by Bell in his book), the distinction between and definitions of the terms "observer", "apparatus", and "system" (where "system" presumably means "object"), all remain quite vague under the Copenhagen interpretation.

It may be noted, in passing, that a single "bit" of information, as may be stored in a computers memory, qualifies as a "quantum object". Bell's Theorem the shows that in a "quantum computer", said bit may not necessarily reside within a single box that sits on one's desk. Indeed , if the quantum computer is part of a quantum-computer network, then the bit's definition may be determined only by joint operations performed on the whole network, but not by any attempt to access it on a single box that is attached to the quantum network.

Question (2) naturally arises when one tries to test the CHSH experimental predictions using optical photons, as Stu Freedman and I had just done. In such an experiment, only about one emitted photon pair in a million is actually detected in coincidence. Most photon pairs emitted by the source are not detected at all. For most of the observed events from the cascade photon source that we used, only one photon of the pair is actually detected. This situation is clearly quite far removed from that of the idealized configuration originally considered by Bell in his 1964 paper, wherein both particles of each and every emitted pair are subsequently detected, and where the pair is known to have been emitted (as per the discussion of "event-ready" detectors in [1]). Mike Horne and I realized that the associated enormous unobserved volume of phase space in such experiments leaves a lot of room for significant sampling biases to occur. Unlike the CHSH inequality, however, the CH inequality's prediction is carefully derived so as to render it unaffected by these biases. The experimental requirements for testing the CH inequality are also derived in the CH paper. Unfortunately, the consequence of using the CH inequality (instead of the CHSH inequality) for experimental tests is that an experiment performed with photons can no longer, by itself, provide a direct test of local realism.[17] Moreover, designing and carrying out a practical experiment that actually meets the CH inequality's experimental requirements is not an easy task. *Fry*, *Walther*, and *Li* [30] have recently offered such a design, and its experimental realization is currently in progress. As of this writing, no direct tests of the full CH inequality have yet been performed. Experimental tests of the CHSH inequality, rather than the CH inequality, are now commonly said to leave open the "detector-efficiency loophole".

Question (3) concerns the relationship between the experimental predictions made by the CHSH and CH inequalities. Since experiments performed with photons, as a general rule, do not violate the CH inequality, but can and do violate the CHSH inequality, it is natural to ask what additional assumptions can (or must) be added to allow the CHSH inequality and experiments using photons to still give reasonably persuasive (but perhaps not totally compelling) conclusions about the truth or falsity of local realism. That is, what is the best we can do with contemporary technology? To a similar end, CHSH had earlier derived their inequality by making an auxiliary assumption regarding the behavior of a photon at a polarizer that then allows an experimentally testable prediction to obtain. Thus, CHSH assume that whenever a photon encounters a polarizer, then the photon is either transmitted by the polarizer or not transmitted by the polarizer. Furthermore, given that two photons have emerged from their respective polarizers, their detection probability is assumed by CHSH to be then independent of the analyzer-pair settings. Here, CHSH have taken the concepts of "passage" and "non-passage" to be primitive, and thereby effectively also assume that photons are particle-like, for the purposes of the experiment. The CHSH auxiliary assumption obviously does not apply to a semi-classical radiation theory (such as Jaynes' "neo-classical radiation theory", see below). In such a theory, each of the individual "photons" that comprise a pair of "coincident photons", must then be represented as a short pulse (wave packet) of classical electromagnetic radiation. Since each pulse is now classical, each "photon" is then partially passed and partially transmitted by a polarizer. Nonetheless, "photons" in such a theory are clearly "objects", under the CH definition.

The CHSH auxiliary assumption is thus deficient, in that it does not apply to this important class of local hidden-variable theories. CH noted, however, that even though the CHSH assumption does not apply to these theories, the CHSH inequality apparently does still apply, and Bell's Theorem clearly does constrain semi-classical radiation theories. To remedy this deficiency, CH provide a significantly improved auxiliary assumption that is also much weaker than the CHSH auxiliary assumption. It is called the "no-enhancement" assumption. It simply requires that the probability of an object's detection, after it has passed through (or partially passed through)

[17] This fact is still true, even for photon pairs that are generated by a parametric down-conversion source, as with most recent experimental tests of Bell's Theorem.

an analyzer, can then be no greater than the probability would have been if the particle had not passed through the analyzer, i.e. greater than if the analyzer had been absent. This assumption clearly holds, for example, for semi-classical radiation theories when the objects are classical electromagnetic pulses and the analyzers are optical polarizers. Furthermore, it seems to be highly plausible and very difficult to avoid for almost any local realistic theory. Then, for all local realistic theories for which the "no-enhancement" assumption also holds, the CH inequality's prediction reduces to the CHSH inequality's prediction. In such cases, (except as noted below) experiments such as that by Freedman and Clauser then convincingly refute all such theories.

Question (4) arises in a quest for finding *both* necessary and sufficient conditions for the CH inequality to constrain local realism. Towards this end, CH provide an *ad-hoc* counterexample, that is consistent with local realism, that *does not* violate the CH inequality, and that *does* violate the CHSH inequality. While the existence of this counterexample does not prove that the "no-enhancement" assumption is "necessary", it does show that at least some additional assumption is required for one to use the CHSH inequality to test local realism. Use of the CHSH inequality (as opposed to the CH inequality) is now commonly said to leave open the "detector-efficiency loophole".

## 6.11 Common Confusion About Count-Rate Normalization

For testing either the CH or CHSH inequalities, a normalization is required for the coincidence count rates as a function of relative polarizer orientation. A valid test of the CHSH inequality's prediction (as obtained from either the original CHSH auxiliary assumption and/or the CH no-enhancement assumption) calls for coincidence rate measurements to be taken, both with polarizers in place and with polarizers absent. The latter measurements then provide a normalization for the former. Here one calculates ratios of coincidence rates to coincidence rates, and since both rates have the same order of magnitude in photon counting experiments, a valid test is straightforward to obtain.

The primary reason that direct tests of the CH inequality are much more difficult is that here, the polarizers are not removed. Instead, this normalization is done via the detector singles count rates (with the polarizers in place). For low detector efficiencies and/or a high particle loss rate via the broad angular correlation associated with a three-body-decay emission, the singles rates are generally much larger than the coincidence rates, the associated ratios are very small, and no violation of the CH inequality, by itself, can then occur. For this reason, a two-body-decay is required for such a test.

Thus, obtaining a correct normalization for the coincidence count rates is always crucial for a proper and valid test to be performed. Unfortunately, this requirement is often either overlooked or significantly downplayed in its importance in many recent experiments. Furthermore, unless "event-ready detectors" are used, testing of either the CH or CHSH inequality is most readily done with only one detector following each analyzer. Indeed, the superiority of two-channel experiments (relative to four-channel experiments) is discussed, at length, in both the CH paper (Appendix B), and in the *Clauser–Shimony* review article [1].

Note that various other more recent workers have been tempted to use an alternative scheme that is more along the lines of Bohm's original *Gedanken-experiment*, and thus also along the lines of the idealized experiment considered by Bell in his original 1964 paper. In these four-channel schemes, two counting channels follow each polarizer. Workers who succumb to this temptation generally ignore the need for event ready detectors, and commonly reason that an alternative and easier to implement normalization method will suffice. In such schemes, the polarizers are never removed. Instead, the observed coincidence rates are normalized to the sum of the four possible coincidence rates in all such channels. This alternative normalization method is *adamantly not* the CHSH (nor the CH) normalization method, although it is frequently (and incorrectly) attributed as such.[18]

A four channel experiment was first used by *Aspect*, *Grangier* and *Roger* [31]. For that experiment, Aspect et al. necessarily rely on a very different set of auxiliary assumptions to allow a comparison of their results to Bell's Theorem. However, their set of assumptions is clearly much stronger than the CH no-enhancement assumption, and thus, allow a much larger "detector-efficiency loophole". The relative strength of their alternative auxiliary assumptions with the no-enhancement assumption is readily seen by noting that it is straightforward to build a rather simple counterexample that violates the CHSH inequality (configured as per the no-enhancement assumption), but that does not violate the count-rate inequality used by Aspect et al. For example, consider a simple local hidden-variable theory that is similar to the CH counterexample, but that does not include the enhancement used by that counterexample. This theory then predicts an angular dependence that has a reduced coincidence-rate sinusoidal amplitude along with a proportionately reduced constant offset from zero. An experiment that uses the Aspect et al. normalization will then incorrectly conclude that this hidden-variable theory is refuted, and that quantum mechanics is correct. However, an experiment performed using the CHSH/CH normalization method will reach the opposite (but correct) conclusion.

## 6.12 Bell's Response to CH and "Local Beables"

Prompted by the CH paper's results, Bell's own fuller discussion of the fundamental basis for Bell's Theorem appeared two years later, but then

[18] In the future, authors are hereby requested to please stop misquoting us!

only as a preprint [32].[19] It is titled *The theory of local beables*, and is based on the CH paper's results,[20] which Bell therein attempts to extend. In this extension, Bell follows a line of reasoning that is essentially the same as that of an early rejected draft of the CH paper. In preparing the CH manuscript, Mike and I had been convinced by our discussions with Abner Shimony that such an extension does not work, and correspondingly we did not use it. Thus, Abner, Mike and I were together confident that Bell had pushed his arguments too far. Whereas the CH paper maintains the observer as free-willed and as a separate entity from the "objects" being observed, in the "Beables" paper Bell further includes the observer as part of the objective reality being constrained. So doing, he is then able to hang the whole argument on locality considerations alone, whereupon considerations of objective reality are then no longer needed. But in doing so, the observer's free will to choose analyzer orientations is lost. Given that the observer unwittingly may be a preprogrammed automaton, his programming then defeats the locality considerations, and the arguments then leading to Bell's Theorem no longer follow. With *Abner Shimony* now taking the initiative, we three proceeded with a public dialog with *Bell* over this point in the abovementioned "quantum subculture" newspaper, "Epistemological Letters" [33].[21] In this dialog, Bell effectively concedes our point. The dialog is also noteworthy in that many other important related issues are also discussed in detail therein, including the important concepts of "parameter independence" and "outcome independence". [34]

## 6.13 The Quantum-Optics Community Encounters Related Problems

Despite the grand successes of quantum electrodynamics (QED) in accurately predicting the Lamb Shift, g-2 for an electron, etc. it was quite common in the 1960s and 1970s for atomic physicists to use semi-classical radiation theory (SCT) to explain their results, since doing so was much easier than using QED. However, any author who did so, also generally included an apology, conceding, "of course a *correct* treatment of this problem requires QED". Nonetheless, atomic physicists found that, with very few exceptions,

[19] Perhaps, Bell's reason for not publishing this result in 1975 is that he may have been afraid that doing so might warrant his being branded a quack, given the then existing stigma against such work.

[20] Bell therein acknowledges our prior work, by stating
"*As regards the literature on the subject, I am particularly conscious of having profited from the paper of Clauser, Horne, Holt, and Shimony which gave the prototype of (16), and from that of Clauser and Horne. As well as a general analysis of this topic this last paper contains a valuable discussion of how best to apply the inequality in practice.*"

[21] For a republication of this interchange that also includes Bell's "Beables" paper.

SCT accurately predicts observed experimental results. As time progressed, the number of exceptions kept shrinking. It then became sort of a game among these workers to see just how often QED effects can be ignored. For example, as part of this game, *Willis Lamb* and *Marlan Scully* [35], and *Peter Franken* [36] showed that important observable aspects of the photoelectric effect can be derived without quantizing the radiation field, and Franken made a similar demonstration for resonance fluorescence. In fact, the use of SCT became so common[22] and gave results in agreement with atomic physics experiments so often, that various workers such as Franken began to jest that QED may be totally unnecessary! The game began to get serious, however, when iconoclast Ed Jaynes said that he was not joking, and formulated a new semi-classical theory as a serious alternative to QED. He referred to it as neoclassical radiation theory (NCT). Furthermore, he formally challenged the quantum-optics community to refute NCT with direct experimental evidence. Talk about a break from the "standard religion" (regarding acceptance of a theory without compelling experimental evidence) – Wow!

Jaynes' NCT assumes that (1) Schrödinger's equation governs the evolution of an atom's wave function, (2) the absolute square of an atom's wave function provides a physically real electric charge density, (3) Maxwell's equations (without field quantization) describe the physically real electromagnetic radiation field,[23] and (4) radiation reaction effects are included, so that radiation emitted by an atom affects the atom's own evolution. Using only these elements, Jaynes and his students found that (without QED) they could predict absorption of radiation, spontaneous and stimulated emission of radiation, the Lamb shift, and the black body radiation spectrum [37, 38].

Jaynes' work, however, had crossed a magic line in the sand and obviously had to be viewed as heresy. Heretofore, it was a firmly held belief that all of these latter effects do require a quantization of thc radiation field, as is done in QED. But Jayne's' work now cast serious doubts on these cherished beliefs. Nonetheless, it would seem that if quantum theory rests on a solid foundation, then Jaynes' challenge should be easy and straightforward to meet. But Jaynes' challenge went unanswered for several years. A small crisis was thus at hand within the quantum-optics community.

In a historical context, Jaynes was iconoclastically following a path similar to that taken by Bohm, de Broglie, and Schrödinger. While Jaynes did not

---

[22] Strictly speaking a semi-classical treatment of such systems *is* a "hidden-variables" treatment of these systems, and, of course, such a treatment represents "forbidden thinking" under the canons of the "standard religion". However, it is well known in matters of both politics and religion, that issues of practicality commonly override such canons, via the use of "creative ambiguity", especially when appropriate obsequious apologies are also offered. Indeed, "creative ambiguity" allows the present existence only one China in the world.

[23] Elements (1)–(3) will be recognized as together constituting the so-called "old Schrödinger interpretation". It is noteworthy that these elements are still essential primitives for X-ray crystallography.

refer to NCT as a hidden-variables theory (perhaps because of the stigma against such theories), he did recognize that he was indeed tinkering with very fundamental issues within quantum theory. Unlike the theory's founders, however, Jaynes based his work on a very broad range of experimental evidence, most of which postdated the work by the theory's founders. Unfortunately, he was not yet familiar with the power of Bell's Theorem and the associated peculiarities of entanglement.

Upon reading Jaynes' work, I quickly realized that NCT is clearly a local realistic theory. The electromagnetic field and the charge density in NCT are, in principle, all measurable objectively real quantities. Photons (or whatever the stuff is that gives rise to the coincidences in the Freedman–Clauser experiment) within NCT thus qualify as objects, as per the above discussion. Moreover, to refute NCT, the full power of Bell's Theorem is not even needed, and a much simpler argument is sufficient to do so. I thus accepted Jayne's challenge [39].

The 1972 Third Rochester Conference on Coherence and Quantum Optics became a watershed for discussions of this issue, and many of the papers and discussions at this conference were directed toward resolving the issues that Jaynes had raised. At this conference, I pointed out to the participants that if NCT is applied to the EPR configuration, it essentially becomes an example of the Schrödinger–Furry hypothesis. Given the above-mentioned argument by Bohm and Aharonov, it is clear that the Wu–Shacknov experiment already refutes this hypothesis. I also pointed out that the Kocher–Commins experiment provides additional refutation, and offered a slightly more general proof of this fact than that given by Bohm and Aharonov. In his own written contribution to the Conference's Proceedings [40], Jaynes' publicly concedes that my arguments do indeed refute NCT. He soon dropped any further pursuit of this theory, and in a personal letter to me, both praised my work and admitted that his own efforts now "lay in ruins". In defense of Jaynes' valiant efforts, I contend that his efforts provide a heuristically valuable and therefore worthwhile exercise that is highly beneficial to our overall basic understanding of physics.

## 6.14 Splitting Photons?

To obtain an experimentally testable prediction, CHSH originally assumed that whenever a photon (or at least its detectable component part[24]) encounters a polarizer, then it is either transmitted by the polarizer or not transmitted by the polarizer. That is, CHSH, with their auxiliary assumption require that photons are always particle-like for the purposes of the experiment. While the CH no-enhancement assumption now obviates the

---

[24] In a "pilot-wave" theory, separate component parts of a quantum particle exist that then respectively account for the particle-like and the wave-like aspects of its propagation.

CHSH auxiliary assumption, it was then still an open and relevant question as to whether or not there was any compelling *à priori* experimental evidence that justifies the CHSH assumption. That is, do photons clearly demonstrate an unambiguous particle-like behavior, at any time? If so, then under the CH definition of an object, this behavior may be used to constrain further the dimensions of the box that is needed to surround them, and thereby to limit further the number of viable counterexamples to the Freedman–Clauser experiment.[25] It is clear from the simple fact that one observes temporal coincidences between a sequentially emitted pair of cascade photons, that at least the second photon of a cascade must be temporally localized during its propagation. However, this observation, by itself, does not imply that the same photon is also spatially localized (i.e. confined within a box that does not always also still include the emitting atom), and especially spatially localized in directions perpendicular to its propagation direction.

*Josef Jauch*, who had earlier coauthored a standard textbook on QED [41], expresses a view that photons do indeed exhibit an unambiguous particle-like character [42]. He bases this view on an experiment that was originally proposed by Schrödinger. Earlier, Schrödinger had provided a causal interpretation of his own equation – the so-called "old" Schrödinger interpretation – wherein a propagating electron therein has objectively real charge density waves associated with its propagating wave function. For the Compton scattering process, *Schrödinger* [43] shows that these waves can act as moving matter-wave gratings that Bragg-scatter classical electromagnetic wave packets in a manner that exactly predicts the results observed by Compton and Simon. Schrödinger was thus not convinced of the pivotal importance of the Compton–Simon experiment (as von Neumann apparently was) in compelling an acceptance of the Copenhagen interpretation. As an original heretic to the standard religion, he thus persuaded *Ádám*, *Jánnosy* and *Varga* (ÁJV) [44] to actually perform his proposed experiment. In their experiment, two independent photo-detectors are placed respectively in the transmitted and reflected beams of a half-silvered mirror. If photons have a particle-like character, i.e. if their detectable components are always spatially bounded and well localized, then photons impinging on the half-silvered mirror will not be split in two at this mirror. On the other hand, if they are purely wave-like in nature, (as with classical waves) then they can and will be split into two independent classical wave packets at this mirror. This fact then implies that if they are purely wave-like (in this classical sense), then the two detectors will show coincidences when a single temporally localized photon is directed

[25] The huge detector efficiency loss that is due to the small solid-angles subtended by photon detectors in typical cascade photon experiments, can then be rendered effectively harmless by said significantly reduced box volume. This fact is used explicitly (but without citing my experimental justification for it) by Bell in the *Gedankenexperiment* that he depicts on p.107 of his book, Speakable and Unspeakable in Quantum Mechanics.

at said mirror. One of these independent wave packets will be transmitted to illuminate the first detector, and the other will be reflected to illuminate the second detector, and both detectors will then have a finite probability of detecting the same photon (classical wave packet). This latter possibility, however, violates the predictions by QED, which prohibits such coincidences. ÁJV thus searched for anomalous coincidences between photomultiplier tubes that viewed the reflected and transmitted beams behind a half-silvered mirror. Unfortunately, their mirror was illuminated by a temporally continuous beam of light, and not by "tagged" single photons.[26] They found no such anomalous coincidences, and thereby concluded that, consistently with QED, photons do not split at the mirror and thus do exhibit a particle-like character.

A critical parameter in the ÁJV experiment is the efficiency of their photon detectors. Its value then determines whether or not the sought-after anomalous coincidences can be distinguished from the large background of omnipresent "accidental" coincidences. Upon reading their paper, I was astonished by their claim to have achieved 10% photon detection efficiency. Stu Freedman and I had struggled with much better equipment to get an efficiency of about 0.1%. In fact, we actually measured this efficiency using the above-mentioned photon-tagging scheme, by calculating the ratio of the coincidence count rate to the singles count rate for first photons of the cascade. How could ÁJV possibly claim to have achieved this absurdly high efficiency? To do so, they had ignored the solid-angle loss, and had quoted only the so-called photomultiplier quantum efficiency. But it is clear that for a worst-case spherically expanding wave, it is the overall photon detection efficiency that matters to the argument, and *not* the photomultiplier's quantum efficiency. Further study of their result and of similar repetitions of their experiment convinced me that none of these experiments convincingly demonstrated a particle-like behavior for photons. With low detector efficiencies, all of these experiments fell far short of having the required precision to do so. At best, they could only observe coincidences between two photons that accidentally were temporally overlapped. Viewed in this light, the ÁJV experiment was effectively a null experiment that, at best, sets an upper limit to the anomalous coincidence rate that might be present. Moreover, the upper limit that they obtain does not then even usefully constrain semi-classical radiation theories.

I thus decided to perform an experiment that has convincingly efficient detectors, and also that is configured, not as a null experiment, but instead as one that gives a definitive answer. Its results were published in 1974 [45].

---

[26] A "tagged" single photon is readily produced by a cascade photon source, by using a coincidence gate to only look for a sequentially emitted second single photon from an atom that is undergoing a cascade decay, that immediately follows the detection of the first photon of the decay. Given a detection of the former, one then knows with a very high probability that the associated second photon has been emitted by the atom, within roughly an intermediate-state lifetime, following the first photon's emission.

This experiment, unlike that by ÁJV, exploits the temporal localization already exhibited by "tagged" photons. Also, unlike the ÁJV experiment, the experiment employs four photomultiplier tubes, operating pair-wise in coincidence, and two half-silvered mirrors. By a judicious application of the Cauchy–Schwarz inequality to four different measured coincidence rates, it is then possible to obtain the desired definitive (non-null) prediction for this configuration. The experiment's results show that both quantum mechanics and quantum electrodynamics hold true, and photons do not split at a half-silvered mirror. This experiment is further noteworthy, in that it represents the first observed violation of a Cauchy–Schwarz inequality, as derived from classical optics principles, and thus also represents the first observation of sub-Poissonian statistics for light.

Curiously, the results of this experiment also resolve another untidy issue left behind by quantum theory's founders. That issue is whether or not energy conservation in the photoelectric effect is point-wise exact, or only holds as a statistical average, as suggested earlier by Bohr, Kramers and Slater. (Recall that von Neumann cited the Compton–Simon experiment as a refutation of this theory.) Schrödinger, in his above-cited works, is critical of this conclusion, and reopens the question. As a logical follow-on, Schrödinger then worried about the particle-like character of photons and, as noted above, went on to propose the ÁJV experiment as a crucial test. Schrödinger (as a theorist), however, never carefully scrutinized ÁJV's experimental assumptions and parameters to note the difference between ÁJV's incorrect use of the photomultiplier quantum efficiency in place of the (correct) photon detection efficiency. As a result, no prior experiments had ever carefully and conclusively resolved this issue, that instead, heretofore had been simply dismissed by others on purely theoretical grounds. (Here we see yet another critical gap between theorists and experimentalists.)

The issue arises from the fact that there are really two independent definitions of electromagnetic energy – one from quantum mechanics ($h\nu$) and one from classical mechanics (integral over all space of the electromagnetic field energy). Quantum electrodynamics is formally based on an assumed strict equality of these two different independently defined quantities, and thus requires a strict point-wise conservation of energy. But, prior to my experimental result, this crucial fundamental assumption had not yet been directly experimentally demonstrated. My improved version of the ÁJV experiment now finally does just that. *Grangier*, *Roger*, and *Aspect* [46] have since performed a somewhat similar experiment that also observes sub-Poissonian statistics and thereby confirms my results.

## 6.15 Remaining Locality Loopholes

It should be noted that an experiment that is shown to violate the CH inequality then rules out all local realistic theories, except for one remaining special class of theories. These are theories for which Bell (following Bohm and Aharonov) noted in his original 1964 paper that an experiment performed with static analyzers is, itself, logically flawed, since it does not strictly enforce locality. Indeed, it is logically possible that in such an experiment the two different particle detectors, located on opposites sides of one's laboratory, are conspiratorially communicating with each other, with the specific motive of defeating the experimental test. To close this so-called "locality loophole", Bohm and Aharonov proposed that one should rapidly reorient the analyzers while the photons are in flight, and Bell echoes and promotes this suggestion in his 1964 paper. However, Bohm and Aharonov's artifice clearly does not close this loophole! Instead, it simply shifts the burden of perpetrating said conspiracy from one apparatus component to another.

Although logically, it must be admitted that such theories are indeed possible, in a sense they seem to be rather improbable, indeed almost "pathological". To accept the locality loophole as a reasonable possibility, it is first necessary to believe that a pair of detectors and analyzers that are several meters apart are somehow conspiring with each other, so as to defeat the experimenter, and thus somehow are converting a non-sinusoidal coincidence rate dependence on relative polarizer orientation into a sinusoidal dependence. However, once that one is willing to concede that these apparatus components are indeed capable of perpetrating such a conspiracy, then it seems to be equally reasonable (and similarly logically possible) to assume further that the electronics that control a pair of rapidly reorienting analyzers are (perhaps unwittingly) also equally participating in the same conspiracy, also then to defeat the experimenter's efforts with similarly devious tricks. To be sure, it may seem that a belief in such logically possible conspiracies requires a certain degree of paranoia. A logical consequence of admitting to this paranoia[27] is then an associated recognition that every piece of equipment that has been in your laboratory for more than a few nanoseconds might possibly participate in such a grand conspiracy against you, whereupon no effort on your part can rule out this logical possibility! However, if all experimentalists were similarly paranoid, then experimental physics, in general, would seem to be a pointless endeavor.

Without attributing paranoid tendencies for these workers, it is noteworthy that nearly a decade after the publication of the Freedman–Clauser results, *Aspect* et al. [47] took the possible existence of such conspiracies seriously, and following Bohm and Aharonov's original suggestion, performed a second-generation experiment with dynamic (time-varying) analyzers. Their experiment then rules out detector-analyzer conspiracies. Other workers are currently considering performing experiments that might rule out even more insidious higher-level conspiracies.

[27] By similar reasoning, the simple fact that you are paranoid doesn't mean that *they* are not actually trying to kill you!

## 6.16 Conclusions

Clearly, quantum theory's predictive power and Bohr's Complementarity have both proven to be exceedingly robust while under attack by what is probably the most powerful challenge they have ever had to face – Bell's Theorem. History now shows that Bohr's views are probably the correct ones to hold. However, in retrospect, those who held to Einstein's view can, with reasonable justification, claim that Bohr, himself, was just lucky, in that he then had no real compelling (experimentally justifiable) reason to hold his view. Similarly, those who smugly held to Bohr's view now can say, also with a similarly reasonable justification, "See, we told you so!" Nonetheless, it appears reasonable to believe that neither camp can justifiably argue that the resolution, via Bell's Theorem, of the issues separating them was a waste of time. A lot of new physics was discovered in the process.

Given historical hindsight, I assert that our basic understanding of quantum theory has been significantly improved via Bell's Theorem and via its associated experimental testing, long after it was confidently asserted by many textbooks to be well understood. It is truly amazing that so many "killer" details slipped through cracks that existed between experimentalists and theorists. It is clearly of continuing importance for experimentalists and theorists to scrutinize each other's work with great care to try to eliminate such cracks. Given such hindsight, I also assert that it is clearly counterproductive to scientific progress for one camp smugly to hold to a belief that all problems are solved in any given area. It is even more counterproductive for this camp then further to rely on this belief to formulate a religious stigma against others who do not share their cherished belief. Indeed, history also shows that a prohibition against open discussion and experimental testing of the foundations of quantum theory, in turn, led to a significant delay of the discovery of important new applications of these foundations. Quantum cryptography, distributed entanglement, etc. undoubtedly would never have been envisioned without the intellectual challenges posed by Bell's Theorem.

My own final conclusion is that the only real looser here has been the "stigma", itself. I hope that John Bell would have agreed.

## References

1. J.F. Clauser, A. Shimony: Rep. Prog. Phys. **41**, 1881 (1978)
2. J. von Neumann: *Mathematische Grundlagen der Quantenmechanik* (Springer, Berlin 1932). English translation: *Mathematical Foundations of Quantum Mechanics* (Princeton Univ. Press, Princeton 1955)
3. A. Einstein, B. Podolsky, N. Rosen: Phys. Rev. **47**, 777 (1935)
4. N. Bohr: Phys. Rev. **48**, 696 (1935); Nature **136**, 65 (1935)
5. E. Schrödinger: Proc. Cambridge Philos. Soc. **31**, 555 (1935)
6. W.H. Furry: Phys. Rev. **49**, 393 (1936); Phys. Rev. **49**, 476 (1936)
   The second paper emphasizes the differences between his and Schrödinger's views of this result

7. A.H. Compton, A.W. Simon: Phys. Rev. **26**, 289 (1925)
8. N. Bohr, H.A. Kramers, J.C. Slater: Philos. Mag. **47**, 785 (1924)
9. J.S. Bell: *Speakable and Unspeakable in Quantum Mechanics* (Cambridge Univ. Press, Cambridge 1987) Chapt. 18
10. J.S. Bell: Phys. World, Aug. 33–40 (1990)
11. D. Bohm: *Quantum Theory* (Prentis Hall, Englewood Cliffs, NJ 1950)
12. D. Bohm, Y. Aharonov: Phys. Rev. **108**, 1070 (1957)
13. C.S. Wu, I. Shaknov: Phys. Rev. **77**, 136 (1950)
14. D. Bohm: Phys. Rev. **85**, 169 (1952). See also, D. Bohm, J.P. Vigier: Phys. Rev. **96**, 208 (1954)
15. L. de Broglie: *Nonlinear Wave Mechanics* (Elsevier, Amsterdam 1960) *Ondes Electromagnetiques et Photons* (Gauthier-Villars, Paris 1968) *Introduction to the Vigier Theory of Elementary Particles* (Elsevier, Amsterdam 1963)
16. J.S. Bell: Revs. Mod. Phys. **38**, 447 (1965)
17. J.S. Bell: Physics **1**, 195 (1964)
18. A. Peres, P. Singer: Nuovo Cimento **17**, 964 (1960)
19. C.A. Kocher, E.D. Commins: Phys. Rev. Lett. **18**, 575 (1969)
20. D. Wick: *The Infamous Boundary – Seven Decades of Controversy in Quantum Physics* (Birkhäuser, Boston 1995)
21. J.F. Clauser: Bull. Am. Phys. Soc. **14**, 578 (1969)
22. J.F. Clauser, M.A. Horne, A. Shimony, R.A. Holt: Phys. Rev. Lett. **23**, 880 (1969)
23. S.J. Freedman, J.F. Clauser: Phys. Rev. Lett. **28**, 938 (1972)
24. J.F. Clauser: Phys. Rev. Lett. **36**, 1223 (1976)
25. J.F. Clauser: Nuovo Cimento B **33**, 740 (1976)
26. E.S. Fry, R.C. Thompson: Phys. Rev. Lett. **37**, 465 (1976)
27. J.F. Clauser, M.A. Horne: Phys. Rev. D **10**, 526–535 (1974)
28. J.S. Bell: Introduction to the hidden-variables question, in *Foundations of Quantum Mechanics*, Proceedings of the International School of Physics, "Enrico Fermi", ed. by B. d'Espagnat (Academic, New York 1971) pp. 170–194
29. J.S. Bell: Science **177**, 880 (1972)
30. E.S. Fry, T. Walther, S. Li: Phys. Rev. A **52**, 4381 (1995)
31. A. Aspect, P. Grangier, G. Roger: Phys. Rev. Lett. **49**, 91 (1982)
32. J.S. Bell: *Communication at the 6th Gift Conf. Jaca, June 1975*, Res Th 2053-CERN
33. J.S. Bell, A. Shimony, M.A. Horne, J.F. Clauser: Dialectica **39**, 85 (1985)
34. A. Shimony: *Search for a Naturalistic World* (Cambridge Univ. Press, Cambridge 1993); More recently, the importance of these concepts was independently rediscovered by J. Jarrett [Noûs **18**, 569 (1984)].
35. W.E. Lamb, M. Scully: in *Polarization: Matière et Rayonnement* (Presses Universitaires de France, Paris 1969), ed. by Société Française de Physique
36. P.A. Franken: in *Atomic Physics* Proceedings of the 1st International Conference on Atomic Physics, ed. by V. Hughes et al. (Plenum, New York 1968), p.377
37. E.T. Jaynes, F.W. Cummings: Proc. IEEE **51**, 89 (1963); C.R. Stroud, Jr., E.T. Jaynes: Phys. Rev. A **1**, 106 (1970); M.D. Crisp, E.T. Jaynes: Phys. Rev. **179**, 1253 (1969); **185**, 2046 (1969)
38. P.A. Nesbet: Phys. Rev. A **4**, 259 (1971)

39. J.F. Clauser: Phys. Rev. A **6**, 49 (1972)
40. The proceedings of this conference are in L. Mandel, E. Wolf (eds.), *Coherence and Quantum Optics* (Plenum Press, New York 1973)
41. J.M. Jauch, F. Rohrlich: *The Theory of Photons and Electrons* (Addison-Wesley, Reading, MA 1955)
42. J.M. Jauch: *Dialogue on the Question ARE QUANTA REAL?* (Univ. of Geneva, preprint, 1971)
43. E. Schrödinger: Physikalische Zeits. **23**, 301 (1922); Die Naturwissenschaften **12**, 720 (1924); Nuovo Cimento **9**, 162-170 (1958)
44. A. Ádám, L. Jánnosy, P. Varga: Acta Phys. Hung. **4**, 301 (1955); Ann. Phys. **16**, 408 (1956)
45. J.F. Clauser: Phys. Rev. D **9**, 853 (1974)
46. P. Grangier, G. Roger, A. Aspect: Europhys. Lett. **20**, 1061 (1986)
47. A. Aspect, J. Dalibard, G. Roger: Phys. Rev. Lett. **49**, 1804 (1982)

# 7 On Four Decades of Interaction with John Bell

Michael Horne

In the late 1960s Abner Shimony, on his first day as my thesis advisor at Boston University, gave me two papers by *John Bell*: "On the Problem of Hidden Variables in Quantum Mechanics" [1] and "On the EPR Paradox" [2]. These are, respectively, the first and second papers in John's 1987 reprint collection, *Speakable and Unspeakable in Quantum Mechanics* [3]. Both papers are masterpieces. In the first, John reviews and often simplifies earlier proofs that quantum mechanics is incompatible with the existence of dispersion-free states (hidden variables). John then points out a significant flaw in the physical assumptions behind each argument. The second paper is a follow-up in which he provides a flawless proof, via the now famous inequality, of the incompatibility of quantum mechanics and any *local* hidden variable theory, i.e. any theory that assumes the existence of EPR's local elements of reality. The argument is made in the context of Bohm's version of an EPR situation, i.e. two spatial separated yet entangled spin - 1/2 particles.

Abner saw the significance of John's two papers before most physicists knew of their existence and he suggested as a thesis topic that I try to design a realizable Bohm-type EPR experiment. Because of Bell's inequality theorem, the experiment would either violate the quantum mechanical predictions or show that Einstein's elements of reality do not exist. What a thesis topic! Within less than two years a doable experiment was designed and a proposal published [4], in collaboration with John Clauser who independently had the same idea. A referee report made excellent suggestions that improved and strengthened the published paper. One could tell from the quality and style of the report that this referee was John.

In addition to motivating an experimental proposal, John's two papers had other significant effects on me. They showed that elementary quantum mechanics itself can be a worthy subject of study. They showed that even decades after the initial discovery of the quantum rules, important new things can still be uncovered about the content of these rules, i.e. that the subject is not "over". They showed that new aspects and consequences of the quantum rules are most likely to be found if past work is first critically examined, simplified, boiled down to its physical essence, and freed of any unnecessary layers of mathematics. The two papers are models of clear, concise, and focused thinking and writing. In brief, those two papers taught me what good physics looks like.

In the early 1970s, while Clauser was performing the first experiment confirming the quantum predictions and afterwards, Bell and, independently, Abner, Clauser and I, worked at strengthening the argument, i.e. at deriving a testable inequality from weaker assumptions about local reality [5, 6]. These papers were essentially an exchange, we stimulated by his work and vice versa. Toward the end of this period, Abner, Clauser, and I were convinced that Bell had reached a little too far in these efforts to weaken the assumptions and a published debate developed [7–9]. In the end, the two sides were, as Abner has put it, "converging toward consensus".

The summer of 1976 was a turning point. At a workshop in Sicily on Bell physics, I finally met John for the first time. At that same workshop I also met Anton Zeilinger for the first time and we found that we shared a strong interest in the then new field of thermal neutron interferometry, he then working with Helmut Rauch in Vienna and I with Cliff Shull at MIT. For the next ten years, we concentrated on single-particle interference and I didn't think about two-particle entanglement and Bell physics except to occasionally share what I knew about it with Anton and with Danny Greenberger, who both made many visits, long and short, to Cliff's lab.

The return to two-particle quantum mechanics and renewed contact with John Bell came in 1985. Early that year, Anton and I saw an announcement for a summer symposium in Finland to celebrate the 50th anniversary of the EPR paper. Wanting to attend, but needing some new two-particle effect to present, we started pondering a possible connection between single-particle two-path interference and two-particle two-state entanglement of the Bohm type. This led us to invent a new arrangement [10, 11] which eventually came to be known as two-particle interferometry [12, 13]. The arrangement, although it employs spin-free particles, is a perfect analog to the Bohm arrangement and hence can violate Bell's inequality while employing a state closely related to the original EPR state. Bell was not at the Finland meeting, but I gave him a preprint of the Finland paper when he visited Boston University later in the year.

After a few months, John responded to our Finland preprint. Returning to the original EPR state, he noted [14] that since it has a positive Wigner function, there can be "no locality problem" with the EPR state if we are "restricted to measurements of position only, on two non-interacting spinless particles in free space". This conclusion is consistent with the presence of a locality problem in our arrangement, since the two-particle interferometer, although illuminated by an EPR type state, contains beam splitters and mirrors so that the space of the arrangement is no longer "free". In brief, it is not just the state but the whole apparatus that determines whether there is a locality problem.

Our renewed interest in *two*-particle entanglements led Danny Greenberger to suggest investigating whether there might be something new in *three*-particle entanglements. The end result was the GHZ [15] (Greenberger,

Horne, Zeilinger) theorem showing that the non-existence of Einstein elements of reality is even more dramatic in three-particle entanglements than in Bell's two-particle theorem. In the summer of 1990, Danny, Anton, Abner, and I met with John for the last time at a workshop at Amherst College. Mary Bell was also there. We presented the new three-particle theorem and passed out preprints [16]. Had he lived longer, I'm sure John would have added something new and interesting. He always did.

## References


1. J.S. Bell: On the problem of hidden variables in quantum theory, Rev. Mod. Phys. **38**, 447 (1966). Reprinted as paper 1 in [3]
2. J.S. Bell: On the Einstein–Podolsky–Rosen paradox, Phys. **1**, 195 (1964). Reprinted as paper 2 in [3]
3. J.S. Bell: *Speakable and Unspeakable in Quantum Mechanics* (Cambridge University Press, Cambridge 1987)
4. J. Clauser, M.A. Horne, A. Shimony, R. Holt: Proposed experiment to test local hidden variable theories, Phys. Rev. Lett. **23**, 880 (1969)
5. J.S. Bell: Introduction to the hidden-variable question, Proc. of the International School of Physics 'Enrico Fermi', course IL, in *Foundations of Quantum Mechanics* (Academic, New York 1971) pp. 171–181. Reprinted as paper 4 in [3]
6. J. Clauser, M.A. Horne: Experimental consequences of objective local theories, Phys. Rev. D **10**, 526 (1974)
7. J.S. Bell: The theory of local beables, TH-2053-CERN, July 28 (1975). Presented at the sixth GIFT seminar, Jaca, 2–7 June 1975, and reproduced in Epistemological Letters March 1976. This article also appeared in Dialectica **39** (1985). Reprinted as paper 7 in [3]
8. J. Clauser, M.A. Horne, A. Shimony: Comment on Bell's theory of local beables, Epist. Lett. **13**, (1976). Reprinted in Dialectica **39** (1985)
9. J.S. Bell: Free variables and local causality, Epist. Lett., February (1977). This article also appeared in Dialectica **39** (1985). Reprinted as paper 12 in [3]
10. M.A. Horne, A. Zeilinger: A Bell-type EPR experiment using linear momenta, in *Symposium on the Foundations of Modern Physics*, ed. by P. Lahti, P. Mittelstaedt (World Scientific, Singapore 1985). Conference June 16–20, 1985, Joensuu, Finland
11. M.A. Horne, A. Zeilinger: A possible spin-less experimental test of Bell's inequality, in *Microphysical Reality and Quantum Formalism*, Vol. 2, ed. by A. van der Merwe et al. (Kluwer Academic, Dordrecht 1988), p. 401. Conference Sept. 25 – Oct. 3, 1985, Urbino, Italy
12. M.A. Horne, A. Shimony, A. Zeilinger: Two-particle interferometry, Phys. Rev. Lett. **62**, 2209 (1989)
13. M.A. Horne, A. Shimony, A. Zeilinger: Quantum optics: Two-particle interferometry, Nature **347**, 429 (1990)
14. J.S. Bell: EPR correlations and EPW distributions, in *New Techniques and Ideas in Quantum Measurement Theory*, ed. by D.M. Greenberger (New York Acad. Sci., New York 1986). Reprinted as paper 21 in [3]

15. D.M. Greenberger, M.A. Horne, A. Zeilinger: Going beyond Bell's Theorem, in *Bell's Theorem, Quantum Theory and Conceptions of the Universe*, ed. by M. Kafatos (Kluwer Academic, Dordrecht 1989) pp. 69–72. Conference Oct. 21–22, 1988 at George Mason University, Fairfax, VA
16. D.M. Greenberger, M.A. Horne, A. Shimony, A. Zeilinger: Bell's Theorem without inequalities, Am. J. Phys. **58**, 1131 (1990)

# 8 Atom Based Tests of the Bell Inequalities – the Legacy of John Bell Continues ...

Edward S. Fry and Thomas Walther

After a brief historical review including a discussion of loopholes associated with previous tests of the Bell inequalities, we review the loophole-free atom based experiment we have underway. It tests the strong Bell–Clauser–Horne inequality and is based on the photo-dissociation of $^{199}Hg_2$ dimer that is in a nuclear spin-1/2 entangled state. This experiment will be compared to other recent atom based tests.

This paper is dedicated to John Bell whose work not only inspired the experiments discussed in this paper but has also had a far-reaching influence on our understanding of the interpretation of quantum mechanics.

## 8.1 Historical Overview

In a period of 30 years at the beginning of the last century the world of Physics was revolutionized. Now, one hundred years after Planck's quantum hypothesis, we continue to be puzzled by the wonders of quantum mechanics. We have learned to apply concepts such as entanglement to problems in quantum information [1,2]. Other examples of emerging applications that have arisen in the last couple of years include quantum cryptography [3] and teleportation [4–6]. Nevertheless, the article published in 1935 by *Albert Einstein*, *Boris Podolsky*, and *Nathan Rosen* (generally referred to as EPR) continues to fascinate us [7]. EPR argued by means of a *gedanken*experiment that quantum mechanics is not a "complete" theory. This incomplete description could presumably be avoided by postulating the presence of some hidden variables (HV) that would permit deterministic predictions for microscopic events. The relation of quantum mechanics to HV could be considered analogous to the relation of thermodynamics to statistical mechanics. For example, in a gas the thermodynamic quantities, temperature and pressure, can be understood in terms of microscopic (hidden) values of the speeds and directions of the individual atoms. In addition, the HV would eliminate concerns about non-locality.

For many years the discussion was purely philosophical in nature. However, in 1964, *John Bell* made a giant step forward [8]. Bell revisited the EPR experiment and considered a version based on the entanglement of spin-1/2 particles that was introduced by *Bohm* [9]. The latter has conceptual advantages in comparison to the original EPR experiment that involves entanglement in position and momentum. The wave function of such a system can be written as

$$|\Psi\rangle = \frac{1}{\sqrt{2}}\Big(|\uparrow\rangle_1|\downarrow\rangle_2 - |\downarrow\rangle_1|\uparrow\rangle_2\Big). \tag{8.1}$$

Thus, for a given two particle system, if particle one is found to have spin up (↑) in some spatial direction, particle two must necessarily have spin down (↓) in that direction and vice versa. Specifically, quantum mechanics predicts that the total spin of this two particle system is zero, and that for measurement of both components in one specific direction, the results will be opposite with unit probability. However, quantum mechanically it is not possible to predict the absolute orientation of either spin, or to even predict absolute values for two of the components of the spin of a particle.

John Bell showed: (1) Hidden-variable theories can exist in the context of Bohm's version of the EPR experiment. (2) *Any* hidden-variable theory satisfying a physically reasonable condition of locality (LHV theory, e.g. a classical theory) will yield statistical predictions which must satisfy restrictions for certain correlated phenomena such as those involving entangled states (8.1). In other words, classical theories put upper bounds on measurements of the statistical correlations between the spin components of the two particles. These restrictions can be cast into the form of inequalities, which are now generally referred to as Bell inequalities. Thus, at least in principle, Bell's work made it possible for the first time to experimentally distinguish between the LHV and quantum mechanical pictures. It should be emphasized that no specifics of an LHV theory are involved, only its existence. Thus a test of a Bell inequality is general and leads to discrimination between any LHV and quantum mechanics.

Bell's original formulation of the inequalities was idealized and not readily suited to realistic experimental conditions. Several researchers have since formulated other versions of the Bell inequalities [10] that are more experimentally amenable. Well known examples are the *Clauser–Horne–Shimony–Holt* (CHSH) inequality [11] and the *Bell–Clauser–Horne* (BCH) inequality [12]. The CHSH inequality is

$$S(\theta_1,\theta_2,\theta_1',\theta_2') = |E(\theta_1,\theta_2) - E(\theta_1',\theta_2)| + E(\theta_1,\theta_2') + E(\theta_1',\theta_2') \leq 2\,, \tag{8.2}$$

where $\theta_1$, $\theta_1'$ are two values of angles for the first spin analyzer and $\theta_2$, $\theta_2'$ are two values of angles for the second spin analyzer. In (8.2) the quantities $E(\theta_1,\theta_2)$ are expectation values,

$$E(\theta_1,\theta_2) = \frac{N_{\uparrow\uparrow}(\theta_1,\theta_2) - N_{\uparrow\downarrow}(\theta_1,\theta_2) - N_{\downarrow\uparrow}(\theta_1,\theta_2) + N_{\downarrow\downarrow}(\theta_1,\theta_2)}{N_{\mathrm{total}}}\,, \tag{8.3}$$

where $N_{\uparrow\downarrow}(\theta_1,\theta_2)$ is the number of pairs of particles for which analyzer 1 (oriented in direction $\theta_1$) would give spin up for particle 1 and analyzer 2 (oriented in direction $\theta_2$) would give spin down for particle 2. Definitions for the other $N$'s are analogous. $N_{\text{total}}$ is the total number of pairs of particles. The $N$'s denote the number of pairs of particles rather than actual detection events. Or in other words, the $N$'s mean emergence from the analysers, not detection. In this regard (8.2) is not an experimentally testable inequality. However, if the detectors have 100% efficiency, detection and emergence from the analyzer are identical and the inequality is testable, e.g. the Boulder experiment [13]. By making the CHSH assumption (the detection efficiency is independent of analyzer orientation), the $E(\theta_1, \theta_2)$ can be formulated in terms of experimental coincidence detection rates.

One can also make the alternative CH assumption that the probability of detection with the analyzer in place is less than or equal to the corresponding probability with the analyzer removed. From a practical standpoint, employing either the CHSH or the CH assumption enables one to obtain inequalities that depend only on coincidence rates ("simultaneous" detection of both members of an entangled pair of particles) and that do not depend on the singles rates for detection of particles at only one of the detectors. As a consequence, these inequalities are independent of detection efficiencies. *Freedman* has obtained such an inequality in a particularly simple form [14],

$$\frac{|R(22.5°) - R(67.5°)|}{R_0} \leq \frac{1}{4}\,, \tag{8.4}$$

where $R(22.5°)$ and $R(67.5°)$ are coincidence rates with angles 22.5° and 67.5°, respectively, between the polarization transmission directions of the polarizers for each photon, and $R_0$ is the coincidence rate with both polarizers removed. By contrast, the BCH inequality is

$$\begin{aligned} &S(\theta_1, \theta_2, \theta_1', \theta_2') \\ &\quad = \frac{R_{\uparrow\uparrow}(\theta_1, \theta_2) - R_{\uparrow\uparrow}(\theta_1, \theta_2') + R_{\uparrow\uparrow}(\theta_1', \theta_2) + R_{\uparrow\uparrow}(\theta_1', \theta_2')}{R_{1\uparrow}(\theta_1') + R_{2\uparrow}(\theta_2)} \leq 1\,, \end{aligned} \tag{8.5}$$

where a typical $R_{\uparrow\uparrow}(\theta_1, \theta_2)$ denotes the experimental coincidence count rates when both particles are detected with spin up in the directions $\theta_1$ and $\theta_2$, respectively; $R_{1\uparrow}(\theta_1')$ and $R_{2\uparrow}(\theta_2)$ are singles count rates at detectors 1 and 2, respectively. The BCH inequality is especially important because it provides a direct constraint on the experimentally accessible (observed) detection count rates; it does not require any additional assumptions for experimental implementation. However, it depends on the ratio of coincidence rates to singles rates and is therefore proportional to the detector efficiency. (By comparison, (8.4) depends on the ratio of coincidence rates to coincidence rates and thus the detector efficiency in the numerator and denominator of this ratio cancel). Consequently, for low detection efficiency, the inequality (8.5) is always satisfied by the quantum mechanical predictions; this is also recognized by noting that the singles rates in the denominator are large compared to the coincidence rates in the numerator. Therefore, (8.5) can only be used for a definitive test if the detectors have very high efficiencies. Note also, that it is not necessary to simultaneously measure both projections in the case of the BCH inequality.

Both (8.2) and (8.5) apply to deterministic as well as stochastic (a surprising result) theories, that are based on local hidden variables [10]. For a definitive loophole free experiment, the BCH inequality [12] should be tested. Since the BCH inequality contains no additional assumptions, it provides the strongest possible test. It is remarkable that this inequality has actually not yet been tested (cf. Sect. 8.3).

The principle problem for experimental tests of the BCH inequality lies in the fact that experimental imperfections generally preclude a definitive test, i.e. they tend to shift the result such that the Bell inequalities are no longer violated by the quantum mechanical predictions.

## 8.2 The Bell–Clauser–Horne Inequality

Since our experiment will test the BCH inequality, we will briefly discuss the quantum mechanical prediction for this inequality as it applies to the experimentally relevant situation of our experiment (cf. Sect. 8.4.3).

The test of the strong BCH-inequality requires the measurement of coincidence rates $\mathrm{R}_{\uparrow\uparrow}(\theta_1,\theta_2)$ for the simultaneous detection of one particle of the entangled pair at detector 1 with spin-up ("$\uparrow$") in the direction $\theta_1$ and of the other atom at detector 2 with spin-up ("$\uparrow$") in the direction $\theta_2$; the singles rates $\mathrm{R}_{i\uparrow}(\theta_i)$, which are defined as the rate of detection of particles with spin-up in the direction $\theta_i$ at detector $i$, where $i$=1 or 2 must also be measured. The parameter for which the pair is entangled (i.e. polarization, time-position, or spin) or the type of particle (i.e. photon or atom) that is entangled is not significant to the argument. However, we will concentrate on atomic systems with an entanglement of the spin according to (8.1).

The BCH inequality is formulated in terms of the ratio of coincidence rates to singles rates for four different combinations of angles for the spin measurement [10, 12]. The quantum mechanical predictions including experimental imperfections for the BCH inequality is given by [15, 16]:

$$R_{\uparrow\uparrow}(\theta_1,\theta_2) = \eta^2 fgN\frac{1}{4}\left[\varepsilon_+^2 - \varepsilon_-^2 F\cos(\theta_1-\theta_2)\right] , \tag{8.6}$$

$$R_{1\uparrow}(\theta_1) = R_{2\uparrow}(\theta_2) = \frac{\eta fN}{2}\varepsilon_+ , \tag{8.7}$$

where $\eta$ is the detector efficiency for the atoms and is assumed to be the same for both detectors; $\varepsilon_+$ and $\varepsilon_-$ are defined as $\varepsilon_+ = \varepsilon_M + \varepsilon_m$ and $\varepsilon_- = \varepsilon_M - \varepsilon_m$ , respectively, where $\varepsilon_M$ is the transmission of the analyzers for one spin component and $\varepsilon_m$ the leakage through the analyzer for the other spin component; $\varepsilon_m$ and $\varepsilon_M$ are positive definite and are assumed to be identical for both analyzers. Specifically, $\varepsilon_+$ and $\varepsilon_-$ are measures of the capability to discriminate between the two spin components in a measurement. For optimum discrimination, $\varepsilon_+$ and $\varepsilon_-$ are equal to unity. The two detectors are assumed symmetric; they have identical geometries and $f$ is the detector acceptance solid angle; $g$ is the conditional probability that if one of the atoms from a dissociated dimer enters the aperture of one detector, then the other atom from that dimer enters the aperture of the other detector; $N$ is the total number of dissociating dimers per unit time; $F$ is a measure for the purity of the entangled state.

In an actual experiment the goal will be to choose the angles $\theta_1$, $\theta_2$, $\theta_1'$, and $\theta_2'$ such that when the quantum mechanical predictions (8.6) and (8.7) are used in (8.5), the resulting $S_{\mathrm{QM}}$ provides a maximum violation of the BCH inequality. For $\theta_1$, $\theta_2$, $\theta_1'$, $\theta_2'$ equal to 135°, 0°, 225°, and 90°, respectively, we find:

$$S_{\mathrm{QM}}(135°, 0°, 225°, 90°) = \frac{1}{2}\eta g \varepsilon_+ \left[1 + \sqrt{2}F\left(\frac{\varepsilon_-}{\varepsilon_+}\right)^2\right] . \tag{8.8}$$

In order to obtain the largest possible violation of the BCH inequality by the quantum mechanical prediction, the right hand side of (8.8) must exceed one, and the parameters $\eta$, $g$, $\varepsilon_M$, $F$ should be as close to unity as possible while $\varepsilon_m$ should be as small as possible.

Finally, any test of the Bell inequalities is statistical in nature; specifically, count rates have statistical errors and the measured quantities are ratios of coincidence count rates to singles count rates. These statistical errors must be sufficiently small so that the resulting error in the measured value of $S$ is much less than the magnitude of the violation (the amount by which $S_{\mathrm{QM}}$ exceeds one). Specifically, it is not the magnitude of the violation, but rather its comparison to the error limits of an experimentally determined value of $S$ that are a measure of the significance of the violation. As an aside it should be noted that some experimental parameters must be measured (and therefore have errors associated with them) in order to evaluate $S_{\mathrm{QM}}$; therefore, the quantum mechanical prediction $S_{\mathrm{QM}}$ will have errors associated with it.

## 8.3 Loopholes

Despite a substantial number of experimental tests of the Bell inequalities, no experiment to date has been entirely loophole free [10, 12, 17–20]. Specifically, one or more of the following loopholes were present: (1) the spatial correlation loophole (2) the detection efficiency loophole, and (3) the enforcement of locality (communication loophole).

### 8.3.1 The First Loophole

The spatial correlation loophole relates to the ideal case in which the two-particle entangled state is generated by a two-body decay that involves only the two particles that are entangled. In this case, momentum conservation ensures a strong spatial correlation that is necessary since the experimenter must make sure that both particles of an entangled pair end up in the detectors. This loophole is effectively eliminated by employing experiments based on two-body processes. The original Bell inequality experiments employing two photon atomic cascades suffered from spatial correlation. Specifically, the two entangled particles were the cascade photons but they were produced in the presence of a third particle, the atom, that carried off some momentum and smeared out the spatial correlation between the two photons.

### 8.3.2 The Second Loophole

The detection efficiency loophole originates from experimental imperfections. Low detection efficiencies reduce the observable correlation, since the lower the detection efficiency, the lower the probability that both partners of an entangled pair will be detected. In particular, the quantum mechanical expectation value $S_{\mathrm{QM}}$ (8.8) of the BCH inequality (8.5) involves the detection efficiency $\eta$. Clearly, if $\eta$ drops below some critical level the quantum mechanical prediction will always drop below one. Thus, a distinction between quantum mechanics and LHV theories is no longer possible. For the case of a symmetric entangled state as in (8.1) this critical value is $\eta_{\mathrm{crit}} = 0.82$ assuming all other parameters being perfect. However, it has been shown that for asymmetric entangled states (i.e. states for which the two components have weights different from $1/\sqrt{2}$) an efficiency of $\eta_{\mathrm{crit}} = 0.67$ would be sufficient [21]. The great benefit of atom based tests is that photo-ionization schemes can provide high detection efficiencies for almost all elements of the periodic table [22].

Since photon detectors had relatively low detection efficiencies, the CSHS and CH assumptions (cf. Sect. 8.1) were introduced in order to obtain experimentally testable inequalities. These assumptions allowed one to obtain inequalities that depend on ratios of coincidence rates to coincidence rates (8.4) so that the detector efficiency did not appear in the quantum mechanical predictions for the inequality.

As pointed out by *Santos* [19] and as evident in the quantum mechanical prediction (8.8), high detection efficiency alone is not enough to test the BCH inequality. High detection efficiencies must also be accompanied by a high probability of actually detecting both particles of an entangled pair in their respective detectors, i.e. a high conditional probability $g$ of finding particle 2 in detector 2 provided that particle 1 entered detector 1. This also relates to the first loophole, spatial correlation.

### 8.3.3 The Third Loophole

The enforcement of the locality condition is also known as the communication loophole. It requires that the choice of correlation measurement for the two particles be completely independent at each analyzer. Specifically, one must guarantee no communication between the two analyzers during a measurement, i.e. the time interval from choices of analyzer "orientations" to detection must be outside each others space-time light cone.

This means that the times required for the two detection events must be short compared to the time required for a light signal to propagate from one analyzer/detector to the other. This involves not only a large separation between the two analyzer/detectors, but also the requirement for fast detection schemes and electronics. The time of a detection event as defined in this context is the time it takes to choose and set the analyser settings in a random fashion, analyze the spin state, and complete and record the outcome of the experiment.

The first attempt to enforce the locality condition was pursued by *Aspect* and coworkers [23]. However, their innovative experimental scheme did not allow a strict enforcement of the locality condition; in particular, they used periodic switching of the polarizer orientation rather than random orientations [18]. Very recently *Zeilinger* et al. did succeed in rigorously closing this loophole [24]. Zeilinger's success represents a very important step in the clarification of the EPR argument. However, since the experiment was performed with a detection efficiency of approximately 5%, the detection loophole is not yet closed. The ultimate definitive test must be one that enforces locality and simultaneously closes all other loopholes associated with previous tests.

## 8.4 Atom Based Experiments

Significant progress in the experimental tests of Bell inequalities has been achieved. Most of the experiments have been performed using photons [23–33]. However, in this article we will concentrate on atom based experiments; these have features that make them distinctly different from their photon based counterparts [34]. For example, there are fermions as well as bosons, massive versus massless particles, long lived entangled states, and entanglement produced by direct manipulation of the atomic degrees of freedom. We will briefly review recent atom based tests of Bell inequalities and then conclude with a summary of the key ideas of the experiment which is underway at Texas A&M University.

### 8.4.1 The Paris Experiments

In the Paris experiments, entanglement between atoms is produced using a high-Q micro-wave cavity, specifically, through the interaction of two completely independent Rydberg atoms with a common radiation field in the cavity. The two Rydberg atoms pass one after another through the cavity.

With a proper choice of the interaction time between the atoms and the cavity field, entanglement between the two atoms is achieved [35].

While the detection efficiencies in this experiment are potentially high, the actual current implementation still has low detection efficiencies and relatively poor discrimination. Although these could be overcome, a principle difficulty for tests of the Bell inequalities based on entanglement in a micro-maser is that the generation of the entanglement is a sequential process, which requires precise control of the experimental parameters such as interaction time of the atoms with the cavity fields, separation of atoms etc. The enforcement of Einstein locality would also be a problem since the detection of the atoms occurs sequentially.

### 8.4.2 The Boulder Experiments

In the Boulder experiment, *Wineland* and coworkers [13] are the first group to close the detection efficiency loophole. They prepared two $Be^+$-ions in an ion trap in an entangled state between two hyperfine Zeeman substates. The two substates can be identified with spin-up and spin-down and are therfore equivalent to an entangled state as depicted by (8.1).

The determination of the spin components is considerably more subtle than in the usual case of spin-1/2 particles. The measurement of a spin component with respect to a certain quantization axis is essentially equivalent to a classical manipulation of the spin, i.e. a rotation onto the quantization axis. The two possible results are parallel or anti-parallel, which then can be identified as spin-up or down with respect to the quantization axis. In the Boulder experiment, the classical manipulation occurs by irradiating the ion with two Raman beams, which prepare the ions in a "bright" and "dark" superposition of the two possible Zeeman substates. The phase of the Raman lasers determine the two phases. The actual measurement occurs by scattering photons off the ions. The "bright" state does scatter photons, while the "dark" state does not. The up-up or down-down outcome as defined in (8.3) corresponds to both ions bright, or both dark, respectively. Equivalently, an up-down or down-up outcome corresponds to one ion dark and the other bright. By manipulation of the trap potential, the ions can be shifted within the trap without influencing the internal degrees of freedom, i.e. without affecting the entanglement. This shift in position then leads to a change in the relative phase of the Raman beams as seen by the two ions, and thereby enables the measurement of "spin" components in arbitrary directions.

Due to the proximity of the two ions in the trap, it is impossible in these experiments to enforce the locality condition. In fact, *Vaidman* has argued that the situation is even worse in these experiments [36]. Specifically, due to the detection of the state by the scattering of many photons, both detection events and the choice of both analyzer orientations overlap throughout a relatively long period of time (orders of magnitude longer than the time required for light to travel between the two ions).

### 8.4.3 The Texas A&M Experiment

The need for a two-particle system in an entangled state in the test of Bell inequalities suggests that a diatomic molecule might be an ideal starting ground for such a test. In fact, assuming a nuclear spin 1/2 for each atom, the nuclear wave function of the molecule in the separated atom basis can be written as in (8.1). This state is identical to that of the two spin 1/2 particles in Bohm's classic version of the EPR *gedanken*experiment [9, 10]. The question is now whether it is possible to prepare a diatomic molecule in such a state. This is, in fact, possible, and in addition to our experiment, an approach using $Na_2$ dimer has been proposed by *Shimony* et al. [37]. Our experiment is essentially an implementation of Bohm's version of an EPR experiment (cf. Fig 8.1). For a more detailed discussion of the experiment, the reader is referred to the following set of references [15, 16, 38–42]. The experiment is based on $^{199}$Hg, which has nuclear spin $I = 1/2$; it is the first attempt to actually test the BCH inequalities (8.5).

In our experiment $^{199}Hg_2$ dimers are generated in a supersonic jet expansion and are photo-dissociated using two laser beams in a stimulated Raman transition. The two wavelengths are at approximately 266 nm and 355 nm. Due to the difference in the photon energies, the molecule will be pumped onto a dissociating part of the potential energy surface of the ground electronic state. This leads to the dissociation of the molecule and thus to the spatial separation of the two atoms in the entangled state. The energy difference between a photon from each laser beam is distributed equally to the kinetic energies of the two atoms. Due to momentum conservation the

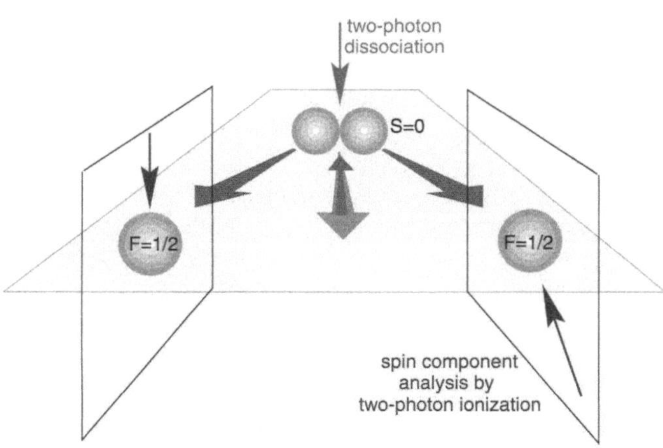


**Fig. 8.1.** Schematic of the Texas A&M experiment, which is a realization of Bohm's spin-1/2 particle EPR experiment. A $^{199}Hg_2$ molecule is dissociated into an entangled pair of atoms by a two-photon dissociation process. After the separation of the $^{199}$Hg atoms their spin components are measured by a two-photon ionization process.

atoms will fly apart in exactly opposite directions (180°) in the center-of-mass frame. However, as a consequence of the initial velocity of the dimers in the supersonic jet, the dissociated atoms will separate at an angle of approximately 130° in the laboratory frame.

The preparation of the entangled state given by (8.1) requires selecting a molecular state with a total nuclear spin $I = 0$. This is possible due to the specific symmetry rules for the total wavefunction of a homonuclear diatomic molecule consisting of two fermions. Thorough analysis of these symmetries [16] shows that only the entangled nuclear spin singlet state is dissociated if the excitation transition at 266 nm starts from states whose rotational quantum number $N$ is even.

Determination of the correlation in the spin components of the two entangled atoms as well as detection of the atoms is achieved via a state-selective two-photon excitation-ionization scheme. Based on the selection rules for electronic transitions, one can photo-ionize only those atoms which are in the state $m_F = +1/2$, i.e. spin up along the quantization axis (cf. Fig. 8.2). Since the quantization axis is determined by the propagation direction of

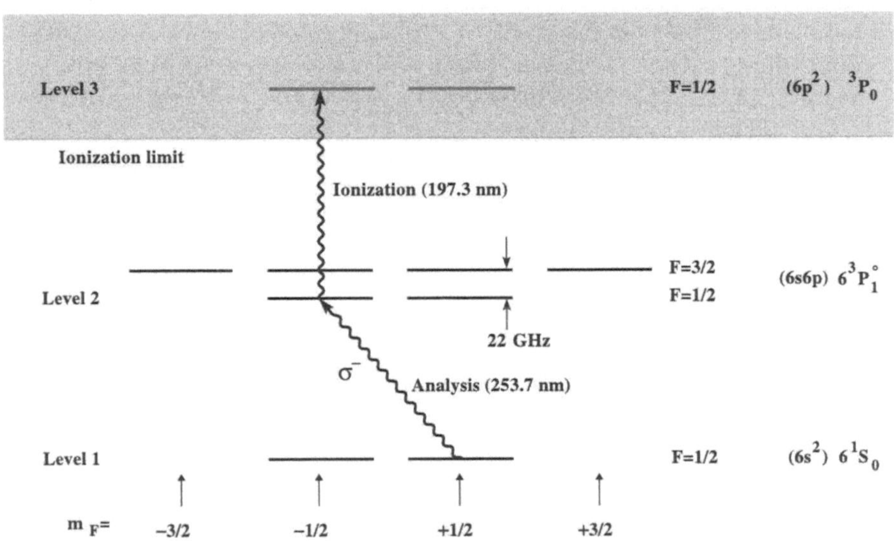


**Fig. 8.2.** Photo-ionization scheme used for spin analysis of the Hg atoms. The quantization axis is determined by the propagation direction of the left hand circular ($\sigma^-$) polarized 253.7 nm laser beam. The atom has been analyzed as "spin-up" ($m_F = +1/2$) if it is ionized in the two-photon process, 253.7 nm + 197.3 nm [via the $6^3\mathrm{P}_1^0$ ($F = 1/2$) state]. Excitation/ionization transitions from $m_F = -1/2$ correspond to a "spin-down" analysis; by the selection rule for $\sigma^-$, $\Delta m_F = -1$, this transition must go via the $6^3\mathrm{P}_1^0$ ($F = 3/2$) state and is highly suppressed due to its 22 GHz detuning from the $F = 1/2$ state

the left circular polarized laser beam, the spin component in any direction is measured by orienting the laser beam to propagate in that direction.

The ionized atoms are detected in two independent ways, i.e. by observing both the photo-ions and the photo-electrons. The photo-electrons are accelerated and guided by static electric fields onto a secondary electrode (CuBe surface), where they produce secondary electrons; the latter are then detected by a Channeltron multiplier. The photo-ions are focused, accelerated through 25 kV, and directed onto a slanted Al surface where they produce many secondary electrons that are also detected in a Channeltron multiplier.

Discussions of critical aspects of this experiment follow:

#### 8.4.3.1 Purity of the State

A high purity selection of the entangled state 8.1 to be dissociated is ensured since the different molecular rotational levels are spectroscopically well separated (typically $> 500$ MHz). Furthermore the use of nuclear spins ensures that the entanglement is relatively immune to external influences such as stray magnetic fields. Specifically, since the nuclear magnetic moment is very small (approximately 1800 times smaller than the electron magnet moment), its interactions with the external environment are extremely weak. These factors lead to a very robust entangled state with very high purity.

#### 8.4.3.2 Spatial Correlation

In order to be able to measure correlations with high efficiency, one must ensure that, with high probability, both of the entangled atoms from a dissociated dimer enter their respective analyzers. Since the photo dissociation process constitutes a two-body decay, the two atoms will move apart with a 180° separation in the center-of-mass (COM) frame; i.e. there is perfect spatial correlation in the COM. Consequently, if the particles are at rest, and are also generated in a point-like dissociation volume, the conditional probability $g$ defined in Sect. 8.2 would be unity. However, in practice the spatial correlation is not perfect and $g$ is less than unity due to two factors. First, the dissociation volume is not a point; it is macroscopic and is formed by the common intersection of the supersonic jet and the two lasers. Second, the laboratory velocity of an atom is the vector addition of the dimer velocity with its COM velocity. Consequently, due to the small spread of the dimer velocities in the supersonic beam, there is a slight smearing of the spatial distribution of dissociated atom velocities in the laboratory frame. The smearing is significantly reduced by using the Doppler effect on the 266 nm transition to spectroscopically select and dissociate a narrow velocity subgroup from the initial dimer velocity distribution. Monte-Carlo simulations show that in our experiment values of $g$ greater than 0.94 can be expected based on the geometry of the beams and the velocity selection capabilities of our 266 nm laser system. This value of $g$ is dominated by the velocity spread. If the

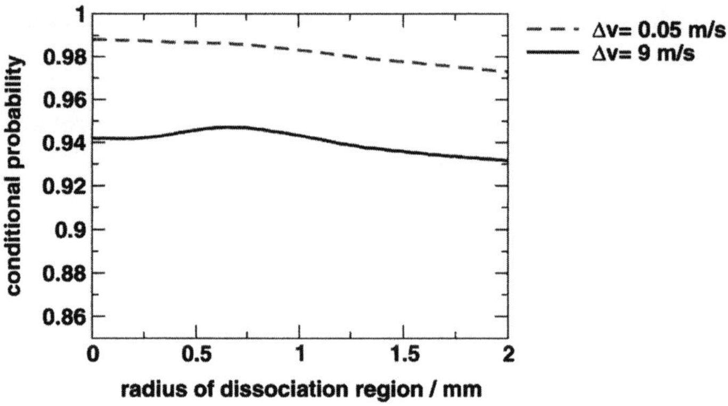


**Fig. 8.3.** Conditional probability $g$ as a function of the radius of the 1.5 mm long cylindrical dissociation region for two different velocity spreads of the dissociating dimers

velocity spread in our system is reduced to $\Delta v = 0.05\,\mathrm{m/s}$, the value of $g$ (greater than $\approx 0.98$) is dominated by the size of the dissociation region, see Fig. 8.3.

#### 8.4.3.3 Analyzer Quality

Two basic processes can reduce the quality of the spin analyzer. First, a Hg atom could be erroneously detected as spin up when it should have been detected as spin down, or vice-versa. Second, an atom is missed, i.e. it should have been detected as spin up, but no detection event is observed. The first error constitutes leakage, while the second corresponds to non perfect transmission through the analyzers.

Theoretical analysis of the two-photon ionization process shows that, for a carefully chosen time separation between the analysis and ionization laser pulses, transmission in excess of 99% is possible. For 8 ns laser pulses with near-Fourier transform limited linewidths, this requires a relative time delay of $2-3$ ns. This specification has been achieved using a Ti:Sapphire laser operating simultaneously at 761.1 nm (frequency tripled to yield 253.7 nm photons) and 789.9 nm (frequency quadrupled to yield 197.3 nm photons) [43]. Two factors can contribute to leakage in our analysis scheme, non-perfect polarization of the lasers, and excitation/ionization transitions through the $6^3\mathrm{P}_1^0$ $(F = 3/2)$ state. Our simulations show that the latter is the most significant contribution to the leakage, and is on the order of $10^{-3}$. This leakage occurs despite the relatively large 22 GHz detuning of the $6^3\mathrm{P}_1^0$ $(F = 3/2)$ state from the correct $6^3\mathrm{P}_1^0$ $(F = 1/2)$ state.

#### 8.4.3.4 Detector Design and Efficiency

After the spin analysis and photo-ionization, the $\mathrm{Hg}^+$ and/or its associated photo-electron must be detected with high efficiency and extremely low background. The latter is particularly important since the signal level must be reduced so that the probability of more than one Hg atom in the detection region at a time is negligible (otherwise coincidences between uncorrelated atoms from different dissociating molecules will be observed). Reducing photoelectron background signals to a negligible level is especially difficult in a vacuum system in the presence of ultraviolet laser beams that produce copious numbers of photoelectrons at walls and surfaces. Background signals associated with the ion detector have been easy to suppress to negligible levels.

Both the photo-ion and the photo-electron are observed with independent detectors. For confirmation of the detection of one of the atoms from a dissociated dimer, there are four choices. One can use the signal from the photo-ion detector alone; one can use the signal from the photo-electron detector alone; one can require a signal from both detectors (an AND configuration); or one can require a signal from either detector (an OR configuration). The AND configuration has the lowest overall detection efficiency but is least vulnerable to background signals. The OR configuration has the highest overall detection efficiency but is most vulnerable to background signals. In any case all four signals (an ion and electron detector signal for each atom of an entangled pair) are recorded and the inequality can be tested using each of the choices. This is an example of both the redundancy and the extensive systematic checks available with this experiment.

By observing both the photo-ion and the photo-electron, the measurement data inherently include measurements of the detection efficiencies, since the photo-electron and photo-ion are perfectly correlated. Knowing the detector efficiencies, one can use the measured coincidence rates for entangled atom pairs to also determine the conditional probability $g$ [42]. No *a priori* measurements of detector efficiencies are required; the same data used to test the Bell inequality can also be used to determine the efficiencies of the detectors while the data is being collected.

The acceptance solid angle of the detectors should be as large as possible, not only to maximize the signals (the usual situation), but also to maximize the conditional probability $g$.

Overall detection efficiencies close to unity for both the electron and the ion should be possible and have been confirmed by preliminary measurements [38].

#### 8.4.3.5 The Locality Loophole

In our experiment the enforcement of Einstein locality, also known as the communication loophole, can be implemented by employing electro-optic modulators (EOM). Specifically, the EOM together with a beam splitting polarizer can, in a couple nanoseconds, change the propagation direction of the excitation laser beam and hence the component of nuclear spin angular momentum being observed. A separation between our detectors of approximately 12 m will be necessary in order to close the locality loophole. This estimate includes allowances for the selection of a random number, switching the EOM, firing the nanosecond-detection lasers, the run-time of the electrons through the detection system including the Channeltron, as well as digitizing the detector output.

## 8.5 Summary

John Bell's work on the foundations of quantum mechanics continues to influence and inspire physics. The quest for a final answer in the test of Bell inequalities and thus the answer to the question whether or not quantum mechanics is a complete theory creates new ideas for more and more refined experiments. Simultaneously closing the existing loopholes has proven difficult. However, the atom based experiments that have emerged in the last few years are especially promising.

## Acknowledgments

This research was supported by the Robert A. Welch Foundation grant No. A-1218, the Texas Advanced Research Program grant 010366-0367-1999, and by the National Science Foundation grant PHY-9732459. The authors thank John Clauser and Marlan Scully for many helpful discussions on the theory and encouragement as well as Joel Bondurant, Robert Kenefick, Cechan Tian and Seiichirou Yokoo for their experimental contributions and active participation.

## References

1. A. Steane: Rep. Prog. Phys. **61**, 117 (1998)
2. P. Rodgers: Phys. World (special issue) **11**, 33 (1998)
3. W. Buttler, R. Hughes, P. Kwiat, S. Lamoreaux, G. Luher, G. Morgan, J. Nordjolt, C. Peterson, C. Simmons: Phys. Rev. Lett. **81**, 3283 (1998)
4. D. Bouwmeester, J.-W. Pan, K. Mattle, M. Eible, H. Weinfurter, A. Zeilinger: Nature **390**, 575 (1997)
5. D. Boschi, S. Branca, F. DeMartini, L. Hardy, S. Popescu: Phys. Rev. Lett. **80**, 1121 (1998)
6. A. Furusawa, J. Sørensen, S. Braunstein, C. Fuchs, H. Kimble, E. Polziak: Science **282**, 706 (1998)
7. A. Einstein, B. Podolsky, N. Rosen: Phys. Rev. **47**, 777 (1935)
8. J. Bell: Physics **1**, 195 (1964), reprinted in J.S. Bell, *Speakable and Unspeakable in Quantum Mechanics* (Cambridge Univ. Press, Cambridge 1987)
9. D. Bohm: *Quantum Physics* (Prentice Hall, New York 1951)
10. J. Clauser, A. Shimony: Rep. Prog. Phys **41**, 1881 (1978)
11. J. Clauser, M. Horne, A. Shimony, R. Holt: Phys. Rev. Lett. **23**, 880 (1969)

12. J. Clauser, M. Horne: Phys. Rev. D **10**, 526 (1974)
13. M. Rowe, D. Kielpinski, V. Meyer, C. Sackett, W. Itano, C. Monroe, D. Wineland: Nature **409**, 791 (2001)
14. S. Freedman: Rep. No. LBL 191 (Lawrence Livermore Lab., 1972)
15. E. Fry, T. Walther, S. Li: Phys. Rev. A **52**, 4381 (1995)
16. T. Walther, E. Fry: Z. Naturforsch. **52a**, 20 (1997), Proc. of the workshop in honor of E.C.G. Sudarshan
17. P. Pearle: Phys. Rev. D **2**, 1418 (1970)
18. A. Zeilinger: Phys. Lett. A **118**, 1 (1986)
19. E. Santos: Phys. Lett. A **212**, 10 (1996)
20. N. Gisin, B. Gisin: Phys. Lett. A **260**, 323 (1999)
21. P. Eberhard: Phys. Rev. A **47**, 747 (1993)
22. G. Hurst, M. Payne, S. Kramer, J. Young: Rev. Mod. Phys. **51**, 767 (1979)
23. A. Aspect, J. Dalibard, G. Roger: Phys. Rev. Lett. **49**, 1804 (1982)
24. G. Weihs, T. Jennewein, C. Simon, H. Weinfurter, A. Zeilinger: Phys. Rev. **81**, 5039 (1998)
25. S. Freedman: J. Clauser, Phys. Rev. Lett. **28**, 938 (1972)
26. J. Clauser: Phys. Rev. Lett. **36**, 1223 (1976)
27. E. Fry, R. Thompson: Phys. Rev. Lett. **37**, 465 (1976)
28. A. Aspect, P. Grangier, G. Roger: Phys. Rev. Lett. **49**, 91 (1982)
29. C. Alley, Y.H. Shih: in *Proc. Int. Symp. Foundations of Quantum Mechanics*, ed. by M. Namiki (1987), p. 47
30. Z. Ou, L. Mandel: Phys. Rev. Lett. **61**, 50 (1988)
31. P. Kwiat, K. Mattle, H. Weinfurter, A. Zeilinger, A. Sergienko, Y. Shih: Phys. Rev. Lett. **75**, 4337 (1995)
32. P. Kwiat, E. Waks, A. White, I. Appelbaum, P. Eberhard: quant-ph/9810003 (1998)
33. W. Tittel, J. Brendel, H. Zbinden, N. Gisin: Phys. Rev. Lett. **81**, 3563 (1998)
34. R. Nicolaescu, T. Walther, E.S. Fry, M. Muendel: Appl. Optics **38**, 1784 (1999)
35. E. Hagley, X. Maître, G. Nogues, C. Wunderlich, M. Brune, J. Raimond, S. Haroche: Phys. Rev. **79**, 1 (1997)
36. L. Vaidman: quant-ph/0102139 (2001)
37. T. Lo, A. Shimony: Phys. Rev. A **23**, 3003 (1981)
38. E. Fry, T. Walther, R. Kenefick: Phys. Scripta **T76**, 47 (1998)
39. E. Fry, T. Walther: in *Experimental Metaphysics – Quantum Mechanical Studies for Abner Shimony, Vol. I*, ed. by R. Cohen, J. Stachel (Kluwer Academic, Dordrecht 1996)
40. E. Fry, T. Walther: in *Quantum Interferometry*, ed. by F.D. Martini, G. Denardo, Y. Shih (VCH, Weinheim, 1996)
41. E. Fry, T. Walther: Adv. At. Mol. Opt. Phys. B **42**, 1 (1999)
42. T. Walther, E. Fry: in *New Developments on Fundamental Problems in Quantum Physics*, ed. by M. Ferrero, E. Santos, S. Huelga, Fundamental Theories of Physics, 81 (Kluwer Academic, Dordrecht, 1997)
43. C. Tian, T. Walther, R. Nicolaescu, X. Pan, Y. Liao, E. Fry: Opt. Lett. **24**, 1496 (1999)

# 9 Bell's Theorem: The Naive View of an Experimentalist

Alain Aspect

## 9.1 Introduction

It is a very emotional experience to contribute to this book commemorating of John Bell. I first met John in 1975, a few months after reading his famous paper [1]. I had been so strongly impressed by this paper, that I had immediately decided to do my "thèse d'état" – which at that time, in France, could be a really long work – on this fascinating problem. I definitely wanted to carry out an experiment "in which the settings are changed during the flight of the particles", as suggested in the paper, and I had convinced a young professor at the Institut d'Optique, Christian Imbert, to support my project and to act as my thesis advisor. He had advised me to go first to Geneva, to discuss my proposal with John Bell. I got an appointment without delay, and I showed up in John's office at CERN, quite nervous. While I explained my planned experiment, he listened silently. Eventually, I stopped talking, and the first question came: "Have you a permanent position?" After my positive answer, he started talking of physics, and he definitely encouraged me, making it clear that he would consider the implementation of variable analysers a fundamental improvement. Remembering this first question reminds me both of his celebrated sense of humour and of the general atmosphere at that time about raising questions on the foundations of quantum mechanics. Quite frequently there was open hostility, and in the best case, irony: "quantum mechanics has been vindicated by such a large amount of work by the smartest theorists and experimentalists; how can you hope to find anything with such a simple scheme, in optics, a science of the 19th century?" In addition to starting the experiment, I had then to develop a line of argument to try to convince the physicists I met (and among them some had to give their opinion about funding my project). After some not so successful attempts at quite sophisticated pleas, I eventually found out that it was much more efficient to explain the very simple and naive way in which I had understood Bell's theorem. And to my great surprise, that simple presentation was very convincing even with the most theoretically inclined interlocutors. I was lucky enough to be able to present it in front of John Bell himself, and he apparently appreciated it. I am therefore going to explain now how I understood Bell's theorem twenty five years ago, and I hope to be able to communicate the shock I received, which was so strong that I spent eight years of my life working on this problem.

This contribution is partly based on a paper that was published two decades ago as a proceedings of a conference, not so easy to find nowadays [2]. The first part of the paper aims at explaining what are Bell's theorem and Bell's inequalities, and why I find them so important. It is followed by a rapid review of the *first generation of experimental tests of Bell's inequalities with correlated photons*, carried out between 1971 and 1976. I give then a more detailed description of the three *second-generation experiments* that we performed at the Institut d'Optique d'Orsay between 1976 and 1982, with a dramatically improved source of pairs of entangled photons, using a nonlinear two-photon laser excitation of atomic radiative cascades. The last part gives an overview of the *third-generation experiments*, developed since the late 1980s and carried out with pairs of correlated photons produced in nonlinear parametric down conversion: these experiments can close most of the loopholes still left open by the second-generation experiments. I am deliberately concentrating on optics experiments, since they are at the present time the most convincing and the closest to the ideal Gedankenexperiment, but the interested reader should be aware that other systems do exist, in other domains of physics, that may offer the possibility to perform different experiments.

In Sects. 9.2 to 9.6, we will see that Bell's inequalities provide a quantitative criterion to test "reasonable" supplementary-parameters theories versus quantum mechanics. Following Bell, I will first explain the motivations for considering supplementary-parameter theories: the argument is based on an analysis of the famous Einstein–Podolsky–Rosen (EPR) Gedankenexperiment [3]. Introducing a reasonable locality condition, we will then derive Bell's theorem, which states:

- that local supplementary-parameter theories are constrained by Bell's inequalities;
- that certain predictions of quantum mechanics violate Bell's inequalities, and therefore that quantum mechanics is incompatible with local supplementary-parameter theories.

We will then discuss the fundamental assumption for this conflict is the *locality assumption.* And we will show that in a more sophisticated version of the EPR thought experiment ("timing experiment"), the locality condition may be considered a *consequence of Einstein's causality*, preventing faster-than-light interactions.

The purpose of the first part is to convince the reader that the *formalism leading to Bell's inequalities is very general and reasonable.* What is *surprising* is that such a reasonable formalism *conflicts with quantum mechanics.* In fact, situations exhibiting a conflict *are very rare*, and quantum optics is the domain where the most significant tests of this conflict have been carried out (Sects. 9.7 to 9.11).

## 9.2 Why Have Supplementary Parameters? The Einstein–Podolsky–Rosen–Bohm Gedankenexperiment

### 9.2.1 Experimental Scheme

Let us consider the optical variant of *Bohm*'s version [4] of the EPR Gedankenexperiment (Fig. 9.1). A source, $S$, emits a pair of photons with different frequencies, $\nu_1$ and $\nu_2$, counterpropagating along $Oz$. Suppose that the polarization part of the state vector describing the pair is

$$|\Psi(\nu_1,\nu_2)\rangle = \frac{1}{\sqrt{2}}\left(|x,x\rangle + |y,y\rangle\right) , \tag{9.1}$$

where $|x\rangle$ and $|y\rangle$ are linear polarizations states. This state is remarkable: it cannot be factorized into a product of two states associated with each photon, so we cannot ascribe any well-defined state to each photon. As a consequence, we cannot assign any polarization to each photon. Such a state describing a system of several objects that can only be thought of globally is an *entangled state*.

We perform linear polarization measurements on the two photons, with analysers I and II. Analyser I, in orientation $\boldsymbol{a}$, is followed by two detectors, giving results + or −, corresponding to a linear polarization found parallel or perpendicular to $\boldsymbol{a}$. Analyser II, in orientation $\boldsymbol{b}$, acts similarly[1].

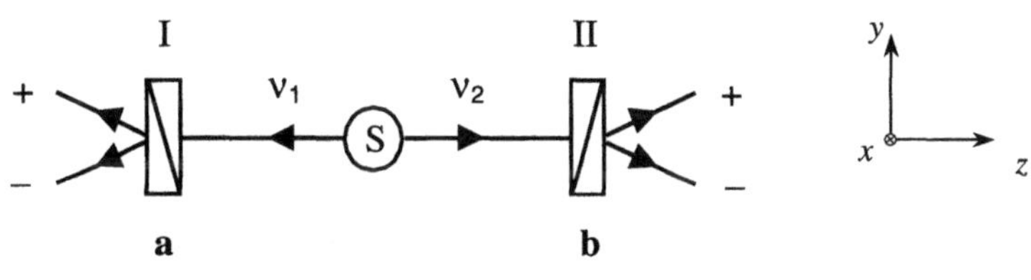


**Fig. 9.1. Einstein–Podolsky–Rosen–Bohm Gedankenexperiment with photons.** The two photons $\nu_1$ and $\nu_2$, emitted in the state $|\Psi(1,2)\rangle$ of (9.1), are analyzed by linear polarizers in orientations $\boldsymbol{a}$ and $\boldsymbol{b}$. One can measure the probabilities of single or joint detections in the output channels of the polarizers

It is easy to derive the quantum-mechanical predictions for these measurements of polarization, single or in coincidence. Consider first the single probabilities $P_\pm(\boldsymbol{a})$ of getting the results ± for photon $\nu_1$, and similarly, the single probabilities $P_\pm(\boldsymbol{b})$ of obtaining the results ± for photon $\nu_2$. Quantum

[1] There is a one-to-one correspondence with the EPR–Bohm Gedankenexperiment dealing with a pair of spin 1/2 particles, in a singlet state, analysed by two orientable Stern–Gerlach filters.

mechanics predicts the following:

$$\begin{aligned} P_+(\boldsymbol{a}) &= P_-(\boldsymbol{a}) = 1/2 \,, \\ P_+(\boldsymbol{b}) &= P_-(\boldsymbol{b}) = 1/2 \,. \end{aligned} \tag{9.2}$$

These results are in agreement with the remark that we cannot assign any polarization to each photon, so that each individual polarization measurement gives a random result.

Let us now consider the probabilities $P_{\pm\pm}(\boldsymbol{a}, \boldsymbol{b})$ of joint detections of $\nu_1$ and $\nu_2$ in the channels + or − of polarizers I or II, in orientations $\boldsymbol{a}$ and $\boldsymbol{b}$. Quantum mechanics predicts

$$\begin{aligned} P_{++}(\boldsymbol{a}, \boldsymbol{b}) &= P_{--}(\boldsymbol{a}, \boldsymbol{b}) = \frac{1}{2} \cos^2(\boldsymbol{a}, \boldsymbol{b}) \,, \\ P_{+-}(\boldsymbol{a}, \boldsymbol{b}) &= P_{-+}(\boldsymbol{a}, \boldsymbol{b}) = \frac{1}{2} \sin^2(\boldsymbol{a}, \boldsymbol{b}) \,. \end{aligned} \tag{9.3}$$

We are going to show that these quantum-mechanical predictions have far reaching consequences.

### 9.2.2 Correlations

Consider first the particular situation, $(\boldsymbol{a}, \boldsymbol{b}) = 0$, where polarizers are parallel. The quantum-mechanical predictions for the joint detection probabilities (9.3) are:

$$\begin{aligned} P_{++}(\boldsymbol{a}, \boldsymbol{a}) &= P_{--}(\boldsymbol{a}, \boldsymbol{a}) = \frac{1}{2} \,, \\ P_{+-}(\boldsymbol{a}, \boldsymbol{a}) &= P_{-+}(\boldsymbol{a}, \boldsymbol{a}) = 0 \,. \end{aligned} \tag{9.4}$$

According to this result, and taking into account (9.2), we conclude that when photon $\nu_1$ is found in the + channel of polarizer I, $\nu_2$ is found *with certainty* in the + channel of II (and similarly for the − channels). For parallel polarizers, there is thus a *total correlation* between the individually random results of measurements of polarization on the two photons $\nu_1$ and $\nu_2$.

A convenient way to measure the amount of correlation between random quantities is to calculate the correlation coefficient. For the polarization measurements considered above, it is equal to

$$E(\boldsymbol{a}, \boldsymbol{b}) = P_{++}(\boldsymbol{a}, \boldsymbol{b}) + P_{--}(\boldsymbol{a}, \boldsymbol{b}) - P_{+-}(\boldsymbol{a}, \boldsymbol{b}) - P_{+-}(\boldsymbol{a}, \boldsymbol{b}) \,. \tag{9.5}$$

Using the prediction (9.3) of quantum mechanics, we find the correlation coefficient

$$E_{\text{QM}}(\boldsymbol{a}, \boldsymbol{b}) = \cos 2(\boldsymbol{a}, \boldsymbol{b}) \,. \tag{9.6}$$

In the particular case of parallel polarizers $[(\boldsymbol{a}, \boldsymbol{b}) = 0]$, we find $E_{\text{QM}}(0) = 1$; this confirms that the correlation is total.

In conclusion, the quantum-mechanical calculations suggest that, although each individual measurement gives random results, these random results are correlated, as expressed by (9.6). For parallel (or perpendicular) orientations of the polarizers, the correlation is total ($|E_{\mathrm{QM}}| = 1$).

### 9.2.3 Difficulty of an Image Derived from the Formalism of Quantum Mechanics

As a naive physicist, I would like to raise the question of whether a simple image can be found to understand these strong correlations. The most natural way to find an image may seem to follow the quantum-mechanical calculations leading to (9.3). In fact, there are several ways to do this calculation. A very direct one is to project the state vector (9.1) onto the eigenvector corresponding to the relevant result. This immediately gives the joint probabilities (9.3). However, since this calculation bears on state vectors describing globally the two photons, I do not know how to build a picture in our ordinary space.

In order to overcome this problem, and to identify separately the two measurements happening at the two ends of the experiment, we can split the joint measurement in two steps. Suppose for instance that the measurement on photon $\nu_1$ takes place first, and gives the result +, with the polarizer I in orientation $\boldsymbol{a}$. The + result (associated with the polarization state $|\boldsymbol{a}\rangle$) has a probability of 1/2. To proceed with the calculation, we must then use the postulate of reduction of the state vector, which states that, after this measurement, the new state vector $|\Psi'(\nu_1, \nu_2)\rangle$ describing the pair is obtained by projection of the initial state vector $|\Psi(\nu_1, \nu_2)\rangle$ (9.1) onto the eigenspace associated with the result +: this two-dimensional eigenspace has a basis $\{|\boldsymbol{a}, x\rangle, |\boldsymbol{a}, y\rangle\}$. Using the corresponding projector, we find after a little algebra

$$|\Psi'(\nu_1, \nu_2)\rangle = |\boldsymbol{a}, \boldsymbol{a}\rangle \; . \tag{9.7}$$

This means that immediately after the first measurement, photon $\nu_1$ takes the polarization $|\boldsymbol{a}\rangle$; this is obvious because it has been measured with a polarizer oriented along $\boldsymbol{a}$, and the result + has been found. More surprisingly, the distant photon $\nu_2$, which has not yet interacted with any polarizer, has also been projected into the state $|\boldsymbol{a}\rangle$ with a well-defined polarization, parallel to the one found for photon $\nu_1$. This surprising conclusion however leads to the correct final result (9.3), since a straightforward application of the Malus law shows that a subsequent measurement performed along $\boldsymbol{b}$ on photon $\nu_2$ will lead to

$$P_{++}(\boldsymbol{a}, \boldsymbol{b}) = \frac{1}{2} \cos^2(\boldsymbol{a}, \boldsymbol{b}) \; . \tag{9.8}$$

The calculation in two steps therefore gives the same result as the direct calculation. But, in addition, it suggests a picture for the two-step measurement:

- Photon $\nu_1$, which did not have a well-defined polarization before its measurement, takes the polarization associated with the obtained result, at the moment of its measurement. This is not surprising.
- When the measurement on $\nu_1$ is done, photon $\nu_2$, which did not have a well-defined polarization before this measurement, is projected into a state of polarization parallel to the result of the measurement on $\nu_1$. This is very surprising, because this change in the description of $\nu_2$ happens instantaneously, whatever the distance between $\nu_1$ and $\nu_2$ at the moment of the first measurement.

This picture seems in contradiction with relativity. According to Einstein, what happens in a given region of space–time cannot be influenced by an event happening in a region of space–time that is separated by a space-like interval. It therefore not unreasonable to try to find more acceptable pictures for "understanding" the EPR correlations. It is such a picture that we consider now.

### 9.2.4 Supplementary Parameters

Correlations between distant measurements on two separated systems that had previously interacted are common in the classical world. For instance, if a mechanical object with a null linear (or angular) momentum is split in two parts by some internal repulsion, the linear (or angular) momentum of the two separated parts remain equal and opposite in the case of a free evolution. In the general case, where each fragment is submitted to some interaction, the two momenta remain correlated, since they are at each moment determined by their initial values, which had a perfectly defined sum.

It is tempting to use such a classical picture to render an account of the EPR correlations, in terms of common properties of the two systems. Let us consider again the perfect correlation of polarization measurements in the case of parallel polarizers $(\boldsymbol{a}, \boldsymbol{b}) = 0$. When we find + for $\nu_1$, we are sure to also find + for $\nu_2$. We are thus led to admit that there is some property (Einstein said "an element of physical reality") pertaining to this particular pair and determining the result ++. For another pair, when the results is $--$, we can similarly invoke a common property, determining the result $--$. It is then sufficient to admit that half the pairs are emitted with the property ++ and half with the property $--$, to reproduce all the results of measurement in this configuration. Note, however, that such properties, differing from one pair to another one, are not taken into account by the quantum-mechanical state vector $|\Psi(\nu_1, \nu_2)\rangle$, which is the same for all pairs. This is why we can conclude with Einstein that *quantum mechanics is not complete.* And this is why such additional properties are referred to as *supplementary parameters*, or *hidden-variables.*[2]

[2] Einstein actually did not "speak of hidden variables" or "supplementary parameters", but rather of "elements of physical reality". Accordingly, many au-

As a conclusion, it seems possible to "understand" the EPR correlations by such a classical-looking picture, involving supplementary parameters differing from one pair to another one. It can be hoped that the statistical quantum-mechanical predictions are recovered when averaging over the supplementary parameters. It seems that this was also Einstein's position [5–7]. Note that, at this stage of the reasoning, a commitment to this position does not contradict quantum mechanics; there is no logical problem in fully accepting the predictions of quantum mechanics *and* in invoking supplementary parameters giving an acceptable picture of the EPR correlations.

## 9.3 Bell's Inequalities

### 9.3.1 Formalism

Three decades after the EPR paper, Bell translated into mathematics the consequences of the preceding discussion, and he explicitly introduced supplementary parameters, denoted $\lambda$. Their distribution on an ensemble of emitted pairs is specified by a probability distribution, $\rho(\lambda)$, such that

$$\begin{aligned} \rho(\lambda) &\geq 0 \, , \\ \int \mathrm{d}\lambda \rho(\lambda) &= 1 \, . \end{aligned} \tag{9.9}$$

For a given pair, characterized by a given supplementary parameter, $\lambda$, the results of measurements are given by the bivalued functions

$$\begin{aligned} A(\lambda, \boldsymbol{a}) &= \pm 1 \quad \text{at analyzer I (in orientation } \boldsymbol{a}) \, , \\ B(\lambda, \boldsymbol{b}) &= \pm 1 \quad \text{at analyzer II (in oreintation } \boldsymbol{b}) \, . \end{aligned} \tag{9.10}$$

A particular supplementary-parameter theory is completely defined by the explicit form of the functions $\rho(\lambda)$, $A(\lambda, \boldsymbol{a})$ and $B(\lambda, \boldsymbol{b})$. It is then easy to express the probabilities of the various results of measurements. For instance, noting that the function $[A(\lambda, \boldsymbol{a}) + 1]/2$ assumes the value $+1$ for the $+$ result, and 0 otherwise (and similarly $[1 - B(\lambda, \boldsymbol{b})]/2$ assumes the value $+1$ for the $-$ result, and 0 otherwise), we can write

$$\begin{aligned} P_{+}(\boldsymbol{a}) &= \int \mathrm{d}\lambda \rho(\lambda) \frac{[A(\lambda, \boldsymbol{a}) + 1]}{2} \, , \\ P_{+-}(\boldsymbol{a}, \boldsymbol{b}) &= \int \mathrm{d}\lambda \rho(\lambda) \frac{[A(\lambda, \boldsymbol{a}) + 1]}{2} \frac{[1 - B(\lambda, \boldsymbol{b})]}{2} \, . \end{aligned} \tag{9.11}$$

Similarly, the correlation function assumes the simple form

$$E(\boldsymbol{a}, \boldsymbol{b}) = \int \mathrm{d}\lambda \rho(\lambda) A(\lambda, \boldsymbol{a}) B(\lambda, \boldsymbol{b}) \, . \tag{9.12}$$

---

thors refer to "realistic theories" rather than to "hidden-variable theories" or to "supplementary-variable theories".

### 9.3.2 A (Naive) Example of the Supplementary-Parameter Theory

As an example of the supplementary-parameter theory, we present a model where each photon travelling along O$z$ is assumed to have a well-defined linear polarization, determined by its angle ($\lambda_1$ or $\lambda_2$) with the $x$-axis. In order to account for the strong correlation, we assume that the two photons of a pair are emitted with the same linear polarization, defined by the common angle $\lambda$ (Fig. 9.2).

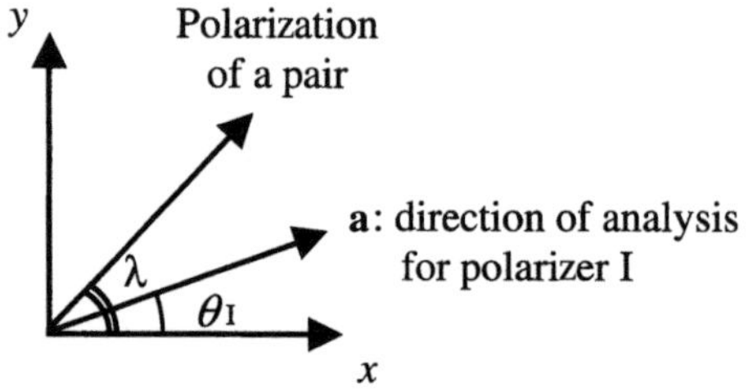


**Fig. 9.2. The naive example.** Each pair of photons has a "direction of polarization", defined by $\lambda$, which is the supplementary parameter of the model. Polarizer I makes a polarization measurement along $\boldsymbol{a}$, at an angle $\theta_{\mathrm{I}}$ from the $x$-axis

The polarization of the various pairs is randomly distributed, according to the probability distribution, $\rho(\lambda)$, which we take to be rotationally invariant:

$$\rho(\lambda) = \frac{1}{2\pi} \,. \tag{9.13}$$

To complete our model, we must give an explicit form for the functions $A(\lambda, \boldsymbol{a})$ and $B(\lambda, \boldsymbol{b})$. We take the following form:

$$\begin{aligned} A(\lambda, \boldsymbol{a}) &= \text{sign}\left[\cos 2(\theta_{\mathrm{I}} - \lambda)\right] \,, \\ B(\lambda, \boldsymbol{b}) &= \text{sign}\left[\cos 2(\theta_{\mathrm{II}} - \lambda)\right] \,, \end{aligned} \tag{9.14}$$

where the angles $\theta_{\mathrm{I}}$ and $\theta_{\mathrm{II}}$ indicate the orientations of the polarizers. Note that these forms are very reasonable: $A(\lambda, \boldsymbol{a})$ assumes the value $+1$ when the polarization of photon $\nu_1$ makes an angle less than $\pi/4$ with the direction of analysis $\boldsymbol{a}$, and $-1$ for the complementary case (polarization closer to perpendicular to $\boldsymbol{a}$).

With this explicit model, we can use (9.11) to calculate the probabilities of the various measurements. We find, for instance, single probabilities,

$$P_+(\boldsymbol{a}) = P_-(\boldsymbol{a}) = P_+(\boldsymbol{b}) = P_-(\boldsymbol{b}) = \frac{1}{2} \,, \tag{9.15}$$

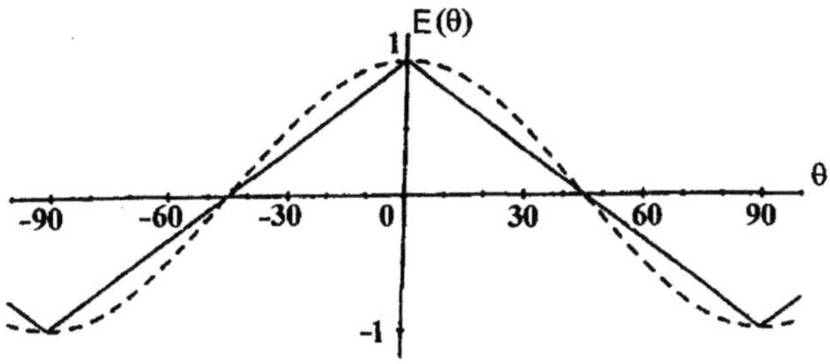


**Fig. 9.3.** Polarization correlation coefficient as a function of the relative orientation of the polarizers: *Dotted line*: quantum-mechanical prediction; *solid line*: our naive model

identical to the quantum-mechanical results. The model also allows us to calculate the joint probabilities, or equivalently the correlation function, and we find, using (9.12):

$$
\begin{aligned}
&E(\boldsymbol{a},\boldsymbol{b}) = 1 - 4\frac{|\theta_{\mathrm{I}} - \theta_{\mathrm{II}}|}{\pi} = 1 - 4\frac{|(\boldsymbol{a},\boldsymbol{b})|}{\pi} \\
&\text{for} \quad -\frac{\pi}{2} \leq \theta_{\mathrm{I}} - \theta_{\mathrm{II}} \leq \frac{\pi}{2} \,.
\end{aligned}
\tag{9.16}
$$

This is a remarkable result. Note first that $E(\boldsymbol{a},\boldsymbol{b})$ depends only on the relative angle $(\boldsymbol{a},\boldsymbol{b})$, as for the quantum-mechanical prediction (9.6). Moreover, as shown in Fig. 9.3, the difference beween the predictions of the simple supplementary-parameter model and the quantum-mechanical predictions is always small, and the agreement is exact for the angles 0 and $\pi/2$, i.e. cases of total correlation. This result, obtained with an extremely simple supplementary-parameter model, is very encouraging, and it might be hoped that a more sophisticated model would reproduce exactly the quantum-mechanical predictions. *Bell's discovery is the fact that the search for such models is hopeless*, as we are going to show now.

### 9.3.3 The Inequalities

There are many different forms and demonstrations of Bell's inequalities. We give here a very simple demonstration leading to a form directly applicable to experimentation.[3]

[3] It is important to distinguish between inequalities which show a mathematical contradiction with quantum mechanics, but without the possibility of an experimental test with (necessarily) imperfect apparatus, and inequalities allowing an experimental test provided that the experimental imperfections remain in certain limits.

Let us consider the quantity

$$\begin{aligned} s &= A(\lambda, \boldsymbol{a}) \cdot B(\lambda, \boldsymbol{b}) - A(\lambda, \boldsymbol{a}) \cdot B(\lambda, \boldsymbol{b}') + A(\lambda, \boldsymbol{a}') \cdot B(\lambda, \boldsymbol{b}) \\ &\quad + A(\lambda, \boldsymbol{a}') \cdot B(\lambda, \boldsymbol{b}') \\ &= A(\lambda, \boldsymbol{a}) \left[B(\lambda, \boldsymbol{b}) - B(\lambda, \boldsymbol{b}')\right] + A(\lambda, \boldsymbol{a}') \left[B(\lambda, \boldsymbol{b}) + B(\lambda, \boldsymbol{b}')\right] . \end{aligned} \tag{9.17}$$

By remembering that the four numbers $A$ and $B$ only take the values $\pm 1$, a simple inspection of the second line of (9.17) shows that

$$s(\lambda, \boldsymbol{a}, \boldsymbol{a}', \boldsymbol{b}, \boldsymbol{b}') = \pm 2 . \tag{9.18}$$

The average of $s$ over $\lambda$ is therefore between $+2$ and $-2$:

$$-2 \leq \int \mathrm{d}\lambda \rho(\lambda) s(\lambda, \boldsymbol{a}, \boldsymbol{a}', \boldsymbol{b}, \boldsymbol{b}') \leq 2 . \tag{9.19}$$

According to (9.12), we can rewrite these inequalities

$$-2 \leq S(\lambda, \boldsymbol{a}, \boldsymbol{a}', \boldsymbol{b}, \boldsymbol{b}') \leq 2 \tag{9.20}$$

with

$$S(\boldsymbol{a}, \boldsymbol{a}', \boldsymbol{b}, \boldsymbol{b}') = E(\boldsymbol{a}, \boldsymbol{b}) - E(\boldsymbol{a}, \boldsymbol{b}') + E(\boldsymbol{a}', \boldsymbol{b}) + E(\boldsymbol{a}', \boldsymbol{b}') . \tag{9.21}$$

These are BCHSH inequalities, i.e. Bell's inequalities generalized by *Clauser*, *Horne*, *Shimony*, and *Holt* [8]. They bear upon the combination, $S$, of the four polarization correlation coefficients, associated with the two directions of analysis for each polarizer ($\boldsymbol{a}$ and $\boldsymbol{a}'$ for polarizer I, $\boldsymbol{b}$ and $\boldsymbol{b}'$ for polarizer II). Note that they apply to any supplementary-parameter theory of the very general form defined in Sect. 9.3.1 [(9.9), (9.10) and (9.12)], of which our naive model is only an example.

## 9.4 Conflict with Quantum Mechanics

### 9.4.1 Evidence

We can use the predictions (9.6) of quantum mechanics for EPR pairs to evaluate the quantity $S(\boldsymbol{a}, \boldsymbol{a}', \boldsymbol{b}, \boldsymbol{b}')$ defined by (9.21). For the particular set of orientations shown in Fig. 9.4, the result is

$$S_{\mathrm{QM}} = 2\sqrt{2} . \tag{9.22}$$

This quantum-mechanical prediction definitely conflicts with the Bell's inequality (9.20) which is valid for any supplementary-parameter theory of the general form defined in Sect. 9.3.1.

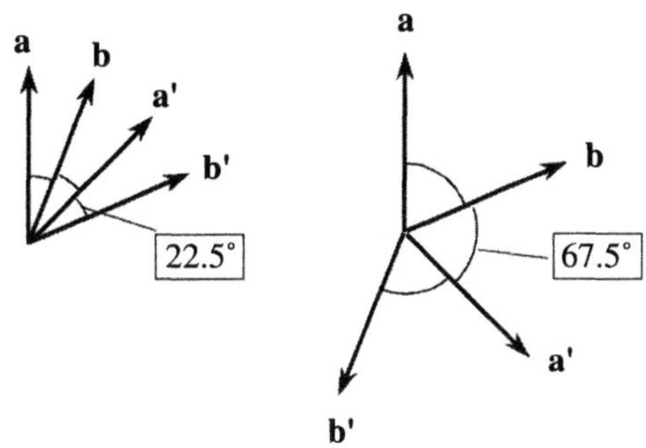


**Fig. 9.4.** Orientations yielding the largest conflict between Bell's inequalities and quantum mechanics

We have thus found a situation where the quantum-mechanical predictions cannot be reproduced (mimicked) by supplementary-parameter theories. This is the essence of Bell's theorem: it is impossible to find a supplementary-parameter theory, of the general form defined in Sect. 9.3.1, that reproduces *all* the predictions of quantum mechanics. This statement is the generalisation of what appears in Fig. 9.3, for the particular supplementary-parameter model considered in Sect. 9.3.2: the model exactly reproduces the predictions of quantum mechanics for some particular angles $(0, \pi/4, \pi/2)$, but it somewhat deviates at other angles. The importance of Bell's theorem is that it is not restricted to a particular supplementary-parameter model, but it is general.

### 9.4.2 Maximum Conflict

It is interesting to look for the maximum violation of Bell's inequalities by the quantum-mechanical predictions. Let us take the quantum-mechanical value of $S$:

$$S_{\mathrm{QM}}(\boldsymbol{a}, \boldsymbol{b}, \boldsymbol{a}', \boldsymbol{b}') = \cos(\boldsymbol{a}, \boldsymbol{b}) - \cos(\boldsymbol{a}, \boldsymbol{b}') + \cos(\boldsymbol{a}', \boldsymbol{b}) + \cos(\boldsymbol{a}', \boldsymbol{b}') \ . \tag{9.23}$$

It is a function of the three independant variables $(\boldsymbol{a}, \boldsymbol{b})$, $(\boldsymbol{b}, \boldsymbol{a}')$, and $(\boldsymbol{a}', \boldsymbol{b}')$. Note that

$$(\boldsymbol{a}, \boldsymbol{b}') = (\boldsymbol{a}, \boldsymbol{b}) + (\boldsymbol{b}, \boldsymbol{a}') + (\boldsymbol{a}', \boldsymbol{b}') \ .$$

In order to find the extremum values of $S_{\mathrm{QM}}$, we make the three partial derivatives null, and we find

$$(\boldsymbol{a}, \boldsymbol{b}) = (\boldsymbol{b}, \boldsymbol{a}') = (\boldsymbol{a}', \boldsymbol{b}') = \theta \tag{9.24}$$

and

$$\sin\theta = \sin 3\theta \ . \tag{9.25}$$

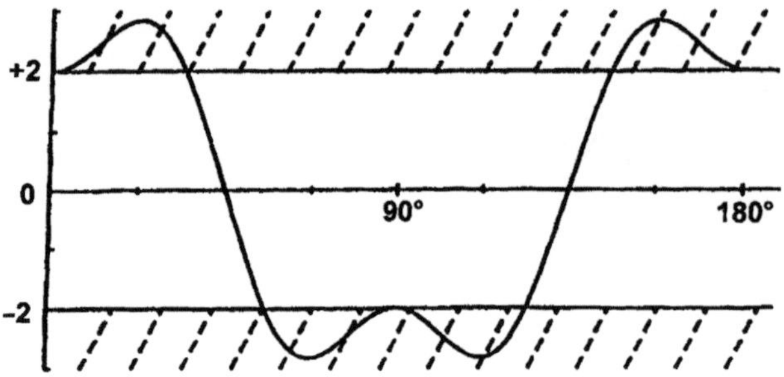


**Fig. 9.5.** $S(\theta)$ as predicted by quantum mechanics for EPR pairs. The conflict with Bell's inequalities happens when $|S|$ is larger than 2, and it is maximum for the sets of orientations of Fig. 9.4

The function $S_{\mathrm{QM}}(\theta)$ evaluated in the case of condition (9.24) is plotted in Fig. 9.5. It shows that the absolute maximum and minimum of $S_{\mathrm{QM}}$ are

$$S_{\mathrm{MQ}} = 2\sqrt{2} \qquad \text{for} \quad \theta = \pm\frac{\pi}{8}\,, \tag{9.26}$$

$$S_{\mathrm{MQ}} = -2\sqrt{2} \qquad \text{for} \quad \theta = \pm\frac{3\pi}{8}\,. \tag{9.27}$$

These values are solutions of (9.25). The corresponding sets of orientations are displayed in Fig. 9.4. They give the maximum violations of Bell's inequalities.

More generally, Fig. 9.5 shows that there is a full range of orientations leading to a conflict with Bell's inequalities. However, it is also clear that there are many sets of orientations for which there is no conflict.

## 9.5 Discussion: The Locality Condition

We have now established Bell's theorem: quantum mechanics conflicts with any supplementary-parameter theory as defined in Sect. 9.3.1, since it violates a consequence (Bell's inequalities) of any such theory. Clearly, it is interesting at this stage to look for the hypotheses underlying the formalism introduced in Sect. 9.3.1. One may then hope to point out a specific hypothesis responsible for the conflict. We therefore examine now the various hypotheses underlying the supplementary-parameter theories introduced in Sect. 9.3.1.

A first hypothesis is the existence of supplementary parameters. As we have seen, they have been introduced in order to render an account of the correlations at a distance. This hypothesis is strongly linked to a conception of the world, as expressed by Einstein, where the notion of separated physical realities for separated particles is meaningful. It is even possible to derive the existence of supplementary parameters from general statements about physical reality, in the spirit of Einstein's ideas [9]. An hypothesis in this spirit seems absolutely necessary to obtain inequalities conflicting with quantum mechanics.

The second considered hypothesis is determinism. As a matter of fact, the formalism of Sect. 9.3.1 is deterministic: once $\lambda$ is fixed, the results $A(\lambda, \boldsymbol{a})$ and $B(\lambda, \boldsymbol{b})$ of the polarization measurements are certain. One has speculated that it may be a good reason for a conflict with the non deterministic formalism of quantum mechanics. In fact, as first shown by *Bell* [10], and subsequently developed [11], it is easy to generalize the formalism of Sect. 9.3.1 to *stochastic* supplementary-parameter theories, where the deterministic measurement functions $A(\lambda, \boldsymbol{a})$ and $B(\lambda, \boldsymbol{b}')$ are replaced by probabilistic functions. One then finds that Bell's inequalities still hold and that the conflict does not disappear. It is therefore generally accepted that the deterministic character of the formalism is not the reason for the conflict [12].[4]

The most important hypothesis, stressed by Bell in all his papers, is the local character of the formalism of Sect. 9.3.1. We have indeed implicitly assumed that the result $A(\lambda, \boldsymbol{a})$ of the measurement at polarizer I does not depend on the orientation $\boldsymbol{b}$ of the remote polarizer II, and vice versa. Similarly, it is assumed that the probability distribution, $\rho(\lambda)$ (i.e. the way in which pairs are emitted), does not depend on the orientations $\boldsymbol{a}$ and $\boldsymbol{b}$. This *locality assumption* is crucial: Bell's inequalities would no longer hold without it. It is indeed clear that the demonstration of Sect. 9.3.3 fails with quantities such as $A(\lambda, \boldsymbol{a}, \boldsymbol{b})$ or $\rho(\lambda, \boldsymbol{a}, \boldsymbol{b})$.

To conclude, there are two hypothesis that seem to be necessary to obtain Bell's inequalities and, consequently, a conflict with quantum mechanics:

- Distant correlations can be understood by the introduction of supplementary parameters carried along by the separated particles, in the spirit of Einstein's ideas that separated objects have separated physical realities.
- The quantities $A(\lambda)$, $B(\lambda)$, and $\rho(\lambda)$ obey the *locality condition*, i.e. they do not depend on the orientations of the distant polarizers.

This is why one often claims that quantum mechanics conflicts with local realism.

## 9.6 Gedankenexperiment with Variable Analyzers: The Locality Condition as a Consequence of Einstein's Causality

In static experiments, in which the polarizers are held fixed for the whole duration of a run, the locality condition must be stated as an assumption. Although highly reasonable, this condition is not prescribed by any fundamental physical law. To quote *Bell* [1] "the settings of the instruments are made sufficiently in advance to allow them to reach some mutual rapport by

[4] This conclusion is not shared by all authors. For instance, it has been argued that the stochastic theories of *Bell* [10] or of *Clauser* and *Horne* [11] achieve no further generality, since they can be mimicked by a deterministic theory.

exchange of signals with velocity less than or equal to that of light." In that case, the result $A(\lambda)$ of the measurement at polarizer I could depend on the orientation $\boldsymbol{b}$ of the remote polarizer II, and vice versa. The locality condition would no longer hold, nor would Bell's inequalities.

Bell thus insisted upon the importance of "experiments of the type proposed by *Bohm* and *Aharonov* [5], in which the settings are changed during the flight of the particles."[5] In such a *timing experiment*, the *locality condition would become a consequence of Einstein's causality, which prevents any faster-than-light influence.*

As shown in our 1975 proposal [13], it is sufficient to switch each polarizer between two particular settings ($\boldsymbol{a}$ and $\boldsymbol{a}'$ for polarizer I, $\boldsymbol{b}$ and $\boldsymbol{b}'$ for polarizer II). It then becomes possible to test experimentally a large class of supplementary-parameter theories: those obyeing Einstein's causality. In such theories, the response of polarizer I at time $t$ is allowed to depend on the orientation $\boldsymbol{b}$ (or $\boldsymbol{b}'$) of polarizer II at times anterior to $t - L/c$ ($L$ is the distance between the polarizers). A similar retarded dependence is considered for the probability distribution, $\rho(\lambda)$, i.e. the way in which pairs are emitted at the source. For random switching times, with both sides uncorrelated, the predictions of these more general *separable supplementary-parameter theories* are constrained by *generalized Bell's inequalities* [13], based on Einstein's causality and not on Bell's locality condition.

On the other hand, it is easy to show [14] that the polarization correlations predicted by quantum mechanics depend only on the orientations $\boldsymbol{a}$ and $\boldsymbol{b}$ at the time of the measurements and do not involve any retardation terms, $L/c$. For a suitable choice of the set of orientations ($\boldsymbol{a}$, $\boldsymbol{a}'$, $\boldsymbol{b}$, $\boldsymbol{b}'$) – for instance the sets displayed in Fig. 9.4 – the quantum-mechanical predictions still conflict with generalized Bell's inequalities.

In an experiment with time-varying polarizers, Bell's theorem therefore states that quantum mechanics is incompatible with supplementary-parameter theories obeying Einstein's causality. Note that Einstein's causality already played an important role in the discussions leading to the notion of supplementary parameters, or equivalently of an independant physical reality for each separated subsystem [6]. It therefore does not seem exaggerated to conclude that, in a scheme with time-varying polarizers, Bell's theorem establishes a *contradiction between quantum mechanics and a description of the world in the spirit of Einstein's ideas.* Note, however that *Einstein did not know Bell's theorem*, and he could logically think that his world view was compatible with all the algebraic predictions of quantum mechanics. It is impossible to know what would have been his reaction to the contradiction revealed by Bell's theorem.

[5] The idea had already been expressed in Bohm's book [4].

## 9.7 From Bell's Theorem to a Realistic Experiment

### 9.7.1 Experimentally Testing Bell's Inequalities

With Bell's theorem, the debate on the possibility (or necessity) of completing quantum mechanics changed dramatically. It was no longer a matter of philosophical position (realism versus positivism) or of personal taste. It became possible to settle the question by an experiment. If one can produce pairs of photons (or of spin 1/2 particles) in an EPR state and measure the four coincidence rates, $N_{\pm\pm}(\boldsymbol{a},\boldsymbol{b})$, with detectors (Fig. 9.1) in the output channels of the polarizers (or Stern–Gerlach filters), one obtains the polarization correlation coefficient, for polarizers in orientations $\boldsymbol{a}$ and $\boldsymbol{b}$:

$$E(\boldsymbol{a},\boldsymbol{b}) = \frac{N_{++}(\boldsymbol{a},\boldsymbol{b}) - N_{+-}(\boldsymbol{a},\boldsymbol{b}) - N_{-+}(\boldsymbol{a},\boldsymbol{b}) + N_{--}(\boldsymbol{a},\boldsymbol{b})}{N_{++}(\boldsymbol{a},\boldsymbol{b}) + N_{+-}(\boldsymbol{a},\boldsymbol{b}) + N_{-+}(\boldsymbol{a},\boldsymbol{b}) + N_{-}(\boldsymbol{a},\boldsymbol{b})} \,. \tag{9.28}$$

By performing four measurements of this type in orientations $(\boldsymbol{a},\boldsymbol{b})$, $(\boldsymbol{a},\boldsymbol{b}')$, $(\boldsymbol{a}',\boldsymbol{b})$, and $(\boldsymbol{a}',\boldsymbol{b}')$, one obtains a measured value $S_{\mathrm{exp}}(\boldsymbol{a},\boldsymbol{a}',\boldsymbol{b},\boldsymbol{b}')$ for the quantity $S$ defined in (9.21). Choosing a situation where quantum mechanics predicts that this quantity violates Bell's inequalities (9.20), we have a test allowing one to discriminate between quantum mechanics and any local supplementary-parameter theory. If, in addition, we use a scheme with variable polarizers, we even test the more general class of *separable* (or *causal* in the relativistic sense) supplementary-parameter theories.

### 9.7.2 Sensitive Situations Are Rare

Quantum mechanics has been upheld in such a great variety of experiments that Bell's theorem might just appear as a proof of the impossibility of supplementary parameters. However, situations in which the conflict revealed by Bell's inequalities arises (sensitive situations) are so rare that in 1965 none had been realized.

To better appreciate this point, let us first note that Bell's inequalities are compatible with all classical physics, namely classical (relativistic) mechanics and classical electrodynamics, since this formalism can be imbedded into the supplementary-parameter formalism obeying Einstein's causality. For instance, in classical mechanics, the $\lambda$'s would be the initial positions and velocities of the particles, from which the future evolution can be derived. Similarly, in classical electrodynamics, the $\lambda$'s would be the trajectories of the charges in the sources, from which one can deduce the electromagnetic fields, and their action on the measuring apparatus.

Moreover, in situations usually described by quantum mechanics, it does not often happen that there is a conflict with Bell's inequalities. More precisely, for situations in which one looks for correlations between two separated subsystems (that may have interacted in the past), we can point out two conditions necessary for the possibility of a conflict with Bell's inequalities:

- The two separated subsystems must be in an *entangled state*, non factorizable, such as (9.1) (or the singlet state for two spin 1/2 particles).
- For each subsystem, it must be possible to choose the measured quantity among at least two non commuting observables (such as polarization measurements along directions $\boldsymbol{a}$ and $\boldsymbol{a}'$, neither parallel nor perpendicular).

Even in such cases, we have seen that the conflict exists only for well-chosen measured quantities (sets of orientations). But, as shown in Fig. 9.5, there are many orientations sets for which the quantum-mechanical predictions do not violate Bell's inequalities.

It was realized in 1965 that there was no experimental evidence of a violation of Bell's inequalities. Since these inequalities are derived from a very reasonable hypothesis, one could consider the possibility that the violation of Bell's inequalities would indicate a situation in which quantum mechanics fails. It was thus tempting to design a *sensitive experiment*, i.e. an experiment in which the predictions of quantum mechanics for the real situation definitely violate Bell's inequalities. The experiment would then give a clearcut result between quantum mechanics and supplementary-parameter theories obeying Bell's locality condition.

### 9.7.3 Production of Pairs of Photons in an EPR State

As pointed out by *Clauser* et al. [8], pairs of photons emitted in suitable atomic radiative cascades are good candidates for a sensitive test. Consider for instance a $J = 0 \rightarrow J = 1 \rightarrow J = 0$ atomic cascade (Fig. 9.6). Suppose that we select, with the use of wavelength filters and collimators, two plane waves of frequencies $\nu_1$ and $\nu_2$ propagating in opposite directions along the $z$-axis (Fig. 9.7).

It is easy to show, by invoking parity and angular-momentum conservation, that the polarization part of the state vector describing the pair $(\nu_1, \nu_2)$ can be written as follows:

$$\Psi(\nu_1, \nu_2) = \frac{1}{\sqrt{2}} (|R, R\rangle + |L, L\rangle) \ , \tag{9.29}$$

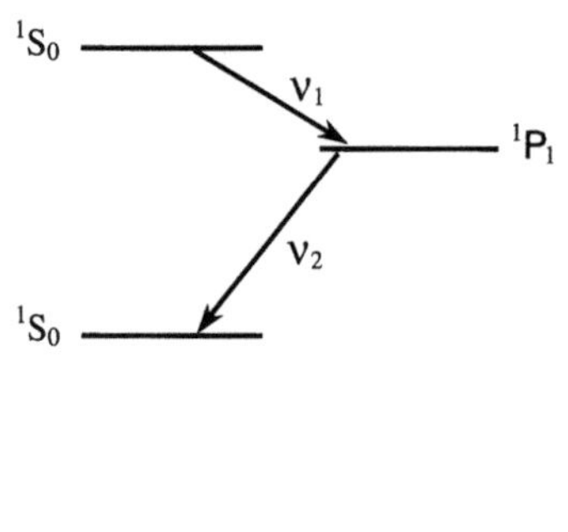


**Fig. 9.6.** Radiative cascade emitting pairs of photons correlated in polarization

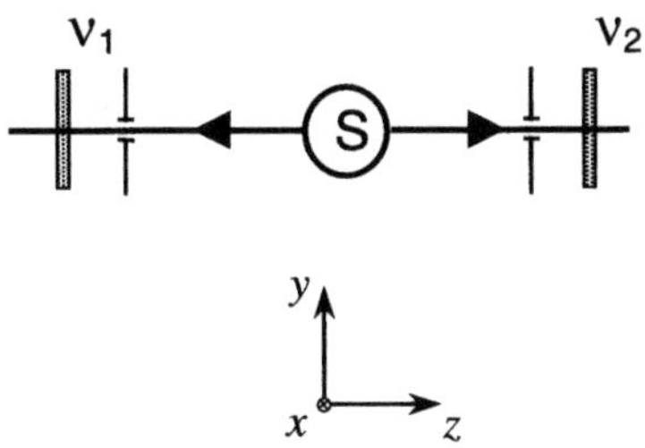


**Fig. 9.7.** Ideal configuration (infinitely small solid angles)

where $|R\rangle$ and $|L\rangle$ are circularly polarized states. By expressing $|R\rangle$ and $|L\rangle$ on a linear polarization basis, we obtain the state (9.1)

$$|\Psi(\nu_1, \nu_2)\rangle = \frac{1}{\sqrt{2}} \left(|x, x\rangle + |y, y\rangle\right) .$$

With this entangled EPR state, one can envisage a sensitive experiment.

### 9.7.4 Realistic Experiment

The real experiment differs from the ideal one in several respects. For instance, the light should be collected in finite angles, $2u$, which should be as large as possible (Fig. 9.8). In this situation, one can show [15] that the contrast of the correlation function decreases, since (9.6) is replaced by:

$$E_{\mathrm{QM}}(\boldsymbol{a}, \boldsymbol{b}) = F(u) \cos 2(\boldsymbol{a}, \boldsymbol{b}) , \tag{9.30}$$

where $F(u) \leq 1$. Figure 9.9 displays $F(u)$ for a $J = 0 \to J = 1 \to J = 0$ cascade. Fortunately, one can use large angles without great harm. For $u = 32°$ (our experiments), one has $F(u) = 0.984$.

All experimental inefficiencies (polarizers defects, accidental birefringences, etc.) will similarly lead to a decrease in the correlation function, $E(\boldsymbol{a}, \boldsymbol{b})$. The function $S_{\mathrm{QM}}(\Theta)$ (Fig. 9.5) is then multiplied by a factor of less than 1, and the conflict with Bell's inequalities decreases and may even vanish. Therefore, an actual experiment must be carefully designed and every auxiliary effect must be evaluated. All relevant parameters must be perfectly controlled, since a forgotten effect could similarly lead to a decrease in the conflict. For instance, a hyperfine structure dramatically decreases $F(u)$, so that only even isotopes should be used [15].

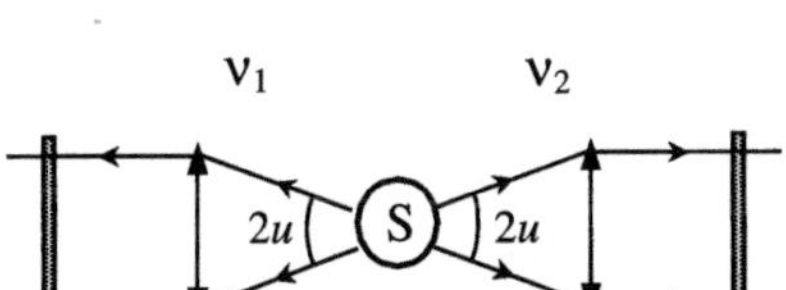


**Fig. 9.8.** Realistic configuration, with finite solid angles

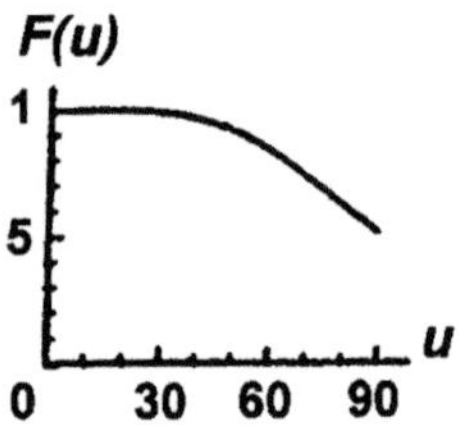


**Fig. 9.9.** Reduction factor, $F(u)$, for a $J = 0 \to J = 1 \to J = 0$ cascade

### 9.7.5 Timing Conditions

As we have seen in Sect. 9.6, Bell's locality condition may be considered to be a consequence of Einstein's causality, if the experiment fulfills requirements that can be split in two conditions:

- The distant *measurements* on the two subsystems must be *space-like separated.*
- The *choices* of the quantities measured on the two separated subsystems must be made at *random*, and must be *space-like separated.*

The second condition is obviously more difficult to fulfill.

## 9.8 First-Generation Experiments

The *Clauser* et al. paper [8], published in 1969, had shown the possibility for realistic, sensitive experiments with correlated photons produced in certain atomic cascades. Two groups, one at Berkeley, one at Harvard, started an experiment. After their conflicting results, a third experiment was carried out at College Station (Texas). All three experiments used a simplified experimental scheme, somewhat different from the ideal one, since it involved one-channel polarizers.

### 9.8.1 Experiments with a One-Channel Polarizer

In this simplified experimental scheme, one uses polarizers that transmit light polarized parallel to $\boldsymbol{a}$ (or $\boldsymbol{b}$), but block that which is orthogonal. Compared to the scheme in Fig. 9.1, one can thus only detect the + results, and the coincidence measurements only yield the coincidence rates, $N_{++}(\boldsymbol{a}, \boldsymbol{b})$, between the + channels. In order to recover the missing data, auxiliary runs are performed with one or both polarizers removed (the "orientation" of a removed polarizer is conventionally denoted $\infty$). We can then write relations between the *measured* coincidence rates, $N_{++}(\boldsymbol{a}, \boldsymbol{b})$, $N_{++}(\boldsymbol{a}, \infty)$, and $N_{++}(\infty, \boldsymbol{b})$, and the coincidence rates which are not measured:

$$\begin{aligned} N_{++}(\infty, \infty) &= N_{++}(\boldsymbol{a}, \boldsymbol{b}) + N_{-+}(\boldsymbol{a}, \boldsymbol{b}) + N_{-+}(\boldsymbol{a}, \boldsymbol{b}) + N_{--}(\boldsymbol{a}, \boldsymbol{b}) \, , \\ N_{++}(\boldsymbol{a}, \infty) &= N_{++}(\boldsymbol{a}, \boldsymbol{b}) + N_{+-}(\boldsymbol{a}, \boldsymbol{b}) \, , \\ N_{++}(\infty, \boldsymbol{b}) &= N_{++}(\boldsymbol{a}, \boldsymbol{b}) + N_{-+}(\boldsymbol{a}, \boldsymbol{b}) \, . \end{aligned} \tag{9.31}$$

By substitution into (9.28), an expression of the polarization correlation coefficient, and into inequalities (9.21), one can eliminate all the quantities which are not measured and obtain new BCHSH inequalities:

$$-1 \leq S' \leq 0 , \tag{9.32}$$

where

$$S' = \frac{N(\boldsymbol{a},\boldsymbol{b}) - N(\boldsymbol{a},\boldsymbol{b}') + N(\boldsymbol{a}',\boldsymbol{b}) + N(\boldsymbol{a}',\boldsymbol{b}') - N(\boldsymbol{a}',\infty) - N(\infty,\boldsymbol{b})}{N(\infty,\infty)} \tag{9.33}$$

is expressed as a function of *measured* coincidence rates only (we have omitted the implicit subscripts ++ in the expression above).

For the orientation sets shown in Fig 9.4, the quantum-mechanical predictions violate Bell's inequalities (9.32):

$$\begin{aligned} S'^{\mathrm{Max}}_{\mathrm{QM}} &= \frac{\sqrt{2}-1}{2} \qquad \text{for} \quad \theta = \pi/8 , \\ S'^{\mathrm{Min}}_{\mathrm{QM}} &= \frac{-\sqrt{2}-1}{2} \qquad \text{for} \quad \theta = 3\pi/8 . \end{aligned} \tag{9.34}$$

It is therefore also possible to make a sensitive test with one-channel polarizers.

Note however that the derivation of Bell's inequalities (9.32) requires a *supplementary assumption.* Since the detection efficiencies are low (due to a small collection angle and low photomultiplier efficiencies), the probabilities involved in the expression of $E(\boldsymbol{a},\boldsymbol{b})$ must be redefined on the ensemble of pairs that would be detected with polarizers removed. This procedure is valid only if one assumes a reasonable hypothesis about the detectors. The CHSH assumption [8] states that, "given that a pair of photons emerges from the polarizers, the probability of their joint detection is independent of the polarizer orientations" (or of their removal). *Clauser* and *Horne* [11] have shown another assumption, leading to the same inequalities. The status of these assumptions has been thoroughly discussed in [16].

### 9.8.2 Results

In the Berkeley experiment [17], *Freedman* and *Clauser* built a source where calcium atoms were excited to highly lying states by ultraviolet radiation. Each atom would then decay, and among the various deexcitation routes, it had some probability to emit a couple of green- and violet-correlated photons ($4p^{2\,1}S_0 \rightarrow 4s4p^1P_1 \rightarrow 4s^{2\,1}S_0$ radiative cascade). Since the signal was weak, and spurious cascades occurred, it took more than 200 h of measurement for a significant result. The results were found to be in agreement with quantum mechanics, and a violation of the relevant Bell's inequalities (9.32) was observed (by 5 standard deviations).

At the same time, at *Harvard*, *Holt* and *Pipkin* [18] found a result in disagreement with quantum mechanics, and in agreement with Bell's inequalities. Their source was based on the $9^1P_1 \to 7^3P_1 \to 6^3P_0$ cascade of mercury (isotope 200), excited by electron bombardment. The data accumulation lasted 150 h. *Clauser* subsequently repeated their experiment, but with mercury 202. He found an agreement with quantum mechanics, and a significant violation of Bell's Inequalities [19].

In 1976, at *Houston*, *Fry* and *Thompson* [20] built a much improved source of correlated photons, emitted in the $7^3S_1 \to 6^3P_1 \to 6^3S_0$ cascade in mercury 200. This is a $J = 1 \to J = 1 \to J = 0$ cascade, a priori not as favorable as a $J = 0 \to J = 1 \to J = 0$ cascade, but they could selectively excite the upper level of the cascade, by use of a cw single-line tunable laser (a quite rare instrument at that time). The signal was several orders of magnitude larger than in previous experiments, allowing them to collect the relevant data in a period of 80 min. Their result was in excellent agreement with quantum mechanics, and they found a violation, by 4 standard deviations, of the Bell's inequalities (9.32) specific to single-channel-polarizer experiments.

## 9.9 Orsay Experiments (1980–1982) [14]

### 9.9.1 The Source

From the beginning of our programme, our goal was to implement more sophisticated experimental schemes [13], so we devoted a lot of effort to developing a high-efficiency, stable, and well-controlled source of correlated photons. This was achieved (Fig. 9.10) by a two-photon selective excitation [21] of the $4p^2\ ^1S_0 \to 4s4p\ ^1P_1 \to 4s^2\ ^1S_0$ cascade of calcium that had been used by Clauser and Freedman. This cascade is very well suited to coincidence experiments, since the lifetime, $\tau_r$, of the intermediate level is rather short (5 ns). If one can reach an excitation rate of about $1/\tau_r$, then an optimum signal-to-noise ratio for coincidence measurements with this cascade is reached.

We have been able to obtain this optimum rate with the use of a krypton ion laser ($\lambda_K = 406\,\text{nm}$) and a tunable dye laser ($\lambda_D = 581\,\text{nm}$) tuned to resonance for the two-photon process. Both lasers are single-mode operated. They are focused onto a calcium atomic beam (laser beam waists about 50 μm). Two feedback loops provide the required stability of the source (better than 0.5% for several hours): the first loop controls the wavelength of the tunable laser to ensure the maximum fluorescence signal; a second loop controls the angle between the laser polarizations and compensates for all the fluctuations of the cascade rate. With a few tens of milliwatts from each laser, the cascade rate is about $N = 4 \times 10^7\,\text{s}^{-1}$. An increase beyond this rate would not significantly improve the signal-to-noise ratio for coincidence counting, since the accidental coincidence rate increases as $N^2$, while the true coincidence rate increases as $N$. At this cascade rate, the coincidence

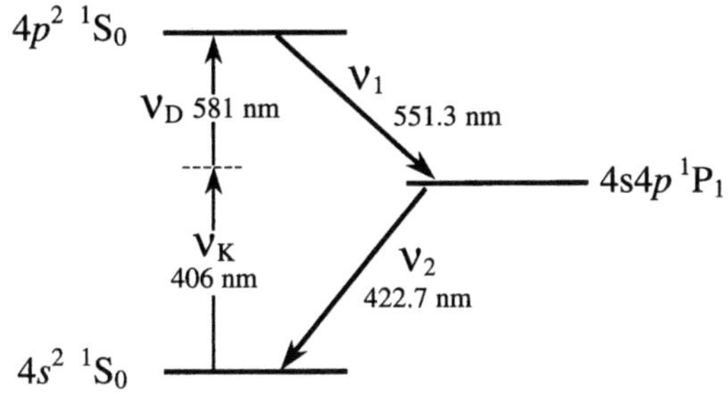


**Fig. 9.10.** Two-photon selective excitation of the $4p^{21}S_0$ state of calcium with a krypton ion laser and a tunable dye laser. From this state, the atom radiative decay can only deliver the pair of entangled photons $(\nu_1, \nu_2)$

rate with parallel polarizers was about $10^2\,\mathrm{s}^{-1}$, several orders of magnitude larger than in the first experiments. A statistical accuracy of 1% could then be achieved in each individual run of duration 100 s.

### 9.9.2 Detection–Coincidence Counting

The fluorescence light is collected by two large-aperture aspherical lenses ($u = 32°$, as defined in Fig. 9.8), followed in each leg by an interference filter (at 551.3 nm and 422.7 nm respectively), a transport optical system, a polarizer, and a photomuliplier tube. The photomultipliers feed the coincidence-counting electronics, which includes a time-to-amplitude converter and a multichannel analyzer, yielding the time-delay spectrum of the two-photon detections (Fig. 9.11). This spectrum first shows a flat background due to acci-

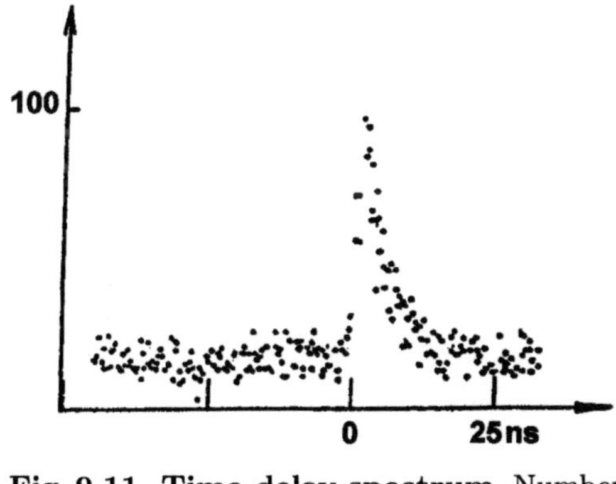


**Fig. 9.11. Time-delay spectrum.** Number of detected pairs as a function of the delay between the detections of two photons. The flat background corresponds to accidental coincidences between uncorrelated photons emitted by different atoms, and scales as $N^2$. The peak, whose area scales as $N$, corresponds to correlated photons and gives the coincidence rate to be measured

dental coincidences (between photons emitted by different atoms). The true coincidences (between photons emitted by the same atom) are displayed in the peak rising at zero-delay and exponentially decaying with a time constant, $\tau_r = 5\,\text{ns}$ (lifetime of the intermediate state of the cascade). The measured coincidence signal is thus the area of the peak.

Additionally, a standard coincidence circuit with a 19 ns coincidence window monitors the rate of coincidences around zero delay, while a delayed-coincidence channel monitors the accidental rate. It is then possible to check that the true coincidence rate obtained by subtraction is equal to the signal in the peak of the time-delay spectrum.

In the second and third experiments, we used a fourfold coincidence system, involving a fourfold multichannel analyzer and four double-coincidence circuits. The data were automatically gathered and processed by a computer.

### 9.9.3 Experiment with One-Channel Polarizers [22]

Our first experiment was carried out using one-channel, pile-of-plates polarizers, made of ten optical-grade glass plates at Brewster's angle, ensuring excellent rotational invariance. For fully polarized light, the maximum and minimum transmission were $0.975 \pm 0.005$ and $0.030 \pm 0.005$ respectively.

Thanks to our high-efficiency source, allowing us to achieve excellent statistical accuracy in 100 s runs, we could perform various statistical checks, as well as physical checks, for instance, on the rotational invariance of the signals (for all these measurements, the long-term stability of the source, at the level of 0.5%, was found to be crucial).

A direct test of the Bell's inequalities for single-channel polarizers (9.32) was performed. We found for the quantity $S'$ (9.33),

$$S'_{\text{exp}} = 0.126 \pm 0.014 \,, \tag{9.35}$$

violating inequalities (9.32) by 9 standard deviations and in good agreement with the quantum-mechanical predictions for our polarizer efficiencies and lens aperture angles:

$$S'_{\text{QM}} = 0.118 \pm 0.005 \,. \tag{9.36}$$

The error in $S'_{\text{QM}}$ accounts for the uncertainty in the measurements of the polarizer efficiencies.

The agreement between the experimental data and the quantum-mechanical predictions has been checked in a full 360° range of orientations (Fig. 9.12).

We repeated these measurements with the polarizers moved to 6.5 m from the source. At such a distance (four coherence lengths of the wave packet associated with the lifetime $\tau_r$) *the detection events are space-like separated* (we therefore fulfill the first time condition of Sect. 9.7.5). No modification of the experimental results was observed, and Bell's inequality was violated by the same amount.

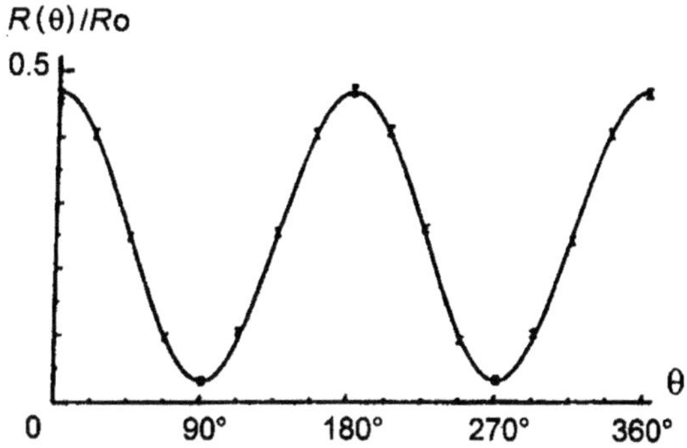


**Fig. 9.12. Experiment with one-channel polarizers.** Normalized coincidence rate as a function of the relative polarizers orientation. Indicated errors are $\pm 1$ standard deviation. The *solid curve* is not a fit to the data but the prediction by quantum mechanics for the actual experiment

### 9.9.4 Experiment with Two-Channel Analyzers [23–25]

With single-channel polarizers, the measurements of polarization are inherently incomplete. When a pair has been emitted, if no count is obtained at one of the photomultipliers, there is no way to know if "it has been missed" by the detector or if it has been blocked by the polarizer (only the latter case corresponds to a $-$ result for the measurement). This is why one had to resort to auxiliary experiments, and indirect reasoning using supplementary assumptions, in order to test Bell's inequalities.

With the use of two-channel polarizers, we have performed an experiment following much more closely the ideal scheme of Fig. 9.1. Our polarizers were polarizing cubes with dielectric layers transmitting one polarization and reflecting the orthogonal one.[6] Such a polarization splitter and the two corresponding photomultipliers are fixed on a rotatable mount. This device (polarimeter) yields the $+$ and $-$ results for a linear polarization measurement. It is an optical analog of a Stern–Gerlach filter for a spin 1/2 particle.

With polarimeters I and II in orientations $\boldsymbol{a}$ and $\boldsymbol{b}$, and a fourfold coincidence counting system, we are able to measure in a single run the four coincidence rates, $N_{\pm\pm}(\boldsymbol{a}, \boldsymbol{b})$, and to obtain directly the polarization correlation coefficient, $E(\boldsymbol{a}, \boldsymbol{b})$, by plugging the numbers into (9.28). It is then sufficient to repeat the same measurement for a sensitive set of four orientations, and the ideal Bell's inequality (9.20) can be directly tested.

Because this experimental scheme is much closer to the ideal scheme of Fig. 9.1 than the previous experiments with one-channel polarizers, we do not need the strong supplementary assumption on the detectors. However, the detection efficiency in each channel is well below unity, first because of

[6] A similar experiment, using calcite two-channel polarizers, had been considered at the University of Catania [26].

the limited solid angle of collection and second because of the efficiency of the photomultiplier. An advocate of hidden-variable theories could then argue that we are not sure that the sample on which the measurement bears remains the same when the orientations of the polarimeters are changed. In order to be logically allowed to compare our measurements to Bell's inequalities, we therefore also need a supplementary assumption: we must assume that the ensemble of actually detected pairs is independent of the orientations of the polarimeters. This assumption is very reasonable with our symmetrical scheme, where the two orthogonal output channels of a polarizer are treated in the same way (the detection efficiencies in both channels of a polarimeter are equal). Moreover, we have experimentally checked that the sum of the four coincidence rates, $N_{\pm\pm}(\boldsymbol{a}, \boldsymbol{b})$, remains constant when the orientations change, although each coincidence rate is 100% modulated. This shows that the size of the selected sample of pairs is constant. Of course, it is not proof of the validity of the assumption, but at least it is consistent with it. Note that it is possible to use a stronger assumption, the *fair-sampling assumption*, in which one assumes that the ensemble of detected pairs is a fair sample of the ensemble of all emitted pairs. The assumption above is a logical consequence of the fair-sampling assumption, which is stronger. On the other hand, the fair-sampling assumption is very reasonable and easy to express.

The experiment was done at the sensitive set of orientations in Fig. 9.4a, for which maximum conflict is predicted. We found

$$S_{\text{exp}} = 2.697 \pm 0.015 \ , \tag{9.37}$$

violating the inequalities (9.21) ($|S| \leq 2$) by more than 40 standard deviations! Note that this result is in excellent agreement with the predictions of quantum mechanics for our polarizer efficiencies and lens apertures:

$$S_{\text{QM}} = 2.70 \pm 0.05 \ . \tag{9.38}$$

The uncertainty indicated for $S_{\text{QM}}$ accounts for the slight lack of symmetry of the two channels of a polarizer ($\pm 1\%$). The effect of these dissymmetries has been computed and cannot create a variation of $S_{\text{QM}}$ greater than 2%.

We have also performed measurements of the polarization correlation coefficient, $E(\boldsymbol{a}, \boldsymbol{b})$, in various orientations, for a direct comparison with the predictions of quantum mechanics (Fig. 9.13). The agreement is clearly excellent.

These measurements can be presented in a different way, to emphasize the relevance to the test of Bell's inequalities. In Fig. 9.14, we show the measured quantity $S(\theta)$, as defined in Sect. 9.4.2. The violation of Bell's inequalities is clear around 22.5° [which corresponds to the result (9.37)] and 67.5°, but one also sees, as already emphasized, that there are many situations where there is no conflict with Bell's inequalities.

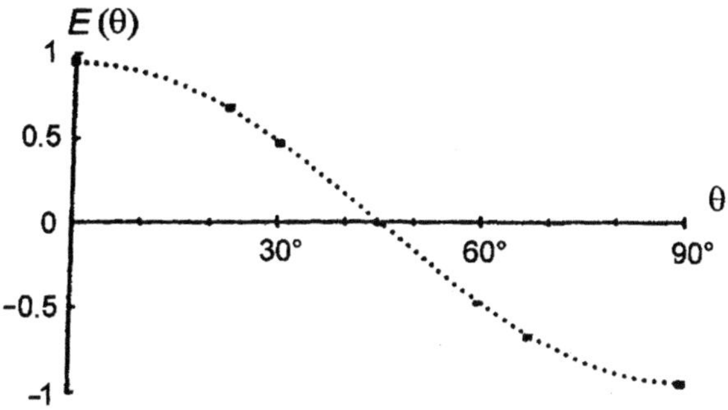


**Fig. 9.13. Experiment with two-channel polarizers**. Polarization correlation as a function of the relative angle of the polarimeters. The indicated errors are $\pm 2$ standard deviations. The *dotted curve* is not a fit to the data, but quantum-mechanical predictions for the actual experiment. For an ideal experiment, the *curve* would exactly reach the values $\pm 1$

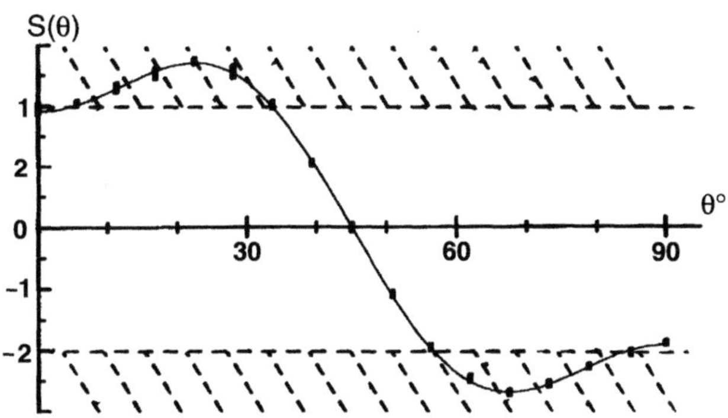


**Fig. 9.14. Experiment with two-channel polarizers**. Quantity $S(\theta)$, to be tested by Bell's inequalities ($-2 \leq S \leq +2$), as a function of the relative angle of the polarimeters. The indicated errors are $\pm 2$ standard deviations. The *solid curve* is not a fit to the data, but quantum-mechanical predictions for the actual experiment. For an ideal experiment, the *curve* would exactly reach the values $\pm 2.828$

### 9.9.5 Timing Experiment [27]

As stressed in Sects. 9.6 and 9.7.5, an ideal test of Bell's inequalities would involve the possibility of switching at random times the orientation of each polarizer [13], since the locality condition would become a consequence of Einstein's causality. We have made a step towards such an ideal experiment by using the modified scheme shown in Fig. 9.15.

Each (single-channel) polarizer is replaced by a setup involving a switching device followed by two polarizers in two different orientations: $\boldsymbol{a}$ and $\boldsymbol{a}'$ on side I and $\boldsymbol{b}$ and $\boldsymbol{b}'$ on side II. The optical switch $C_1$ is able to rapidly redirect the incident light either to the polarizer in orientation $\boldsymbol{a}$ or to the polarizer in orientation $\boldsymbol{a}'$. This setup is thus equivalent to a variable polarizer switched

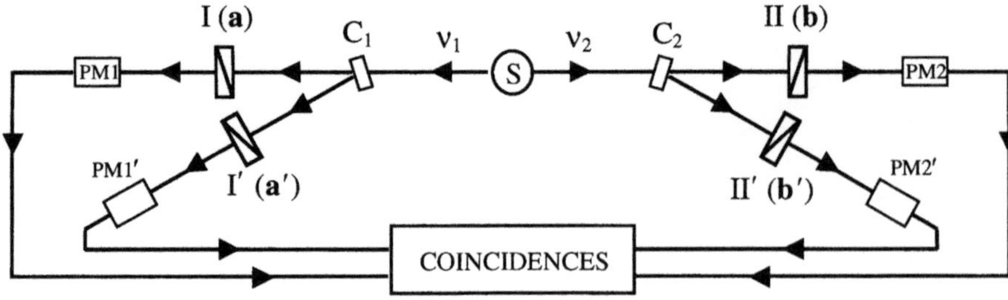


**Fig. 9.15. Timing experiment with optical switches** ($C_1$ and $C_2$). Switch $C_1$ followed by the two polarizers in orientations $\boldsymbol{a}$ and $\boldsymbol{a}'$ is equivalent to a single polarizer switched between orientations $\boldsymbol{a}$ and $\boldsymbol{a}'$. A switch occurs approximately every 10 ns. A similar setup, independently driven, is implemented on the second side

between the two orientations $\boldsymbol{a}$ and $\boldsymbol{a}'$. A similar setup is implemented on the other side and is equivalent to a polarizer switched between the two orientations $\boldsymbol{b}$ and $\boldsymbol{b}'$. In our experiment, the distance, $L$, between the two switches is 13 m, and $L/c$ has a value of 43 ns.

The switching of the light is done by home-built devices, based on the acousto-optical interaction of the light with an ultrasonic standing wave in water. The incidence angle (Bragg angle) and the acoustic power, are adjusted for a complete switching between the zeroth and first order of diffraction. The switching function is then of the form $\sin^2\left(\frac{\pi}{2}\cos\Omega_{\mathrm{a}}t\right)$, with the acoustic frequency $\Omega_{\mathrm{a}}/2\pi$ of the order of 25 MHz. The change of orientation of the equivalent variable polarizer then occurs after inequal intervals of 6.7 ns and 13.37 ns. Since these intervals as well as the delay between the emissions of the two photons of a pair (average value of $\tau_{\mathrm{r}} = 5$ ns), are small compared to $L/c$ (43 ns), a detection event on one side and the corresponding change of orientation on the other side are separated by a space-like interval. The first time condition is clearly fulfilled. The second time condition is also fulfilled, except for the fact that the switching was not truly random, but rather quasiperiodic (we discuss this point below).

The experiment was far from ideal on other points. First, in order to match the photon beams to the aperture of the switches, we had to reduce their size by a factor of three, entailing a reduction of the coincidence rates by one order of magnitude. As a consequence, to achieve significant statistical accuracy, the duration of data accumulation was much longer than in previous experiments, and we had to face problems of drifts. It was then necessary to average out the various measured quantities. Second, for not infinitely small beams, the commutation by the switches is incomplete, because the incidence angle is not exactly the Bragg angle for all rays. In our experiment, the minimum of the light intensity in each channel was 20%, so that not all photons were submitted to forced switching. Third, in this experiment, we used single-channel polarizers, which allowed us to use the same fourfold coincidence system as in the static experiment of Sect. 9.9.4.

Our test of Bell's inequalities involved a total of 8000 s of data accumulation with the 4 polarizers in the orientations of Fig. 9.4. A total of 16 000 s was devoted to the measurements with half or all the polarizers removed. In order to compensate for the effects of systematic drifts, the data accumulation was alternated between the various configurations each 400 s, and the data were averaged out. We finally obtained

$$S'_{\text{exp}} = 0.101 \pm 0.020 \,, \tag{9.39}$$

violating the upper limit of the Bell's inequality (9.32) by 5 standard deviations and in good agreement with the quantum-mechanic predictions,

$$S'_{\text{QM}} = 0.113 \pm 0.005 \,. \tag{9.40}$$

Other measurements of the coincidence rate have been carried out, for a comparison with quantum mechanics at different angles. As shown in Fig. 9.16, the results are in good agreement with the predictions of quantum mechanics.

According to these results, supplementary-parameter theories obeying Einstein's causality seem to be untenable. However, as indicated earlier, our experiment was not ideal, from several points of view, and several loopholes were left open for a strict advocate of hidden-variable theories. First, because we used single-channel polarizers, the experiment is significant only if one accepts a strong version of the fair-sampling assumption. Addressing more specifically the timing aspect of this experiment, an advocate of hidden-variable theories might argue that the switching was not complete, and that only the pairs undergoing forced switching must obey Bell's inequalities. But

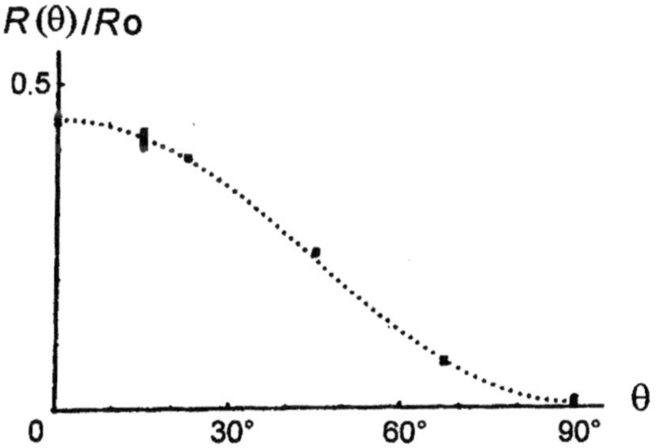


**Fig. 9.16. Timing experiment**. Average normalized coincicence rate as a function of the relative orientation of the polarizers. Indicated errors are ±1 standard deviation. The *dashed curve* is not a fit to the data but the predictions by quantum mechanics for the actual experiment

since these pairs represent a large fraction of the total number, it is hard to believe that we would not have observed a significant discrepancy between our results and the quantum-mechanical predictions.

The most important point to discuss is the fact that the switches were not truly random, since the acousto-optical devices were driven by periodic generators. Note, however, that the two generators on the two sides were functioning in a completely uncorrelated way, since they were operated by different RF generators at different frequencies (23.1 MHz and 24.2 MHz), with uncorrelated frequency drifts. Moreover, another random feature is provided by the random delay between the two photons of a pair (exponentially decaying distribution of constant, $\tau_{\mathrm{r}} = 5\,\mathrm{ns}$, as shown in Fig. 9.11), which are distributed on an interval larger than the time between two successive switchings.

In conclusion, this experiment, which was until 1998 the only one involving forced fast changes of the settings of the analysers, had enough imperfections to leave open the possibility of ad hoc supplementary-parameter models fulfilling Einstein's causality. However, several models that we have tried have been eliminated by our experimental results, which are constituted not only by the measured value (9.39) of $S'$, but also by the time-delay spectra without any accident observable on the exponential decay, and with areas in good agreement with quantum mechanics as shown in Fig. 9.15. We think that these data should be taken into account by any advocate of local, hidden-variable theories trying to build a model compatible with experimental observations.

## 9.10 Third Generation: Experiments with Pairs of Photons Produced in Parametric Down Conversion

As we have already noted, the calcium radiative cascade that was used in our experiments was excited to an optimum rate, beyond which there is not much of a possibility for gain in signal-to-noise ratio. Since the lifetime of the intermediate stage is quite short (5 ns) the situation is very favourable for coincidence counting, and there was not much room left for improvement with sources based on atomic radiative cascades[7].

In the late 1980s, new sources of pairs of correlated photons were developed simultaneously by two groups [29, 30]. In these sources, a pair of red photons is produced by parametric down conversion of a UV photon. Because of the phase-matching condition in the nonlinear crystal used for

[7] Note however the work of *Perrie* et al.[28] in which the intermediate lifetime is much shorter. Unfortunately, for other experimental reasons, the statistical significance of the results was limited as a test of Bell's inequalities. This experiment was however a remarkable "tour de force" in atomic physics.

this process, there is a strong correlation between the directions of emission of the two photons of a pair, so that, by spatial selection with two diaphragms positioned in conjugate positions, one can in principle be sure to get the two photons of a pair. This is in contrast with the atomic radiative cascades which produce photons only weakly correlated in direction [15]: since each photon is collected in solid angle $\Omega$, small compared to $4\pi$, the probability of obtaining the second photon of a pair, once the first one is detected, is of the order of $\Omega/4\pi$, so that the sample of detected pairs is smaller by this factor than the sample of selected pairs. The new scheme with photons correlated in direction allows one to get rid of this reduction factor, and this has far reaching consequences, both practical and fundamental. On the practical side, it allows larger coincidence rates to be obtained, for similar cascade rates: in the most favourable case [31] the coincidence rate may be more than one order of magnitude larger than in our best experiments (Sect. 9.9)[8]. Moreover, such large coincidence rates can be obtained with narrow photon beams (with a small Fresnel number). Such beams can easily be matched into small optical components, or even optical fibers, which opens up many new possibilities.

These new sources can produce pairs or photons correlated in polarization [29–32], in states analogous to (9.1). But they can also produce entangled states exhibiting EPR-type correlations between observables other than polarization. An interesting case [33] considers pairs of photons where each photon is emitted "at two different times". Here, the relevant observable is the time of emission of the two photons of the pair, and the conjugate is the energy (wavelength). Corresponding experiments have been carried out [34–36]. Note that this scheme, where polarization is not the relevant observable, is especially interesting for experiments with optical fibers, in which polarization control may be a crucial issue. Another interesting scheme considers the directions of emission as observables [37]: each photon of an entangled pair involves two different directions of emission, strongly correlated to two directions of emission for the second photon. An experiment of this type has also been carried out [38].

As emphasised in [37], all these new schemes can be embedded in the general framework of "two-particles interferences"; indeed, the joint measurement probabilities are the square of a sum of two amplitudes (each involving the two photons), with a relative phase that can be controlled experimentally. Although it was not pointed out in *Horne* et al. [37], the original EPRB scheme is a very clear example of this situation. For instance, for the polarization entangled state of Sect. 9.2, the state (9.1) can also be rewritten [see 9.29] as the superposition of a state, $|L, L\rangle$, where both photons have

[8] Note that the reported violation of Bell's inequalities by 100 standard deviations relies on a stronger version of the "fair sampling hypothesis" than our second experiment (Sect. 9.9.4), since this experiment uses one channel polarizers and not two channel polarizers.

a left-handed helicity, and a state, $|R,R\rangle$, with two right-handed helicities. For each of these two states, the amplitude for being detected in any couple of output channels behind the linear polarizers (see Fig. 9.1) results from the addition of the amplitudes associated with $|L,L\rangle$ and $|R,R\rangle$. It leads to an interference term responsible for the sinusoidal variations of the joint probabilities (9.3) when the orientations change.

These new sources and schemes led to a series of tests of Bell's inequalities, *which have all confirmed the predictions of quantum mechanics.* Clear violations of Bell's inequalities were found, under the assumption that the fair-sampling hypothesis holds. Among these, it is worth pointing out a violation of Bell's inequalities by 100 standard deviations in a few minutes only [31]. Note also an experiment [35] where a clear violation of Bell's inequalities was observed with one leg of the apparatus made of 4 km of optical fiber. More recently, EPR correlations have been observed with photons propagating in several tens of kilometers of commercial telecommunication fibers [39].

These third-generation experiments should eventually lead to an ultimate experiment where there would be no remaining loophole left open. First, the perfect correlation between the directions of emission offers the possibility of closing the loophole related to the low detection efficiency [11] when photon detectors with quantum efficiency close to unity are available[9].

The second class of fundamental improvements is related to the "timing experiments" (Sects. 9.7.5 and 9.9.5). Ideally [13], one needs polarizers that can be independently reorientated at random times, with a reorientation autocorrelation time shorter than the space separation, $L/c$, between the polarizers. Our third experiment (Sect. 9.9.5), which was the first attempt in this direction, was basically limited by the wide size of the beams carrying the correlated photons: this prevented us from using small electro-optical devices suitable for random switching of polarization. With the new schemes using optical fibers, it becomes possible to work with small integrated electro-optical devices. Moreover, with use of optical fibers, the detectors can be kilometers apart. At such separations (several microseconds), the time conditions become less stringent, and truly random active operation of the polarizers become possible at this time scale [41]. An experiment of this type has been completed in the group led by *Anton Zeilinger* [42]. From the point of view of the timing condition, one can say that this experiment meets all the criteria of an ideal experiment [43].

[9] Note that a recent experiment bearing on two entangled ions has achieved large enough detection efficiencies to observe a violation of Bell's inequalities without any supplementary hypothis on the detectors [40]. Note however that the two ions in the same trap are very close to each other, so that the detection events are not space-like separated, and it seems difficult to fulfill the timing conditions in experiments following this scheme.

## 9.11 Conclusion

We have nowadays an impressive number of sensitive experiments in which Bell's inequalities have been clearly violated. Moreover, the results are in excellent agreement with the quantum-mechanical predictions including all the known features of the real experiment. Each of the remaining loopholes has been separately closed [40, 42], and although yet more ideal experiments are still desirable [44], it is legitimate to discuss the consequences of the rejection of supplementary-parameter theories obeying Einstein's causality.

It may be concluded that quantum mechanics has some nonlocality in it, and that this nonlocal character is vindicated by experiments [45]. It is very important, however, to note that such a nonlocality has a very subtle nature, and in particular that it cannot be used for faster-than-light telegraphy. It is indeed simple to show [46] that, in a scheme where one tries to use EPR correlations to send a message, it is necessary to send complementary information (about the orientation of a polarizer) via a normal channel, which of course does not violate causality. This is similar to the teleportation schemes [47] where a quantum state can be teleported via a nonlocal process provided that one also transmits classical information via a classical channel. In fact, there is certainly a lot to understand about the exact nature of nonlocality, by a careful analysis of such schemes [48].

When it is realized that this quantum nonlocality does not allow one to send any useful information, one might be tempted to conclude that in fact there is no real problem and that all these discussions and experimental efforts are pointless. Before rushing to this conclusion, I would suggest an ideal experiment done in the following way is considered (Fig. 9.17): On each side of the experiment of Fig. 9.1, using variable analysers, there is a monitoring system that registers the detection events in channels + or − with their exact dates. We also suppose that the orientation of each polarizer is changed at random times, also monitored by the system of the corresponding side. It is only when the experiment is completed that the two sets of data, separately collected on each side, are brought together in order to extract the correlations. Then, looking into the data that were collected previously and that correspond to paired events that were space-like separated when they happened, one can see that indeed the correlation did change at the very moment when the relative orientation of the polarizers changed.

So when one takes the point of view of a delocalized observer, which is certainly not inconsistent when looking into the past, it must be acknowledged that there is nonlocal behaviour in the EPR correlations. Entanglement is definitely a feature going beyond any space–time description à la Einstein: a pair of entangled photons must be considered to be a single global object that we cannot consider to be made of individual objects separated in space–time with well-defined properties.

For many years, I have been quoting the scheme of Fig. 9.17 as a Gedanken-experiment useful for the sake of the discussion. Nowadays, we are lucky that

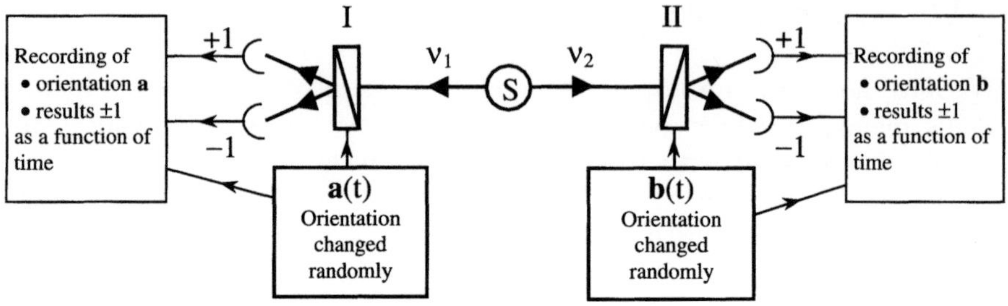


**Fig. 9.17.** Ideal timing experiment. Each polarizer is randomly reoriented during the propagation of photons between the source and the polarizers. On each side, one records the orientation of the polarizer as well as the results of polarization measurements as a function of time. When a run is completed, the two data sets from the two sides are brought together, and one can determine the value of the correlation as a function of the relative orientation at the moment of measurement

this experiment has been done in the real world: the experiment of *Zeilinger* and *Weihs* [42], sketched in Fig. 9.18, exactly follows [43] the scheme of Fig. 9.17. Once again, the EPR problem has switched from fundamental discussions bearing on a Gedankenexperiment to real experiments. We must thank John Bell for having shown us that philosophical questions about the nature of reality could be translated into problems for physicists, where naive experimentalists can contribute.

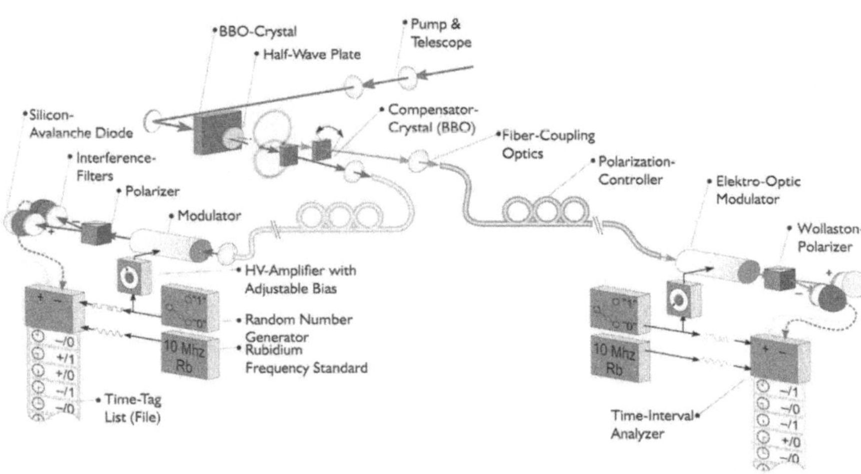


**Fig. 9.18.** The timing experiment of Weihs et al. This experiment follows closely the ideal scheme of Fig. 9.17, since the two ends of the experiment are totally independent. It is only after completion of a run that the data from the two sides are compared, in order to extract the correlation coefficient. Note also that the orientations of the polarizers are changed randomly during the photon propagation from the source to the polarizers

## References


1. J.S. Bell, On the Einstein–Podolsky–Rosen Paradox, Phys. **1**, 195 (1964)
2. A. Aspect, *Experimental tests of Bell's inequalities in atomic physics*, in: *At. Phys.* **8**, *Proceedings of the Eighth International Conference on Atomic Physics*, I. Lindgren, A. Rosen, S. Svanberg (eds.) (1982)
3. A. Einstein, B. Podolsky, N. Rosen, Can a quantum-mechanical description of physical reality be considered complete? Phys. Rev. **47**, 777 (1935)
   See also Bohr's answer: N. Bohr, Can a quantum-mechanical description of physical reality be considered complete? Phys. Rev. **48**, 696 (1935)
4. D. Bohm, *Quantum Theory* (Prentice-Hall, Englewoods Cliffs 1951). Republished by Dover, New York (1989)
5. D. Bohm, Y. Aharonov, Discussion of experimental proof for the paradox of Einstein, Rosen and Podolsky, Phys. Rev. **108**, 1070 (1957)
6. P.A. Schilp (ed.), *A. Einstein Philosopher Scientist* (Open Court and Cambridge Univ. Press, Cambridge 1949)
7. *Correspondence* between A. Einstein and M. Born. French translation: (Seuil, Paris, 1972)
8. J.F. Clauser, M.A. Horne, A. Shimony, R.A. Holt, Proposed experiment to test local hidden-variable theories, Phys. Rev. Lett. **23**, 880 (1969)
9. B. d'Espagnat, Use of inequalities for the experimental test of a general conception of the foundation of microphysics, Phys. Rev. D **11**, 1424 (1975)
   L. Hardy, Quantum mechanics, local realistic theories, and Lorentz-invariant realistic theories, Phys. Rev. Lett. **68**, 2981 (1992)
10. J.S. Bell, 'Introduction to the hidden-variable question', in: *Foundations of Quantums Mechanics*, B. d'Espagnat (ed.), (Academic, New York 1972)
11. J.F. Clauser, M.A. Horne, Experimental consequences of objective local theories, Phys. Rev. D **10**, 526 (1974)
12. A. Fine, *Hidden variables*, Joint probability, and the Bell inequalities, Phys. Rev. Lett. **48**, 291 (1982)
13. A. Aspect, Proposed experiment to test separable hidden-variable theories, Phys. Lett. A **54**, 117 (1975)
    A. Aspect, Proposed Experiment to test the nonseparability of Quantum Mechanics, Phys. Rev. D **14**, 1944 (1976)
14. A. Aspect, Trois tests expérimentaux des inégalités de Bell par mesure de corrélation de polarization de photons, thèse d'Etat, Orsay (1983)
15. E.S. Fry, Two-photon correlations in atomic transitions, Phys. Rev. A **8**, 1219 (1973)
16. J.F. Clauser, A. Shimony, Bell's theorem: Experimental tests and implications, Rep. Prog. Phys. **41**, 1881 (1978)
17. S.J. Freedman, J.F. Clauser, Experimental test of local hidden-variable theories, Phys. Rev. Lett. **28**, 938 (1972)
18. F.M. Pipkin, *Atomic physics tests of the basics concepts in quantum mechanics*, in: *Advances in Atomic and Molecular Physics*, D.R. Bates and B. Bederson, (eds.), (Academic, New York 1978)
19. J.F. Clauser, Experimental investigation of a polarization correlation anomaly, Phys. Rev. Lett. **36**, 1223 (1976)
20. E.S. Fry, R.C. Thompson, Experimental test of local hidden-variable theories, Phys. Rev. Lett. **37**, 465 (1976)

21. A. Aspect, C. Imbert, G. Roger, Absolute measurement of an atomic cascade rate using a two photon coincidence technique. Application to the $4p^2\,{}^1S_0 - 4s4p\,{}^1P_1 - 4s^2\,{}^1S_0$ cascade of calcium excited by a two photon absorption, Opt. Commun. **34**, 46 (1980)
22. A. Aspect, P. Grangier, G. Roger, Experimental Tests of Realistic Local Theories via Bell's Theorem, Phys. Rev. Lett. **47**, 460 (1981)
23. A. Aspect, P. Grangier, G. Roger, Experimental realization of Einstein–Podolsky–Rosen–Bohm gedankenexperiment: A new violation of Bell's inequalities, Phys. Rev. Lett. **49**, 91 (1982)
24. P. Grangier, Thèse de troisième cycle, Orsay (1982)
25. A. Aspect, P. Grangier, About resonant scattering and other hypothetical effects in the Orsay atomic-cascade experiment tests of Bell inequalities: A discussion and some new experimental data, Lett. Nuovo Cim. **43**, 345 (1985)
26. A. Garrucio, V.A. Rapisarda, Nuovo Cim. A **65**, 269 (1981)
27. A. Aspect, J. Dalibard, G. Roger, Experimental test of Bell's inequalities using variable analyzers, Phys. Rev. Lett. **49**, 1804 (1982)
28. W. Perrie, A.J. Duncan, H.J. Beyer, H. Kleinpoppen, Polarization correlation of the two photons emitted by metastable atomic deuterium: A test of Bell's inequality, Phys. Rev. Lett. **54**, 1790 (1985); **54**, 2647 (1985)
29. Y.H. Shih, C.O. Alley, New type of Einstein–Podolsky–Rosen–Bohm experiment using pairs of light quanta produced by optical parametric down conversion, Phys. Rev. Lett. **61**, 2921 (1988)
30. Z.Y. Ou, L. Mandel, Violation of Bell's inequality and classical probability in a two-photon correlation experiment, Phys. Rev. Lett. **61**, 50 (1988)
31. P.G. Kwiat, K. Mattle, H. Weinfurter, A. Zeilinger, New high-intensity source of polarization-entangled photon-pairs, Phys. Rev. Lett. **75**, 4337 (1995)
Note that the reported violation of Bell's inequalities by 100 standard deviations relies on a stronger version of the "fair sampling hypothesis" than our second experiment (Sect. 9.9.4), since this experiment uses one channel polarizers and not two channel polarizers
32. T.E. Kiess, Y.H. Shih, A.V. Sergienko, C.O. Alley, Einstein–Podolsky–Rosen–Bohm experiment using pairs of light quanta produced by type-II parametric down-conversion, Phys. Rev. Lett. **71**, 3893 (1993)
33. J.D. Franson, Bell inequality for position and time, Phys. Rev. Lett. **62**, 2205 (1989)
34. J. Brendel, E. Mohler, W. Martienssen, Experimental test of Bell's inequality for energy and time, Europhys. Lett. **20**, 575 (1992)
35. P.R. Tapster, J.G. Rarity, P.C.M. Owens, Violation of Bell's inequality over 4 km of optical fiber, Phys. Rev. Lett. **73**, 1923 (1994)
36. W. Tittel, J. Brendel, T. Herzog, H. Zbinden, N. Gisin, Non-local two-photon correlations using interferometers physically separated by 35 meters, Europhys., Lett. **40**, 595 (1997)
37. M.A. Horne, A. Shimony, A. Zeilinger, Two-particle interferometry, Phys. Rev. Lett. **62**, 2209 (1989)
38. J.G. Rarity, P.R. Tapster, Experimental violation of Bell's inequality based on phase and momentum, Phys. Rev. Lett. **64**, 2495 (1990)
39. W. Tittel, J. Brendel, H. Zbinden, N. Gisin, Violation of Bell inequalities by photons more than 10 km apart, Phys. Rev. Lett. **81**, 3563 (1998)

40. M.A. Rowe, D. Kielpinsky, V. Meyer, C.A. Sackett, W.M. Itano, D.J. Wineland, Experimental violation of a Bell's inequality with efficient detection, Nature **409**, 791 (2001)
41. The interest of an active rather than a passive switching of the polarizers is questioned in: N. Gisin, H. Zbinden, Bell inequality and the locality loophole: active versus passive switches, Phys. Lett. A **264**, 103 (1999)
42. G. Weihs, T. Jennewein, C. Simon, H. Weinfurter, A. Zeilinger, Violation of Bell's inequality under strict Einstein locality condition, Phys. Rev. Lett. **81**, 5039 (1998)
43. A. Aspect, Bell's inequality test: more ideal than ever, Nature **398**, 189 (1999)
44. For a scheme that could be ideal, using entangled atoms, see the paper of E. Fry in this book
45. J.S. Bell, Atomic cascade photons and quantum-mechanical nonlocality, Comm. Atom. Mol. Phys. **9**, 121 (1981)
46. A. Aspect, Expériences basées sur les inégalités de Bell, J. Phys. Coll. C **2**, 940 (1981)
47. C.H. Bennet, G. Brassard, C. Crépeau, R. Josza, A. Peres, W.K. Wootters, Phys. Rev. Lett. **70**, 1895 (1993)
    D. Bouwmeester, J.-W. Pan, K. Mattle, M. Eibl, H. Weinfurter, A. Zeilinger, Experimental quantum teleportation, Nature **390**, 575 (1997)
    D. Boschi, S. Branca, F. De Martini, L. Hardy, S. Popescu, Experimental realization of teleporting an unknown pure quantum state via dual classical and Einstein–Podolsky–Rosen channels, submitted to Phys. Rev. Lett. (1997)
    A. Furusawa, J.L. Sorensen, S.L. Braunstein, C.A. Fuchs, H.J. Kimble, E.S. Polzik, Unconditional quantum teleportation, Science **282**, 706 (1998)
48. S. Popescu, Bell's inequalities versus teleportation: what is non-locality? Phys. Rev. Lett. **72**, 797 (1994)

# 10 Bell's Theorem for Space-Like Separation

Gregor Weihs

Soon after John Bell published his famous theorem first experiments [1–3] were performed to test whether nature was compliant with local realistic theories or not. Now, more than 30 years later various groups are still trying to do better and better tests of the same sort, and indeed techniques have come remarkably close to ideal. So, why bother?

The answer is that Bell's theorem is an extremely stringent criterion to assess the viability of a large class of physical models. In its abstract form it relates quantities drawn from simple thought experiments. If one now tries to apply this to real experiments a lot of conditions on the various components of the apparatus appear. Up to now no experiment could stand up to meet all of these requirements simultaneously, leaving so-called loopholes open for local realistic theories to escape their refutation by experimental data.

The usual way to show the physical importance of such an experiment is to specify some additional assumptions – as weak as possible – about the models to be tested. Using these assumptions one derives a theorem in the spirit of Bell's theorem, which is then applicable to the experiment in question. This kind of reasoning amounts to a restriction of the class of local realistic models under test.

So, however reasonable these additional assumptions may seem in a specific context, it is still desirable to eventually get rid of them and to test a so-called strong Bell inequality, i.e. one which does not need any other premise than locality and realism.

## 10.1 From Gedanken to Real Experiments

The Gedanken experiment employed by Bell to formulate his theorem can not precisely be reconstructed from a single source in the literature. The clearest description is given in [4]. Therein, a schematic of a general EPR setup, similar to Fig. 10.1 is used to describe the basic features that are necessary, without any reference to terms as "particle", "analyzer", or "source". A central "go" signal is fed into the long box-like machine, soon after which the machine outputs a positive or negative answer to the "go" signal. This makes us ready to feed two parameter-control signals "a" and "b" locally into the remote parts of the machine. Then again shortly after that the machine will output an answer at each side and we may — after a number of runs — proceed to calculate probabilities and correlations.

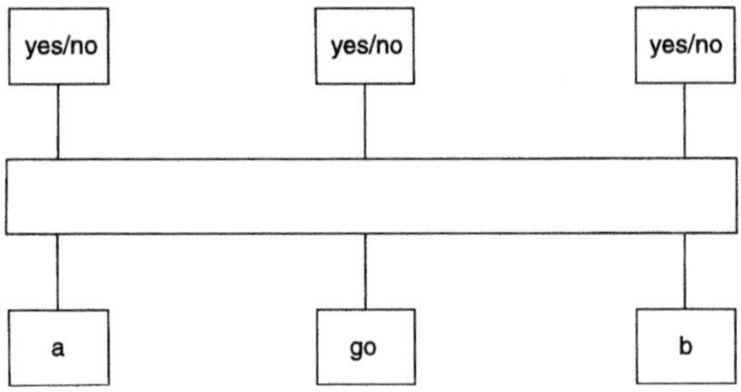


**Fig. 10.1.** A general schematic for an EPR/Bell experiment. Below are three inputs (a,b,go), above are three outputs

This scheme however does not account for the cases in which there is no output at all on at least one of the two sides. Such a lossy, inefficient machine calls for a treatment of the unpaired events. Usually, an assumption is invoked which allows to extrapolate from the available correct measurements to the ideal case.

Also the long box-like machine only gives a vague account of what we have to consider under the notion of locality. Bell says that the signals "a" and "b" are injected shortly before the output is to be expected. He continues in saying that we can arrange our actions such that the time between injection and output is much shorter than the signaling delay between opposite ends of the machine. The apparent difficulty here is to specify where the interface between the inside and outside of the machine would be located in a real experiment and, consequently, how to fix the critical time mentioned.

This setup also seems to imply that it would generally be necessary to know in advance when output is to be expected, a feature also known as "event-ready"-ness. I think that this is just one aspect of the experimental definition of locality that could in principle be traded for a measurement of the time between the injection of the parameter control signals and the output.

## 10.2 Efficiency

The first of the two foremost issues in the relationship between Bell's theorem and real experiments is the problem of the notoriously low detection efficiencies. Although very recently this problem has been solved employing a trapped two-ion system [5], which will be described later, it is interesting to discuss the general problem.

In the Gedankenexperiment from which Bell's theorem is derived only pairs are emitted and detected. With the noteable exception of [5] this has not been true for any experiment be it optical or with massive particles. In all these tests many particles were lost on their way from the source to the measurement apparatus or in the detectors themselves.

To estimate the efficiency that is needed to still get a violation of Bell's inequality — using the quantum physical probabilities — despite the loss of particles, we assume that both our apparatus will detect a fraction $\eta$ of all incoming particles on each side. The probability to detect a pair at analyzer angles $\alpha$ and $\beta$ is given by

$$p_{\mathrm{AB}}^{\mathrm{ex}} = \frac{1}{2}\eta^2 \cos^2(\beta - \alpha) \ , \tag{10.1}$$

for an entangled system of two particles emerging from their source in the state

$$\Phi = \frac{1}{\sqrt{2}}\left[|0\rangle_1|0\rangle_2 + |1\rangle_1|1\rangle_2\right] \ , \tag{10.2}$$

where 0 and 1 are the two basis states used. The probabilities for a count on either side are analogously

$$p_{\mathrm{A}}^{\mathrm{ex}} = p_{\mathrm{B}}^{\mathrm{ex}} = \frac{1}{2}\eta \ . \tag{10.3}$$

Inserting these expressions into the *Clauser–Horne* [6] version of Bell's inequality we calculate the minimum efficiency $\eta_{\mathrm{min}}$ that is necessary to achieve a violation in the case of ideal experimental contrast.

$$\begin{aligned} &\tfrac{1}{4}\eta^2\left(\cos^2 22.5^\circ - \cos^2 67.5^\circ + \cos^2 22.5^\circ + \cos^2 22.5^\circ\right) - \eta\left(\tfrac{1}{2} + \tfrac{1}{2}\right) = \\ &= \tfrac{1}{2}\eta^2\left(1+\sqrt{2}\right) - \eta \le 0 \\ &= \eta \le \tfrac{2}{1+\sqrt{2}} = 0.828 = \eta_{\mathrm{min}} \ . \end{aligned}$$

In the early days of optical experiments using atomic cascade transitions it was not even in principle possible to capture all pairs. This is because cascades are three-body decay processes and therefore the emitted photons do not possess any directional correlation. Without directional correlation the practical collection efficiency was so low that even with perfect detectors it would have been impossible to capture a large enough fraction. Parametric down-conversion sources, in contrast, yield nearly perfect momentum correlation for the emitted photons and consequently we can place our detectors such that whenever we find one photon in one detector its partner will be found in the other one.

In the following I will summarize a theoretical step and experimental improvements that have been undertaken to get closer to the above "threshold".

### 10.2.1 Reducing the Required Efficiency

In 1993 *Eberhardt* [7] found a way to reduce, or better, shift the experimental requirements. His idea was to use quantum mechanical states for a Bell test that are not maximally entangled. In general such a state can be described by

$$\Phi = \frac{1}{\sqrt{a^2+b^2}}\left[a|0\rangle_1|0\rangle_2 + b|1\rangle_1|1\rangle_2\right], \tag{10.4}$$

By choice of the coefficients $a$ and $b$ one can manipulate the magnitudes of the terms appearing in the above inequality such that one can in principle achieve a violation even if the overall efficiency is as low as 67%. Still, in order to be able to go below the naive limit of 82.2% one needs to have less than about 2% background, which doesn't make the experimentalists' lives much easier.

### 10.2.2 State of the Art and Development of Optical Experiments

Parametric down-conversion sources have the great advantage that the emission directions of two photons forming a pair are strongly correlated. This gives us the chance of catching every photon's partner as well. Therefore we are only limited here by the transmission of optical elements and by the quantum efficiency of our detectors. By careful engineering of the optical setup recently an overall efficiency of 28.6% has been achieved [8], thereby getting very close to the intrinsic efficiency limit of the detector used. Obviously this value — while being the record for optical experiments — is far from touching the critical boundaries at 67% or even 83%.

However, there have been reports [9, 10] saying that much better detectors exist. These so-called Solid-State-Photomultipliers (SSPM) or Visible-Light-Photon-Counters (VLPC) are notoriously hard to operate, but at least one group managed to achieve proper shielding from thermal radiation while maintaining extremely high efficiency for visible light [11–13]. They now pursue an experiment in which they want to close the detection efficiency loophole. So, combined with the above techniques there is a good chance of success for the optical domain, which at the same time leaves a realistic possibility to simultaneously close all the loopholes in a single experiment.

### 10.2.3 A Proposal Using Mercury Atoms

A very interesting proposal was given by *Fry* et al. [14] suggesting to use laser assisted dissociation of Mercury dimers to produce spin entangled states of mercury atoms, which have been experimentally proven to yield detection efficiencies of close to 100%. Since there are remarkable experimental difficulties that have to be overcome it will probably take much more development until this program can be carried out.

### 10.2.4 Closure of the Detection Efficiency Loophole with an Ion Trap Experiment

In a very recent experiment *Rowe* et al. [5] have managed to do a test of Bell's inequality with perfect detection efficiency. Their experimental set-up differs fundamentally from all previous work. Pairs of ions are captured in a microscopic trap and prepared in any desired state, particularly in an entangled state. Once they have been prepared a transformation is carried out on each individual ion in order to set the basis for the subsequent measurement.

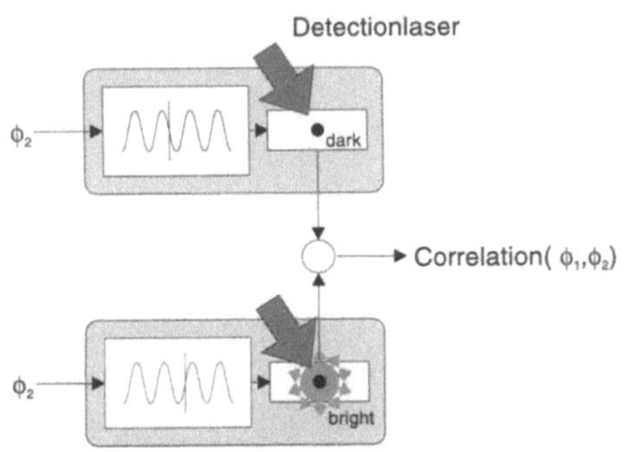


**Fig. 10.2.** The Bell inequality test by *Rowe* et al. [5]. Two ions are prepared in an entangled state and measured after having undergone a certain phase rotation. The measurement signal is the summed fluorescence intensity of both ions from which the correlation can immediately be extracted

Measurements are performed by looking at light which is scattered from each ion in a state dependent fashion. The big advantage over all other schemes is that the particles themselves are not destroyed by the measurement but recycled for the next run. In this way Rowe et al. made sure that not a single measurement "failed". The only imperfection was the finite accuracy in discriminating the various measurement outcomes. Still, the method yielded a high enough contrast and a clear violation of Bell's inequality.

This new approach has one little drawback: The ions are separated by not more than 3 μm prohibiting any discrimination between the results of the individual particles not to speak of any more demanding locality issues.

## 10.3 Locality

In commenting upon the loophole discussion John S. Bell himself expressed: "Of more importance, in my opinion, is the complete absence of the vital *time* factor in existing experiments. The analyzers are not rotated during the flight of the particles." [4]. Although Bell did not specify a clear concept of what he would have considered an appropriate experiment including the "time factor", there seems to be some consensus that the two measurements that contribute to the correlation should be separated by a space-like interval. This is sometimes called Einstein locality.

It becomes immediately clear that a space-like interval can only be defined if the notion of time is introduced. However, there is absolutely no unique way to ascribe a certain time-interval to a measurement, neither from quantum physics, nor from general local realism. In a phenomenological sense one could try to deduce a time scale from some decoherence model for the detector used. This should yield a time scale comparable to the time resolution of that detector. In this way the detector time resolution would determine the bare minimum time interval that has to be included.

The above argumentation does not at all consider the analyzers. If one wants to include these into the decisive time scale – as John S. Bell did (see above) – things become considerably more involved. Definitely it is the analyzer settings where locality is introduced into the derivation of Bell's inequality. If the analyzer setting of one side has to be kept a local secret, the only generally acknowledged way is to change it in a way that can not be controlled or observed by the other side within the time that is relevant for each measurement.

Lots of suggestions have been made on how to accomplish this task but none is completely satisfactory. A remarkable suggestion says that one should use the light from stars at opposite ends of the galaxy as a source of randomness on each side of the experiment. As a consequence no information could have reached the other side early enough to violate the locality condition. An approximation to this concept is the use of independent physical random generators at both observer locations.

Frequently, in discussing the above issues the topic of complete determinism comes up, the proponents of which assume that everything – including random generators and even distant stars – could be controlled by a hidden-variable machinery that makes everything look quantum in a superficial way. A funny twist would be put to this argumentation, if one imagines human observers choose the analyzer settings at their own free will. Then, if one still pursues the complete determinist's viewpoint, consequently they have to arrive at the conclusion that they themselves are being controlled by that machinery.

### 10.3.1 Experiments with Large Separation Between Measurement Stations

One of the basic ingredients for any experiment trying to tackle the locality loophole is spatial separation between the measurement stations. The record so far for an experiment showing a violation of Bell's inequality is held by *Tittel* et al. [15] in Geneva. Using optical fibers at telecommunication wavelengths they managed to separate their observers by more than 10 km.[1] There was no (fast) variation in the analyzers, but the mere fact that quantum physical correlations can be kept up over seemingly arbitrary distance is not only stunning but has also led people to think of new applications for these phenomena.

### 10.3.2 Analyzer Switching

The first experiment trying to include some fast variation of the analyzers was done by *Aspect* et al. [16]. In this extremely difficult test the distance between the detectors was about 12 m and the switch rise time 2 ns, corresponding to 60 cm at the vacuum speed of light. The variation was done in a periodic way with different frequencies independently on both sides. A periodic variation is at least in principle predictable and therefore the setting at any time in the future is determined. In this sense also no specific time could have been associated with an individual measurement rendering a spacetime analysis impossible.

**Fig. 10.3.** A test of Bell's inequality was performed for observers separated by 360 m with random switching of the analyzers within a time less than a tenth of $360\,\mathrm{m}/c_{\mathrm{vacuum}}=1.2\,\mu\mathrm{s}$

In an effort to go beyond Aspect et al.'s work we constructed an experiment that unites the aspects of large spatial separation (about 360 m) and fast switching [17]. Further we added randomness and total independence of the two measurement apparatus by devising a novel scheme of data collection that disposed of any link between the stations.[2] The randomness was drawn from fast physical random number generators with extremely short autocorrelation

[1] Meanwhile they extended their studies in the direction of measurements with moving detectors in order to check whether there exist preferred frames or any other connection between entanglement and relativity. Up to now these experiments did not show any measurable effect or deviation from ordinary quantum theory.

[2] This also removes a minor loophole that is opened by the "common context", which is created by an electronic link between the stations.

time. Here a definable time-scale comes into play that enables one to assess the relativistic ordering of events with reference to a certain "memory" of the observing stations.

Taking into account a few multiples of the autocorrelation time and all deterministic delays that occur in electronics, switches, and registration we were safe to say that our measurement was faster than a tenth of the time it would have taken any signal to travel to the other station at the vacuum speed of light.

In conclusion I think that it is safe to say that today no loopholes are open in the experimental tests of Bell's theorem, although no experiment succeeded in closing all of them simultaneously. Given the enormous interest that we experience now for entanglement it might well be that new experiments will try to push the limits ever more into the extreme.

## References


1. S.J. Freedman, J.F. Clauser, Phys. Rev. Lett. **28**, 938 (1972)
2. R.A. Holt, F.M. Pipkin, Harvard Univ. preprint (1974)
3. E.S. Fry, R.C. Thompson, Phys. Rev. Lett. **37**, 465 (1976)
4. J.S. Bell, J. Phys. (Paris) **42**, C2 41 (1981)
5. M.A. Rowe et al., Nature **409**, 791 (2001)
6. J.F. Clauser, M.A. Horne, Phys. Rev. D **10**, 526 (1974)
7. P.H. Eberhard, Phys. Rev. A **47**, R747 (1993)
8. C. Kurtsiefer, M. Oberparleiter, H. Weinfurter, Phys. Rev. A **64**, 023802 (2001)
9. P.G. Kwiat et al., Phys. Rev. A **48**, R867 (1993)
10. P.G. Kwiat et al., Appl. Opt. **33**, 1844 (1994)
11. J. Kim, Y. Yamamoto, H.H. Hogue, Appl. Phys. Lett. **70**, 2852 (1997)
12. S. Takeuchi, J. Kim, Y. Yamamoto, H.H. Hogue, Appl. Phys. Lett. **74**, 1063 (1999)
13. J. Kim, S. Takeuchi, Y. Yamamoto, H.H. Hogue, Appl. Phys. Lett. **74**, 902 (1999)
14. E.S. Fry, T. Walther, S. Li, Phys. Rev. A **52**, 4381 (1995)
15. W. Tittel, J. Brendel, H. Zbinden, N. Gisin, Phys. Rev. Lett. **81**, 3563 (1998)
16. A. Aspect, J. Dalibard, G. Roger, Phys. Rev. Lett. **49**, 1804 (1982)
17. G. Weihs et al., Phys. Rev. Lett. **81**, 5039 (1998)

# 11 The EPR Paradox in Massive Systems or about Strange Particles

Reinhold A. Bertlmann, Walter Grimus and Beatrix C. Hiesmayr

We give an introduction to an entangled massive system, specifically the neutral-kaon system, which has similarities to the entangled two-photon system, but also challenging differences.

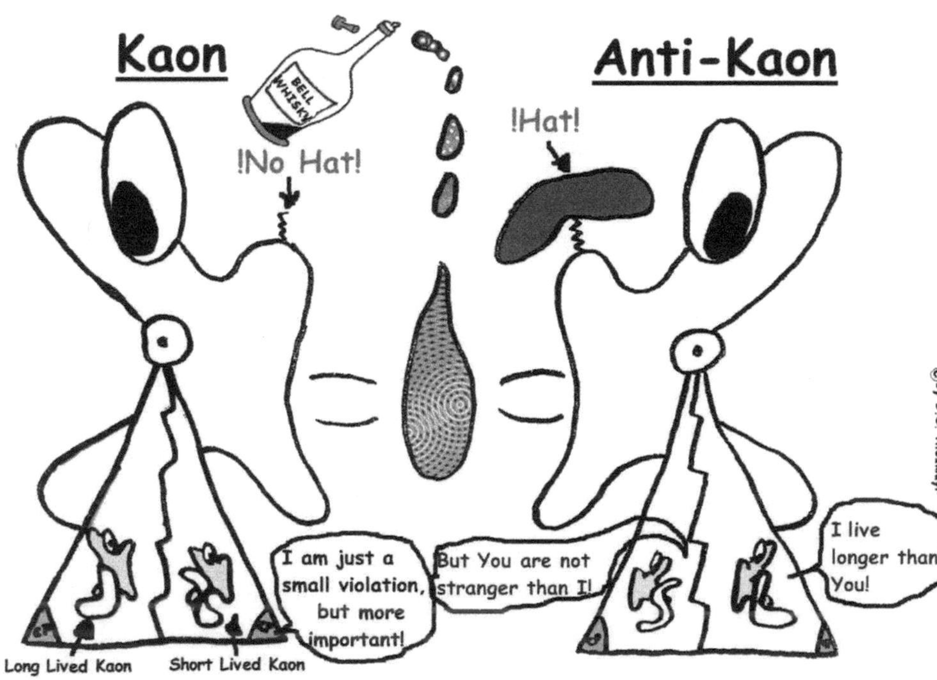


This shows the production of two entangled neutral kaons. The hat symbolizes the strangeness eigenstate, thus it is a particle or an anti-particle. The ties symbolize the superposition of a strangeness eigenstate of two mass eigenstates, the short-lived one and the long-lived one. The little triangles in the ties stand for the $CP$ violation which occurs in this system. The whisky bottle is copied from [1], which was a response to the article by John S. Bell, "Bertlmann's socks and the nature of reality" [2]

## 11.1 Introduction

As it is well known John Bell's questioning of orthodox quantum mechanics was just his "hobby", and it is this hobby *John Bell* is most famous for [3, 4]. Though he was an authority at his working place, CERN, with his hobby he was rather isolated there. It took about 30 years before Bell inequalities were investigated in the particle-physics community. His broad knowledge and his deep understanding of physics were very impressive. He wrote many classical papers in different areas, as was witnessed at the Bell Conference 2000, held in Vienna in honour of him, or as can be witnessed by the book you hold in your hands. One of his classical papers, written in 1965 together with *Jack Steinberger* [5], was about $CP$ violation of the neutral kaon system, and curiously, nowadays precisely this system yields an opportunity to investigate Bell inequalities in massive systems. Compared to photons, kaons are decaying and have $CP$ violation, which gives new features to the original EPR paradox.

## 11.2 The Bell–CHSH Inequality for Photons and for Kaons

*Before we explain the formalism of neutral kaons, we compare the Bell–Clauser–Horne–Shimony–Holt inequality for photons with the one for neutral kaons. We emphasize why the neutral-kaon system can be considered to have an EPR-like correlation, but also show the differences.*

The neutral kaon and its anti-particle can be distinguished by the strangeness number, $S$. This quantum number was introduced by Gell–Mann and Nishijima in 1953 to solve the strange behaviour of these particles. They are produced nearly as often as pions, but they live long enough to travel a measurable distance – several centimeters. This new quantum number is conserved by the strong interaction, but violated by the weak interaction, which is responsible for the decay of the kaons. For this section it is important to know that due to the strong interaction we can distinguish between the particle $K^0$, having $S = 1$, and its anti-particle $\bar{K}^0$, having $S = -1$. A further nice feature of the neutral-kaon system is that one of the four Bell states, the antisymmetric spin-singlet state or the polarisation $|\psi^-\rangle \sim |H\rangle|V\rangle - |V\rangle|H\rangle$ state, can be produced:

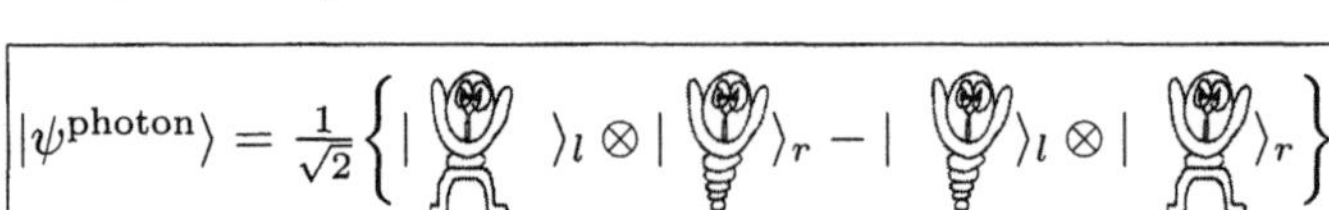
,

.

Starting from that initial state for the photons we want to derive the probability of detecting on the left ($l$) side a photon behind a linear polarisator, whose optical axis is rotated by the angle $\alpha$, and on the right ($r$) side a photon, whose optical axis is rotated by the angle $\beta$. For neutral kaons we then similarly have to derive the probability of finding a neutral kaon, $K^0$, on the left side at time $t_l$ and a $K^0$ on the right side but at time $t_r$.

|  $P(\alpha,\beta)$ $= \frac{1}{4}\{1-\cos[2(\beta-\alpha)]\}$ , | $P(K^0 t_l, K^0 t_r)$ $= \frac{1}{8}\Big\{e^{-\gamma_S t_l - \gamma_L t_r} + e^{-\gamma_L t_l - \gamma_S t_r}$ $-2\cos[\Delta m(t_l - t_r)]\cdot e^{-\gamma(t_l+t_r)}\Big\}$  |
|---|---|

Where $\gamma_S$ and $\gamma_L$ are the decay widths of the two mass eigenstates $\left[\gamma = \frac{1}{2}(\gamma_S+\gamma_L)\right]$ and $\Delta m = m_L - m_S$ is the mass difference of these states. We see immediately that if we forbid the kaons to decay we obtain the following probability:

|  $P(\alpha,\beta)$ $= \frac{1}{4}\{1-\cos[2(\beta-\alpha)]\}$ , | No decays allowed: $\gamma_S = \gamma_L = 0$ , $P(K^0 t_l, K^0 t_r)$ $= \frac{1}{4}\{1-\cos[\Delta m(t_l - t_r)]\}$  |
|---|---|

Thus the probability of finding a $K^0$ on each side at the same time $t_l = t_r$ is zero, whereas the probability of finding a $K^0$ on one side and a $\bar{K}^0$ on the other is one, completely analog to the photon case if we choose $\alpha = \beta$. This is the reason why we can speak of an EPR-like correlation of this massive system, but note that it is only true for equal times on the left and right sides.

*So we learn that the time difference, $\Delta t = t_l - t_r$, in the kaon system plays a role similar to the angle difference, $\phi_{\alpha,\beta} = \beta - \alpha$, in the photon system. For $\Delta t = 0$ we have this peculiar strong EPR correlation; upon finding a kaon on the right side, a measurement at the same time on the left side will definitely indicate an anti-kaon (of course, only if this kaon did not yet decay).*

We have seen the similarities of the photon and neutral-kaon systems. So now, a natural question arises:

*Can we find a Bell–CHSH inequality for the kaons – similar to that for the photons – in order to distinguish in particle physics between local realistic theory and quantum theory?*

The derivation of the Bell–CHSH inequality based on local realistic theory (LRT) in the kaon case is quite similar to that for photons; however, to then calculate the quantum-mechanical expectation value is not so straightforward, because one has to include the decay states. We will discuss this in the Sect. 11.5. Here only the result and its interesting implications are presented.

The CHSH inequality for the photon system is [20, 21]

$$S^{\text{photon}}(\phi_{\alpha,\beta}, \phi_{\alpha,\gamma}, \phi_{\delta,\beta}) = |\cos(\phi_{\alpha,\beta}) - \cos(\phi_{\alpha,\gamma})| + |\cos(\phi_{\delta,\beta}) + \cos(-\phi_{\alpha,\beta} + \phi_{\alpha,\gamma} + \phi_{\delta,\beta})| \leq 2\,,$$

with $\phi_{\alpha,\beta} = 2(\beta - \alpha)$.

The CHSH inequality for the kaon system is [23]

$$S^{\text{kaon}}(t_a, \phi_{a,b}, \phi_{a,c}, \phi_{d,b}) = |\cos(\phi_{a,b}) \cdot \mathrm{e}^{-\gamma(t_a+t_b)} - \cos(\phi_{a,c}) \cdot \mathrm{e}^{-\gamma(t_a+t_c)}| + |\cos(\phi_{d,b}) \cdot \mathrm{e}^{-\gamma(t_d+t_b)} + \cos(-\phi_{a,b} + \phi_{a,c} + \phi_{d,b}) \cdot \mathrm{e}^{-\gamma(t_d+t_c)}| \leq 2\,,$$

with $\phi_{a,b} = \Delta m\,(t_b - t_a)$.

Now let us analyze these two $S$ functions. Note that in the kaon case we actually have four free parameters to choose. To see if the left-hand side of the inequality, the $S$ function, becomes bigger than two, all we have to do is to find the maximums. The maximal violation of the $S^{\text{photon}}$ function is $2\sqrt{2}$. But what about the kaon system? The result for the following choice of parameters is

$$S^{\text{photon}}(\tfrac{3\pi}{4}, \tfrac{\pi}{4}, \tfrac{\pi}{4}) = 2.828\,, \qquad S^{\text{kaon}}(0, \tfrac{3\pi}{4}, \tfrac{\pi}{4}, \tfrac{\pi}{4}) = 0.426\,.$$

Thus in this case we do not find any violation (larger $t_a$'s make the result even worse). As the $S^{\text{kaon}}$ function is too complicated to analyze analytically, one has to handle it numerically [23, 14]. One finds the highest value for the following choices

$$S^{\text{photon}}(\tfrac{3\pi}{4}, \tfrac{\pi}{2}, 0) = 2.414\,, \qquad S^{\text{kaon}}(0, \tfrac{3\pi}{4}, \tfrac{\pi}{2}, 0) = 1.362\,.$$

But this is still not bigger than two! There is no choice for the four free parameters which lets $S^{\text{kaon}}$ become bigger than two; thus there is no way to distinguish between an LRT and quantum theory. It turns out because the quantity $x = 2\Delta m/\gamma_S$ is about one in the kaon system and not about a factor of 4.3 bigger, *no* violation of the CHSH inequality is possible, just for theoretical reasons[1].

[1] The quantity $x$ expresses the interplay of the strangeness oscillation, $\Delta m$, and the decay constant, $\gamma_S$ (see Sect. 11.3).

*Is there really no way to distinguish between a LRT and quantum mechanics?*

As we show in this paper, there is another way to obtain an answer to that tricky question, namely through a decoherence approach. In Sect. 11.4 we will develop that approach in detail. There also exist different Bell inequalities for neutral kaons, using other properties of the neutral-kaon system, for instance, the $CP$ violation; this we show in Sect. 11.5. Last but not least we connect both approaches in Sect. 11.6. However, to understand the following sections additional information on kaons and their properties is needed; the next section gives such an overview.

## 11.3 A Neutral Kaon Introduces Itself

*Here we are going to learn something about the strange particles we have been talking about.*

There exist two neutral kaons $K^0$ and $\bar{K}^0$ in addition to the two charged kaons $K^{\pm}$ with $S = \pm 1$, which form the following isospin doublets:

| | $I_3$ | |
|---|---|---|
| $S$ | $+\frac{1}{2}$ | $-\frac{1}{2}$ |
| $+1$ | $K^+$ | $K^0$ |
| $-1$ | $\bar{K}^0$ | $K^-$ |

.

Thus the neutral kaon is not its own anti-particle; the strong interaction can be used to distinguish between the particle and anti-particle.

If one studies the strong interactions of the neutral kaons, one finds that $K^0$ and $\bar{K}^0$ are pseudoscalar particles; hence the parity operator $P$ acts on the neutral kaon as follows:

$$P\,|K^0\rangle = -\,|K^0\rangle\,,$$
$$P\,|\bar{K}^0\rangle = -\,|\bar{K}^0\rangle\,. \tag{11.1}$$

The charge conjugation, $C$, transforms a neutral kaon, $K^0$, to its anti-particle, $\bar{K}^0$; the phases can be defined as follows:

$$C\,|K^0\rangle = |\bar{K}^0\rangle\,,$$
$$C\,|\bar{K}^0\rangle = |K^0\rangle\,. \tag{11.2}$$

For the product $CP$ one has

$$CP\,|K^0\rangle = -\,|\bar{K}^0\rangle\,,$$
$$CP\,|\bar{K}^0\rangle = -\,|K^0\rangle\,. \tag{11.3}$$

The decay of the $K$-meson is a weak process. We know that the weak interactions neither conserve the strangeness, $S$, nor the parity, $P$, nor the charge conjugation, $C$. In the most cases the neutral kaons decay into two pions. Both $K^0$ as well as $\bar{K}^0$ can produce the same final state:

$$\begin{aligned} K^0 &\Rightarrow \pi^+\pi^-,\ \pi^0\pi^0, \\ \bar{K}^0 &\Rightarrow \pi^+\pi^-,\ \pi^0\pi^0. \end{aligned} \tag{11.4}$$

*From that moment on it was clear that the weak interactions induce higher-order transitions between $K^0$ and $\bar{K}^0$!*

Thus the decay process of the neutral kaons has to be handled as a two-state system, $K^0 - \bar{K}^0$. A formalism for the decay mechanism of an unstable state which is degenerate with one or more other states is called the Wigner–Weisskopf approximation (see for example [6, 7]).
We deal with an effective Schrödinger equation,

$$\mathrm{i}\frac{\partial}{\partial t}\,|\psi(t)\rangle = \mathcal{H}\,|\psi(t)\rangle\,, \tag{11.5}$$

where $\mathcal{H}$ is an operator in the two-dimensional space of the neutral kaons and is in general non-Hermitian. One can show that this operator can be separated into Hermitian and non-Hermitian parts. These matrices are the generalization of the mass and decay-width of a decaying particle without state-mixing.

Now we consider the eigenvalue problem of $\mathcal{H}$. The eigenvectors will be called $|K_S\rangle$ and $|K_L\rangle$ with the eigenvalues $\lambda_S$, $\lambda_L$:

$$\begin{aligned} \mathcal{H}\,|K_S\rangle &= \lambda_S\,|K_S\rangle\,, \\ \mathcal{H}\,|K_L\rangle &= \lambda_L\,|K_L\rangle\,. \end{aligned} \tag{11.6}$$

Because $\mathcal{H}$ is non-Hermitian, so neither are the eigenvalues real nor are the eigenstates necessarily orthogonal, one denotes

$$\begin{aligned} \lambda_S &= m_S - \frac{\mathrm{i}}{2}\gamma_S\,, \\ \lambda_L &= m_L - \frac{\mathrm{i}}{2}\gamma_L\,, \end{aligned} \tag{11.7}$$

where $m_S, m_L, \gamma_S, \gamma_L$ are all real. The time evolution in this basis is exponential:

$$\begin{aligned} |K_S(t)\rangle &= \mathrm{e}^{-\mathrm{i}\lambda_S t}\,|K_S\rangle\,, \qquad \text{with} \qquad \lambda_S = m_S - \frac{\mathrm{i}}{2}\gamma_S\,, \\ |K_L(t)\rangle &= \mathrm{e}^{-\mathrm{i}\lambda_L t}\,|K_L\rangle\,, \qquad \text{with} \qquad \lambda_L = m_L - \frac{\mathrm{i}}{2}\gamma_L\,. \end{aligned} \tag{11.8}$$

The experimental values of these quantities are

$$\begin{aligned} \tau_S &= \frac{1}{\gamma_S} = (0.8935 \pm 0.0008) \times 10^{-10}\mathrm{s}\,, \\ \tau_L &= \frac{1}{\gamma_L} = (5.17 \pm 0.04) \times 10^{-8}\mathrm{s}\,, \\ \Delta m &= m_L - m_S = (0.5300 \pm 0.0012) \times 10^{10}\mathrm{s}^{-1}\,. \end{aligned} \tag{11.9}$$

There exist two different mass eigenstates, the short-lived eigenstate, $|K_S\rangle$, and the long-lived eigenstate, $|K_L\rangle$, which lives about 600 times longer than the short-lived kaon, $|K_S\rangle$. The mass difference, $\Delta m$, is responsible for the strangeness oscillation; thus an initial kaon transforms with a certain probability into an anti-kaon.

In (11.3) we see that both the kaon and the anti-kaon transform odd under the $CP$ operation. Now it is easy to construct the $CP$ eigenstates:

$$|K_1^0\rangle = \frac{1}{\sqrt{2}}\left(|K^0\rangle - |\bar{K}^0\rangle\right) ,$$
$$|K_2^0\rangle = \frac{1}{\sqrt{2}}\left(|K^0\rangle + |\bar{K}^0\rangle\right) . \tag{11.10}$$

In the decay processes it was observed that the short-lived kaon decayed into 2 pions and the long-lived kaon decayed into 3 pions. The two pions are in a $CP = +1$ state, and the three-pion state has $CP = -1$. So it was natural to identify the short-lived state, $|K_S\rangle$, with the $CP = +1$ state, $|K_1^0\rangle$, and the long-lived state, $|K_L\rangle$, with $|K_2^0\rangle$.

However, in 1964 the famous experiment by Christenson, Cronin, Fitch and Turlay [8] demonstrated that the long-lived kaon also decays into two pions.

*This means $CP$ symmetry is broken!*

Anyway the value of the $CP$ symmetry, the $CP$ parameter, $\varepsilon$, is small, about $10^{-3}$.

For the neutral kaons this means that we have the following physically important quasispin eigenstates[2]:

$$\begin{aligned} &|K^0\rangle , && |\bar{K}^0\rangle , \\ &|K_S\rangle = \frac{1}{N}\left(p|K^0\rangle - q|\bar{K}^0\rangle\right) , && |K_L\rangle = \frac{1}{N}\left(p|K^0\rangle + q|\bar{K}^0\rangle\right) , \\ &|K_1^0\rangle = \frac{1}{\sqrt{2}}\left(|K^0\rangle - |\bar{K}^0\rangle\right) , && |K_2^0\rangle = \frac{1}{\sqrt{2}}\left(|K^0\rangle + |\bar{K}^0\rangle\right) , \end{aligned} \tag{11.11}$$

with $p = 1 + \varepsilon, q = 1 - \varepsilon$ and $N^2 = |p|^2 + |q|^2$.

## 11.4 The Experiment at CERN and Possible Decoherence

*In this section we introduce the CPLEAR experiment performed at CERN. We describe the hypothesis that spontaneous decoherence of the wave function takes places and illustrate the consequences. With the help of the data from*

[2] The generalized definition of a quasispin eigenstate is an arbitrary superposition of the strangeness eigenstates.

*the CPLEAR experiment, we estimate the possible values for the "decoherence parameter", $\zeta$.*

This experiment [9] was performed in 1998. The $K^0\bar{K}^0$ pairs were produced in a $J^{PC} = 1^{--}$ state – the one corresponding to the $|\psi^-\rangle$ state of the photons – by proton–anti-proton annihilation. So the initial state of that production is

$$|\psi(t=0)\rangle = \frac{1}{\sqrt{2}}\left(|K^0\rangle_l \otimes |\bar{K}^0\rangle_r - |\bar{K}^0\rangle_l \otimes |K^0\rangle_r\right) , \tag{11.12}$$

and with the information from Sect. 11.3 we can rewrite this state in the mass eigenstate basis:

$$|\psi(t=0)\rangle = \frac{1}{2\sqrt{2}pq}\left(|K_S\rangle_r \otimes |K_L\rangle_l - |K_L\rangle_r \otimes |K_S\rangle_l\right) . \tag{11.13}$$

The CPLEAR group constructed two different setups, shown in Fig. 11.1.

The first one is called the $C(0)$ configuration; here each kaon travels 2 cm, thus the time difference, $\Delta t$, of the two flying paths is zero. This is the EPR-like case; by detecting on the right side a kaon with $S = +1$ implies that on the left side there is no such kaon with $S = +1$. The second configuration is called $C(5)$, because now the time difference is about 5 cm; one kaon is detected after a flying path of 2 cm, and the second one after 7 cm.

What the CPLEAR group measured was the difference between the like-strangeness events and the unlike-strangeness events, i.e., how often they detected the two kaons with $S = 1$ or with $S = -1$ (like-strangeness events) and how often they detected one kaon with $S = 1$ and the other one with $S = -1$ (unlike-strangeness events).

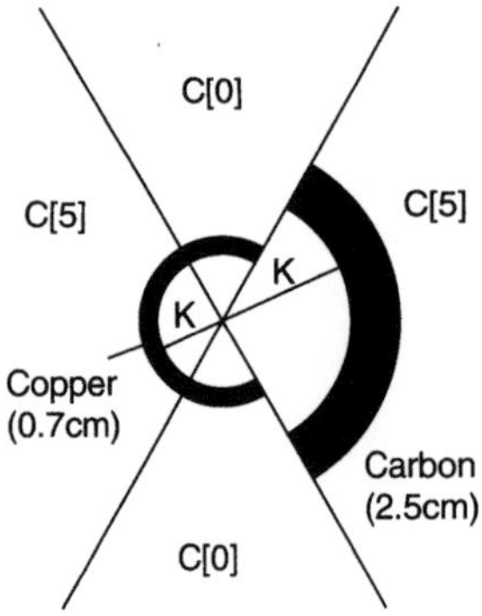


**Fig. 11.1.** A section through the experimental construction of the CPLEAR experiment at CERN in 1998. The $C(0)$ region shows the EPR-like configuration. Both kaons have nearly equal proper times when they interact with the copper absorber which is about 2 cm away from the creation point. The $C(5)$ configuration shows a proper time difference of $\Delta t = 5$ cm; the carbon absorber is about 7 cm away

This quantity is normalized by the sum of these two probabilities and is called the asymmetry term:

$$A^{\mathrm{QM}}(t_l, t_r) = \frac{P_{\mathrm{unlike}}(t_l, t_r) - P_{\mathrm{like}}(t_l, t_r)}{P_{\mathrm{unlike}}(t_l, t_r) + P_{\mathrm{like}}(t_l, t_r)}. \tag{11.14}$$

In quantum mechanics such a like-strangeness probability, $P_{\mathrm{like}}(t_l, t_r)$[3], is calculated in a straightforward way:

$$\begin{aligned} P_{\mathrm{like}}(t_l, t_r) &= ||\langle K^0|_l \otimes \langle K^0|_r \, |\psi(t_l, t_r)\rangle||^2 \\ &= ||\langle \bar{K}^0|_l \otimes \langle \bar{K}^0|_r \, |\psi(t_l, t_r)\rangle||^2 \\ &= \frac{1}{8}\Big[\mathrm{e}^{-\gamma_S t_l - \gamma_L t_r} + \mathrm{e}^{-\gamma_L t_l - \gamma_S t_r} \\ &\qquad -2\cos(\Delta m \Delta t)\cdot \mathrm{e}^{-\gamma(t_l + t_r)}\Big], \end{aligned} \tag{11.15}$$

with $\Delta t = t_l - t_r$ [for a definition of $\psi(t_l, t_r)$ see (11.31)]. Measuring the same strangeness on both sides at the same time, $t_l = t_r$, has probability zero. Similarly the unlike-strangeness probability yields the result

$$\begin{aligned} P_{\mathrm{unlike}}(t_l, t_r) &= ||\langle K^0|_l \otimes \langle \bar{K}^0|_r \, |\psi(t_l, t_r)\rangle||^2 \\ &= ||\langle \bar{K}^0|_l \otimes \langle K^0|_r \, |\psi(t_l, t_r)\rangle||^2 \\ &= \frac{1}{8}\Big[\mathrm{e}^{-\gamma_S t_l - \gamma_L t_r} + \mathrm{e}^{-\gamma_L t_l - \gamma_S t_r} \\ &\qquad +2\cos(\Delta m \Delta t)\cdot \mathrm{e}^{-\gamma(t_l + t_r)}\Big]. \end{aligned} \tag{11.16}$$

We note that the interference term, the strangeness oscillation, has changed its sign; so for equal time measurements this unlike-strangeness probability is maximal. Inserting the two probabilities in our asymmetry term (11.14) gives

$$A^{\mathrm{QM}}(\Delta t) = \frac{\cos(\Delta m \Delta t)}{\cosh(\frac{1}{2}\Delta\gamma \Delta t)}, \qquad \text{with} \quad \Delta\gamma = (\gamma_L - \gamma_S). \tag{11.17}$$

The asymmetry term depends only on the time difference, $\Delta t$, and is direct proportional to the interference term, the strangeness oscillation. This was the quantity which was tested in the CPLEAR experiment. The experimental results for the two configurations of the asymmetry term are as follows:

| | Experiment | Theory (corrected) |
|---|---|---|
| $C(0)$ | $0.81 \pm 0.17$ | 0.93 |
| $C(5)$ | $0.48 \pm 0.12$ | 0.56 |

[3] The $CP$ violation is neglected in the whole section, because it gives only corrections of the order $10^{-3}$ and this is far removed from the experimental accuracy.

They are compared with the theoretical values corrected according to the experimental configuration. The experimental values agree within one standard deviation with quantum theory. And the peculiar quantum entanglement seems to be confirmed.

*But is it really that way?*

Imagine that the initial state (11.13) collapses spontaneously into its components immediately after its creation, e.g.,

$$
\begin{aligned}
|\psi\rangle \quad &\Longrightarrow \quad |K_S\rangle_l \otimes |K_L\rangle_r \qquad \text{in half of the cases,} \\
|\psi\rangle \quad &\Longrightarrow \quad |K_L\rangle_l \otimes |K_S\rangle_r \qquad \text{in the other half.}
\end{aligned}
\tag{11.18}
$$

This factorization of the wave function is called Furry's hypothesis[4] [10].

We will now modify the calculation of the probabilities by considering that we have on one hand the quantum-mechanical probability and on the other hand Furry's hypothesis or spontaneous factorization of the initial wave function. For this we again look at the derivation of the quantum-mechanical like-strangeness probability (11.15) and modify it in the following way [13, 14]:

$$
\begin{aligned}
&P_{\text{like}}(t_l, t_r) = \| \langle K^0|_l \otimes \langle K^0|_r \, |\psi(t_l, t_r)\rangle \|^2 \quad \longrightarrow \quad P^{\zeta}_{\text{like}}(t_l, t_r) \\
&= \frac{1}{2}\Big[ \mathrm{e}^{-\gamma_S t_l - \gamma_L t_r} |\langle K^0|K_S\rangle_l|^2 \, |\langle K^0|K_L\rangle_r|^2 \\
&+ \mathrm{e}^{-\gamma_L t_l - \gamma_S t_r} |\langle K^0|K_L\rangle_l|^2 \, |\langle K^0|K_S\rangle_r|^2 \\
&-2 \underbrace{(1-\zeta)}_{\text{modification}} \, \mathrm{Re}\left( \langle K^0|K_S\rangle_l^* \, \langle K^0|K_L\rangle_r^* \langle K^0|K_L\rangle_l \, \langle K^0|K_S\rangle_r \mathrm{e}^{+\mathrm{i}\Delta m \Delta t} \right) \cdot \mathrm{e}^{-\gamma(t_l + t_r)} \Big] \\
&= \frac{1}{8}\Big[ \mathrm{e}^{-\gamma_S t_l - \gamma_L t_r} + \mathrm{e}^{-\gamma_L t_l - \gamma_S t_r} - 2 \underbrace{(1-\zeta)}_{\text{modification}} \cos(\Delta m \Delta t) \cdot \mathrm{e}^{-\gamma(t_l + t_r)} \Big] .
\end{aligned}
\tag{11.19}
$$

Thus we have multiplied the interference term of the decay amplitudes by the factor $(1-\zeta)$. For $\zeta = 0$ we have the quantum-mechanical expression for the like-strangeness probability, but if $\zeta = 1$, the quantum-mechanical interference term of the decay amplitudes vanish, and the wave function factorizes as shown in (11.18). We actually found a way to have both the quantum-mechanical result on one hand and the factorization on the other hand, just depending on the choice of the decoherence parameter, $\zeta$.

All we have now to do is to recalculate the asymmetry term (11.14) with this simple modification and compare it with the measurement results of the CPLEAR experiment.

---

[4] Actually it should be called Schrödinger's hypothesis, because first he stated it a full year earlier [11] and second he remarked that it could really happen, to be re-read in [12]. Clauser in his contribution to this book refers to it as Schrödinger–Furry hypothesis.

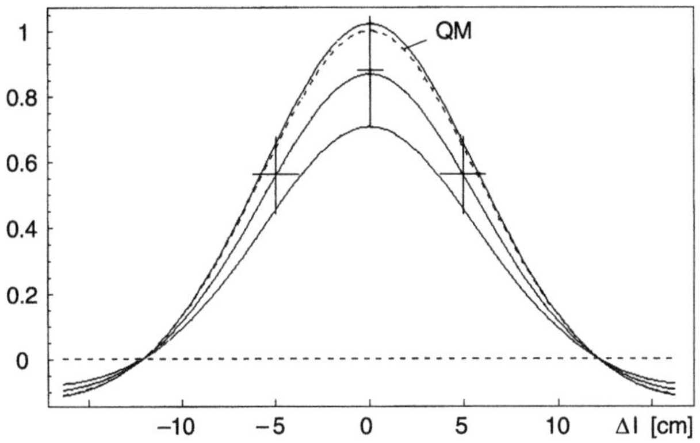


**Fig. 11.2.** The asymmetry (11.20) as a function of the difference in the distances traveled by the kaons to the points where their strangeness is measured. The *dashed curve* corresponds to quantum mechanics with the decoherence parameter, $\zeta = 0$, whereas the *solid curves* correspond to the values of $\zeta$ obtained by the fit (11.21) to the CPLEAR data. The two data points represented by the *crosses* have been taken from [9]. The *horizontal dashed line* indicates zero asymmetry for $\zeta = 1$, the consequence of Furry's hypothesis with respect to the $K_S K_L$ basis

The asymmetry term reads

$$A_\zeta(\varDelta t) = A^{\mathrm{QM}}(\varDelta t) \cdot (1 - \zeta)\,; \tag{11.20}$$

thus it is only linearly affected by our modification. Obviously, for $\zeta$ equal to zero, we have the quantum-mechanical result, but for $\zeta$ equal to one, the asymmetry term (11.20) vanishes for all time differences. We have already considered the results of the CPLEAR experiment and have seen that they did not measure values equal to zero; however, if we compare both the experimental values and their corresponding uncertainties with the modified asymmetry term, we obtain the following fit result[5]:

$$\bar{\zeta} = 0.13^{+0.16}_{-0.15}\,. \tag{11.21}$$

The result is shown in Fig. 11.2.

Now we could conclude that no such factorization is possible, due to the data from the CPLEAR experiment.

*But is this really true?*

Lets go back to (11.18); why should it actually factorize this way, why not:

$$\begin{aligned} |\psi\rangle \quad &\Longrightarrow \quad |K^0\rangle_l \otimes |\bar{K}^0\rangle_r \qquad \text{in half of the cases,} \\ |\psi\rangle \quad &\Longrightarrow \quad |\bar{K}^0\rangle_l \otimes |K^0\rangle_r \qquad \text{in the other half.} \end{aligned} \tag{11.22}$$

[5] Of course fitting the experimental results with such a modified theory requires more considerations, which have been made and explained in [13, 14]; however, the correct result with a confidence level (CL) of 97% is given here.

In quantum theory one has a free choice for the basis, so it is natural to also consider such a case. For that we have to start from the initial state in the $K^0\bar{K}^0$ basis choice (11.12) and then recalculate the modified probabilities:

$$
\begin{aligned}
&P_{\text{like}}(t_l,t_r) = \|\langle K^0|_l \otimes \langle K^0|_r \, |\psi(t_l,t_r)\rangle\|^2 \quad \longrightarrow \qquad\qquad (11.23)\\
&P^{\zeta}_{\text{like}}(t_l,t_r) = \\
&\quad \frac{1}{2}\Big[|\langle K^0|K^0(t_l)\rangle_l|^2\,|\langle K^0|\bar{K}^0(t_r)\rangle_r|^2 + |\langle K^0|\bar{K}^0(t_l)\rangle_l|^2\,|\langle K^0|K^0(t_r)\rangle_r|^2\\
&\quad -2\underbrace{(1-\zeta)}_{\text{modification}}\mathrm{Re}\big(\langle K^0|K^0(t_l)\rangle_l^*\langle K^0|\bar{K}^0(t_r)\rangle_r^*\langle K^0|\bar{K}^0(t_l)\rangle_l\langle K^0|K^0(t_r)\rangle_r\big)\Big]\,.
\end{aligned}
$$

It is clear that the decay amplitudes starting with the $K_S K_L$ basis differ from the ones with the $K^0\bar{K}^0$ basis, since the interference term of (11.19) differs from the one in (11.23). Or put simply, the factorization in the $K_S K_L$ basis (11.18) is not equal to the factorization with the $K^0\bar{K}^0$ basis (11.22). In this sense the decoherence approach is basis dependent.

Now we can derive the asymmetry term with the decoherence parameter, $\zeta$, with the $K^0\bar{K}^0$ basis choice; we find

$$
A_{\zeta}^{K^0\bar{K}^0}(t_l,t_r) = \frac{\cos(\Delta m \Delta t) - \frac{1}{2}\zeta\,\{\cos(\Delta m\Delta t) - \cos\left[\Delta m(t_l+t_r)\right]\}}{\cosh(\frac{1}{2}\Delta\gamma\Delta t) - \frac{1}{2}\zeta\left\{\cosh(\frac{1}{2}\Delta\gamma\Delta t) - \cosh\left[\frac{1}{2}\Delta\gamma(t_l+t_r)\right]\right\}}\,. \tag{11.24}
$$

This asymmetry term is a little more complicated; it depends not only on the time difference, $\Delta t$, but also on the sum of the two involved times, and, moreover, the dependence on $\zeta$ is not linear. Again we compare this asymmetry term with the experimental data, the result[6] is

$$
\bar{\zeta} \sim 0.4 \pm 0.7. \tag{11.25}
$$

But this is curious! The quantum-mechanical result and Furry's hypothesis are within one standard deviation of each other. With this basis choice we cannot distinguish between quantum mechanics and a spontaneous factorization of the wave function!

### Résumé

*We have learned that Furry's hypothesis is basis dependent and so is our decoherence approach. With help of the CPLEAR data, we could restrict the degree of decoherence, although an arbitrariness of the basis choice remains.*

*However, what is essential is the existence of a basis where the $K^0\bar{K}^0$ system is far removed from total decoherence and the corresponding $\zeta$ is*

[6] The CL is about 67%, see [13, 14].

*close to zero, in agreement with quantum mechanics. We have seen that the "best basis" in this respect is the $K_S K_L$ basis. So in this sense we have demonstrated the existence of interference effects of massive particles over macroscopic distances (remember they were separated by several centimeters).*

Now naturally the question arises, of how this decoherence parameter is really connected to hidden-variable theories. Another question arises, regarding whether we somehow can use the other properties of the neutral kaons, which have no similarities to the photon system, and construct Bell inequalities. As curious as it is, the answer is yes, but with some limitations, as we will see now.

## 11.5 The Generalized Bell Inequality and Unitary Time Evolution

*Here we will derive a Bell–CHSH inequality based on locality, realism and induction for different times (this is then the one analyzed in Sect. 11.2) but also vary the quasispin eigenstates. We analyze this generalized Bell inequality for the different choices.*

When testing quantum mechanics against LRT we find analogies but also important physical differences between kaons and spin 1/2 particles[7] (for an alternative view see [15, 16]):

- While in the spin 1/2 case one can test whether a system is in an arbitrary state $\alpha|\Uparrow_n\rangle + \beta|\Downarrow_n\rangle$, there is no analogous way to test the arbitrary superposition $\alpha|K^0\rangle + \beta|\bar{K}^0\rangle$. However, as in [17, 18], we will assume that the mass and $CP$ eigenstates (11.11) can be measured by a Gedankenexperiment.
- While in the spin 1/2 case the direct product space $H^l_{\text{spin}} \otimes H^r_{\text{spin}}$ is sufficient to account for all spin properties of the entangled system, it is *not* true for the neutral-kaon case. Indeed, we want to emphasize that due to the unitary time evolution of the states one has to include the decay-product spaces which are orthogonal to the product space, $H^l_{\text{kaon}} \otimes H^r_{\text{kaon}}$. This leads to additional terms in the resulting Wigner-type inequality compared to other works [18, 19, 16] and compared to the photon case.

In the case of spin variables one can derive the common Bell–CHSH inequality [20, 21] for the averaged spin values along arbitrary quantization directions $n$ and $m$. The analogue in the kaon case is the free choice of the times at which measurements are performed and in addition the freedom of choosing the quasispin state of the meson, the strangeness eigenstate, the mass eigenstate or the $CP$ eigenstate.

[7] We will now talk about the spin 1/2 systems, because of the analogy to kaons; but the reader common with photons can just image the photon system.

*How can one derive a Bell–CHSH inequality in LRT for neutral kaons?*

Bell's locality assumption requires then that the results on one side be completely independent of the choice of the time and the choice of the quasispin eigenstates on the other side. To define the appropriate correlation functions to be used in a Bell inequality, we consider an observable, $O^l(k_n, t_a)$, on the left side, which has the value +1 if the measurement at time point $t_a$ gives the quasispin $k_n$ and the value −1 if the quasispin $k_n$ is not found. In terms of such an observable we can define the correlation function, $O(k_n t_a; k_m t_b)$, which takes the value +1 both when a $k_n$ at $t_a$ and a $k_m$ at $t_b$ are detected and when no $k_n$ and no $k_m$ are detected. In the case where only one of the desired quasispin eigenstates has been found, no matter on which side, the correlation function takes the value −1.

The locality assumption implies then that $O(k_n t_a, k_m t_b)$, in a specific individual experiment, equals the product of $O^l(k_n, t_a)$ and $O^r(k_m, t_b)$:

$$O(k_n t_a; k_m t_b) = O^l(k_n, t_a) \cdot O^r(k_m, t_b) \,. \tag{11.26}$$

It implies

$$\begin{aligned} &|O(k_n t_a; k_m t_b) - O(k_n t_a; k_{m'} t_c)| \\ &\qquad + |O(k_{n'} t_d; k_{m'} t_c) + O(k_{n'} t_d; k_m t_b)| = 2 \,, \end{aligned} \tag{11.27}$$

with $k_n, k_m, k_{m'}$ and $k_{n'}$ being arbitrary quasispin eigenstates of the meson and $t_a, t_b, t_c$ and $t_d$ four different times.

Let us now consider a sequence of $N$ identical measurements, and let us denote by $O_i$ the value taken by $O$ in the $i$th experiment. The average is given by

$$M(k_n t_a; k_m t_b) = \frac{1}{N} \sum_{i=1}^{N} O_i(k_n t_a; k_m t_b) \tag{11.28}$$

and satisfies the Bell–CHSH inequality

$$\begin{aligned} &|M(k_n t_a; k_m t_b) - M(k_n t_a; k_{m'} t_c)| + |M(k_{n'} t_d; k_{m'} t_c) + M(k_{n'} t_d; k_m t_b)| \\ &\leq \frac{1}{N} \sum_{i=1}^{N} \Big[ |O_i(k_n t_a; k_m t_b) - O_i(k_n t_a; k_{m'} t_c)| \\ &\quad + |O_i(k_{n'} t_d; k_{m'} t_c) + O_i(k_{n'} t_d; k_m t_b)| \Big] = 2 \,. \end{aligned} \tag{11.29}$$

Note that, by setting $M(k_n t_a; k_m t_b)$ equivalent to $M(\boldsymbol{a}, \boldsymbol{b})$, (11.29) reads exactly the same as the CHSH inequality for photons.

*How can one derive the appropriate quantum probabilities for neutral kaons?*

As we have emphasized, a unitary time evolution for the neutral kaons is necessary. Note that in the photon system we automatically have unitarity. Such a unitary time evolution for neutral kaons looks like

$$U(t,0)|K_{S,L}\rangle = \mathrm{e}^{-\mathrm{i}\lambda_{S,L} t}|K_S\rangle + |\Omega_{S,L}(t)\rangle \,, \tag{11.30}$$

where $|\Omega_{S,L}(t)\rangle$ describes the decay products of the neutral kaons and operates in a Hilbert space orthogonal to $H_{\text{kaon}}$. Thus we operate in a complete Hilbert space, analogous to the photon case. The time evolution of the initial state $|\psi(0)\rangle$ (11.13) is (for details please see [22])

$$\begin{aligned}|\psi(t_l,t_r)\rangle &= U_l(t_l)\otimes U_r(t_r)\,|\psi(0)\rangle \\ &= \left[\mathrm{e}^{-\mathrm{i}\lambda_S t_l}|K_S\rangle_l + |\Omega_S(t_l)\rangle_l\right]\otimes\left[\mathrm{e}^{-\mathrm{i}\lambda_L t_r}|K_L\rangle_r + |\Omega_L(t_r)\rangle_r\right] \\ &\quad - \left[\mathrm{e}^{-\mathrm{i}\lambda_L t_l}|K_L\rangle_l + |\Omega_L(t_l)\rangle_l\right]\otimes\left[\mathrm{e}^{-\mathrm{i}\lambda_S t_r}|K_S\rangle_r + |\Omega_S(t_r)\rangle_r\right]\,. \end{aligned} \tag{11.31}$$

Now the only thing we need to do to obtain, for example, the probability of finding the a special quasispin state, $|k_n\rangle$, on the left side at $t_l$ (a yes $[Y]$ event) and another quasispin state, $|k_m\rangle$, on the right side at $t_r$ ($[Y]$ event), is to apply the corresponding projection operators to that state (11.31) and square it:

$$P_{k_n,k_m}(Yt_l,Yt_r) = ||P_l(k_n)\otimes P_r(k_m)\,|\psi(t_l,t_r)\rangle||^2\,. \tag{11.32}$$

We can also ask for the probability of finding on the left side a $|k_n\rangle$ at $t_l$ and on the right side *no* $|k_m\rangle$ at $t_r$ (no $[N]$ event). This means for the experimenter that she has a detector for observing a $|k_m\rangle$ state, but she finds no such state at time $t_r$:

$$P_{k_n,k_m}(Yt_l,Nt_r) = ||P_l(k_n)\otimes(1-P_r(k_m))\,|\psi(t_l,t_r)\rangle||^2. \tag{11.33}$$

In this way our quantum-mechanical expectation value is given by

$$\begin{aligned}&M^{\mathrm{QM}}(k_nt_a;k_mt_b) = \\ &P_{n,m}(Yt_a,Yt_b)+P_{n,m}(Nt_a,Nt_b)-P_{n,m}(Yt_a,Nt_b)-P_{n,m}(Nt_a,Yt_b)\,.\end{aligned} \tag{11.34}$$

Further we can use the fact that the sum of the probabilities of the results $(Y,Y)$, $(N,N)$, $(Y,N)$ and $(N,Y)$ is unity for all times; so (11.34) can be rewritten as

$$M^{\mathrm{QM}}(k_nt_a;k_mt_b) = -1+2\left[P_{n,m}(Yt_a,Yt_b)+P_{n,m}(Nt_a,Nt_b)\right]. \tag{11.35}$$

By setting this expression into the Bell–CHSH inequality (11.29), we obtain the following inequality:

$$\begin{aligned}&S^{\mathrm{kaon}}(k_nt_a;k_mt_b;k_{n'}t_c;k_{m'}t_d) \\ &= |P_{n,m}(Yt_a,Yt_b)+P_{n,m}(Nt_a,Nt_b) \\ &\quad -P_{n,n'}(Yt_a,Yt_c)-P_{n,n'}(Nt_a,Nt_c)| \\ &\quad +|-1+P_{m',m}(Yt_d,Yt_b)+P_{m',m}(Nt_d,Nt_b) \\ &\quad +P_{m',n'}(Yt_d,Yt_c)+P_{m',n'}(Nt_d,Nt_c)|\;\leq\;1\,.\end{aligned} \tag{11.36}$$

This is our generalized Bell–CHSH inequality for neutral kaons! Now we have a lot of possibilities to choose considering that there are 8 parameters. Let us first choose them in a way such that we obtain the Bell–CHSH inequality of Sect. 11.2.

### 11.5.1 The Choice of the Strangeness Eigenstate

In this case we have to choose all quasispin eigenstates to be the same, e.g., the anti-kaon[8] [23]

$$k_n = k_m = k_{n'} = k_{m'} = \bar{K}^0 \,. \tag{11.37}$$

By calculating now all of the involved probabilities and inserting them into (11.36), we obtain the following result[9]:

$$\begin{aligned} &|\mathrm{e}^{-\frac{\gamma_S}{2}(t_a+t_b)} \cos[\Delta m(t_a - t_b)] - \mathrm{e}^{-\frac{\gamma_S}{2}(t_a+t_c)} \cos[\Delta m(t_a - t_c)]| \\ &+|\mathrm{e}^{-\frac{\gamma_S}{2}(t_d+t_b)} \cos[\Delta m(t_d - t_b)] + \mathrm{e}^{-\frac{\gamma_S}{2}(t_d+t_c)} \cos[\Delta m(t_d - t_c)]| \le 2 \,. \end{aligned} \tag{11.38}$$

Unfortunately, – as discussed in Sect. 11.2 – inequality (11.38) *cannot* be violated for any choice of the four (positive) times, $t_a, t_b, t_c$, and $t_d$, due to the interplay between the kaon decay width and the strangeness oscillation.

### 11.5.2 The Choice Sensitive to the $CP$ Violating Parameter, $\varepsilon$

Now we will set $t_a = t_b = t_c = t_d = 0$ and further choose $n' = m'$; the remaining states we choose in the following way:

$$\begin{aligned} |k_n\rangle &= |K_S\rangle \,, \\ |k_m\rangle &= |\bar{K}^0\rangle \,, \\ |k_{n'}\rangle &= |K_1^0\rangle \,, \end{aligned} \tag{11.39}$$

and if we denote the probabilities $P_{K_S,\bar{K}^0}(t_a = 0, t_b = 0)$ by $P(K_S, \bar{K}^0)$ etc. the generalized Bell–CHSH inequality (11.36) gives the following inequality:

$$P(K_S, \bar{K}^0) \le P(K_S, K_1^0) + P(K_1^0, \bar{K}^0) \,. \tag{11.40}$$

This is a Wigner-type inequality, originally found by *Uchiyama* [17] by a set theoretical approach. The interesting point here is its connection to a physical parameter, the $CP$ violating parameter, $\varepsilon$ (for a connection to the leptonic $CP$ asymmetry, see [24]). By calculating the above probabilities, this inequality is turned into an inequality for $\varepsilon$:

$$\mathrm{Re}(\varepsilon) \le |\varepsilon|^2. \tag{11.41}$$

The experimental value of $\varepsilon$, measured in experiments which had nothing to do with entangled states, has an absolute value of about $10^{-3}$ and a phase of about 45°. A quick calculation shows that — yes — this inequality is

---

[8] The particular choice of the anti-kaon has experimental reasons; this strangeness state reacts with matter stronger than the one with $S = 1$.

[9] $CP$ violation is neglected, and we have made $\gamma_L = 0$.

violated! So we have found a way of distinguishing between LRT and quantum mechanics.

But first we have to see if this inequality holds for different times. For $t = 0$ the situation is not really physical, the particles have not yet flown apart. So what happens if we let the system evolve in time?

Let us first set all times equal, $t_a = t_b = t_c = t$; we obtain the following inequality:

$$\begin{aligned} e^{-2\gamma t}\, P(K_S, \bar{K}^0) \leq\ & e^{-2\gamma t}\, P(K_S, K_1^0) + e^{-2\gamma t}\, P(K_1^0, \bar{K}^0) \\ & + h(K_S, \bar{K}^0, K_1^0; t)\,, \end{aligned} \tag{11.42}$$

where

$$\begin{aligned} h(K_S, \bar{K}^0, K_1^0; t) = & -P_{K_S,\bar{K}^0}(Nt; Nt) + P_{K_S,K_1^0}(Nt; Nt) \\ & + P_{K_1^0,\bar{K}^0}(Nt; Nt) + P_{K_1^0,K_1^0}(Nt; Nt)\,. \end{aligned} \tag{11.43}$$

The function $h$ is missing if one does not consider a unitary time evolution; the exponential factors, $e^{-2\gamma t}$, can be divided out of (11.42), and one would wrongly conclude that (11.41) holds for all times.

However, it turns out that (11.42) is only for times, $t < 8 \cdot 10^{-4} \tau_S$, violated due to the fast damping of the probabilities. Thus for larger times we again cannot distinguish between LRT and quantum mechanics.

But, fortunately, there exist certain cases where the situation is better. We can avoid a fast increase in the function $h$ by taking $t_a = t_c$ and $t_a \leq t_b$. Then a violation of the Bell–CHSH inequality occurs which is strongest for $t_a \approx 0$; in this case $t_b$ can be chosen up to $t_b \leq 4\tau_S$, which is really quite large.

The reader would surely agree that considering Bell inequalities for neutral kaons is a very strange thing!

## 11.6 Connection of the Bell Inequality and the Decoherence Approach

We have seen in Sect. 11.4 that with a simple modification of quantum theory we can achieve continuously the factorization of the wave function. What has this approach to do with LRTs, i.e., Bell inequalities? It is clear that for $\zeta = 0$ — thus quantum theory — Bell inequalities may not be fulfilled, but for $\zeta = 1$, which is clearly a local situation — no interference term exists between the two amplitudes — Bell inequalities are certainly satisfied.

*What can we say for $\zeta$ values between 0 and 1?*

Let's consider again (11.40) and recalculate it with the modified probabilities, then we will find a bound on $\zeta$. In this way we can relate the decoherence approach to an LRT.

The result in our 'best' basis choice $K_S K_L$ is [25]

$$\frac{\mathrm{Re}\,(\varepsilon) - |\varepsilon|^2}{\mathrm{Re}\,(\varepsilon) + 4\mathrm{Re}^2\,(\varepsilon) + |\varepsilon|^2} = 0.987 \leq \zeta\,. \tag{11.44}$$

So the decoherence parameter, $\zeta$, has to be close to one; hence, Furry's hypothesis or spontaneous factorization has to take place totally. This means in our case that the created initial state vector (11.13) factorizes in 50% of the cases in a short-lived state on the left side and in a long-lived state on the right side, and in the other 50% of the cases vice versa.

Intuitively, we would have expected that there exist LRTs which allow an interference term at least partially (see for instance [26, 27]). Our result demands a vanishing interference term; hence, the locality assumption underlying this inequality forces the $K_S K_L$ interference term to vanish.

But on the other hand we can compare this result with the experimental $\bar{\zeta}^{K_S K_L} = 0.13^{+0.16}_{-0.15}$ (11.21), where $\zeta = 1$ is excluded by many standard deviations.

*This means that for experimental reasons a local realistic variable theory equivalent to the $K_S K_L$ basis choice is excluded!*

However, the situation changes when using the $K^0 \bar{K}^0$ basis; then we cannot discriminate between quantum mechanics and Furry's hypothesis (for details see Ref. [25]).

## 11.7 Final Remark

The authors hope that they could give a short introduction into the EPR–Bell-like correlations in particle physics, demonstrating their similarities to the photon system, but more challenging the differences (for an analogous treatment of the beauty system see, for instance, [28, 29]). We want to emphasize that this field is a very young one and many experiments are still missing; however, the latter will be done in the near future.

## Acknowledgement

The authors want to thank the Austrian FWF, project P14143-PHY, and the Austrian–Czech Republic Scientific Collaboration, project KONTAKT 2001-11, for their support.

## References

1. R.A. Bertlmann, Found. Phys. **20**, 1191 (1990)
2. J.S. Bell, Bertlmann socks and the nature of reality, Article 16 of [3]

3. J.S. Bell, *Speakable and Unspeakable in Quantum Mechanics* (Cambridge Univ. Press, Cambridge 1987)
4. J.S. Bell, Phys. **1**, 195 (1964)
5. J.S. Bell, J. Steinberger, *Weak Interactions of Kaons*, Proceedings of the Oxford International Conference, 10–25 September 1965, reprinted in: *Quantum Mechanics, High Energy Physics and Accelerators*, ed. by. M. Bell, K. Gottfried, M. Veltman (World Scientific, Singapore 1995) p. 280
6. O. Nachtmann, *Phänomene und Konzepte der Elementarteilchenphysik* (Vieweg, Braunschweig 1986)
7. P.K. Kabir, *The CP Puzzle; Strange Decays of the Neutral Kaon* (Academic, New York 1968)
8. J.H. Christenson, J.W. Cronin, V.L. Fitch, R. Turlay, Phys. Rev. Lett. **13**, 138 (1964)
9. A. Apostolakis et al., CPLEAR Coll., Phys. Lett. B **422**, 339 (1998)
10. W.H. Furry, Phys. Rev. **49**, 393 (1936); ibid. **49**, 476 (1936)
11. E. Schrödinger, Naturwissenschaft **23**, 807, 823, 844 (1935)
12. J.E. Clauser, A. Shimony, Rep. Prog. Phys. **41**, 1881 (1978)
13. R.A. Bertlmann, W. Grimus, B.C. Hiesmayr, Phys. Rev. D **60**, 114 032 (1999)
14. B.C. Hiesmayr, Diploma Thesis, University of Vienna (1999)
15. B. Ancochea, A. Bramon, M. Nowakowski, Phys. Rev. D **60**, 094 008 (1999); A. Bramon, M. Nowakowski, Phys. Rev. Lett. **83**, 1 (1999)
16. N. Gisin, A. Go, Am. J. Phys. **69**, 264 (2001)
17. F. Uchiyama, Phys. Lett. A **231**, 295 (1997)
18. A. Di Domenico, Nucl. Phys. B **450**, 293 (1995)
19. F. Benatti, R. Floreanini, Eur. Phys. J. C **13**, 267 (2000)
20. J.S. Bell, Introduction to the hidden-variable question, Article 4 of [3]
21. J.E. Clauser, M.A. Horne, A. Shimony, R.A. Holt, Phys. Rev. D **10**, 880 (1969)
22. R.A. Bertlmann, B.C. Hiesmayr, Bell inequalities for entangled kaons and their unitary time evolution, Phys. Rev. A **63**, 062 112 (2001), hep-ph/0101356
23. G.C. Ghirardi, R. Grassi, T. Weber, in Proc. of Workshop on *Physics and Detectors for DA$\Phi$NE, the Frascati $\Phi$ Factory*, April 9–12, 1991, ed. by G. Pancheri (INFN - Laboratori Nazionali di Frascati, Frascati 1991)
24. R.A. Bertlmann, W. Grimus, B.C. Hiesmayr, Phys. Lett. A **289**, 21 (2001), quant-ph/0107022
25. B.C. Hiesmayr, Found. Phys. Lett. **14**, 231 (2001)
26. J. Six, Phys. Lett. A **150**, 243 (1990)
27. P. Privitera, F. Selleri, Phys. Lett. B **296**, 261 (1992); R. Foadi, F. Selleri, Phys. Lett. B **461**, 123 (1999); A. Afriat, F. Selleri, *The Einstein, Podolsky, and Rosen Paradox in Atomic, Nuclear and Particle Physics*, (Plenum, New York 1999)
28. R.A. Bertlmann, W. Grimus, Phys. Rev. D **58**, 034 014 (1998)
29. R.A. Bertlmann, W. Grimus, A model for decoherence of entangled beauty, Phys. Rev. D **64**, 056 004 (2001)

# Part III

# Quantum Information

# 12 Are There Measurements?

Stig Stenholm

This paper discusses the special role measurements play in quantum mechanics. It is argued that no such special events may occur in nature, but they must be part of our interpretation of experience. No formal frameworks, be they sophisticated mathematical constructions or human-based information interpretations, can have any power to structure reality; they affect only our symbolic mode of representation of experience. But this representation works to guide our actions in reality, and hence it must convey some knowledge about what there really is. Thus we end up with an old dualiy: the unreachable thing in itself and our man-made descriptions thereof. It is argued that such an approach is inherent in the view of Niels Bohr, who separated classical objects from quantum entities. The Copenhagen interpretation is thus valid within it own positivistic framework, but it is not satisfactory as a description of reality. Today we want to go beyond this stage and let new experiments guide our understanding of physical reality through suitable extensions and elaborations of the language of description.

## 12.1 An Encounter with John Bell

Only once did I have the opportunity to meet John Bell. He came to Helsinki and gave a seminar titled: *What on earth is quantum mechanics about?* After the talk I had the pleasure of putting my objections to him, and listening to his retorts. Calmly, almost amused, he reiterated his arguments why quantum theory, as we know it, cannot be a satisfactory description of reality. Later, I have realized that all my arguments must have been well known to him; he knew the standard arguments in favor of the received view of quantum mechanics as well as anyone. He had not yet coined the phrase FAPP (=for all practical purposes), but that describes well the attitude he took to my arguments. What he certainly did not learn, was my own critical attitude to the received view; I only wanted to experience his more informed objections. My own arguments were far less developed and lacked adequate verbal form. On that occasion, of course, I did not anticipate that I would not have any opportunity to take up our dialogue later. However, his influence as an informed critic of the received view has ensured a continuing discussion characterized by tolerance and open minds.

So "What on earth is quantum mechanics about?". The paradoxical situation is that quantum theory is the best theory we have of nature, its predictions have been verified over an immense range of phenomena. But yet, we do not know what it speaks about; it does not answer the fundamental question "what there is" [1]. As it stands, quantum mechanics does not tell which concepts are real, i.e. are "elements of reality". Physical quantum entities appear and disappear in most discomforting ways; conserved entities seeming more real than others. The only object we are left with is the wave function, but is that real? It is not a description of an ensemble, but it clings to the individual system in some way. Many would say that it describes our knowledge about the system. But how come then that it is supposed to be objective, that it governs the outcomes of our observations? In a cynical mood one may say that a minimal credo of the quantum physicist is that the theory describes statistical properties of lists of recorded data. But as *Bell* states [2]:

"*But experiment is a tool. The aim remains: to understand the world.*"

Quantum theory as we know it is singularly unhelpful to endeavors towards this goal. We do really desire some understanding of the fabric of reality, which implies knowing the essence and being of the basic building blocks. According to *Bell* [3]:

"*The theory should not be about observables but about beables.*"

The world where our most successful theories work is not a world where we can conduct our everyday life.

This was recognized by *Niels Bohr*, who took for granted the existence of macroscopic entities, which behave in the way our classical human environment behaves, and which are described by our intuitive classical concepts [4]:

"*The main argument was that unambiguous communication of physical evidence demands that the experimental arrangement as well as the recording of the observations be expressed in common language, suitably refined by the vocabulary of classical physics.*"

Thus the elements of reality, which we can "speak about", are the preparation setups and the results of our observations. The rest of the formalism is taken to be just a tool to reach these statements we can grasp. The view is very much in tune with the positivistic picture of reality dominating the first half of the twentieth century. It is continued in the semantic approach to theories, where ontology is relegated to include those entities which are needed to "make sense of the discourse" only [5].

But a working physicist is a realist, even if the exact character of this realism has been debated incessantly. We do believe that we are making sense of external reality, and that our predictions will be verified in a real laboratory, whoever does the test. We have to insist that two separate entities exist, the external world and our description of it. This legacy of Bohr may be uncomfortable for a philosopher, because it leads to endless troubles, but to deny the existence of the problem makes a mockery of our effort to understand the activity of physics, and clearly even the everyday experience of humans. To me, a philosopher trying to convince me of a nonrealistic interpretation of reality, is just as incomprehensible as the proverbial solipsist surprised at others not sharing his view.

So what is quantum mechanics about, if we are not satisfied with only describing the outcomes of measurements? In order to approach this problem, we need to know what a measurement is. Bohr did neither want nor need any theoretical measurement description, but such formal mechanisms have been developed. Some working physicists find them highly satisfactory, but here we want to analyze them to see if we cannot only understand them but put them in their place. Whatever that place is!

## 12.2 Measurements or Not?

So what is a measurement? Simply stated, it is a process whereby the state of a physical system influences a recording device so that reading its state conveys some knowledge of the original system to the observer. Information about the unobservable thing in itself is transferred to a place where information can be read. According to the rules of quantum mechanics, the amount of information transferrable in this way is limited by quantum uncertainty, and the full knowledge of the original state eludes us. This lies in an essential way in the concept of quantum information. Only information residing in Bohr's macroscopic device is fully accessible. All the rest of the machinery lacks realistic interpretation. *Bohr*'s view is adequately expressed in his answer to the EPR paper in 1935 [6]:

"*The necessity to discriminate in each experimental arrangement between those parts of the physical system considered, which are to be treated as measuring instruments and those which constitute the object under investigation may indeed be said to form a principal distinction between classical and quantum mechanical description of physical phenomena.*

*Its fundamental importance in quantum theory has its roots in the indispensible use of classical concepts in the interpretation of all proper measurements.*

*In each experimental arrangement we have only a free choice of the place of discrimination between object and measuring agent within the region where the quantum mechanical description of the process concerned is effectively equivalent with the classical description.*"

Recently this train of arguments has been reversed; our concept of information has been projected back into reality, and all of physics is interpreted in terms of its information contents. The mathematical formalism of information exchange has played a central role in communication technology ever since its introduction by Shannon in 1948. It is natural to extend the realm of this concept in an age which is awed by the growth of information distribution. The new science of *quantum information* is born, and it is making progress both as a language for describing laboratory progress and as a mathematical disciple with more or less relevance for applications. The mirage of *quantum computing* is providing both much of the motivation and many of the challenges. What fruits this branch of the three of knowledge will bring is still an open question.

One eager advocate of the information approach to reality has been Rolf Landauer; his thesis "Information in inevitably physical" [7], has made a lasting mark on the discussion. With his extensive background in analyzing the physics of information processing, his point of view carries real weight. However, he served as a stern critic of the unrestrained extension of classical concepts into the field of quantum computing. Thus he, with his experience, could sense the dangers of uncritical extension of concepts beyond their original area of relevance. Unfortunately he is no longer with us, and the theory evolves without his keen analysis of its results.

Personally I find it hard to ascribe all we encompass with the word information to the physical environment. Of course, all information not constituting eerie visions only, must be found on a material substrate. However, I can tell you about the star constellations Cassiopeia and Southern Cross including their use in finding the way on open sea. But I cannot really imagine that these man-made concepts have any real existence; the star configurations do not form natural entities. This becomes manifest when I want to process the information. Then its material character will determine which physical occurrences are possible and how to proceed to effect them. We know of no physical process to make a cross out of Cassiopeia! Thus I conclude: the processing of information is inevitably physical, but the interpretation of the information contents is happening in our minds only. To quote *Quine* [8]:

"*The totality of our so-called knowledge or beliefs, from the most casual matters of geography and history to the profoundest laws of atomic physics or even of pure mathematics and logic, is a man-made fabric which impinges on experience only along the edges.*"

Quine proceeds from this to question the difference between analytic truths based on our theoretical frameworks and contingent facts conditioned on the structure of reality. His problem is just the one we discuss her, but his conclusions are, like most conclusions of philosophical discourse, rather unhelpful in providing satisfaction to physicists.

So what is a measurement; thereupon hangs the tale? We have to find out this by ourselves, we do not believe that any amount of formal structure can answer the question if we do not know what we want to do. Only a pure Platonist could imagine that a formalism can contain a unique recipe how it relates to experienced reality. Experiments are tools. Recent developments in mathematical physical may suggest that, if a structure is logically possible, it must become manifest in nature. Most such attempts have failed before, general relativity possibly providing an exception, so we cannot really relay on this. *Bell*'s arguments against measurements [2], indicate that no particular events in reality should deserve the designation "measurement" in contrast to other manifestations of the interactions. The only serious attempt to integrate an interpretation into the formalism, the *Everett* many-world interpretation [9], fails just in this respect. If we do not single out certain events as observations, the wave functions will branch continuously in space and time thus multiplying the worlds to a cardinality even overwhelming the imagination of the most eager supporter of Cantor.

The information approach to knowledge has proved exceedingly successful in classical communication and computing. Its extension to quantum theory provides a fertile ground for research [10] and an efficient method for structuring experiments [11]. However, if we claim that its concepts have implications forced on reality, we indulge in a hubris which mistakes our man-made structure for reality. *Wittgenstein* warned against this nearly one hundred years ago [12]:

"*The whole modern conception of the world is founded on the illusion that the so-called laws of nature are the explanations of natural phenomena.*" (§ 6.371)

As all information is information for us, we know no mechanism whereby it would have the force to impose anything on reality except to the extent that reality was contributing to its construction. To assert, that the external world is a materialization of our information, is to propose idealism beyond that ascribed to Bishop Berkeley.

But we do perform experiments, we do acquire knowledge and we do communicate the results we have learned. In order to understand how that is possible we have to look at the problem from a different angle.

## 12.3 Describing a Measurement

Bohr was totally unwilling to consider a formal theory of measurements, he was probably opposed to such a formalization of the acquisition of knowledge by humans. *Von Neumann* [13] took exactly the opposite view. His description of measurements has been the main inspiration for later works, but it has also mislead the development into a direction which may well have prevented a more balanced view to emerge over the years. Textbooks usually state that " ... a measurement always causes the system to jump into an eigenstate of the dynamical variable being measured ... " [14]. This is called the von Neuman projection postulate. In addition to being oversimplified and often incorrect, this statement does not adequately describe von Neumann's view. He states namely:

"*In the measurement we cannot observe the system* **S** *by itself, but must rather investigate the system* **S** + **M**, *in order to obtain (numerically) its interaction with the measuring apparatus* **M**. *The theory of the measurement* *is a statement concerning* **S** + **M**, *and should describe how the state of* **S** *is related to certain properties of the state of* **M** *(namely, the positions of a certain pointer, since the observer reads these)*." ([13], Chap. V.1).

We conclude that von Neuman was well aware of the fact that the measurement interaction is a physical process. He only postulated the instantaneous reduction principle because of our inability to follow the process in detail; after all his aims were logical consistency not realism. The description of a measurement should thus start from the change of state of a system $M$ conditioned on the state of another system $S$. This is simply transfer of quantum information. The assumption that the system $M$ is such that its state can be observed, i.e. we can read the value of the meter, is an example of the separation of the world into microscopic and macroscopic entities, as introduced by Bohr. This is obviously a tricky point, which Bell terms "a shifty split" of the universe. We will analyze the meaning of this below, but no resolution of the enigma will be found.

Even before the publication of von Neumann's measurement analysis, *Landau* and *Peierls* [15] in 1931 published their penetrating analysis of the measurement process. They set out to learn about the properties of the quantum state $|\psi\rangle$ describing the system $S$. They formulate a principle called the "predictable measurement", which implies that there are states $|\psi\rangle = |v_n\rangle$ such that we know with certainty which outcome to expect. They then prove that the states $\{|v_n\rangle\}$ must form a complete orthonormal set. Thus we expand

$$|\psi\rangle = \sum_n a_n |v_n\rangle \ , \tag{12.1}$$

with uniquely determined coefficients $a_n$, which give the probability $|a_n|^2$ of the outcome $|v_n\rangle$. The statement that choosing the observable implies choosing an orthonormal basis instead of an operator is re-expressed by *Wigner* [16]. If we now choose the state of the meter $M$ to be expressed in terms of the orthonormal basis $\{|\varphi_n\rangle\}$, we assume that the uniquely distinguishable states $|\varphi_n\rangle$ constitute pointers which can be read; initially the pointer is introduced in a fiducial state $|\varphi_0\rangle$. Then the measurement transformation is given by

$$|\psi\rangle\,|\varphi_0\rangle = \sum_n a_n |v_n\rangle\,|\varphi_0\rangle \Rightarrow \sum_n a_n |u_n\rangle\,|\varphi_n\rangle. \tag{12.2}$$

The new states of the system $|u_n\rangle$ need not in general have any simple relations to $|v_n\rangle$ except being uniquely determined by (12.2). In particular the observation of a photon polarization is likely to replace the state by the vacuum whatever the outcome of the measurement is. I quote the paper [15]:

"*In a measurement which occupies a short time it can easily be shown that the* $u_n$ *are identical with the* $v_n$ *only when the corresponding operator commutes with the energy of interaction between the system and the apparatus. In wave mechanics, this interaction energy is always a function of the co-ordinate. The only quantity for which a repeatable measurement is possible is therefore the co-ordinate. Measurements of the co-ordinates always actually have this property. It is also seen that the* $u_n$ *need not in general be orthogonal, i.e. the measurement does not in general diagonalizes an operator. This physical circumstance also is usually overlooked in the presentation of transformation theory.*"

If we assume that our states are numbered in a cyclic manner, we can easily see that the transformation (12.2) is effected by the operator

$$\hat{M} = \sum_{n,m} |\varphi_{n+m}\rangle\, |u_n\rangle\langle v_n|\langle\varphi_m|. \tag{12.3}$$

This operator is not unitary unless the states $\{|u_n\rangle\}$ form a complete set; the von Neumann measurement with $|u_n\rangle \equiv |v_n\rangle$ constitutes an example. On the other hand, if we assume that the states $\{|u_n\rangle\}$ are orthonormal, it was shown by *Schrödinger* [17] that the states $\{|\varphi_n\rangle\}$ are eigenstates of the reduced density matrix in the system $M$; see also [18]. Then the final state in (12.2) is a Schmidt decomposition.

In passing, we note that the mapping (12.2) and (12.3) can be interpreted as effecting a quantum computation, namely the mapping $M : v_n \Rightarrow \varphi_n \equiv \varphi[v_n]$. If we want a reversible evaluation, the states $\{|u_n\rangle\}$ must form an orthonormal set; usually they are assumed to be identical with $\{\ |v_n\rangle\}$, but this is not necessary.

Next we formalise the von Neumann concept of readability of the pointer states. This implies that the meter $M$ is unable to retain coherence between its various states. Thus we have a further transformation occurring on the level of the density matrix of the combined system

$$\begin{aligned}\hat{\rho} &= \sum_{n,m} a_n a_m^* |u_n\rangle\langle u_m| \otimes |\varphi_n\rangle\langle\varphi_m| \\ &\Rightarrow \sum_n |a_n|^2 |u_n\rangle\langle u_n| \otimes |\varphi_n\rangle\langle\varphi_n|. \end{aligned}\tag{12.4}$$

Now the density matrix describes an ensemble where the state $|u_n\rangle\,|\varphi_n\rangle$ occurs with probability $|a_n|^2$; notice that the transformation (12.4) does not affect the reduced density matrices in $S$ and $M$. Thus the information available in each subsystem is unaffected by this transformation.

But now we are where we want to be. The observation of the state of the pointer $|\varphi_n\rangle$ is uniquely correlated with the initial state $|v_n\rangle$ and the state of the system is given by $|u_n\rangle$; if these are identical, we have a von Neumann measurement. The combined process of the measurement interaction and the emergence of the final state can be described using the operators

$\hat{A}_n = |u_n\rangle\langle v_n|$. In the system $S$ this effects the operation

$$|\psi\rangle \Rightarrow \frac{\hat{A}_n|\psi\rangle}{||\,\hat{A}_n|\psi\rangle\,||} = |u_n\rangle, \tag{12.5}$$

which occurs with probability $|a_n|^2 = \langle\psi|\hat{A}_n^\dagger \hat{A}_n|\psi\rangle$. These operators satisfy the relation characteristic of positive valued observables

$$\sum_n \hat{A}_n^\dagger \hat{A}_n = \sum_n |v_n\rangle\langle v_n| = \mathbf{1}. \tag{12.6}$$

These relations connect the Landau–Peierls formulation to the more mathematical formulation of the measurement problem [19].

The description of the measurement process is thus reduced to three essential steps:

1. The state of the system is correlated with the state of the meter in a unique way.
2. The meter is supposed to be macroscopic in the sense that it is unable to retain quantum correlations between its states.
3. From the density matrix describing an ensemble, the potential states involved become actual when an observation is performed.

Of these steps, #1 constitutes an ordinary quantum process exchanging information between two systems. The same type of dynamics is utilized in the popular quantum information processing devices. Step #2 involves coupling the meter to a reservoir. The research on this problem constitutes the main part of the history of measurement theory. By now, it is rather well known under which circumstances the density matrix dephases to diagonal form and which time scales are involved. Even this part of the argument is rather standard, and falls within the class of questions which can be investigated utilizing the present formulation of quantum theory.

Only the final step #3 offers a problem. At this stage, it seems rather harmless, but a little thinking shows it to be problematic. We started by a well defined state of system and meter, in the end we find that this has to be characterized by an ensemble; when we learn the value of the pointer, all other potential values become nonexistent. Based on our new knowledge we have to assign new states to the components of the system. This step is a consequence of the intrinsic statistical interpretation of the original theory. As in all probabilistic theories, new information forces us to reassign probabilities to all parts of the system, even to those very far from the place where the information was acquired. This is the ultimate origin of the "spooky action at a distance", which offers the rather paradoxical nonlocal effects in quantum mechanics. Quantum mechanics is weird also locally; vastly separated systems just bring out this weirdness in a manifest manner. All talk about nonlocal interactions caused by measurements are futile, probability does it all.

This is where the current interpretation of quantum mechanics makes its "shifty split". After having prepared the initial state, which is determined in an objective manner by the preparation device, the physicist at some point stops relating the state to the real system and starts interpreting it as a mathematical probability distribution describing our knowledge of the system. That shift if usually introduced in such a sneeky manner that it is not easily noticed. This is possible because of the double role probability plays in quantum theory: as a description of the original ensemble prepared and as a rational expection of future measurement outcomes. This "split" is the one which introduces the transition from a microscopic description to a human interpretation. In quantum information theory a similar procedure is used to relate observed data to the probabilities characterizing the signal source.

The reduction of the state of the system when new information is obtained manifests the mystery of quantum mechanics: How is it possible that a fundamental theory supposed to describe reality contains a probability concept which does not originate in lack of knowledge and cannot be reduced to pure subensembles? That is the question? Some physicists regard the situation as satisfactory; *Belinfante* [20] says that "when the system becomes reducible (to a classical ensemble), we may as well perform the reduction." But that does not solve the problem how the beables "... on the macroscopic level yield an image of the everyday classical world ..." [3].

One of the well known objections to the Copenhagen interpretation is Schrödinger's dislike of quantum jumps. In his papers [21] called "Are there quantum jumps?", he argues against them. As we have seen, the jumps are only necessary when observations are carried out. Left to themselves the quantum states remain harmonious and smoothly behaved. Thus Schrödinger's question can be rephrased as "Are there measurements?" If we succeed in eliminating the measurements, we have solved the jump problem at the same time.

Whatever it means, we asume that we take cognizance of the state of the meter. This knowledge is then incorporated into our common language description of the situation. We would like to understand what this means, but then we need to understand better how we represent the world of reality.

## 12.4 How to Map Reality

The reality of physics or of our lives cannot be identified with the world of the philosophers. Arthur Schopenhauer set the tone when his work "Die Welt als Wille und Vorstellung" appeared in 1818. Admitting the difficulty in rendering the word "Vorstellung" into English, I think the sense of the first sentence of this work is captured in:

"*The world is my experience – this truth holds for every living and feeling being, even if only man can bring it into abstract cognizance.*"

The world of later philosophy is very much based on the same approach. Whatever quantum mechanics is about, this is not to be found in the philosopher's world. This world is very dependent on words, it is a man-made fabric of symbolic representations, which acquires meaning only through the fate and actions of human beings. The mental construction we call the world is a map which helps us find our way through life, but it is neither a true nor a faithful representation of reality.

Like a map the abstract picture we have of reality is based on our ability to represent it. It cannot be a one-to-one mapping of our external environment, because that would cause problems. Already *Plato* stated that a name must differ from the object it represents [22] because otherwise:

"*They would be the doubles of them, and no one would be able to determine which were names and which were realities.*"

*Bertrand Russel* [23] discusses which entities can be given names, and he too concludes that identity is not the signature of correct naming:

"*If there were in New York an Eiffel Tower exactly like the one in Paris, would there be two Eiffel Towers, or one Eiffel Tower in two places?*"

This problem, of course, does not appear with a picture of the tower. Our verbal description of reality is like that. We can only make approximate pictures no exact reproductions. But the pictures we make allow us to find our way in reality, convey information to other humans and record our experiences. This holds true about ordinary language too, and even there the understanding has remained clouded in spite of endless philosophical investigations.

I claim that our best physical theories are just like the ordinary language in this aspect. Following *Bohr*, we simply extend our ordinary language to encompass an ever broadening experience about reality [24]:

"*Notwithstanding refinements of terminology due to the accumulation of experimental evidence and development of theoretical conception, all account of physical experience is, of course, based on common language, adapted to orientation in our surroundings and to tracing relationships between cause and effect.*"

When we learn more, we have to extend our language, but it is still based on the capability of the mind to form symbolic representations. When experience extends beyond the phenomena shaping our everyday life, it should not be a surprise that the results seem unfamiliar and strange. Their character of a road map in reality remains.

In physics applications it is important that the formal structure is not made too precise. If that happens it can convey only logical truth and we are saying nothing about reality. As *Russel* puts it [23]:

"*Thus the more precision we give to a concept, the oftener it can be proved to be inapplicable, and the seldomer it can be proved to be applicable. When it is completely precise, it can never be proved to be applicable.*"

In physics the concepts have to have enough flexibility to retain their place in the formal structure even if new experience makes them change their contents. Thus the concept of energy has survived all the revolutions in physics which have occurred since it was first introduced.

Maps, however, have another property; they namely contain features and limitations, which are inherent in the representation. Most maps have lines indicating the elevations in the terrain; we look in vain for these in nature. In the same way, our best theories will contain features which are not naming any existing elements of reality. The standard discussion of Bell inequalities shows that some classical concepts can not be carried over into quantum theory. But which quantum concepts name real entities and which are only formal parts of our description? Does the wave function name or is it a tool for making predictions only? Which is the role of the probability interpretation inherent in quantum mechanics?

Two-dimensional maps also fail to represent the curved surface of earth faithfully. This is a limitation posed by our mode of symbolic representation. As even our best theories of physics are based on our ability to construct abstract structures, there may be features of reality we cannot capture; the representation is limited by the performance of the human mind. Theoretical physics is aiming at a theory of everything, a theory which can explain the totality of reality. Even if such a theory exists in some Platonic sense, what justifies the hubris which assumes that this theory can be captured by the human mind? May be, the best we can do is to cover various patches of experience by local charts, which work well in their realm of validity, but which cannot be extended to the whole of reality. Only in some minor regions of overlap can we make sense of several theories. May be quantum theory and relativity cannot be proved to be in conflict where both are valid, but we still cannot construct any theory encompassing both globally. We are finite beings, and grasping the multifarious reality in its totality may be beyond our abilities.

On the other hand, we have done well, our best theories of reality have a validity which allows us to penetrate ever further into nature, manipulate its phenomena and govern physical processes in desired directions. The maps we have allow us to navigate over large areas of the ocean of knowledge. This is what our language can do for us when it is suitably refined by the concepts and methods of classical physics. The abstract structures have proved their ability to extend and renew our concepts of reality. Human intuition is capable of remarkable creative feats, but it needs a structure to work with. The progress of theoretical physics is probably no more remarkable than the fact that young humans are able to use a language in creative ways even after exposure to a very limited sample of usage. This gives hope that the progress in experimental investigations will enrich and extend our ability to describe reality.

What we can say about reality is thus based in the capabilities of our minds to form concepts and relations based on experience and experiments. All information carried to the mind of a human being must be founded on such mental representations, and no absolute concept of information is possible. Information must be in the mind of the beholder, and all other statements imply an overvaluation of our understanding of nature and the capacity of our intellects. However productive and challenging the information approach to physics may be, it is only the latest step in the process of construction of an abstract representation of the reality where we live. To overinterpret its significance brings to mind the warning by *Wittgenstein* [12]:

"*All theories that make a proposition of logic appear to have contents are false ... Indeed, the logical propositions acquire all the characteristics of a proposition of natural science and this is a sure sign that it has been construed wrongly.*" (§ 6.111)

## 12.5 Conclusion

In summary, I want to conclude, that the best theories we have are only mental constructions helping us to find our way through reality and utilize its resources to our advantage. We have no absolute information about nature, but as scientists we must accept that it exists independently of our human endeavors. Collecting experience we can extend and generalize the theoretical web of explanations, but we do not know how much it is conditioned by reality and to what extent it derives from our mental setup. All information theoretic interpretations have to operate within this framework. However, that is not too bad, after all the goals of our theories is communication and guidance. Increased experience of reality will provide the symbolic representations to talk about new entities and phenomena.

In the end we cannot escape the mind-body duality. Our symbolic representations exist in the minds, and our bodies live and act in external reality. Any attempt to form a unified view has failed so far. It may be tempting to restrict the world to encompass only our description of it. This philosophical trend easily leads to an anti-realism which provides a closed universe of discourse, but fails to explain what the scientists are actually up to.

Alternatively we may consider a world constructed from elements of information, where we think that our mental activities are only part of the great pattern of information flow constituting reality. This extremely materialistic view of nature could become reality, if we only had the necessary understanding of the material basis of the human mind. However, such understanding is patently lacking today; we do not know what consciousness is or how it relates to the physiological activities in the neural system. If such understanding can be achieved one day, may be then the need to distinguish between mental constructions and physical phenomena will become unnecessary. Until then we have to live with the old dichotomy between matter and mind.

This seems to be very much the attitude of Niels Bohr, who assumed that we can know which variables relate to our immediate cognition. He stubbornly refused to formalize the relation between the abstract description of nature and our knowledge of it. He writes in 1958 [24]:

"*On their side, the quantal features of the phenomena are revealed in the information about the atomic objects derived from the observations ...*

*From the preceding argumentation it will appear, however, that all departures from common language and ordinary logic are entirely avoided by reserving the word 'phenomenon' solely for reference to unambiguously communicable information, in the account of which the word 'measurement' is used in its plain meaning of standardized comparison.*"

This clearly describes a dichotomy between the map of nature in our formalism and the corresponding knowledge in our minds. Nature knows no processes recognizable as measurement, but we are free to apply this term to occurrences which we can formulate in our common language description of reality. A measurement is some process which increases human information.

We have learned much since Bohr; features of quantum mechanics beyond the dreams of the founding fathers have been verified experimentally. But we are no longer satisfied with just acknowledging the situation. The guide should be the experiments; new fundamental measurements should lead the way. I do not believe that any formal machinery can have implications for how we describe reality. However, following the challenge posed by John Bell, we want to progress beyond the Copenhagen interpretation. Only new experience can inspire and form the development of our "common language suitably refined by the terminology of classical physics".

## References


1. W.V. Quine, *On what there is*, Rev. Methaphys. **2**/1 (Sept. 1948). Reprinted in *From a Logical Point of View* (Harvard Univ. Press, Harvard 1953)
2. J. Bell, *Against 'measurements'*, Phys. World 44, (August 1990)
3. J. Bell, Subject and Object, in *Speakable and Unspeakable in Quantum Mechanics* (Cambridge Univ. Press, Cambridge 1987)
4. N. Bohr, *The Solvay Meetings and the Development of Quantum Physics*, Address at the 12th Solvay Meeting 1961. Reprinted in *The Philosophical Writings of Niels Bohr, Vol.III: Essays 1958–1962.* (Ox Bow Press 1987)
5. J. Hintikka, The Semantics of Modal Notions and the Indeterminacy of Ontology, in *The Intentions of Modal Notions and other New Models for Modalities* (Kluwer, Dordrecht 1975)
6. N. Bohr, Can quantum-mechanical description of physical reality be considered complete? Phys. Rev. **48**, 696 (1935)
7. R. Landauer, Information is Inevitably Physical, in *Feynman and Computation*, A.J.G. Hey (ed.) (Perseus Books, 1999)
8. W.V. Quine, Two Dogmas of Empirism, in *From a Logical Point of View* (Harvard Univ. Press, Harvard 1953)

9. H. Everett III, "Relative state" formulation of quantum mechanics, Rev. Mod. Phys. **29**, 454 (1957)
10. M.A. Nielsen, I.L. Chung, *Quantum Computation and Quantum Information*, (Cambridge Univ. Press, Cambridge 2000)
11. D. Bouwmeester, A. Ekert, A. Zeilinger (eds.), *The Physics of Quantum Information* (Springer, Berlin, Heidelberg 2000)
12. L. Wittgenstein, *Tractatus Logico-Philosophicus* (Routledge, London 1961)
13. J. v Neumann, *Mathematical Foundations of Quantum Mechanics* (Princeton Univ. Press, Princeton 1955; German original 1931)
14. P.A.M. Dirac, *Quantum Mechanics*, 3d edn. (Oxford Univ. Press, Oxford 1948)
15. L.D. Landau, R. Peierls, Extension of the uncertainty principle to relativistic quantum theory, Z. Phys. **69**, 56 (1931), translation in *Collected Papers of Landau*, D. ter Haar (ed.) (Gordon Breach, New York 1965)
16. E. Wigner, Review of the quantum measurement problem, in *Quantum Optics, Experimental Gravitation and Measurement Theory*, P. Meystre, M.O. Scully (eds.) (Plenum, New York 1983)
17. E. Schrödinger, Discussion of Probability Relations between Separated Systems, Proc. Cambridge Phil. Soc. **31**, 535 (1935)
18. S. Stenholm, Observations and quantum information, J. Mod. Opt. **47**, 311 (2000)
19. P. Busch, M. Grabowski, P.J. Lahti, *Operational Quantum Physics* (Springer, Berlin, Heidelberg 1997)
20. F.J. Belinfante, *Measurement and Time Reversal in Objective Quantum Theory* (Pergamon, New York 1975)
21. E. Schrödinger, Are there quantum jumps, Br. J. Philos. Sci. **3**, 109 (1952); ibid 233 (1952)
22. Plato, *Cratylus*
23. B. Russel, *An Inquiry into Meaning and Truth*, (Allen, Unwin 1978); originally published 1940
24. N. Bohr, Quantum physics and philosophy, in *The Philosophical Writings of Niels Bohr, Vol.III: Essays 1958–1962* (Ox Bow Press 1987)

# 13 Sundays in a Quantum Engineer's Life

Nicolas Gisin

## 13.1 I am a Quantum Engineer, but on Sundays I Have Principles

*"I am a quantum engineer, but on Sundays I have principles".* This is how John Bell opened his "underground colloquium" in March 1983; words which I will never forget! What! John Bell, the great John Bell, presented himself as an engineer!?! one of those people who make things work without even understanding how they function?!? In contrast I thought of John Bell as one of the greatest theoreticians.[1]

In March 1983 the *Association Vaudoise des Chercheurs en Physique* organized their annual one-week course in Montana, an excellent combination of skiing and physics, on the foundations of quantum mechanics [1]. For one of the reasons peculiar to the community of people interested solely in these foundations, John Bell was invited, but without any time slot for a presentation. With some friends we, i.e. the PhD students, managed to convince him to give us an evening lecture, after dinner, while the professors enjoyed the local wine. At first John declined, arguing that he did not have his transparencies, but when we learned that his wife Mary was to join him during the week, we had the perfect counter-argument: Mary would bring the transparencies and we would organize a room and an overhead projector. The talk took place in the basement; the ceiling was too low, and the students sat on the ground: a perfect underground atmosphere. When John Bell started you could "hear the silence": *I am a quantum engineer, but on Sundays I have principles.*

The next few pages are not supposed to be selfcontained; for the physics the reader is directed to existing literature. These pages simply contain an illustration of my understanding of John's words [2].

## 13.2 Quantum Cryptography on Sundays

Today, quantum engineers have many good professional opportunities, and even more should be expected in the near future. Nowadays, for example,

[1] Today I would add: one of those people who are unable to make anything work, but think they know why it doesn't function!

they work on quantum cryptography [3, 4]. For this purpose they develop techniques to send "single photon pulses" through telecom optical fibers from a sender, conventionally called Alice, to a receiver called Bob. A convenient solution, according to the engineer, consists in mimicking the single photons by very weak laser pulses, so that the probability that a pulse contains more than a photon is negligiable. But on a sunny Sunday the physicist says [5]: "weak pulses are not good, because they may contain two photons." So the engineer works out a better single-photon source, based on a photon-pair source, using one photon as a trigger for the single photon pulse (see Fig. 13.1). The physicist, impressed by the engineer's ability, likes the idea so much that a special issue of the journal of *Pretty Rational Letters* is edited [6–8]. However, on the following Sunday, he realizes that there is no need, in principle, to set the source on Alice's side [9]: the source could as well be at the center! This is much more elegant, because of higher symmetry.

But then, in this symmetric configuration, there is no longer any physical object transmitted from Alice to Bob: Where does the correlation then come from? What then guarantees the confidentiality? The engineer doesn't care (it is not even clear whether he understands the problem). Actually, it is also

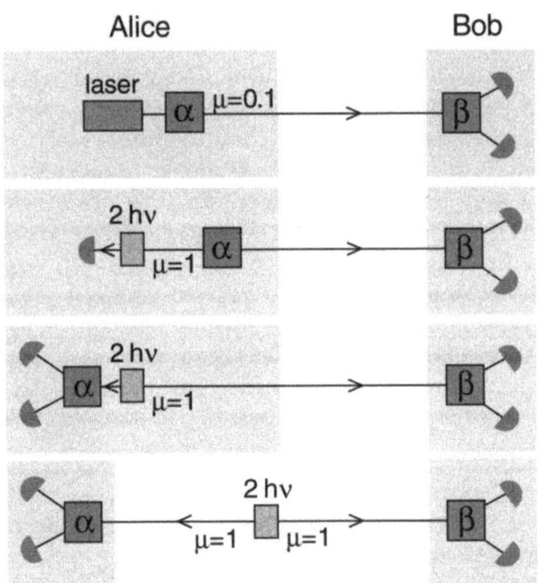


**Fig. 13.1.** Schematics of the relation between the setups using faint laser pulses (top) and those used to test the inequality (bottom). In the upper scheme, each pulse has a mean photon number, $\mu \approx 0.1$. In the second scheme, the source is replaced by a two-photon ($2h\nu$) source; one is used as a trigger, the other is prepared according to the setting $\alpha$ and sent to Bob. In the third scheme, Alice's photon is used to prepare (at a distance) Bob's photon. Finally, the last scheme is completely symmetric between Alice and Bob

unclear to the physicist whether there is a problem or not. Until he discovers *the inequality*[2]. This inequality [10] doesn't explain the correlation observed by the engineer, nor does it say anything about the fragility, and hence about the security of quantum cryptography. The inequality, or more precisely, its violation, says that the correlation will never be explained by any theory based only on local variables (local beables in John's words [12]), i.e. that any description of the "world out there" must incorporate some nonlocal influences.

The quantum engineer enjoys his work. He develops better single-photon sources (i.e. sources coming closer to the ideal case), and he works on better detectors, and higher bit rates, implements efficient error correction and privacy amplification algorithms, etc. During that time, the physicist writes things such as $U_1 \otimes U_2 \Phi^{(+)} = \mathbf{1} \otimes U_2 U_1^t \Phi^{(+)}$ and concludes that the one-photon and two-photon schemes are logically equivalent [13]. He analyses the optimal eavesdropping attack, assuming Eve (the eavesdropper) is limited only by the laws of quantum physics. Thus, he finds that Alice and Bob are guaranteed to share a higher mutual (Shannon) level of information than Eve (with either Alice or Bob) if and only if the error rate between Alice and Bob (something the engineer calls QBER: quantum-bit error rate) is smaller than $\frac{1}{2}(1 - 1/\sqrt{2}) \approx 15\%$ [14, 15]. This result makes the engineer happy; his experimental results are well below this 15% threshold: he is on the safe side. The physicist, however, has mixed feelings: How can it be that this limit, obtained when comparing Shannon information with their logarithms, defines precisely the noise limit above which the inequality can no longer be violated [16]!?![3] Is it a coincidence? Or is it deep [17]? (This is still an open question!)

## 13.3 Let's Assume That the Collapse Is Real

Several Sundays pass. The physicist is fascinated by the connection between the bound derived from the metaphysical assumption of local hidden variables and the security of his engineer friend's quantum crypto device. For the last few weeks he has been interested in the infamous wave packet collapse as a possible interpretation the nonlocal correlation. Suddenly he thinks: "What if the collapse is real? Could it be that the collapse triggered by Alice's measurement really prepares the state of the photon flying to Bob?" The physicist knows, of course, that the collapse and the related measurement problem are

---

[2] How did John Bell refer to his inequality? Very simple, he would just say *the inequality*!

[3] The error rate relates to the two-photon interference visibility, $V$, as follows: $\mathrm{QBER} = (1 - V)/2$ (recall that "outside the visibility" Bob still has one chance out of two of finding the correct result). Hence the QBER threshold mentioned corresponds to a visibility of $1/\sqrt{2}$, well known as the threshold for the Bell–CHSH inequality [11]

notoriously bad questions, since their prediction are indistinguishable from those of quantum mechanics without any collapse. But, the metaphysical assumption of local hidden variables led to interesting physics (and interesting engineering), although the main result is that they do not exist. Maybe it is worth trying some metaphysical assumptions about the collapse and seeing what kind of experiments should be done to test the assumptions?

This was really a nice Sunday, and the physicist thought about testing the speed of what Einstein called the spooky action at a distance! If it is really Alice's measurement that prepares Bob's photon at a distance, says the physicist to the engineer, let's carry out Bob's measurement at precisely the same time as Alice's, so that the nonlocal preparation has no time to operate. According to this assumption, the nonlocal correlation should disappear when proper timing is used. If the correlation remains, then either there is no collapse or the speed of the spooky action is faster than the bound set by the timing accuracy.

The engineer likes the challenge of aligning his system such that both measurements actually take place "at the same time". The solution is far from obvious, knowing that Alice and Bob are connected by almost 20 km of optical fibers over a straight line distance of more than 10 km! But, the engineer has heard of relativity and asks: "In which reference frame should I align the experiment?", "Well, hum, I do not really know!" admits the physicist, "let's try the most obvious choices: the reference frame in which the Swiss Alps[4] are at rest and also the reference frame of the cosmic background radiation [center of mass of the Universe]!"

The engineer smiles, but since it is fun work he is willing to try the experiment (it gives him a break from the task of improving these photon counters, which are so noisy that one needs to cool them, but if they are too cold then the dead time has to be increased because the after pulses take more time to resorb).

"Okay," says the engineer, "What exactly should be aligned? The beam splitters? The detectors? The computers? The observers?"

The physicist is amazed. Now that he dared to consider the assumption that the collapse is real, so many questions arise! And each hypothetical answer can, in principle, be tested! How should one continue?

The following Sunday, our physicist goes for a walk, with his friend David Bohm[5]. They speak about the engineer's question: What should be aligned? Clearly, it should be the device that triggers the collapse. But what could one reasonably assume as the trigger? After a few minutes of silence, the

[4] In other words the reference frame in which all the massive parts of the experiment are at rest (the lab frame, if you prefer)

[5] John Bell always insisted that Bohm's pilot wave model, being experimentally undistinguishable from standard quantum mechanics, should be taught to students at the same level [18]. But who follows this advice?

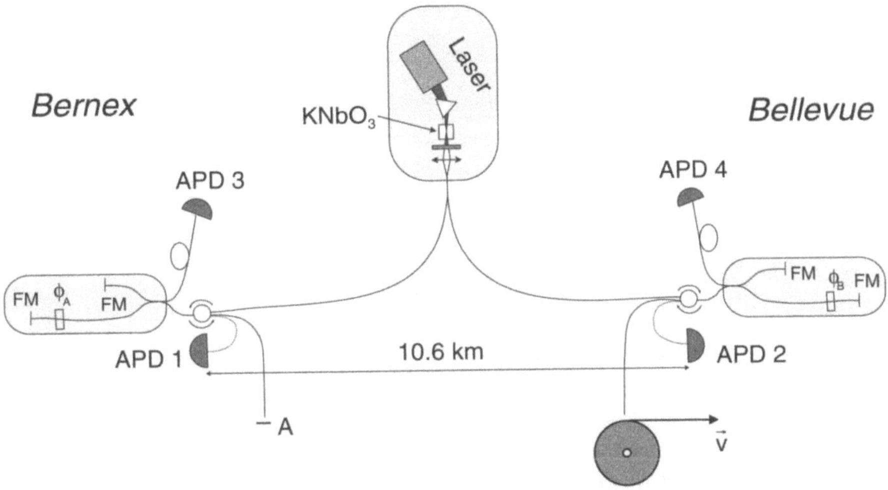


**Fig. 13.2.** Schematic of the Geneva long-distance quantum correlation experiments. When the 4 APDs (photon counters) are connected, this establishes a quantum channel for key distribution (i.e. quantum cryptography). The same setup allows one to test — and violate — Bell's inequality [24]. When APD 1 and 2 are set precisely at the same optical distance from the source, bounds on the speed of the spooky action (nowadays called quantum information!) can be set, as discussed in Sect. 13.3. When APD 2 is replaced by the absorber on the fast rotating wheel, two-photon interferences can be observed between APD 3 and 4. This realises the before–before experiment presented in Sect. 13.4

physicist states: "It must be the detector! That's where the irreversible event happens!"

"Possibly," replies Bohm, "but I bet it is the beam splitters! Because it is there that the particle makes its choice."[6]

On Monday, the engineer starts aligning the detectors. Indeed; to him aligning the beam splitters sounds even stranger than aligning the detectors.

This experiment was actually performed in 1999 in Geneva [20] (Fig. 13.2). The referee reports are interesting, ranging from fascination to desperation![7] There is no doubt that the experiment was a performance (a better than 5 ps alignment over almost 20 km of fibers: $\approx 10^{-7}$ precision). There is no doubt that for many physicists the collapse is taboo (but certainly not for John

[6] Indeed, in Bohm's pilot wave model, the irreversible choice is made at the beam splitters; in this model the detectors merely reveal this choice. This influenced *Antoine Suarez* and *Valerio Scarani* when they developed their proposal [19]

[7] Note that John Bell never published his papers on the foundation of quantum mechanics in regular physics journals! The reason was that he wanted to avoid the too-often sterile discussions with more or less anonymous referees

Bell[8]). There is also no doubt that much remains to be done, both on the theory side and experimentally (an experiment with moving beam splitters is under progress in Geneva).

The experimental result established impressive lower bounds on the speed of the spooky action: $\frac{2}{3}10^7$ and $\frac{3}{2}10^4$ times the speed of light in the "Swiss Alps" and the cosmic background radiation frames, respectively. These numbers are very large, similar to the ratio between the speed of sound in air and that of light (for a long time the speed of sound was the fastest measurable speed, while light was assumed to be instantaneously everywhere).

## 13.4 ... and Relativity?

There comes yet another sunny Sunday. The physicist rests in his armchair and thinks: "All this is quite exciting! But what if we let relativity enter the game even deeper (Fig. 13.3)? What if the detectors are in relative motion such that each detector in its own reference frame analyses its photon before the other? It would seem then that each photon–detector pair must make their choice before the other?!?!?" [19].

"This renews the tension between quantum physics and relativity," says the physicist to himself. Indeed the tension is not new. *Abner Shimony*, a good friend of John Bell, termed this tension "peaceful coexistence", because the tension does not lead to any testable conflict [21]. "However," continues loudly the physicist, although he is alone, "once one assumes that the collapse is a real phenomena, and once one considers specific models, the conflict is real and is testable!"[9] His idea is that the reference frame in which the collapse propagates is not the Swiss Alps frame, nor the cosmic background radiation frame, nor any environmental or universal frame. The intuition is that the frame is determined by the "trigger device", i.e. by the inertial reference frame of the massive device which triggers the collapse. If all trigger devices are at rest in some frame, then it is this frame which is the relevant one.[10] This new assumption opens even clearer tests: If each measurement happens before the other, then the quantum correlation should disappear, however large the speed of the spooky action!

[8] When asked for advice about a research direction on the foundation of quantum mechanics, John Bell always replied: "Do you have a permanent position?" If you have a chance to meet Alain Aspect, ask him about this!

[9] This comes as a surprise because, until recently, all models were developed by people whose primary concern was to avoid any testable difference with quantum mechanics: they were proud that they couldn't be wrong, apparently without realizing that this also implies that they couldn't be right (i.e. be scientifically relevant [22]). Very strange indeed!

[10] If one further assumes that detectors are trigger devices, then the results in the previous section provide a bound on the speed at which entangled objects become separated from one detected

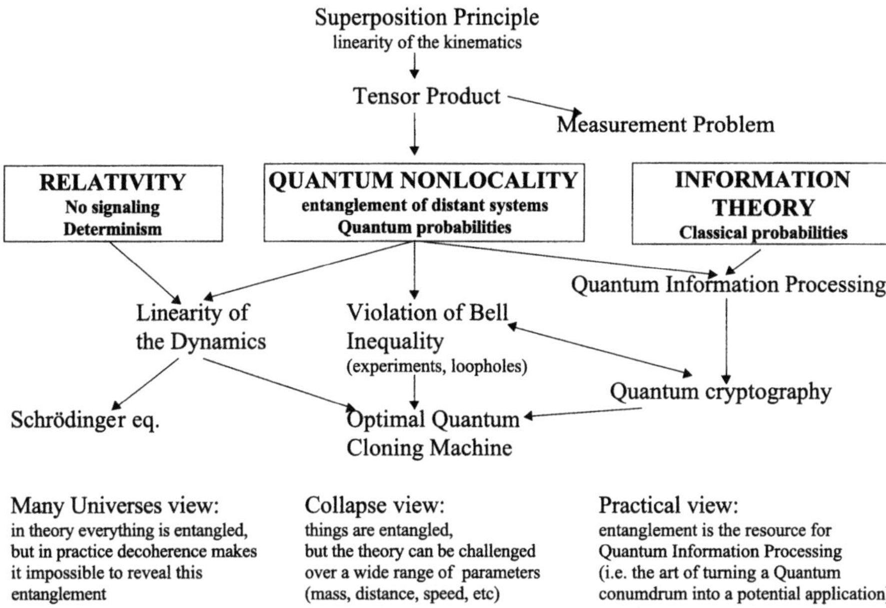


**Fig. 13.3.** Quantum nonlocality is central to today's physics! Logically it follows directly from the superposition principle. This principle leads also to the measurement problem, a dead end since it is not testable (at least in the foreseeable future). In addition quantum nonlocality, relativity and information theory, are the two other main scientific achievements of the first half of the 20th century. Relativity is characterized by determinism, information theory by classical probabilities. From quantum nonlocality and the relativistic no-signaling condition, one can derive the linearity of quantum dynamics, hence the Schrödinger equation [25]. From quantum nonlocality and information theory, one derives quantum cryptography, whose security is intimately connected to Bell's inequality (Sect. 13.2). Finally, the optimal quantum cloning machine can be defined as the eavesdropping strategy on a quantum-cryptographic channel providing Eve with optimal Shannon information and can be derived from the no-signaling condition and the existence of distant entangled states [26]

Very excited, the physicist starts to evaluate some orders of magnitude. Indeed, the experiment he just dreamt up requires that one purposely realizes events in moving reference frames such that the time ordering is altered by relativistic effects. A priori this requires relativistic speeds. But the formula actually leads to much more optimistic numbers [19]: the relative speed, $v$, should satisfy $v > \frac{c^2\delta t}{\ell}$. With $\delta t \approx 5\,\mathrm{ps}$ and the distance $\ell \approx 10\,\mathrm{km}$, one finds $v > 50\,\mathrm{m/s}$. A Ferrari can do it! So, my dear engineer, let's perform the experiment!!

"But," complains the engineer again, "do I really need to activate this bloody moving detector! This requires liquid nitrogen and is a real mess!"

The physicist tries to convince the engineer to use "real detectors". But he has to admit that he can't explain why: If the assumed collapse happens on the first microns of the detector, it should make no difference whether the detector is activated or not!?![11] Moreover, the effect could be seen anyway. Let's arrange the experiment such that one of Bob's detectors is clearly the first one to interact with the photon. Either this detector detects the photon and then the second detector does not see the photon or the first detector triggers a collapse corresponding to "no photon" and then the photon becomes localized in the path to the second detector. Hence, the engineer and physicist compromise: a first experiment without active detectors is planned.

A week later the physicist realizes that the unactivated detector could even be replaced by a simple absorber mounted on a rotating wheel. The engineer is happy: now the experiment is feasible! "Well," thinks the physicist, "this looks too simple. But what could be wrong?"

This experiment was also performed in Geneva, in the spring of 1999 [20]. The two-photon interferences were still visible, independent of the relative velocity between Alice and Bob's reference frames (actually the magnitude of the velocity was constant at 100 m/s, but its orientation varied). What would John Bell have thought of this?

## 13.5 Conclusion

The main lesson I learned from John Bell led me, on the one hand to become a quantum engineer, and, on the other hand, not to forget about the principles. In this way one can make a (very) good living. Simultaneously, one can keep ones fascination for the basic questions without losing ground in endless metaphysical discussions: there is nothing wrong with metaphysical assumptions, but the good ones are those that can be tested [23]. For example, wouldn't it be nicer to dispute the collapse of the wave packet as a physical phenomenon by designing and performing experiments, rather than arguing that the collapse is a metaphysical assumption?

## Acknowledgements

This work was supported by the "Fondation Odier de psycho-physique" and the Swiss NSF. The author is grateful to Cheryl Dotti and Wolfgang Tittel for carefully reading the manuscript.

[11] If it takes place later in the detector, then the time jitter (tens or even hundreds of ps) makes the experiment not yet feasible

## References


1. Ch. Gruber et al. (eds.), *Les fondements de la mécanique quantique*, 25$^{e}$ cours de l'AVCP (CP 101, 1015 Lausanne) 1983
2. J.S. Bell, *Speakable and Unspeakable in Quantum Mechanics: Collected Papers on Quantum Philosophy* (Cambridge Univ. Press, Cambridge 1987)
3. C. Bennett, G. Brassard, 'Quantum cryptography: public key distribution and coin tossing', Int. Conf. Computers, Systems & Signal Processing, Bangalore, Conf. Dig. 175–179 (1984)
4. N. Gisin, G. Ribordy, W. Tittel, H. Zbinden, Rev. Mod. Phys. **74**, 145 (2002)
5. G. Brassard, N. Lütkenhaus, T. Mor, B. Sanders, Phys. Rev. Lett. **85**, 1330 (2000)
6. T. Jennewein, C. Simon, G. Weihs, H. Weinfurter, A. Zeilinger, Phys. Rev. Lett. **84**, 4729 (2000)
7. D. Naik, C. Peterson, A. White, A. Berglund, P. Kwiat, Phys. Rev. Lett. **84**, 4733 (2000)
8. W. Tittel, J. Brendel, H. Zbinden, N. Gisin, Phys. Rev. Lett. **84**, 4737 (2000)
9. A. Ekert, Phys. Rev. Lett. **67**, 661 (1991)
10. J.S. Bell, Physics **1**, 195 (1964). Actually, we shall concentrate on another *Bell inequality*, the so called Bell–CHSH inequality [11]
11. J.F. Clauser, M.A. Horne, A. Shimony, R.A. Holt, Phys. Rev. Lett. **23**, 880 (1969)
12. J.S. Bell, The theory of local beables (1976) reprinted in [2]
13. C. Bennett, G. Brassard, N.D. Mermin, Phys. Rev. Lett. **68**, 557 (1992)
14. C. Fuchs, N. Gisin, R.B. Griffiths, C.-S. Niu, A. Peres, Phys. Rev. A **56**, 1163 (1997)
15. I. Cirac, N. Gisin, Phys. Lett. A **229**, 1997
16. N. Gisin, B. Huttner, Phys. Lett. A **228**, 1997
17. V. Scarani, N. Gisin, Phys. Rev. Lett. **87**, 117 901 (2001); see also Phys. Rev. A **65**, 012 311 (2001)
18. J.S. Bell, Found. Phys. **12**, 989 (1982) reprinted in [2]
19. A. Suarez, V. Scarani, Phys. Lett. A **232**, 9 (1997)
20. H. Zbinden, J. Brendel, W. Tittel, N. Gisin, Phys. Rev. A **63**, 022 111 (2001); V. Scarani, N. Gisin, W. Tittel, H. Zbinden, Phys. Lett. A **276**, 1 (2000); N. Gisin, V. Scarani, W. Tittel, H. Zbinden, Ann. Phys. **9**, 831–842 (2000); H. Zbinden, J. Brendel, W. Tittel, N. Gisin, J. Phys. A **34**, 7103 (2001)
21. A. Shimony, in *Foundations of Quantum Mechanics in the Light of New Technology*, S. Kamefuchi (ed.), (Phys. Soc. Japan, Tokyo, 1983)
22. K.R. Popper, *Logik der Forschung*, 1934; *The Logic of Scientific Discovery* (Hutchinson, London 1959)
23. R.S. Cohen, M. Horne, J. Stachel (eds.), *Experimental Metaphysics* (Kluwer, Dordrecht 1997)
24. W. Tittel, J. Brendel, H. Zbinden, N. Gisin, Phys. Rev. Lett. **81**, 3563 (1998)
25. N. Gisin, Helv. Phys. Acta **62**, 363 (1989); N. Gisin, M. Rigo, J. Phys. A **28**, 7375 (1995) see also quant-ph/0102125
26. N. Gisin, Phys. Lett. A **242**, 1 (1998)

# 14 Secret Sides of Bell's Theorem

Artur Ekert

## 14.1 Is the Bell Theorem of any Practical Use?

In 1935 Albert Einstein together with Boris Podolsky and Nathan Rosen (EPR) published a paper in which they outlined how a 'proper' fundamental theory of nature should look like [1]. The EPR programme required completeness ("In a complete theory there is an element corresponding to each element of reality"), locality ("The real factual situation of the system A is independent of what is done with the system B, which is spatially separated from the former"), and defined the element of physical reality as "If, without in any way disturbing a system, we can predict with certainty the value of a physical quantity, then there exists an element of physical reality corresponding to this physical quantity". EPR then considered a thought experiment on two entangled particles which showed that quantum states cannot in all situations provide a complete description of physical reality. The EPR argument, as subsequently modified by *David Bohm*, goes as follows [2]. Imagine the singlet-spin state of two spin $\frac{1}{2}$ particles

$$| \Psi_- \rangle = \frac{1}{\sqrt{2}} \left( | \uparrow \rangle \, | \downarrow \rangle - | \downarrow \rangle \, | \uparrow \rangle \right) , \tag{14.1}$$

where the single particle kets $| \uparrow \rangle$ and $| \downarrow \rangle$ denote spin up and spin down with respect to some chosen direction. This state is spherically symmetric and the choice of the direction does not matter. In the following we will also use the "binary" notation: $|0\rangle$ for $| \uparrow \rangle$ and $|1\rangle$ for $| \downarrow \rangle$, and refer to particles with such two states as *qubits*. The two particles, which we label $A$ and $B$, are emitted from a source and fly apart. After they are sufficiently separated so that they do not interact with each other we can predict with certainity the $x$ component of spin of particle $A$ by measuring the $x$ component of spin of particle $B$. This is because the total spin of the two particles is zero and the spin components of the two particles must have opposite values. The measurement performed on particle $B$ does not disturb particle $A$ (by locality) therefore the $x$ component of spin is an element of reality according to the EPR criterion. By the same argument and by the spherical symmetry of state $| \Psi \rangle$ the $y$, $z$, or any other spin components are also elements of reality. However, since there is no quantum state of a spin $\frac{1}{2}$ particle in which all components of spin have definite values the quantum description of reality is not complete.

The EPR programme asked for a different description of quantum reality but until John Bell came up with his theorem it was not clear whether such a description was possible and if so whether it would lead to different experimental predictions [3]. Bell showed that the EPR propositions about locality, reality, and completeness are incompatibile with some quantum mechanical predictions involving entangled particles. The contradiction is revealed by deriving from the EPR programme experimentally testable inequalities. Subsequent extension of Bell's original theorem made experimental tests of the EPR programme feasible and quite a few of them have been performed. The experiments have supported quantum mechanical predictions.

However, the impact of Bell's Theorem goes far beyond the refutation of local hidden variables. The very trick used by John Bell to test the conceptual foundations of quantum theory was applied to solve the centuries old problem of perfectly secure communication [4]!

## 14.2 Is There a Perfect Cipher?

Despite a long and colorful history, cryptography became part of mathematics and information theory only last century, in the late 1940s, mainly as a result of the work of Claude Shannon of Bell Laboratories in New Jersey. *Shannon* showed that truly unbreakable ciphers do exist and, in fact, they had been known for over 30 years [5]. The one time pad, devised in about 1918 by an American Telephone and Telegraph engineer Gilbert Vernam, is one of the simplest and most secure encryption schemes. The message, also known as a plaintext, is converted to a sequence of numbers using publicly known digital alphabet (e.g. ASCII code) and then combined with another sequence of random numbers called *a key* to produce a cryptogram. Both sender and receiver must have two exact copies of the key beforehand; the sender needs the key to encrypt the plaintext, the receiver needs the exact copy of the key to recover the plaintext from the cryptogram. For example, if we choose a simple digital alphabet in which we use only capital letters and some punctuation marks such as

| A | B | C | D | E | ... | ... | X | Y | Z | | ? | , | . |
|---|---|---|---|---|---|---|---|---|---|---|---|---|---|
| 00 | 01 | 02 | 03 | 04 | ... | ... | 23 | 24 | 25 | 26 | 27 | 28 | 29 |

then we can illustrate the one-time-pad by the following simple example (we refer to the dietary requirements of 007):

| S | H | A | K | E | N | | N | O | T | | S | T | I | R | R | E | D |
|---|---|---|---|---|---|---|---|---|---|---|---|---|---|---|---|---|---|
| 18 | 07 | 00 | 10 | 04 | 13 | 26 | 13 | 14 | 19 | 26 | 18 | 19 | 08 | 17 | 17 | 04 | 03 |
| 15 | 04 | 28 | 13 | 14 | 06 | 21 | 11 | 23 | 18 | 09 | 11 | 14 | 01 | 19 | 05 | 22 | 07 |
| 03 | 11 | 28 | 23 | 18 | 19 | 17 | 24 | 07 | 07 | 05 | 29 | 03 | 09 | 06 | 22 | 26 | 10 |

In order to obtain the cryptogram $C$(sequence of digits in the bottom row) we add the plaintext numbers $P$ (the top row of digits) to the key numbers $K$ (the middle row of digits), which are randomly selected from between 0 and 29, and take the remainder after division of the sum by 30, that is we perform addition modulo 30. For example, the first letter of the message "S" becomes a number "18"in the plaintext, then we add $18 + 15 = 33$; $33 = 1 \times 30 + 3$, therefore we get 03 in the cryptogram. The encryption and decryption can be written as $P+K \pmod{30} = C$ and $C-K \pmod{30} = P$ respectively. The randomness of the key wipes out various frequency patterns in the cryptogam that are used by code-breakers to crack ciphers. Without the key the cryptogram looks like a random sequence of numbers.

The modern version "one-time pad" is based on binary representation of messages and keys. That is, the message is usually converted into a sequence of 0's and 1's and the key is another sequence of 0's and 1's of the same length. Each bit of the message is then combined with the respective bit of the key by addition in base 2 (logical XOR). As long as the key is truly random, has the same length as the message, and is never reused then the one-time pad is perfectly secure. So, if we have a truly unbreakable system, what is wrong with classical cryptography?

There is a snag, however. All one-time pads suffer from a serious practical drawback, known as the key distribution problem. Potential users have to agree secretly and in advance on the key – a long, random sequence of 0's and 1's. Once they have done this, they can use the key for enciphering and deciphering and the resulting cryptograms can be transmitted publicly such as by radio or in newspaper without compromising the security of messages. But the key itself must be established between the sender and the receiver by means of a very secure channel – for example, a very secure telephone line, a private meeting or hand-delivery by a trusted courier. Such a secure channel is usually available only at certain times and under certain circumstances. So users far apart, in order to guarantee perfect security of subsequent crypto-communication, have to carry around with them an enormous amount of secret and meaningless as such information (cryptographic keys), equal in volume to all the messages they might later wish to send.

Cryptologists and mathematicians tried very hard to eliminate the problem. The 1970s, for example, brought a clever mathematical discovery in the shape of "public key" systems. The two main public key cryptography techniques in use today are the Diffie–Hellman key exchange protocol [6] and the RSA encryption system (named after the three inventors, *Ron Rivest*, *Adi Shamir*, and *Leonard Adleman*) [7]. They were discovered in the academic community in 1976 and 1978, respectively. However, these techniques were known to the British government agencies prior to these dates, although this was not officially confirmed until recently. They were first discovered at the British Government Communication Headquarters in the early 1970s by James Ellis, who called them "Non-Secret Encryption".

In 1973, building on Ellis' idea, C. Cocks designed what we now call RSA, and in 1974 M. Williamson proposed what is essentially known today as the Diffie–Hellman key exchange protocol.

In the public-key systems users do not need to agree on a secret key before they send the message. They work on the principle of a safe with two keys, one public key to lock it, and another private one to open it. Everyone has a key to lock the safe but only one person has a key that will open it again, so anyone can put a message in the safe but only one person can take it out. The systems avoid the key distribution problem but unfortunately their security depends on unproven mathematical assumptions. For example, RSA – probably the most popular public key cryptosystem – gets its security from the difficulty of factoring large numbers. This means that if and when mathematicians or computer scientists come up with fast and clever procedures for factoring the whole privacy and discretion of public-key cryptosystems could vanish overnight.

Indeed, more recent work in quantum computation shows that quantum computers can, at least in principle, factor much faster than classical computers [8]! Thus in one sense public key cryptosystems are already insecure: any RSA-encrypted message that is recorded today will become readable moments after the first quantum computer is switched on, and therefore RSA cannot be used for securely transmitting any information that will still need to be secret on that happy day. Admittedly, that day is probably decades away, but can anyone prove, or give any reliable assurance, that it is? Confidence in the slowness of technological progress is all that the security of the RSA system now rests on.

Mathematics apart one can approach the problem from a different angle. Physicists view the key distribution as a physical process associated with sending information from one place to another and eavesdropping as measurements performed on carries of information. Until now, such eavesdropping has depended on the eavesdropper having the best possible technology. Suppose an eavesdropper is tapping a telephone line. Any measurement on the signal in the line may disturb it and so leave traces. Legitimate users can try to guard against this by making their own measurements on the line to detect the effect of tapping. However, the tappers will escape detection provided the disturbances they cause are smaller than the disturbances that the users can detect. So given the right equipment, eavesdropping can go undetected. Even if legitimate users do detect an eavesdropper, what do they conclude if one day they find no traces of interception? Has the eavesdropping stopped? Or has the eavesdropper acquired better technology? The way round this problem may lie in quantum physics, which brings us to an entirely new way of solving the key distribution problem.

## 14.3 Quantum Key Distribution

The quantum key distribution which I am going to discuss here is based on distribution of entangled particles [4]. The connection between the Bell theorem and cryptography sounds less surprising when one recalls the EPR definition of an element of reality:

*If, without in any way disturbing a system, we can predict with certainty the value of a physical quantity, then there exists an element of physical reality corresponding to this physical quantity.*

If this particular physical quantity is used to encode binary values of a cryptographic key then all an eavesdropper wants is an element of reality corresponding to the encoding observable (well, at least this was my way of thinking about it back in 1990)[1].

Building upon this vague idea one may suggest the key distribution performed via a quantum channel which consists of a source that emits pairs of spin $\frac{1}{2}$ particles in the singlet state as in (14.1). The particles fly apart along the $y$-axis towards the two legitimate users of the channel, Alice and Bob, who, after the particles have separated, perform measurements and register spin components along one of three directions, given by unit vectors $\boldsymbol{a}_i$ and $\boldsymbol{b}_j$ $(i, j = 1, 2, 3)$, respectively for Alice and Bob. For simplicity both $\boldsymbol{a}_i$ and $\boldsymbol{b}_j$ vectors lie in the $x$–$z$ plane, perpendicular to the trajectory of the particles, and are characterized by azimuthal angles: $\phi_1^{\mathrm{a}} = 0, \phi_2^{\mathrm{a}} = \frac{1}{4}\pi, \phi_3^{\mathrm{a}} = \frac{1}{2}\pi$ and $\phi_1^{\mathrm{b}} = \frac{1}{4}\pi, \phi_2^{\mathrm{b}} = \frac{1}{2}\pi, \phi_3^{\mathrm{b}} = \frac{3}{4}\pi$. Superscripts "a" and "b" refer to Alice's and Bob's analysers, respectively, and the angle is measured from the vertical $z$-axis. The users choose the orientation of the analysers randomly and independently for each pair of the incoming particles. Each measurement, in $\frac{1}{2}\hbar$ units, can yield two results, $+1$ (spin up or bit value 0 ) and $-1$ (spin down or bit value 1), and can potentially reveal one bit of information. Alice and Bob keep separate records which list, for each pair of incoming particles, the orientation of the local analyser and the registered bit value.

The quantity

$$E(\boldsymbol{a}_i, \boldsymbol{b}_j) = P_{++}(\boldsymbol{a}_i, \boldsymbol{b}_j) + P_{--}(\boldsymbol{a}_i, \boldsymbol{b}_j) - P_{+-}(\boldsymbol{a}_i, \boldsymbol{b}_j) - P_{-+}(\boldsymbol{a}_i, \boldsymbol{b}_j) \tag{14.2}$$

is the correlation coefficient of the measurements performed by Alice along $\boldsymbol{a}_i$ and by Bob along $\boldsymbol{b}_j$. Here $P_{\pm\pm}(\boldsymbol{a}_i, \boldsymbol{b}_j)$ denotes the probability that result $\pm 1$ has been obtained along $\boldsymbol{a}_i$ and $\pm 1$ along $\boldsymbol{b}_j$. According to the quantum rules

$$E(\boldsymbol{a}_i, \boldsymbol{b}_j) = \langle \boldsymbol{a}_i \cdot \boldsymbol{\sigma} \otimes b_j \cdot \boldsymbol{\sigma} \rangle = -\boldsymbol{a}_i \cdot \boldsymbol{b}_j \ , \tag{14.3}$$

[1] This entanglement based key distribution [4] has been discovered independently from the key distribution based on partial indistinguishibility of non-orthogonal state vectors, which was pioneered by *Wiesner* [9], and subsequently developed into the full fledged key distribution scheme by *Bennett* and *Brassard* [10].

where $\boldsymbol{\sigma}$ represents the three Pauli matrices $\sigma_x$, $\sigma_y$, $\sigma_z$ , and the averaging is performed for the singlet state. For the two pairs of analysers of the same orientation ($\boldsymbol{a}_2, \boldsymbol{b}_1$ and $\boldsymbol{a}_3, \boldsymbol{b}_2$) quantum mechanics predicts total anticorrelation of the results obtained by Alice and Bob: $E(\boldsymbol{a}_2, \boldsymbol{b}_1) = E(\boldsymbol{a}_3, \boldsymbol{b}_2) = -1$.

For the purpose of what follows it is instructive to derive (14.3) by writing the singlet state as the density operator in the $\sigma_a \otimes \sigma_b$ basis $(a, b = x, y, z)$,

$$|\Psi_-\rangle \langle\Psi_-| = \frac{1}{4}(\mathbf{1} - \sigma_x \otimes \sigma_x - \sigma_y \otimes \sigma_y - \sigma_z \otimes \sigma_z) \tag{14.4}$$

and then evaluate

$$\mathrm{Tr}\ [(\boldsymbol{a}_i \cdot \boldsymbol{\sigma} \otimes \boldsymbol{b}_j \cdot \boldsymbol{\sigma})\ |\Psi_-\rangle \langle\Psi_-|] \tag{14.5}$$

using, for example, the identity

$$(\boldsymbol{a} \cdot \boldsymbol{\sigma})(\boldsymbol{b} \cdot \boldsymbol{\sigma}) = \boldsymbol{a} \cdot \boldsymbol{b}\,\mathbf{1} + \mathrm{i}\,(\boldsymbol{a} \times \boldsymbol{b}) \cdot \boldsymbol{\sigma}\ , \tag{14.6}$$

together with $\sigma^2_{x,y,z} = 1$, and $\mathrm{Tr}\ \sigma_{x,y,z} = 0$.

Let us now define quantity $S$ composed of the correlation coefficients for which Alice and Bob used analysers of different orientation

$$S = E(\boldsymbol{a}_1, \boldsymbol{b}_1) - E(\boldsymbol{a}_1, \boldsymbol{b}_3) + E(\boldsymbol{a}_3, \boldsymbol{b}_1) + E(\boldsymbol{a}_3, \boldsymbol{b}_3)\ . \tag{14.7}$$

This is the same $S$ as in the generalised Bell theorem proposed by *Clauser*, *Horne*, *Shimony*, and *Holt* [11] (CHSH). For the singlet state quantum mechanics requires

$$S = -2\sqrt{2}\ , \tag{14.8}$$

and all local theories which attribute elements of reality to measured properties satisfy the CHSH inequality,

$$|S| \le 2\ . \tag{14.9}$$

Let us try to use this inequality as a criterion for a secure key distribution. After the transmission has taken place, Alice and Bob can announce in public the orientations of the analysers they have chosen for each particular measurement (N.B. results of the measurements remain seceret) and divide the measurements into two separate groups: a first group for which they used different orientation of the analysers, and a second group for which they used the same orientation of the analysers. They discard all measurements in which either or both of them failed to register a particle at all. Subsequently Alice and Bob can reveal publicly the results they obtained but within the first group of measurements only. This allows them to establish the value of $S$, which if the particles were not directly or indirectly "disturbed" should reproduce the result of (14.8). This assures the legitimate users that the results they obtained within the second second group of measurements are anticorrelated and can be converted into a secret string of bits – the key.

An eavesdropper, Eve, cannot elicit any information from the particles while in transit from the source to the legitimate users, simply because there is no information encoded there! The information "comes into being" only after the legitimate users perform measurements and communicate in public afterwards. Eve may try to substitute her own prepared data for Alice and Bob to misguide them, but as she does not know which orientation of the analysers will be chosen for a given pair of particles there is no good strategy to escape being detected. In this case her intervention will be equivalent to introducing elements of *physical reality* to the spin components and will lower $S$ below its 'quantum' value. Indeed, suppose that Eve prepares each particle in each pair separately so that each individual particle in the pair has a well defined spin in some direction. These directions may vary from pair to pair so we can say that she prepares with probability $p(\boldsymbol{n}_a, \boldsymbol{n}_b)$ Alice's particle in state $|\,\boldsymbol{n}_a\rangle$ and Bob's particle in state $|\,\boldsymbol{n}_b\rangle$, where $\boldsymbol{n}_a$ and $\boldsymbol{n}_b$ are two unit vectors describing the spin orientations. The density operator for each pair is

$$\rho = \int p(\boldsymbol{n}_a, \boldsymbol{n}_b)\,|\,\boldsymbol{n}_a\rangle\,\langle\boldsymbol{n}_a\,|\otimes|\,\boldsymbol{n}_b\rangle\,\langle\boldsymbol{n}_b\,|\;d\boldsymbol{n}_a\;d\boldsymbol{n}_b\;. \tag{14.10}$$

Equation (14.7) with appropriately modified correlation coefficients reads

$$\begin{aligned} S = \int p(\boldsymbol{n}_a, \boldsymbol{n}_b) d\boldsymbol{n}_a d\boldsymbol{n}_b\; [(\boldsymbol{a}_1 \cdot \boldsymbol{n}_a)(\boldsymbol{b}_1 \cdot \boldsymbol{n}_b) - (\boldsymbol{a}_1 \cdot \boldsymbol{n}_a)(\boldsymbol{b}_3 \cdot \boldsymbol{n}_b) \\ + (\boldsymbol{a}_3 \cdot \boldsymbol{n}_a)(\boldsymbol{b}_1 \cdot \boldsymbol{n}_b) + (\boldsymbol{a}_3 \cdot \boldsymbol{n}_a)(\boldsymbol{b}_3 \cdot \boldsymbol{n}_b)]\;, \end{aligned} \tag{14.11}$$

and leads to

$$S = \int p(\boldsymbol{n}_a, \boldsymbol{n}_b) d\boldsymbol{n}_a d\boldsymbol{n}_b [\sqrt{2}\boldsymbol{n}_a \cdot \boldsymbol{n}_b] \tag{14.12}$$

which implies

$$-\sqrt{2} \le S \le \sqrt{2}\;, \tag{14.13}$$

for any state preparation described by the probability distribution $p(\boldsymbol{n}_a, \boldsymbol{n}_b)$.

This is the case where Eve, who has total control over the state of individual particles, will always have the edge and Alice and Bob should abandon establishing the key; they will learn about it by estimating $|S|$ which in this case will always be smaller than $\sqrt{2}$. However, this is a negative statement – it does not tell us for which values of $S$ Alice and Bob can establish a secret key. Let us investigate if there is any cryptographic meaning to the CHSH threshold, $|S| = 2$.

## 14.4 Eavesdropping Revisited

The eavesdropping analysis presented above is merely a sketch. Clearly Eve can prepare more complicated states. The question is, what kind of states Eve should prepare and what kind of procedures she should implement in order to maximise her chances of guessing the key bits correctly and to minimise the disturbance. Of course Eve is bound to introduce some disturbance if she eavesdropps. Her only chance of avoiding detection is to hide behind what to Alice and Bob may look like an environmental noise in the channel. Let us assume that the noise is symmetrical in the $x$–$z$ plane, i.e. we require that

$$E(\boldsymbol{a},\boldsymbol{b}) = \langle \boldsymbol{a}\cdot\boldsymbol{\sigma}\otimes b\cdot\boldsymbol{\sigma}\rangle = -\eta\, \boldsymbol{a}\cdot\boldsymbol{b}\,, \tag{14.14}$$

for any two unit vectors $\boldsymbol{a}$ and $\boldsymbol{b}$ in the $x$–$z$ plane and for some fixed $0 \le \eta \le 1$. The noise might show asymmetry if the $y$ components of $\boldsymbol{a}$ and $\boldsymbol{b}$ were taken into account, however, Alice and Bob have to follow a prescribed protocol, and this one excludes measurements with non-zero $y$ components of $\boldsymbol{a}$ and $\boldsymbol{b}$. Of course, Alice and Bob may consider including such measurements but this would be a new protocol with a new eavesdropping method.

Equation (14.14) demands that the reduced density operator of the two particles $A$ and $B$ is of the form,

$$\rho = A\,|\Psi_-\rangle\,\langle\Psi_-| + B\,|\Phi_+\rangle\,\langle\Phi_+| + C\frac{1}{4}\mathbf{1}\,, \tag{14.15}$$

where $A+B+C=1$ (N.B. this is not a convex sum, negative values of $A$, $B$, and $C$ are allowed). This form follows from the fact that both $\mathbf{1}$ and the two states

$$|\Psi_-\rangle\,\langle\Psi_-| = \frac{1}{4}(\mathbf{1} - \sigma_x\otimes\sigma_x - \sigma_y\otimes\sigma_y - \sigma_z\otimes\sigma_z)\,, \tag{14.16}$$

$$|\Phi_+\rangle\,\langle\Phi_+| = \frac{1}{4}(\mathbf{1} + \sigma_x\otimes\sigma_x - \sigma_y\otimes\sigma_y + \sigma_z\otimes\sigma_z) \tag{14.17}$$

are invariant under rotations in the $x$–$z$ plane. For state $\rho$ we obtain

$$\langle \boldsymbol{a}\cdot\boldsymbol{\sigma}\otimes\boldsymbol{b}\cdot\boldsymbol{\sigma}\rangle = \mathrm{Tr}\,\rho\,(\boldsymbol{a}\cdot\boldsymbol{\sigma}\otimes b\cdot\boldsymbol{\sigma}) = -\,(A-B)\,\boldsymbol{a}\cdot\boldsymbol{b}\,. \tag{14.18}$$

Eve can prepare the state $\rho$ by preparing the two particles, and an ancilla $E$ in an entangled state,

$$\sqrt{F}\,\frac{1}{\sqrt{2}}\,(|01\rangle\,|E_{01}\rangle + |10\rangle\,|E_{10}\rangle) + \sqrt{D}\,\frac{1}{\sqrt{2}}\,(|00\rangle\,|E_{00}\rangle + |11\rangle\,|E_{11}\rangle)\,, \tag{14.19}$$

where we switched to more convenient notation: $|0\rangle$ for spin up $|\uparrow\rangle$ and $|1\rangle$ for spin down $|\downarrow\rangle$ along *any* direction in the $x$–$z$ plane. Indeed, tracing over the ancilla we obtain $\rho$ as in (14.15) provided that $F = 1/2(1+A-B)$,

$D = 1/2(1 - A + B)$, and that normalised but not necessarily mutually orthogonal states of the ancilla $|E_{ij}\rangle$ satisfy

$$\langle E_{01}|E_{10}\rangle = \frac{A}{F} = \cos\alpha \quad , \quad \langle E_{00}|E_{11}\rangle = \frac{B}{D} = \cos\beta, \tag{14.20}$$

for some $\alpha$ and $\beta$ (this convenient parametrisation is taken from [12]). All the remaining inner products are zero, i.e. states $\{|E_{01}\rangle, |E_{10}\rangle\}$ and $\{|E_{00}\rangle, |E_{11}\rangle\}$ belong to orthogonal subspaces.

The vectors $|E_{ij}\rangle$ change when we move from one basis to another in the $x$–$z$ plane but their inner products $\langle E_{ij}|E_{nm}\rangle$ remain invariant under all rotations in that plane.

Now the eavesdropping proceeds as follows. Eve prepares the state (14.19), sends the particles $A$ and $B$ to Alice and Bob, respectively, and keeps $E$. She then waits for public communication between Alice and Bob. When the orientations of the analysers are revealed Eve follows the algorithm:

- If the orientations are different ignore the ancilla,
- If the orientations are the same identify the state of the ancilla.

The second point is not trivial, however, Eve knows the orientation of the two analysers and therefore she knows that her ancilla is in one of the four states $|E_{00}\rangle$, $|E_{01}\rangle$, $|E_{10}\rangle$, or $|E_{11}\rangle$. She also knows (e.g. from [13]) the optimal measurement that can distinguish between two given non-orthogonal states $|E_{ij}\rangle$ and $|E_{mn}\rangle$ with the minimal probability of error, which is

$$\frac{1}{2}\left(1 - \sqrt{1 - |\langle E_{ij}|E_{mn}\rangle|^2}\right). \tag{14.21}$$

Eve can first check whether the state of the ancilla is in the subspace spanned by $\{|E_{01}\rangle, |E_{10}\rangle\}$ (probability $F$) or in the orthogonal subspace spanned by $\{|E_{00}\rangle, |E_{11}\rangle\}$ (probability $D$). This can be done without any errors. Then she can apply the optimal measurement to distinguish either between $|E_{01}\rangle, |E_{10}\rangle$ or between $|E_{00}\rangle, |E_{11}\rangle$. This procedure gives her the bit values registered by Alice and Bob with the error rate

$$Q_E = F\,\frac{1}{2}\,(1 - \sin\alpha) + D\,\frac{1}{2}\,(1 - \sin\beta)\,. \tag{14.22}$$

Fixing the disturbance of the correlations

$$\eta = A - B = F\cos\alpha - D\cos\beta \tag{14.23}$$

Eve can minimise her error rate $Q_E$ by choosing $\cos\alpha = -\cos\beta$, which gives

$$Q_E = \frac{1}{2}(1 - \sin\alpha) \quad , \quad \eta = \cos\alpha. \tag{14.24}$$

The error rate in the generated key is

$$Q_{AB} = \frac{1}{2}(1 - \cos\alpha), \tag{14.25}$$

and it matches Eve's error $Q_E = Q_{AB}$, for $\cos\alpha = \sin\alpha = 1/\sqrt{2}$, i.e. in terms of the CHSH inequality exactly for

$$|S| = 2. \tag{14.26}$$

Thus the CHSH threshold corresponds to the crossing point of the two error rates. This point is of some significance in cryptanalysis. It is, roughly speaking, the maximal error rate at which Alice and Bob can establish a secure key using some prescribed error correcting codes and without any further communication in public (see, for example, [14]). Thus the positive statement is: Alice and Bob can establish a secret key whenever $|S| > 2$.

## 14.5 Quantum Privacy Amplification

In fact Alice and Bob can establish a secret key even for some values of $S$ which are smaller than 2. For this Alice and Bob may use a *Quantum Privacy Amplification*(QPA) [15]. The essential element of the QPA procedure is the 'entanglement purification' scheme [16]. Without going into technical details one can describe the QPA as an iterative quantum algorithm which, if performed with perfect accuracy, starting with a collection of EPR-pairs in mixed states, would discard some of them and leave the remaining ones in states converging to the pure singlet state. This means that $|S|$ for the remaining pairs will converge to $2\sqrt{2}$. The QPA can be performed by Alice and Bob at distant locations by a sequence of local unitary operations and measurements which are agreed upon by communication over a public channel. It has been shown that any entangled states of two qubits can be purified [17]. Taking the density operator (14.15) and inserting the optimal coefficients we obtain

$$\begin{aligned}\rho(\alpha) = &\frac{1}{2}\cos\alpha(1+\cos\alpha)\,|\Psi_-\rangle\,\langle\Psi_-| \\ &-\frac{1}{2}\cos\alpha(1-\cos\alpha)\,|\Phi_+\rangle\,\langle\Phi_+| + \frac{1}{4}\sin^2\alpha\mathbf{1}.\end{aligned} \tag{14.27}$$

It gives $|S(\alpha)| = \cos\alpha 2\sqrt{2}$. Now, using the partial transposition test [18, 19] we can check that $\rho(\alpha)$ is entangled when $\cos\alpha > \sqrt{2}-1$. This implies that if Alice and Bob are prepared to use the QPA then they can establish a secret key for values $|S| > 2(2-\sqrt{2})$. (N.B. this does not contradict (14.13) where we did not require the rotational symmetry in the $x$–$z$ plane, such a requirement would give $S = 0$.)

I feel I should add here that, unfortunately, the QPA is rather inefficient – many pairs of particles are discarded in the process. One should also mention here that there are classical techniques, such as "advantage distillation", which can supplement quantum key distributions and guarantee its secrecy for some $|S| < 2$ (see for example [20]), however, these techniques are equally inefficient. Thus the CHSH inequality may remain as a clean criterion for an efficient and secure quantum key distribution, at least for all practical purposes.

## 14.6 Concluding Remarks

This brief overview has only scratched the surface of the many activities that are presently being pursued under the heading of quantum cryptography. For example, one may now venture into more complicated security analysis involving methods in which Eve, instead of pair by pair preparations, prepares several pairs of particles in one go, entangles them with more complicated ancilla, and sends them to Alice and Bob. This kind of methods do not lead to significantly different security limits on error rates but are nonetheless interesting from the theoretical point of view. One can also discuss alternative key distribution protocols, or other cryptographic tasks. However, let me stop here hoping that this story about the 'practical' aspect of Bell's theorem shows quite clearly that a border between blue sky and applied research is quite blurred.

## Acknowledgements

The author is supported by the Royal Society, London. This work was supported in part by the European TMR Research Network ERP-4061PL95-1412, EPSRC, and Elsag plc.

## References

1. A. Einstein, B. Podolsky, N. Rosen, Phys. Rev. **47**, 777 (1935)
2. D. Bohm, *Quantum Theory*, (Prentice-Hall, Englewood Cliffs, NJ 1951)
3. J.S. Bell, Physics **1**, 195 (1964); **23**, 880 (1969)
4. A. Ekert, Phys. Rev. Lett. **67**, 661 (1991)
5. C.E. Shannon, Bell. Syst. Tech. J. **28**, 657–715 (1949).
6. W. Diffie, M.E. Hellman, IEEE Trans. Inf. Theory **IT-22**, 644 (1976)
7. R. Rivest, A. Shamir, L. Adleman, On Digital Signatures and Public-Key Cryptosystems, MIT Laboratory for Computer Science, Technical Report, MIT/LCS/TR-212 (January 1979)
8. P.W. Shor, Proc. 35th Annual Symposium on Foundations of Computer Science, Santa Fe, NM, USA, 1994, IEEE Comp. Soc. Press, 124–134 (1994)
9. S. Wiesner, SIGACT News, **15**, 78 (1983); original manuscript written *circa* 1970
10. C.H. Bennett, G. Brassard, Proc. IEEE Int. Conference on Computers, Systems and Signal Processing (IEEE, New York 1984)

11. J.F. Clauser, M.A. Horne, A. Shimony, R.A. Holt, Phys. Rev. Lett. **23**, 880 (1969)
12. J.I. Cirac, N. Gisin, Coherent eavesdropping strategies for the 4 state quantum cryptography protocol, quant-ph/9702002 (1997)
13. C.W. Hellstrom, *Quantum Detection and Estimation Theory* (Academic Press, 1976)
14. I. Csiszár, J. Körner, IEEE Trans. Inf. Theory **IT-24**, 339–348 (1978).
15. D. Deutsch, A. Ekert, R. Jozsa, C. Macchiavello, S. Popescu, A. Sanpera, Phys. Rev. Lett. **77**, 2818 (1996)
16. C.H. Bennett, G. Brassard, S. Popescu, B. Schumacher, J. Smolin, W.K. Wootters, Phys. Rev. Lett. **76**, 722 (1996)
17. M. Horodecki, P. Horodecki, R. Horodecki, Phys. Rev. Lett. **78**, 574 (1997)
18. A. Peres, Phys. Rev. Lett. **77**, 1413 (1996)
19. M. Horodecki, P. Horodecki, R. Horodecki, Phys. Lett. A **223**, 1 (1996)
20. U.M. Maurer, IEEE Trans. Inf. Theory **IT-39**, 733 (1993)

# 15 An Impossible Necklace

Lev Vaidman

A game which a quantum team can win against any classical team is proposed. The game is based on a Zeno-type proof of Bell's inequality. The unusual feature of this game is that a person who does not know about quantum mechanics might be led to belive in the existence of impossible necklaces.

When I approached John Bell in 1987 asking to be his Post-Doctoral student working on foundations of quantum mechanics, his reply was that his main work is particle physics and, therefore, he cannot take anybody who wants to work on foundations. Nevertheless, John Bell played an important role in my research. I find his works to be exceptionally clear and of tremendous importance. I met Bell in the relaxed atmosphere of Erice in 1989. Our discussions there and following e-mail correspondence played a very important role in forming my views. The most important influence on me was, however, made by his works showing the miraculous features of quantum theory. In this brief note I want to present an elaboration of a Bell-type proof [1] which exhibits one of such miracles.

Conceptually, the most simple, surprising, and convincing among the Bell-type experiments is *Mermin's* version [2] of the *Greenberger–Horne–Zeilinger* (GHZ) setup [3]. I find that it can be best explained as a game [4]. A team of three players is allowed to make any preparations before the players are taken to three remote locations. Then, at a certain time, each player is asked one of two possible questions: "What is $X$?" or "What is $Y$?" to which they must quickly give one of the answers: "1" or "$-1$". According to the rules of the game, either all players are asked the $X$ question, or only one player is asked the $X$ question and the other two are asked the $Y$ question. The team wins if the product of their three answers is $-1$ in the case of three $X$ questions and is 1 in the case of one $X$ and two $Y$ questions. It is a simple exercise to prove that if the answer of each player is determined by a local hidden variable theory, then the best strategy of the team will lead to 75% probability to win. However, a quantum team equipped with ideal devices can win with certainty. Each player performs a spin measurement of a spin-1/2 particle: $\sigma_x$ measurement for the $X$ question and $\sigma_y$ measurement for the $Y$ question and gives the answer 1 for spin "up" and $-1$ for spin "down". Quantum theory ensures that if the players have particles prepared in the GHZ state, the team always wins. Actually constructing such devices and seeing that, indeed, the quantum team wins the game with probability significantly larger than 75% will be a very convincing proof of Bell-type inequalities.

Apart from the GHZ game just described there have been several other proposals: an interesting variation of the GHZ game by *Steane* and *van Dam* [5], a game based on the original Bell proof by *Tsirelson* [6], the "quantum cakes" game based on a non-maximally entangled state by *Kwiat* and *Hardy* [7] (see also related experiment [8]). Note also the proposal of *Cabello* [10] for a two-party Bell-inequality proof which can be transformed into a game, too. Let me present here one more game. My game is called an "impossible necklace" and it is based on the Zeno-type Bell inequalities proof [9].

A team of two players wants to persuade a third party, "the interrogator," that they found a secret of making an "impossible necklace". The impossible two-colored neclace has an even number of beads $N$ and all adjacent beads are of different colors except beads 1 and $N$ which are of the same color. The team does not want to reveal the "secret coloring", but the players are ready to reveal the colors of any two adjacent beads of the necklace. They claim to have identical necklaces of this kind, one necklace for each player. The interrogator arranges to ask one player the color of any single bead and ask the other player, at a space-like separated region, the color of one of the adjacent beads. If the team succeeds in giving the correct answers in many repeated experiments (with new necklaces each time), a naive interrogator might be persuaded that the team knows how to make such necklaces. Indeed, if it is a "classical team", and the players decide in advance what answer they will give for every question, then the probability to fail is at least $1/N$. (There are $N$ different pairs and there is no way to arrange that all have correct coloring.) Therefore, the probability to pass the test, say $5N$ times is

$$prob_{\text{classical}} = \left(1 - \frac{1}{N}\right)^{5N} \sim \mathrm{e}^{-5} \sim 0.01. \tag{15.1}$$

The quantum team can do much better. The players do not make any necklaces. Each player takes with him a spin $-\frac{1}{2}$ particle from the EPR (Einstein–Podolski–Rosen) pair. When a player is asked the color of a bead $i$, he measures the spin component in the direction $\hat{\theta}_i$ in the $x$–$z$ plane which makes an angle $\theta_i = \pi i/N$ with the $z$-axis. He says "green" if the result is "up", and "red" if the result is "down". His partners do the same. For all pairs, the measurements are in the directions which differ by the angle $\pi/N$ except for the pair $\{1, N\}$ in which case the angle is $\pi(N-1)/N$. Therefore, the probability to fail the test is $\sin^2(\pi/2N)$. The probability to pass $5N$ tests is

$$prob_{\text{quantum}} = \left(1 - \sin^2 \frac{\pi}{2N}\right)^{5N} \sim \left(1 - \frac{\pi^2}{4N^2}\right)^{5N} \sim \mathrm{e}^{\frac{-5\pi^2}{4N}}. \tag{15.2}$$

For $N = 100$ the quantum team has probability of almost 90% to succeed, compared with 1% of a classical team.

Technological problems will not allow an experiment with a large number $N$ in a near future. Putting aside the attempt to "fool" the interrogator that the team has impossible necklaces, the game can be defined as the competition of two-player teams to pass the interrogator tests a maximal number of times. For any number $N \geq 4$, the quantum team has an advantage over a classical team, so this game is a realistic proposal for demonstrating Bell-type inequalities. A succesful experiment of this type will show that quantum technology is capable of performing communication tasks which are impossible when classical devices are used.

## Acknowledgements

It is a pleasure to thank Hans Christian von Bayer for helpful discussions. This research was supported in part by grant 62/01 of the Israel Science Foundation and the EPSRC grant GR/N33058.

## References

1. J.S. Bell, Physics **1**, 195 (1964)
2. D.N. Mermin, Am. Phys. **58**, 731 (1990)
3. D.M. Greenberger, M.A. Horne, A. Zeilinger, in *Bell Theorem, Quantum Theorey and Conceptions of the Universe*, M. Kafatos, (ed.) (Kluwer Academic, Dordrecht 1989) p.69
4. L. Vaidman, Found. Phys. **29**, 615 (1999)
5. A. Steane, W. van Dam, Phys. Today **53**, 35 (2000)
6. B. Tsirelson, Lecture Notes, Tel-Aviv University (1996)
7. P.G. Kwiat, L. Hardy, Am. J. Phys. **68**, 33 (2000)
8. G. Brida et al., Phys. Lett. A 268 (2000)
9. E.J. Squires, L. Hardy, H.R. Brown, Stud. Hist. Philos. Sci. **25**, 425 (1994)
10. A. Cabello, Phys. Rev. Lett. **86**, 1911 (2001)

# 16 Multi-Photon Entanglement and Quantum Non-Locality

Jian-Wei Pan and Anton Zeilinger

We review recent experiments concerning multi-photon Greenberger–Horne–Zeilinger (GHZ) entanglement. We have experimentally demonstrated GHZ entanglement of up to four photons by making use of pulsed parametric down-conversion. On the basis of measurements on three-photon entanglement, we have realized the first experimental test of quantum non-locality following from the GHZ argument. Not only does multi-particle entanglement enable various fundamental tests of quantum mechanics versus local realism, but it also plays a crucial role in many quantum-communication and quantum-computation schemes.

## 16.1 Introduction

Ever since its introduction in 1935 by *Schrödinger* [1] entanglement has occupied a central position in the discussion of the non-locality of quantum mechanics. Originally the discussion focused on the proposal by *Einstein, Podolsky* and *Rosen* (EPR) of measurements performed on two spatially separated entangled particles [2]. Most significantly John Bell then showed that certain statistical correlations predicted by quantum physics for measurements on such two-particle systems cannot be understood within a realistic picture based on local properties of each individual particle – even if the two particles are separated by large distances [3].

An increasing number of experiments on entangled particle pairs having confirmed the statistical predictions of quantum mechanics [4–6] and have thus provided increasing evidence against local realistic theories. However, one might find some comfort in the fact that such a realistic and thus classical picture can explain perfect correlations and is only in conflict with statistical predictions of the theory. After all, quantum mechanics is statistical in its core structure. In other words, for entangled-particle pairs the cases where the result of a measurement on one particle can definitely be predicted on the basis of a measurement result on the other particle can be explained by a local realistic model. It is only that subset of statistical correlations where the measurement results on one particle can only be predicted with a certain probability which cannot be explained by such a model.

Surprisingly, in 1989 it was shown by *Greenberger*, *Horne* and *Zeilinger* (GHZ) that for certain three- and four-particle states [7, 8] a conflict with local realism arises even for perfect correlations. That is, even for those cases where, based on the measurement on $N-1$ of the particles, the result of the measurement on particle $N$ can be predicted with certainty. Local realism and quantum mechanics here both make definite but completely opposite predictions.

The main purpose of this paper is to present a tutorial review on the recent progress concerning the first experimental realization of three- and four-photon GHZ entanglement [9, 10] and the first experimental test of GHZ theorem [11]. The paper is organized as follows: In Sect. 16.2 we briefly introduce the so-called GHZ theorem.In Sect. 16.3, we show in detail how pulsed parametric down-conversion can be used to generate multi-photon entanglement. In Sect. 16.4, we present the first three-particle test of local realism following from the GHZ theorem. Finally, the possible applications of the techniques developed in the experiments are briefly discussed in Sect. 16.5.

## 16.2 The GHZ Theorem

To show how the quantum predictions of GHZ states are in stronger conflict with local realism than the conflict for two-particle states as implied by Bell's inequalities, let us consider the following three-photon GHZ state:[1]

$$|\Psi\rangle = \frac{1}{\sqrt{2}}(|H\rangle_1 |H\rangle_2 |H\rangle_3 + |V\rangle_1 |V\rangle_2 |V\rangle_3) , \tag{16.1}$$

where $H$ and $V$ denote horizontal and vertical linear polarizations, respectively. This state indicates that the three photons are in a quantum superposition of the state $|H\rangle_1|H\rangle_2|H\rangle_3$ (all three photons are horizontally polarized) and the state $|V\rangle_1|V\rangle_2|V\rangle_3$ (all three photons are vertically polarized), with none of the photons having a well-defined state of its own.

Consider now measurements of linear polarization along directions $H'/V'$ rotated by 45° with respect to the original $H/V$ directions, or of circular polarization, $L/R$ (left-handed, right-handed). These new polarizations can be expressed in terms of the original ones as

$$|H'\rangle = \frac{1}{\sqrt{2}}(|H\rangle + |V\rangle) , \quad |V'\rangle = \frac{1}{\sqrt{2}}(|H\rangle - |V\rangle) , \tag{16.2}$$

$$|R\rangle = \frac{1}{\sqrt{2}}(|H\rangle + \mathrm{i}\,|V\rangle) , \quad |L\rangle = \frac{1}{\sqrt{2}}(|H\rangle - \mathrm{i}\,|V\rangle) . \tag{16.3}$$

Let us denote $|H\rangle$ by matrix $\begin{pmatrix}1\\0\end{pmatrix}$ and $|V\rangle$ by matrix $\begin{pmatrix}0\\1\end{pmatrix}$; they are thus the two eigenstates of the Pauli operator, $\sigma_z$, correspondingly with the eigenvalues $+1$ and $-1$. We can also easily verify that $|H'\rangle$ and $|V'\rangle$ or $|R\rangle$ and $|L\rangle$ are

[1] The same line of reasoning can also be applied to the four-particle case.

two eigenstates for the Pauli operator $\sigma_x$ or $\sigma_y$ with the values +1 and −1, respectively. For convenience we will refer to a measurement of the $H'/V'$ linear polarization as an $x$ measurement and of the $L/R$ circular polarization as a $y$ measurement.

By representing state (16.1) in the new states using (16.2) and (16.3), one obtains the quantum predictions for the measurements of these new polarizations. For example, for the case of the measurement of circular polarization on, say, both photons 1 and 2 and the measurement of linear polarization $H'/V'$ on photon 3, denoted as a $yyx$ experiment, the state may be expressed as

$$|\Psi\rangle = \frac{1}{2}(|R\rangle_1 |L\rangle_2 |H'\rangle_3 + |L\rangle_1 |R\rangle_2 |H'\rangle_3 + |R\rangle_1 |R\rangle_2 |V'\rangle_3 + |L\rangle_1 |L\rangle_2 |V'\rangle_3) . \tag{16.4}$$

This expression implies, first, that any specific result obtained in any individual or in any two-photon joint measurement is maximally random. For example, photon 1 will exhibit polarization $R$ or $L$ with the same probability of 50%, or photons 1 and 2 will exhibit polarizations $RL$, $LR$, $RR$, or $LL$ with the same probability of 25%. Second, given any two results of measurements on any two photons, we can predict with certainty the result of the corresponding measurement performed on the third photon. For example, suppose photons 1 and 2 both exhibit right-handed ($R$) circular polarization. By the third term in (16.4), photon 3 will definitely be $V'$ polarized.

By cyclic permutation, we can obtain analogous expressions for any experiment measuring circular polarization on two photons and $H'/V'$ linear polarization on the remaining one. Thus, in every one of the three $yyx$, $yxy$ and $xyy$ experiments, any individual measurement result – both for circular polarization and for linear $H'/V'$ polarization – can be predicted with certainty for every photon given the corresponding measurement results of the other two.

We now analyze the implications of these predictions from the point of view of local realism. First, note that the predictions are independent of the spatial separation of the photons and independent of the relative time order of the measurements. Let us thus consider the experiment to be performed such that the three measurements are performed simultaneously in a given reference frame, say, for conceptual simplicity, in the reference frame of the source. Thus we can employ the notion of Einstein locality, which implies that no information can travel faster than the speed of light. Hence the specific measurement result obtained for any photon must not depend on which specific measurement is performed simultaneously on the other two or on the outcome of these measurements. The only way then to explain from a local realistic point of view the perfect correlations discussed above is to assume that each photon carries elements of reality for both $x$ and $y$ measurements considered and that these elements of reality determine the specific individual measurement result [7, 8, 12].

Calling these elements of reality, of photon $i$, $X_i$ with values $+1(-1)$ for $H'(V')$ polarizations and $Y_i$ with values $+1(-1)$ for $R(L)$ polarizations we obtain the relations $Y_1Y_2X_3 = -1$, $Y_1X_2Y_3 = -1$ and $X_1Y_2Y_3$ in order to be able to reproduce the quantum predictions of (16.4) and its permutations [12].

We now consider a fourth experiment measuring linear $H'/V'$ polarization on all three photons, that is, an $xxx$ experiment. We investigate the possible outcomes that will be predicted by local realism based on the elements of reality introduced to explain the earlier $yyx$, $yxy$ and $xyy$ experiments.

Because of Einstein locality any specific measurement for $x$ must be independent of whether an $x$ or $y$ measurement is performed on the other photon. As $Y_iY_i = +1$, we can write $X_1X_2X_3 = (X_1Y_2Y_3) \cdot (Y_1X_2Y_3) \cdot (Y_1Y_2X_3)$ and obtain $X_1X_2X_3 = -1$. Thus from a local realistic point of view the only possible results for an $xxx$ experiment are $V_1'V_2'V_3'$, $H_1'H_2'V_3'$, $H_1'V_2'H_3'$, and $V_1'H_2'H_3'$.

How do these predictions of local realism for an $xxx$ experiment compare with those of quantum physics? If we express the state given in (16.1) in terms of $H'/V'$ polarization using (16.2), we obtain

$$
\begin{aligned}
|\Psi\rangle = \frac{1}{2}(&|H'\rangle_1 |H'\rangle_2 |H'\rangle_3 + |H'\rangle_1 |V'\rangle_2 |V'\rangle_3 \\
&+ |V'\rangle_1 |H'\rangle_2 |V'\rangle_3 + |V'\rangle_1 |V'\rangle_2 |H'\rangle_3) \, .
\end{aligned} \tag{16.5}
$$

We conclude that the local realistic model predicts none of the terms occurring in the quantum prediction and vice versa. This implies that, whenever local realism predicts a specific result definitely to occur for a measurement on one of the photons based on the results for the other two, quantum physics definitely predicts the opposite result. For example, if two photons are both found to be $H'$ polarized, local realism predicts the third photon to carry $V'$ polarization while the quantum state predicts $H'$ polarization. This is the GHZ contradiction between local realism and quantum physics.

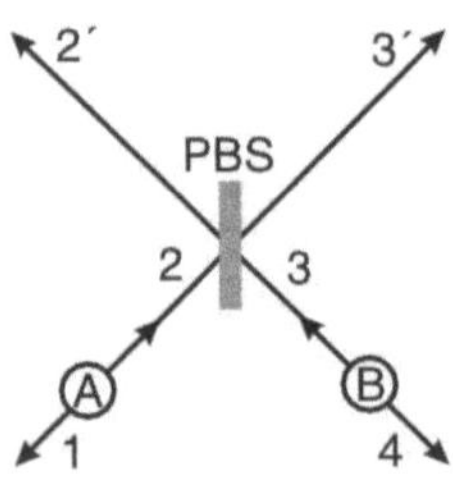


**Fig. 16.1.** Principle for observing three- or four-photon GHZ correlations. Sources A and B each deliver one entangled particle pair. A polarizing beamsplitter (PBS) combines modes 2 and 3. The two photons detected in its output port are either both $H$ (horizontally) or both $V$ (vertically) polarized, projecting the complete four-photon state into a GHZ state

In the case of Bell's inequalities for two photons the conflict between local realism and quantum physics arises for statistical predictions of the theory; but for three entangled particles the conflict arises even for the definite predictions. Statistics now only results from the inevitable experimental limitations occurring in any and every experiment, even in classical physics.

## 16.3 Experimental Multi-Photon GHZ Entanglement

Experimental testing of the GHZ theorem necessitates observations of multi-particle entanglement. The method used here to generate multi-photon GHZ entanglement is a further development of the techniques that have been used in our previous experiments on quantum teleportation [13] and entanglement swapping [14]. The main idea, as was put forward in [15], is to transform two independently created photon pairs into either three- or four-photon entanglement. The working principle is shown in Fig. 16.1.

Suppose that the two pairs are in the state

$$\begin{aligned} |\Psi^i\rangle_{1234} &= \frac{1}{\sqrt{2}}\left(|H\rangle_1|V\rangle_2 - |V\rangle_1|H\rangle_2\right) \\ &\otimes \frac{1}{\sqrt{2}}\left(|H\rangle_3|V\rangle_4 - |V\rangle_3|H\rangle_4\right) , \end{aligned} \tag{16.6}$$

which is a tensor product of two polarization entangled photon pairs.

One photon out of each pair is directed to the two inputs of a polarizing beam-splitter (PBS). Since the PBS transmits horizontal and reflects vertical polarization, coincidence detection between the two PBS outputs implies that photons 2 and 3 are either both horizontally polarized or both vertically polarized, and thus projects (16.6) onto a two-dimensional subspace spanned by $|V\rangle_1|H\rangle_2|H\rangle_3|V\rangle_4$ and $|H\rangle_1|V\rangle_2|V\rangle_3|H\rangle_4$.

After the PBS, the renormalized state corresponding to a four-fold coincidence is

$$|\Psi^f\rangle_{12'3'4} = \frac{1}{\sqrt{2}}\left(|H\rangle_1|V\rangle_{2'}|V\rangle_{3'}|H\rangle_4 + |V\rangle_1|H\rangle_{2'}|H\rangle_{3'}|V\rangle_4\right) , \tag{16.7}$$

which is a GHZ state of four particles.

The scheme described above does not only yield four-particle entanglement but – assuming perfect pair sources and single-photon detectors – could also produce freely propagating three-particle entangled states via so-called entangled entanglement [16]. For example, one could analyze the polarization state of photon $2'$ by passing it through a special PBS that transmits $H'$ polarizations but reflects $V'$ ones. Detecting one photon in one of the two outputs of this PBS makes sure that there will be exactly one photon in each of the outputs 1, $3'$, and 4. Correspondingly, the polarization state of the remaining three photons in modes 1, 3′, and 4 will then be projected onto either the state

$$|\Psi\rangle_{13'4} = \frac{1}{\sqrt{2}}\left(|H\rangle_1 |V\rangle_{3'} |H\rangle_4 + |V\rangle_1 |H\rangle_{3'} |V\rangle_4\right) , \qquad (16.8)$$

if one detects an $H'$ polarized photon in mode $2'$, or the state

$$|\Psi\rangle_{13'4} = \frac{1}{\sqrt{2}}\left(|H\rangle_1 |V\rangle_{3'} |H\rangle_4 - |V\rangle_1 |H\rangle_{3'} |V\rangle_4\right) , \qquad (16.9)$$

if one detects a $V'$ polarized photon in mode $2'$.

Note that, due to the absence of perfect pair sources and perfect single-photon detectors, in our experiments both three- and four-photon entanglements [9, 10] are observed only under the condition that there is one and only one photon in each of the four outputs. As there are other detection events where, for example, two photons appear in the same output port, this condition might raise doubts about whether such a source can be used to test local realism. The same question arose earlier for certain experiments involving photon pairs [17, 18], where a violation of Bell's inequality was only achieved under the condition that both detectors used register a photon. It was often believed [19, 20] that such experiments could never, not even in their idealized versions, be genuine tests of local realism. However, this has been disproved [21]. Following the same line of reasoning, it has been recently shown [22] that our procedure permits a valid GHZ test of local realism. In essence, both the Bell and the GHZ arguments exhibit a conflict between detection events and the ideas of local realism.

We now describe our experimental verification of multi-photon entanglement. Since the methods used in our three- and four-photon experiments are basically the same, in the following we will only present the experimental results on the observation of four-photon entanglement. For details of our three-photon experiment, please see [9].

In our experiment (Fig. 16.2) we create polarization-entangled photon pairs by spontaneous parametric down-conversion from an ultraviolet femtosecond pulsed laser ($\sim 200\,\mathrm{fs}$, $\lambda \simeq 394.5\,\mathrm{nm}$) in a $\beta$-$BaB_3O_6$(BBO) crystal [23]. The laser passes the crystal a second time having been reflected off a translatable mirror. In the reverse pass another conversion process may happen, producing an second entangled pair. One particle of each pair is steered to a polarizing beam-splitter, where the path lengths of each particle have been adjusted such that they arrive simultaneously. On the polarizing beam-splitter a horizontally polarized photon will always be transmitted, whereas a vertically polarized one will always be reflected, both with less than a $10^{-3}$ error rate. The two outputs of the polarizing beam-splitter are spectrally filtered (3.5 nm bandwidth) and monitored by fiber-coupled single-photon counters (D2 and D3). The filtering process stretches the coherent time to about 550 fs, substantially larger than the pump-pulse duration [24]. This

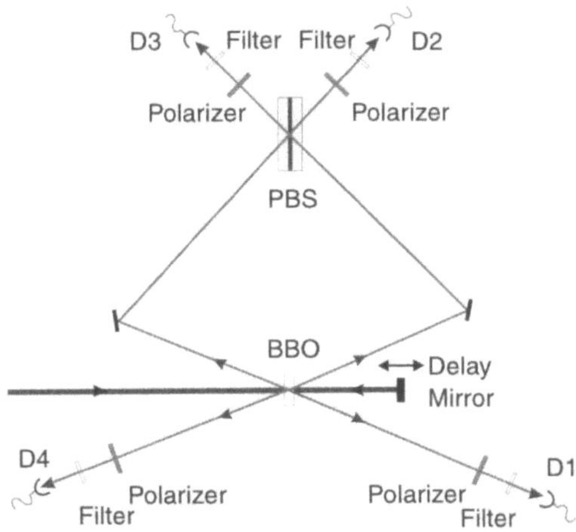


**Fig. 16.2.** Schematic of the experimental setup for the measurement of four-photon GHZ correlations. A pulse of UV light passes a BBO crystal twice to produce two entangled photon pairs. Coincidences between all four detectors, D1–4, exhibit GHZ entanglement

effectively erases any possibility of distinguishing the two photons according to their arrival time and therefore leads to interference.

The remaining two photons – one from each pair – pass identical filters in front of detectors D1 and D4 and are detected directly afterwards. In front of each of the four detectors we may insert a polarizer to assess the correlations with respect to various combinations of polarizer orientations. A correlation circuit extracts only those events where all four detectors registered a photon within a small time window of a few ns. This is necessary in order to exclude cases in which only one pair is created or two pairs in one pass of the pump pulse and none in the other.

To experimentally demonstrate that the state $|\Psi^f\rangle$ of (16.7) has been obtained, we first verified that under the condition of having a four-fold coincidence only the $HVVH$ and $VHHV$ components can be observed, but no others. This was done by comparing the count rates of all 16 possible polarization combinations, $HHHH \ldots VVVV$. The measurement results in the $H/V$ basis (Fig. 16.3) show that the signal-to-noise ratio defined as the ratio of any of the desired four-fold events ($HVVH$ and $VHHV$) to any of the 14 other non-desired ones is about 200:1.

Showing the existence of $HVVH$ and $VHHV$ terms alone is a necessary but not sufficient experimental criterion for the verification of the state $|\Psi^f\rangle$, since the above observation is, in principle, compliant both with $|\Psi^f\rangle$ and with a statistical mixture of $HVVH$ and $VHHV$. Thus, as a further test we have to demonstrate that the two terms $HVVH$ and $VHHV$ are indeed in coherent superposition.

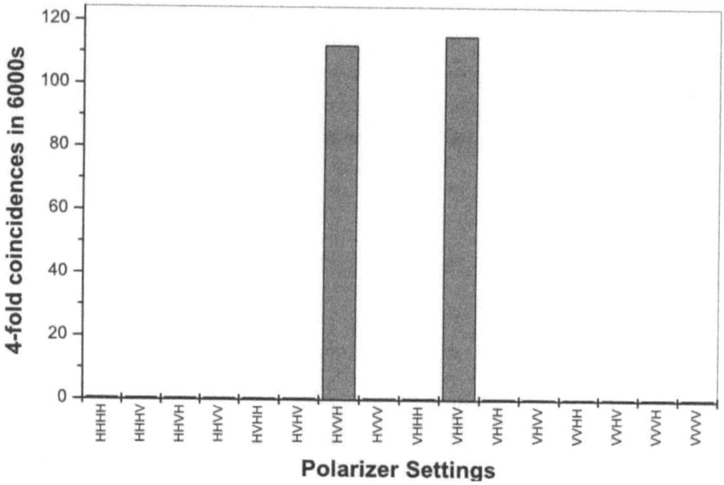


**Fig. 16.3.** Experimental data for horizontal and vertical polarizer settings. Only the two desired terms are present; all other terms which are not part of the state $|\Psi^f\rangle$ (16.7) are so strongly suppressed that they can hardly be discerned in the graph. The number of four-fold coincidences for any of the non-desired terms is 0.5 in 6000 s on average, i.e., seven events for all 14 possibilities

This was done by performing a polarization measurement in the $H'/V'$ basis. Transforming $|\Psi^f\rangle$ to the $H'/V'$ linear polarization basis yields an expression containing eight (out of 16 possible) terms, each with an even number of $|H'\rangle$ components. Combinations with odd numbers of $|H'\rangle$ components do not occur. As a test for coherence we can now check the presence or absence of various components. In Fig. 16.4 we compare the $(H'/H'/H'/H')$ and $(H'/H'/H'/V')$ count rates as a function of the pump delay mirror position. At zero delay – photons 2 and 3 arrive at the PBS simultaneously – the latter component is suppressed with a visibility of $0.79 \pm 0.06$. As explained in [24], many efforts have been made by us to obtain this high visibility reliably. In the experiment we observed that the most important ingredients for a high interference contrast were a high single pair entanglement quality, the use of narrow bandwidth filters, and the high quality of the polarizing beam-splitter.

These measurements clearly show that we obtained four-particle GHZ correlations. The quality of the correlations can be judged by the density matrix of the state

$$\rho = 0.89 \left|\Psi^f\right\rangle \left\langle\Psi^f\right|_{12'3'4} + 0.11 \left|\Phi\right\rangle \left\langle\Phi\right|_{12'3'4} , \qquad (16.10)$$

where $|\Phi\rangle = 1/\sqrt{2}(|HVVH\rangle - |VHHV\rangle)$. This density matrix describes our data under the experimentally well-justified assumption that only phase errors in the $H/V$ basis are present, which appear as bit-flip errors in the $H'/V'$ basis (see Fig. 16.4).

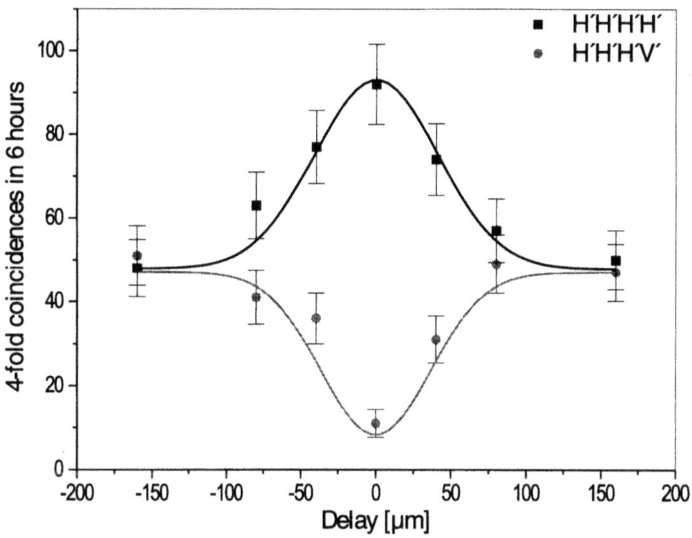


**Fig. 16.4.** Experimental data for 45° polarizer settings. The difference between the four-fold coincidence count rates for $(H'/H'/H'/H')$ and $(H'/H'/H'/V')$ shows that the amplitudes depicted in Fig. 16.3 are in coherent superposition. Maximum interference occurs at zero delay between photons 2 and 3 arriving at the polarizing beam-splitter. The Gaussian curves that roughly connect the data points are only shown to guide the eye. Visibility and errors are calculated only from the raw data

Since performing a polarization decomposition in the $H'/V'$ basis in outputs $2'$ and $3'$ and a subsequent coincidence detection [25] serves exactly the role of Bell-state measurement, we emphasize that our four-photon experiment above can also be viewed as a high-fidelity realization of entanglement swapping, or equivalently teleportation of entanglement. Specifically, the data of Fig. 16.4 indicate that the state of, say, photon 2 was teleported to photon 4 with a fidelity of 0.89. This clearly outperforms our earlier work [14] in this field, and for the first time fully demonstrates the non-local feature of quantum teleportation [26].

## 16.4 Experimental Test of Quantum Non-Locality

Utilizing our source developed for three-photon GHZ entanglement [9], let us now present the first three-particle test of quantum nonlocality [11]. As explained in Sect. 16.2, demonstration of the conflict between local realism and quantum mechanics for three-photon GHZ entanglement consists of four experiments, each with three spatially separated polarization measurements.

First, one performs $yyx$, $yxy$, and $xyy$ experiments. If the results obtained are in agreement with the predictions for a GHZ state, then the predictions for an $xxx$ experiment for a local realist theory are exactly opposite to those for quantum mechanics.

For each experiment we have eight possible outcomes of which ideally four should never occur. Obviously, no experiment either in classical physics or in quantum mechanics can ever be perfect, and therefore, due to principally unavoidable experimental errors, even the outcomes which should not occur will occur with some small probability in any realistic experiment.

All individual fractions which were obtained in our $yyx$, $yxy$ and $xyy$ experiments are shown in Fig. 16.5a–c, respectively. From the data we conclude that we observe the GHZ terms of (16.4) predicted by quantum mechanics in 85% of all cases, and in 15% we observe spurious events.

If we assume the spurious events are just due to experimental errors, we can thus conclude within the experimental accuracy that for each photon, 1, 2 and 3, quantities corresponding to both $x$ and $y$ measurements are elements of reality. Consequently, a local realist, if he accepts that reasoning, would thus predict that for a $xxx$ experiment only the combinations $V'V'V', H'H'V', H'V'H'$, and $V'H'H'$ will be observable (Fig. 16.6b). However, referring back to our original discussion, we see that quantum mechanics predicts that the exact opposite terms should be observed (Fig. 16.6a). To settle this conflict we then perform the actual $xxx$ experiment. Our results, shown in Fig. 16.6c, disagree with the local realism predictions and are consistent with the quantum-mechanical predictions. The individual fractions in Fig. 16.6c clearly show within our experimental uncertainty that only those triple coincidences predicted by quantum mechanics occur and not those predicted by local realism. In this sense, we claim that we have experimentally realized the first three-particle test of local realism following the GHZ argument.

We have already seen that the observed results for an $xxx$ experiment confirm the quantum-mechanical predictions when we assume that deviations from perfect correlations in our experiment, and in any experiment for that matter, are just due to unavoidable experimental errors. However, a local realist might argue against that approach and suggest that the non-perfect detection events indicate that the original GHZ argumentation cannot succeed.

To address this argument, a number of inequalities for $N$-particle GHZ states have been derived [27–29]. For instance, Mermin's inequality for a three-particle GHZ state reads as follows [27]:

$$|\langle\sigma_x\sigma_y\sigma_y\rangle + \langle\sigma_y\sigma_x\sigma_y\rangle + \langle\sigma_y\sigma_y\sigma_x\rangle - \langle\sigma_x\sigma_x\sigma_x\rangle| \leq 2 \,, \tag{16.11}$$

where symbol $\langle\cdot\rangle$ denotes the expectation value of a specific physical quantity. The necessary visibility to violate this inequality is 50%. The visibility observed in our GHZ experiment is $71 \pm 4\%$ and obviously surpasses the 50%

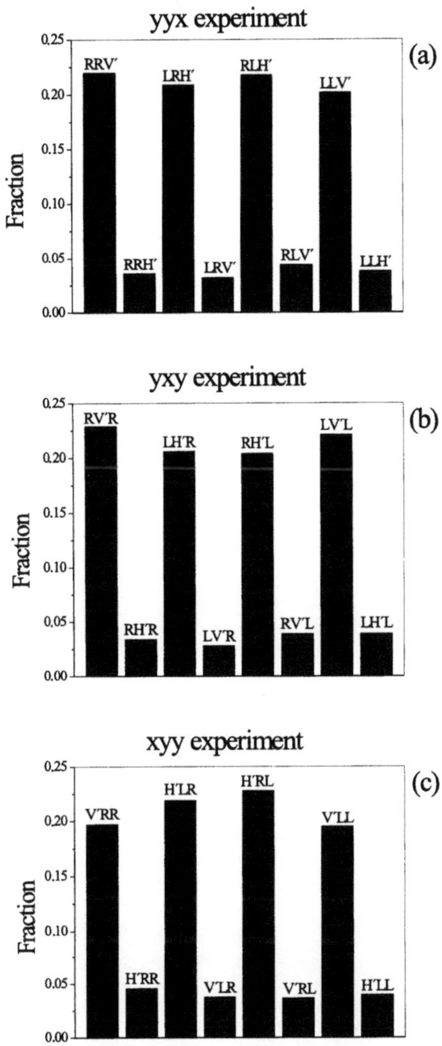


**Fig. 16.5.** Fractions of the various outcomes observed in the (**a**) $yyx$, (**b**) $yxy$, and (**c**) $xyy$ experiments. The experimental data show that we observe the GHZ terms predicted by quantum physics in 85% of all cases and the spurious events in 15%

limitation. Substituting our results measured in the $yyx$, $yxy$ and $xyy$ experiments into the left-hand side of (16.11), we obtain the following constraint:

$$\langle \sigma_x \sigma_x \sigma_x \rangle \leq -0.1 \,, \tag{16.12}$$

by which a local realist can thus predict that in an $xxx$ experiment the probability fraction for the outcomes yielding a $+1$ product, denoted by $P(xxx = +1)$, should be no larger than $0.45 \pm 0.03$ (also refer to the first bar in Fig. 16.7).

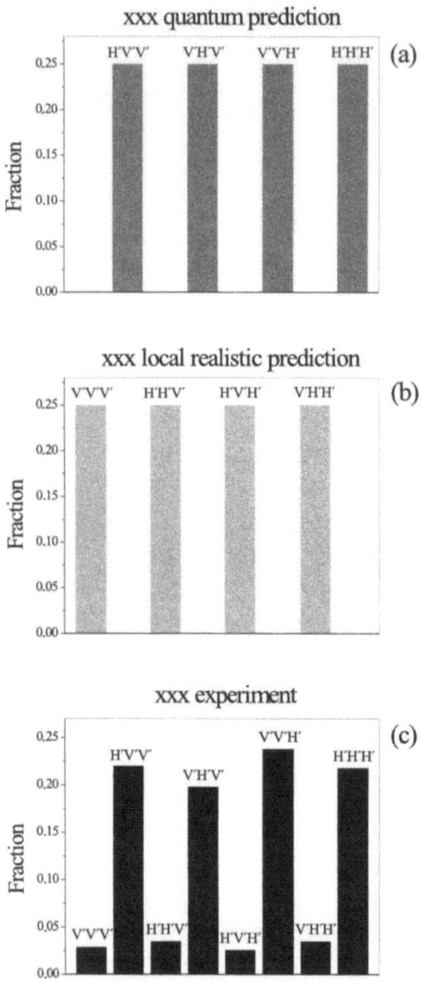


**Fig. 16.6.** The conflicting predictions of (**a**) quantum physics and (**b**) local realism of the fractions of the various outcomes in a $xxx$ experiment for perfect correlations. (**c**) The experimental results are in agreement with quantum physics within experimental errors and in disagreement with local realism

What is the quantum prediction for an $xxx$ experiment following from the $yyx$, $yxy$ and $xyy$ experimental results? Because our experimental visibility is due mainly to the finite width of the interference filters and the finite pulse duration, quantum mechanically it is expected that the same visibility should be observed in an $xxx$ experiment; hence we obtain the quantum prediction as shown in the second bar of Fig. 16.7.

The visibility observed in our $xxx$ experiment is $74 \pm 4\%$, which consequently gives $P(xxx = +1) = 0.87 \pm 0.04$ (shown in the third bar of

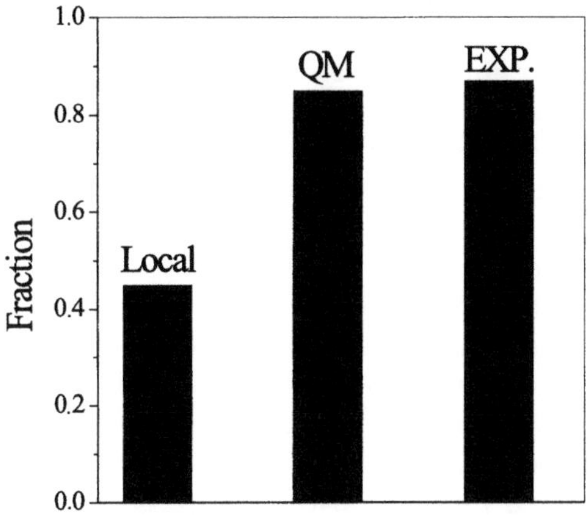


**Fig. 16.7.** Predictions of local realism (Local) and quantum physics (QM) for the probability fraction of the outcomes yielding a +1 product in an $xxx$ experiment based on the real experimental data measured in the $yyx$, $yxy$ and $xyy$ experiments. The experimental results (EXP.) are in good agreement with quantum physics and in distinct conflict with local realism

Fig. 16.7). Comparing the results in Fig. 16.7, we therefore conclude that our experimental results verify the quantum prediction while they contradict the local-realism prediction by over 8 standard deviations; there is no local hidden-variable model which is capable of describing our experimental results.

## 16.5 Discussions and Prospects

The experimental realization of multi-particle GHZ entanglement and high-fidelity teleportation has rather profound implications. First, in higher entangled systems the contradiction with local realism becomes ever stronger, because both the necessary visibility and the required number of statistical tests to reject the local hidden-variable models at a certain confidence level decrease with the number of particles that are entangled [29, 30]. Second, based on the observed visibility of $0.79 \pm 0.06$, one could violate – with an appropriate set of polarization correlation measurements – Bell's inequality for photons 1 and 4, even though these two photons never interact directly. As noted by *Aspect*, "This would certainly help us to further understand nonlocality" [31].

Besides its significance in tests of quantum mechanics versus local realism, the methods developed in the experiment also have many useful applications in the field of quantum information. It was noticed very recently that, while our four-photon setup directly provides a simple way to perform entanglement concentration [32–34], a slight modification of the setup also provides a novel way to perform entanglement purification for general mixed entangled states [10]. Furthermore, following the recent proposal by *Knill* et al. [35], our multi-photon experiment also opens the possibility to experimentally investigate the basic elements of quantum computation with linear optics.

In summary, we have demonstrated a method of creating higher-order entangled states which can, in principle, be extended to any desired number of particles, provided one has efficient pair sources. Given that, more photon-pair sources could be combined with polarizing beam-splitters to yield entangled states of arbitrary numbers of particles. The latest developments in photon-pair sources suggest that it should be possible in the near future to have sources with many orders of magnitude higher emission rates [36]. With these entanglement sources one would be able to implement some quantum-computation algorithms using only entanglement and linear optics [35]. Also, more elaborate entanglement-purification protocols and high-fidelity teleportation over multiple stages as required for the construction of quantum repeaters [37] become possible.

## Acknowledgement

The authors would like to thank Dirk Bouwmeester, Matthew Daniell, Sara Gasparoni, Gregor Weihs, and Harald Weinfurter for fruitful collaborations on various topics in the review. We acknowledge the financial support of the Austrian Science Fund, FWF, project no. F1506, and the European Commission within the IST-FET project "QuComm" and TMR network "The physics of quantum information."

## References

1. E. Schrödinger, Die gegenwärtige Situation in der Quantenmechanik. Naturwissenschaften **23**, 807–812; 823–828; 844–849 (1935)
2. A. Einstein, B. Podolsky, N. Rosen, Can quantum-mechanical description of physical reality be considered complete? Phys. Rev. **47**, 777–780 (1935)
3. J.S. Bell, On the Einstein–Podolsky–Rosen paradox. Physics 1, 195–200 (1964); reprinted in J.S. Bell, *Speakable and Unspeakable in Quantum Mechanics* (Cambridge Univ. Press, Cambridge 1987)
4. S.J. Freedman, J.S. Clauser, Experimental test of local hidden-variable theories, Phys. Rev. Lett. **28**, 938–941 (1972)
5. A. Aspect, J. Dalibard, G. Roger, Experimental test of Bell's inequalities using time-varying analyzers, Phys. Rev. Lett. **47**, 1804–1807 (1982)
6. G. Weihs, T. Jennewein, C. Simon, H. Weinfurter, A. Zeilinger, Violation of Bell's inequality under strict Einstein locality conditions, Phys. Rev. Lett. **81**, 5039–5043 (1998)
7. D.M. Greenberger, M.A. Horne, A. Zeilinger, Going beyond Bell's theorem, in *Bell's Theorem, Quantum Theory, and Conceptions of the Universe*, M. Kafatos (ed.), (Kluwer, Dordrecht 1989) pp. 73–76
8. D.M. Greenberger, M.A. Horne, A. Shimony, A. Zeilinger, Bell's theorem without inequalities, Am. J. Phys. **58**, 1131–1143 (1990)

9. D. Bouwmeester, J.-W. Pan, M. Daniell, H. Weinfurter, A. Zeilinger, Observation of three-photon Greenberger–Horne–Zeilinger entanglement, Phys. Rev. Lett. **82**, 1345–1349 (1999)
10. J.-W. Pan, M. Daniell, S. Gasparoni, G. Weihs, A. Zeilinger, Experimental demonstration of four-photon entanglement and high-fidelity teleportation, Phys. Rev. Lett **86**, 4435–4438 (2001)
11. J.-W. Pan, D. Bouwmeester, M. Daniell, H. Weinfurter, A. Zeilinger, Experimental test of quantum nonlocality in three-photon Greenberger–Horne–Zeilinger entanglement, Nature **403**, 515–519 (2000)
12. N.D. Mermin, What's wrong with these elements of reality? Phys. Today **43**, 9–11 (1990)
13. D. Bouwmeester et al., Experimental quantum teleportation, Nature **390**, 575–579 (1997)
14. J.-W. Pan, D. Bouwmeester, H. Weinfurter, A. Zeilinger, Experimental entanglement swapping: Entangling photons that never interacted, Phys. Rev. Lett. **80**, 3891–3894 (1998)
15. A. Zeilinger, M.A. Horne, H. Weinfurter, M. Zukowski, Three particle entanglements from two entangled pairs, Phys. Rev. Lett. **78**, 3031–3034 (1997)
16. G. Krenn, A. Zeilinger, Entangled entanglement, Phys. Rev. A **54**, 1793–1796 (1996)
17. Z.Y. Ou, L. Mandel, Violation of Bell's inequality and classical probability in a two-photon correlation experiment, Phys. Rev. Lett. **61**, 50–53 (1988)
18. Y.H. Shih, C.O. Alley, New type of Einstein–Podolsky–Rosen–Bohm experiment using pairs of light quanta produced by optical parametric down conversion, Phys. Rev. Lett. **61**, 2921–2924 (1988)
19. P. Kwiat, P.E. Eberhard, A.M. Steinberger, R.Y. Chiao, Proposal for a loophole-free Bell inequality experiment, Phys. Rev. A **49**, 3209–3220 (1994)
20. L. De Caro, A. Garuccio, Reliability of Bell-inequality measurements using polarization correlations in parametric-down-conversion photons, Phys. Rev. A **50**, R2803–2805 (1994)
21. S. Popescu, L. Hardy, M. Zukowski, Revisiting Bell's theorem for a class of down-conversion experiments, Phys. Rev. A **56**, R4353–4357 (1997)
22. M. Zukowski, Violations of local realism in multiphoton interference experiments, Phys. Rev. A **61**, 022109 (2000)
23. P.G. Kwiat, K. Mattle, H. Weinfurter, A. Zeilinger, A.V. Sergienko, Y.H. Shih, New high intensity source of polarization-entangled photon pairs, Phys. Rev. Lett. **75**, 4337–4341 (1995)
24. M. Zukowski, A. Zeilinger, H. Weinfurter, Entangling photons radiated by independent pulsed source, Ann. Acad. Sci. (New York) **755**, 91–102 (1995)
25. J.-W. Pan, A. Zeilinger, Greenberger–Horne–Zeilinger-state analyzer, Phys. Rev. A **57**, 2208–2211 (1998)
26. M. Zukowski, Bell theorem for the nonclassical part of the quantum teleportation process, Phys. Rev. A **62**, 0321011 (2000)
27. N.D. Mermin, Extreme quantum entanglement in a superposition of macroscopically distinct states, Phys. Rev. Lett. **65**, 1838–1841 (1990)
28. S.M. Roy, V. Singh, Tests of signal locality and Einstein–Bell locality for multiparticle systems, Phys. Rev. Lett. **67**, 2761–2764 (1991)
29. M. Zukowski, D. Kaszlikowski, Critical visibility for $N$-particle Greenberger–Horne–Zeilinger correlations to violate local realism, Phys. Rev. A **56**, R1682–1685 (1997)

30. A. Peres, Bayesian analysis of Bell inequalities, Fortschr. Phys. **48**, 531–535 (2000)
31. A. Aspect, Quoted in G.P. Collins, Phys. Today **51**(2), 18–21 (1998)
32. Y. Yamamoto, M. Koashi, N. Imoto, Concentration and purification scheme for two partially entangled photon pairs, Phys. Rev. A **63**, 012304 (2001)
33. Z. Zhao, J.-W. Pan, M.-S. Zhan, Practical scheme for entanglement concentration, Phys. Rev. A **63**, 014301 (2001)
34. J.-W. Pan, C. Simon, C. Brukner, A. Zeilinger, Entanglement purification for quantum communication, Nature **410**, 1067–1070 (2001)
35. E. Knill, R. Laflamme, G. Milburn, A scheme for efficient quantum computation with linear optics, Nature **409**, 46–52 (2001)
36. K. Sanaka, K. Kawahara, T. Kuga, New high-efficiency source of photon pairs for engineering quantum entanglement, Phys. Rev. Lett. **86**, 5620–5623 (2001)
37. H.-J. Briegel, W. Duer, J.I. Cirac, P. Zoller, Quantum repeaters: The role of imperfect local operations in quantum communication, Phys. Rev. Lett. **81**, 5932–5935 (1998)

# 17 Bell's Theorem, Information and Quantum Physics

Anton Zeilinger

## 17.1 Introduction

I had the great fortune to meet John Bell three times in my life. The first encounter was in 1975. I had been working since 1968 at the small Atom Institute in Vienna under Helmut Rauch on applying neutrons to investigate magnetic materials. Then he – together with Wolfgang Treimer and Ulrich Bonse [1] – developed the first crystal-based neutron interferometer as a tool to investigate the foundations of quantum mechanics. Because of my experience with polarized neutrons, I started to think about the role of the spin in such experiments and we were soon able to demonstrate the change of the sign of a spinor wave function upon a complete rotation by $2\pi$ [2]. So it was just in time that I discovered the announcement of an Erice workshop entitled "Thinkshops on Physics", which was organized by John Bell together with Bernard d'Espagnat. The topic of this workshop was exactly experiments on the foundations of Quantum Physics. I went there and reported on our newest experimental results in neutron interferometry. I should mention that this was my first real encounter with the international scientific community. There, I heard for the first time about Bell's theorem, about the Einstein–Podolsky–Rosen paradox, about entanglement, and the like. Needless to say that I did not get any real understanding what this was all about, but I got the hunch that something very important was being discussed. This meeting turned out to be very crucial in my life. There, I met a number of colleagues for the first time, some of whom later became personal friends. These include Mike Horne, Abner Shimony and also Val Telegdi. Mike Horne and Val Telegdi then helped me to get to MIT and work with Cliff Shull at the neutron diffraction laboratory there. At that laboratory I also met Danny Greenberger later, and all this resulted in some of my most important collaborations. So it is fair to say that John Bell's organization of that meeting was very important for me in many different ways.

The next time I met John Bell was in Vienna on the occasion of a conference celebrating the 100th anniversary of Erwin Schrödinger. This was in 1987. At that meeting I had the great honor to be on a panel together with John Bell. The other panel members were K. Baumann (Graz), P. Mittelstaedt (Cologne) and H. Pietschmann (Vienna). This panel discussion took place in one of the gilded state halls of the University of Vienna. And I still remember John Bell's insistence on the desideratum that physics should explain why events happen and that something must be missing in the present formulation of quantum mechanics. A position that already at that time I had the pleasure of disagreeing with. John Bell was probably the most pleasant person to disagree with I ever met. The discussions were always very friendly but to the point and succinct.

The third time I met John Bell was in 1990 at the small workshop entitled "The Amherst Workshop on the Foundations of Quantum Mechanics" which, as the name suggests, took place at Amherst College, June 10–15, 1990 and was organized by George Greenstein and Arthur Zajonc. Two years earlier, Dan Greenberger, Mike Horne and I had found what is called today GHZ states [3], the extreme example of what John Bell had demonstrated, namely, disagreement between quantum mechanics and local realistic theories. To our great joy, John Bell expressed his excitement and pleasure about our work. None of us knew that this was the last time we would meet John, as he passed away a few months later completely unexpectedly for everyone.

## 17.2 Information and Interference

In the famous double-slit experiment [4], which, according to Feynman [5], contains in it the heart of quantum mechanics, a number of deep epistemological questions about quantum mechanics are raised. As is well known, an interference pattern only arises when both slits in the intermediate diaphragm are open (Fig. 17.1). It can be argued that each individual particle has information that both slits are open because all particles avoid the minima in the interference pattern. Yet we still might ask ourselves through which of the two slits does a specific particle pass. Yet, all of the numerous gedanken attempts to demonstrate that one can find out which of the two slits the particle uses, and still obtain the interference pattern, were in vain. Whenever it is possible to determine the path precisely, by whatever means, the interference pattern vanishes. On the other hand, if there is no possibility, not even in principle, to find out which path was taken by a particle registered at the observation screen, the interference pattern arises with perfect visibility. Already here we can note that it is information which plays a crucial role in whether or not interference is observed. Path information and the information contained in the interference pattern exclude the other. It is important to realize that it does not matter whether or not an observer actually takes note of the path the particle took. It is the mere possibility of determining the path which makes it impossible to observe the interference pattern.

The double-slit experiment therefore is a basic example of complementarity in quantum mechanics. Complementarity in general is, as introduced by

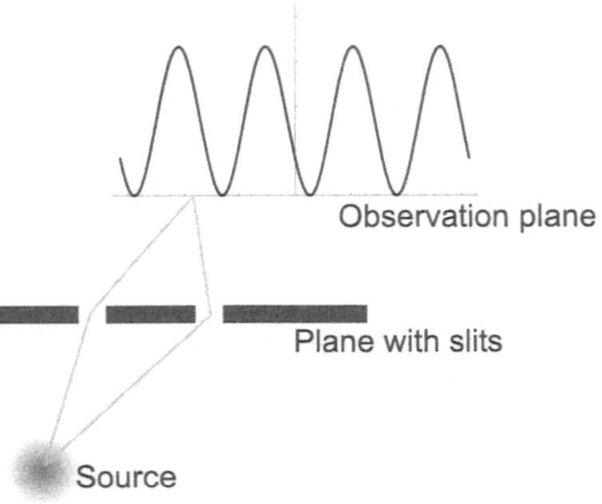


**Fig. 17.1.** The double-slit interference experiment [4]. The interference fringes in the observation plane are collected one particle at a time, each of them passing individually through the plane with the slits

Niels Bohr, the notion that there exist pairs, or combinations of more than two observables, such that if one is determined perfectly, the others are by necessity completely undetermined. For Niels Bohr [6], this was a consequence of the fact that in order to demonstrate mutually complementary observables, one has to use apparata that exclude each other. This might suggest the impossibility of experimental demonstration, or the simultaneous determination of complementary observables, because of the clumsiness of macroscopic apparata which are unsuitable for determining the fine details of quantum objects. Yet, as signified most clearly by the Kochen–Specker theorem [7], as also found by John Bell [7], it is impossible to assign to a quantum system observable properties per se, i.e. independent of considering the apparatus with which the properties will be measured. Thus, we do not in an experiment reveal a pre-existing feature of a quantum system. Therefore, measurement is constitutive in what can be physical reality in the sense that the experimentalist by deciding which apparatus to use chooses which physical observable can become reality. In the case of the double-slit experiment, by deciding to measure the path the particle takes, the experimentalist decides that "path taken" can become an element of reality. On the other hand by deciding to choose an experimental set-up intrinsically unsuitable to determine the path taken, the observer decides that the interference pattern can become reality. This indicates that the observer, by choosing the experiment, can actually choose between different kinds of information that will manifest themselves in the experiment, although the total amount of information is apparently limited.

This has often led to confusion and to the wrong impression that the observer in a quantum measurement has so much influence that she[1] can define reality. Yet, an important point is suggested by the fact that the observer cannot define which particular value will turn out for the observable chosen out of the class of complementary observables. Specifically, the observer has no influence on which of the two paths the particle will be found in if the path is measured. And likewise she has no influence on where exactly in the interference pattern the particle will be found.

This indicates that the observer has a qualitative, but not a quantitative influence on reality. She can define which quality will show up in the experiment, but not the quantity, the exact value, the latter being completely random, except in the rare case when the quantum system is in an eigenstate of the observed quantity.

This small discussion already indicates that information might be at the root of the interpretation of quantum mechanics. In the double-slit experiment, the observer can decide to obtain information either about the path taken or the information contained in the interference pattern. It also turns out that using a new measure of information the total information in a quantum system is a constant [9]. Complementarity then simply is a consequence of the fact that the total information which is represented by a quantum system is finite [10].

## 17.3 Information and Entanglement

Bell's theorem [8] states that any local realistic view of the world is incompatible with quantum mechanics. More precisely, John Bell demonstrates that for entangled states, it is not possible to explain all correlations between two particles using a local realistic model. A crucial assumption of such a model is [11] that a measurement result for each of two entangled particles is independent of whatever measurement is performed on the other particle. While this is often interpreted as demonstrating non-locality in quantum mechanics, there are also alternative viewpoints possible, most notably the assumption that the philosophical notion of counterfactuality does not hold or that the existence of a reality independent of observation makes no sense in physics. By now, the conflict between local realism and quantum mechanics has led to numerous experiments, all of which support quantum mechanics [12–14]. It is therefore safe to assume that the world cannot be understood using the rather intuitively reasonable ideas leading to Bell's inequality.

While entanglement apparently seems to be still posing problems as to understanding its nature, the information-theoretical interpretation of quantum mechanics again leads to a very natural point of view, as we will see now for a specific example [15–17].

---

[1] In some of his papers, John always referred to physicists in the feminine.

Considering just two two-state systems, i.e. two qubits, it is natural to assume that the information carried by the system is one bit of information per qubit. In a classical way of encoding, this would lead to the following factorizable states:

$$
\begin{aligned}
|\Psi_1\rangle &= |0\rangle_1 \, |0\rangle_2 \\
|\Psi_2\rangle &= |0\rangle_1 \, |1\rangle_2 \\
|\Psi_3\rangle &= |1\rangle_1 \, |0\rangle_2 \\
|\Psi_4\rangle &= |1\rangle_1 \, |1\rangle_2 \, .
\end{aligned}
\tag{17.1}
$$

Here the first (second) ket refers to the first (second) system. In that way of encoding, each system is the representative of one well-defined bit of information. In the system of the states represented by (17.1), we indeed have encoded two bits of information, since each of the two particles is represented by a choice of two orthogonal states which easily can be identified. This is just like in classical coding, where we would have two physical bits, each one carrying either the value "0" or the value "1" with four possible combinations corresponding to the four states (17.1). On the other hand, quantum physics is a holistic theory in the sense that a quantum state intrinsically is not limited by space–time allocation. Therefore, two quantum systems can carry two bits of information in such a way that neither of them carries any well-defined information on its own. An example of such maximally entangled states is given by the so-called "Bell-basis" [18].

$$
\begin{aligned}
|\Psi^+\rangle &= \frac{1}{\sqrt{2}}(|0\rangle_1 \, |1\rangle_2 + |1\rangle_1 \, |0\rangle_2) \\
|\Psi^-\rangle &= \frac{1}{\sqrt{2}}(|0\rangle_1 \, |1\rangle_2 - |1\rangle_1 \, |0\rangle_2) \\
|\Phi^+\rangle &= \frac{1}{\sqrt{2}}(|0\rangle_1 \, |0\rangle_2 + |1\rangle_1 \, |1\rangle_2) \\
|\Phi^-\rangle &= \frac{1}{\sqrt{2}}(|0\rangle_1 \, |0\rangle_2 - |1\rangle_1 \, |1\rangle_2) \, .
\end{aligned}
\tag{17.2}
$$

In which way do these four states carry two bits of information in a non-local way? Obviously, one bit of information is encoded in the states $|\Psi^+\rangle$, $|\Psi^-\rangle$ versus $|\Phi^+\rangle$, $|\Phi^-\rangle$. It is the truth-value of the proposition "the two qubits are equal". Apparently, this statement is false for the first two states and correct for the second two states. But where is the other bit of information? This is easily seen if one goes to a conjugate basis by using the transformation

$$
\begin{aligned}
|0\rangle &\rightarrow \frac{1}{\sqrt{2}}(|0\rangle + |1\rangle) \\
|1\rangle &\rightarrow \frac{1}{\sqrt{2}}(|0\rangle - |1\rangle) \, .
\end{aligned}
\tag{17.3}
$$

Then, one will find that in the new basis, the proposition "the two qubits are equal" will now be false for the states $|\Psi^-\rangle$ and $|\Phi^-\rangle$ and correct for the states $|\Psi^+\rangle$ and $|\Phi^+\rangle$. Together, the two statements allow a unique determination of the four non-local states, each one representing a unique combination of the truth values "0", "1" of the two propositions.

The lesson here therefore is that information can be carried by quantum systems in a very non-local way, independent of their spatio–temporal arrangements. One might notice that similar reasoning is possible for higher entangled states, as in GHZ states of the general form [11]

$$|\Psi\rangle = \frac{1}{\sqrt{2}}(|000\rangle + |111\rangle) \tag{17.4}$$

and W-states [19] of the general form

$$|\Psi\rangle = \frac{1}{\sqrt{2}}(|011\rangle + |101\rangle + |110\rangle)\,. \tag{17.5}$$

Here, for example, $|011\rangle = |0\rangle_1|1\rangle_2|1\rangle_3$. In such three-qubit states evidently three propositions define all eight states of the complete basis.

We note that a general criterion exists for quantum non-locality in a two-qubit system, this is that a state violates a Bell inequality if, and only if, more than one bit of information is encoded jointly by the members of the system into the correlations.

These considerations might indicate to the reader that in quantum physics information is a more basic notion than in any classical view of the world.

## 17.4 Bell's Theorem, Quantum Communication and Quantum Information

Most interestingly, essentially in the last decade a fully unexpected and novel development set in. This is based on the realization that using individual quantum systems one can obtain completely novel ways of encoding, transmitting and processing information. These new ways of communication and of computation include as a fundamental concept quantum entanglement. It is safe to say that this very recent development would not have been possible without John Bell's seminal work. Therefore, it might be appropriate to review now a few of the basic procedures.

### 17.4.1 Quantum Dense Coding

It was first suggested by *Bennett* and *Wiesner* [20] that if two players have access to entangled states, Alice can actually encode more than one bit of information into one particle. This can easily be seen by referring to the Bell-basis of (17.2). There one sees that starting, for example, with state $|\Psi^+\rangle$,

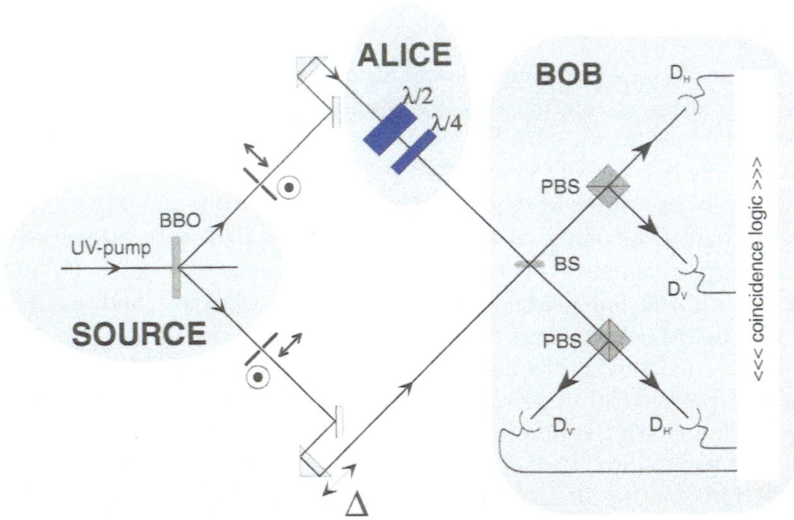


**Fig. 17.2.** Quantum dense coding setup. Entangled photon pairs are produced by spontaneous parametric down-conversion in a BBO crystal. One photon is sent directly to Bob, and the other is sent to Alice, who encodes the three-valued information on this photon by adjusting the $\frac{\lambda}{2}$ and $\frac{\lambda}{4}$ wave plates, and then passes them on to Bob. Bob combines the two photons in his Bell-state analyzer, which allows him to identify three of the four Bell-states [21]

one can easily obtain any of the other Bell-states by just manipulating one of the qubits, say qubit 2. If qubit 2 is flipped, then one obtains the state $|\Phi^+\rangle$, if the phase of qubit 2 is changed by $\pi$, then one obtains $|\Psi^-\rangle$, and if both procedures are applied one obtains $|\Phi^-\rangle$. Thus, Alice can send to Bob more than one bit of information by manipulating just one photon, if Bob has full access to the complete state, i.e. also the other photon, and if he can perform the Bell-state measurement determining which of the four Bell-states characterizes the complete two-qubit system. While at present complete Bell-state analyzers for independent qubits do not exist, the principle of quantum dense coding has been successfully demonstrated in an experiment [21]. In that experiment, see Fig. 17.2, it was possible to determine two Bell-states definitely, and the other two Bell-states gave the same third result. Thus, it is possible to transmit $\mathrm{ld}\, 3 = 1.584...$ bits of information for each qubit manipulated, which is clearly larger than the classical limit of 1. Yet, clearly the real limit of 2 bits per pair is not surpassed. Quantum dense coding may be viewed as a direct application of the fact that information is carried in an entangled state in a holistic or non-local way, and that the experimentalist, by manipulating only part of the state, can change the complete two-qubit state such that it becomes another Bell-state, i.e. qualitatively different. We remark that while Alice only has access to one qubit, Bob needs to have access to both qubits in order to extract the information encoded by Alice.

### 17.4.2 Quantum Teleportation

An extension of these procedures is quantum teleportation, where Alice and Bob initially share an entangled pair of qubits [22, 23]. See Fig. 17.3.

Alice then performs a joint Bell-state measurement on her qubit together with one of the two from the entangled pair. This measurement immediately projects the second qubit from the pair into a specific state directly related to Alice's original. Depending on the result of Alice's Bell-state measurement, Bob then applies a unitary transformation determined by Alice's Bell-state measurement result independent of which state Alice initially had. Thus Bob finally obtains Alice's original state.

It may be worthwhile to analyze this experiment briefly from an information-interpretational point of view [15]. What happens is that initially, Alice and Bob share the non-local information contained in the original entangled quantum state. That is, they know how qubits 2 and 3 relate to each other, should they be measured. By obtaining the Bell-state measurement result, Alice acquires further information on how the original qubit to be teleported and her member of the entangled pair relate to each other. Thus, by a simple logical chain, it is now known how the qubit to be teleported and Bob's entangled member of the pair are related. Thus, one obtains by

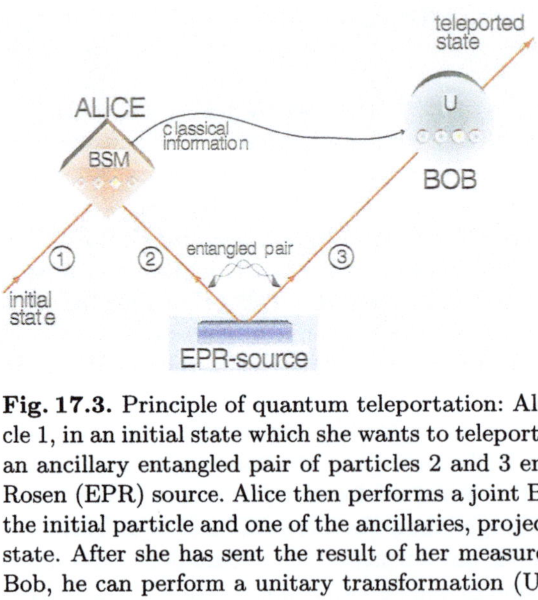


**Fig. 17.3.** Principle of quantum teleportation: Alice has a quantum system, particle 1, in an initial state which she wants to teleport to Bob. Alice and Bob also share an ancillary entangled pair of particles 2 and 3 emitted by an Einstein–Podolsky–Rosen (EPR) source. Alice then performs a joint Bell-state measurement (BSM) on the initial particle and one of the ancillaries, projecting them also onto an entangled state. After she has sent the result of her measurement as classical information to Bob, he can perform a unitary transformation (U) on the other ancillary particle, resulting in it being in the state of the original particle. In the case of quantum teleportation of a qubit, Alice makes a projection measurement onto four orthogonal entangled states (the Bell states) that form a complete basis. Sending the outcome of her measurement, i.e. two bits of classical information, to Bob will enable him to reconstruct the initial qubit

these measurements a sequence of relational statements, and it is therefore uniquely determined through which unitary operation Bob's qubit is related to the original. This simple information-interpretational point of view supersedes all possible paradoxes, as all that changes by Alice's measurement is the quantum state of the total system, that is, the information which the observers have. No action at a distance or other mysterious processes happen.

It is actually interesting to notice that the information contained in Alice's original state is immediately teleported over to Bob's station as soon as Alice's Bell-state measurement is performed. Bob certainly has four different states at hand, depending on the specific result of Alice's measurement. Yet, each of these states is related to the original through a rotation independent of the properties of the original. Thus, in a sense, the information is already there [24], yet Bob cannot really read it out without knowing Alice's result. Interestingly, this has operational consequences. For example, one can start a quantum computer at a time before the classical information has arrived [25] and thus save computational time.

### 17.4.3 Teleportation of Entanglement

A most interesting application of these ideas is the teleportation of entanglement [22, 26, 27], also called entanglement swapping. In that experiment, one teleports a photon which does not enjoy its own quantum state but is still entangled with another one. Thus, one cannot even say here that a quantum state is teleported. In a recent experiment [28], it was possible to perform the teleportation of an entangled photon in such a high quality that the original photon and the teleported one were entangled well enough to observe a violation of Bell's inequality between two independently created photons. This, more than any experiment before, underlines that Bell's theorem is not about properties particles carry, but about the information concerning possible measurement results. See Fig. 17.4.

The teleportation of entanglement also can easily be understood on the basis of an information interpretation of quantum mechanics. This is just one step further than standard teleportation. Alice and Bob initially share two relational statements, each one characterizing each of the two entangled pairs. By performing a Bell-state measurement on one photon from each pair, they obtain further relational statements and thus can conclude the chain of logical links from photon 0 to photon 3. About these two photons, they obtain the same joint statement as about the possible states of the Bell basis of (17.2). Again, these are statements about possible experimental results only, should an experiment actually be performed.

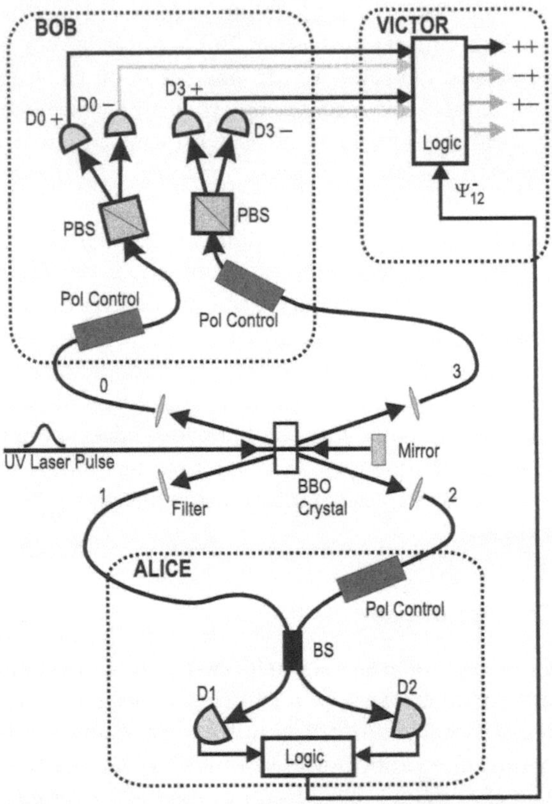


**Fig. 17.4.** Setup of the experiment teleporting an entangled photon. Two entangled photon pairs are produced by down-conversion in a BBO crystal, pumped by femtosecond UV-laser pulses traveling through the crystal in opposite directions. All photons are collected in single-mode optical fibers for further analysis and detection. For performing the Bell-state analysis, photons 1 and 2 interfere at a fiber beam splitter, where one arm contains a polarization controller for compensating the polarization rotation introduced by the optical fibers. Photons 0 and 3 were sent to Bob's two-channel polarizing beam splitters for analysis, and the required orientation of the analyzers was set with polarization controllers in each arm. All photons were detected with silicon avalanche photodiodes, with a detection efficiency of about 40%. Alice's logic circuit detected coincidences between detectors D1 and D2. It is essential that she passes the result as a classical signal to Victor, who determines whether Bob's detection events violate Bell's inequality [28]

### 17.4.4 Quantum Cryptography

The technically most advanced application of fundamental quantum concepts in communication is quantum cryptography. The protocols closest related to Bell's ideas are again based on the use of entangled states, as first suggested by Ekert [30]. Without going into details, one uses an entangled state to create the same identical key by independent measurements by two observers, Alice and Bob. The big advantage of Ekert-type quantum cryptography is that the cryptographic key does not need to be transported from $A$ to $B$, but is really created by measurements in the same basis as a sequence of entangled pairs. Thus, one of the essential security problems of classical cryptography, namely the necessary transportation of a classical key, immediately vanishes. Another advantage of quantum cryptography with entangled states is that any eavesdropper can readily be identified by just observing whether the data measured by Alice and Bob still violate a Bell inequality. If such is the case, then no essential information could have leaked out to an eavesdropper, and Alice and Bob can readily use the key obtained. See Figs. 17.5 and 17.6.

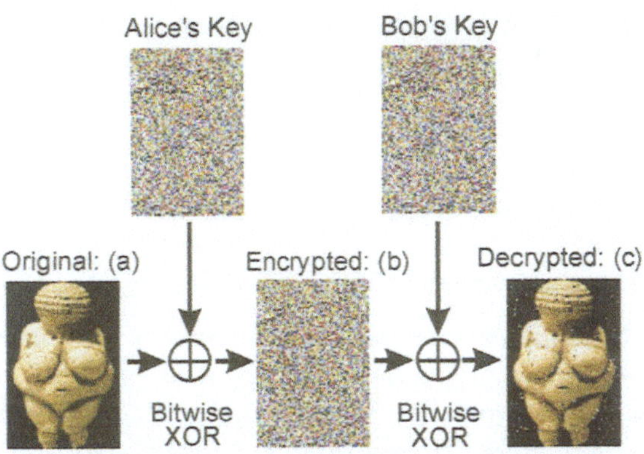


**Fig. 17.5.** In an experimental realization using polarization entangled pairs of photons, a key was created and it was used to actually transmit visual information, in this case a picture of the famous Venus von Willendorf sculpture [29]

There is a very fundamental information-theoretic interpretation of the security of entangled-state quantum teleportation. The basic observation is that a two-qubit system can carry only two bits of information. For a maximally entangled state, three bits are defined non-locally, i.e. the two bits are all used up to define the entanglement. Now, if the eavesdropper extracts any information, this can only be at the expense of the two non-local bits, thus reducing the entanglement.

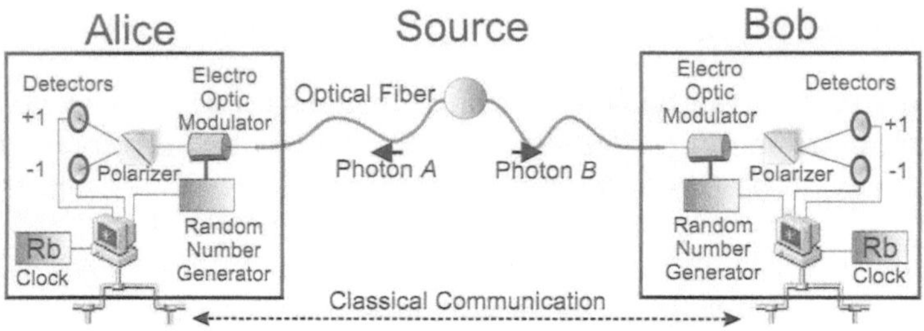


**Fig. 17.6.** Polarization-entangled quantum cryptography [29]. The most salient feature of this experiment is, besides the creation of the entangled pair, that each photon is sent to experimental stations which finally are separated by more than 300 m. In each station, one has an independent Rubidium clock, which simply registers the photon arrival times and two-channel polarizers, where one channel is identified as "0" and the other is identified as "1". If the two polarizers are parallel, the correlations are perfect, and Alice and Bob obtain the same random sequence after eliminating those events where only one photon was registered. In order to identify a possible eavesdropper, both Alice and Bob randomly switch between two bases independent of each other. They then identify those situations where they happen to have the same basis, and they know that in these situations they obtain the key. The situations where they happen to have different bases can be used to check for a possible eavesdropper

## 17.5 John Bell's Desiderata and the Interpretation of Quantum Mechanics

In his famous paper "Against measurement" [31], John Bell suggests:

"*Here are some words which, however legitimate and necessary in application, have no place in a formulation with any pretension to physical precision:* ***system, apparatus, environment, microscopic, macroscopic, reversible, irreversible, observable, information, measurement.***"

By now, the reader might have gathered that the present author does not agree with John Bell's statement. In contrast it is suggested that *information* is the most basic notion of quantum mechanics, and it is *information* about possible *measurement* results that is represented in the quantum states. *Measurement* results are nothing more than states of the classical *apparatus* used by the experimentalist. The quantum *system* then is nothing other than the consistently constructed referent of the *information* represented in the quantum state. In a *measurement*, one of the possible *measurement* results becomes reality with a relative frequency as indicated by the quantum state. For a quantum *system*, the *environment* therefore is just an external bath *information* can leak into, and the situation is *reversible* if this *information* can be recovered somehow from the environment, and *irreversible* if that is not the case.

The point where I agree with John Bell is that *microscopic* and *macroscopic* should not command any fundamental place in any physical theory. Experimental progress will certainly make it possible to push the regime where quantum phenomena have been demonstrated very far into what we would consider *macroscopic*. Thus, while the dichotomy *microscopic–macroscopic* should not have any place in a physical theory, the dichotomy *quantum–classical* is a most fundamental one.

While it is clear that I respectfully disagree with the general philosophical position of John Bell concerning the foundations of quantum mechanics, one cannot but show deep respect for his high intellectual integrity. His way of thinking led him to advocate the quest for more complete theories than quantum mechanics, which, as he hoped, would finally explain why individual events happen. Even if the interpretation of quantum mechanics based on information – which runs completely against his expectation - will turn out to be the correct one, then John Bell will turn out to have raised the correct challenges.

## Acknowledgment

This work was supported by the Austrian Science Foundation, Grant No. SFB 015 "Control and Measurement of Coherent Quantum Systems" and by the European Commission, Grant Nos. IST-1999-10033 "Long Distance Photonic Quantum Communication" and ERBFMRXCT960087 "The Physics of Quantum Information".

## References

1. S.A. Werner, H. Rauch: *Neutron Interferometry: Lessons in Experimental Quantum Mechanics*, Oxford Series on Neutron Scattering in Condensed Matter (Oxford University Press, Oxford 2000)
2. H. Rauch, A. Zeilinger, G. Badurek, A. Wilfing, W. Bauspiess, U. Bonse: Phys. Lett. A **54** (2975) 425
3. D.M. Greenberger, M.A. Horne, A. Zeilinger: 'Going beyond Bell's theorem.' In: *Bell's Theorem, Quantum Theory, and Conceptions of the Universe*, ed. by M. Kafatos (Kluwer Academic, Dordrecht 1989) pp. 73
4. N. Bohr: In: *Albert Einstein: Philosopher-Scientist*, ed. by P.A. Schilpp (The Library of Living Philosophers, Evanston 1949) pp. 200–241
5. R.P. Feynman, R.B. Reighton, M. Sands: *The Feynman Lectures of Physics*, Vol. III (Addison-Wesley, Reading 1965)
6. N. Bohr: Phys. Rev. **48**, 696 (1935)
7. S. Kochen, E.P. Specker: J. Math. Mech. 17, 59 (1967); J.S. Bell: Rev. Mod. Phys. 38, 447 (1966)
8. J.S. Bell: Physics (New York) **1**, 195 (1964). Reprinted in J.S. Bell: *Speakables and Unspeakables in Quantum Mechanics* (Cambridge University Press, Cambridge 1987)

9. Č. Brukner, A. Zeilinger: e-print quant-ph/0201026 (to appear in Proc. Roy. Soc. Lond. A)
10. Č. Brukner, A. Zeilinger: Phys. Rev. Lett. **83**, 354 (1999)
11. D.M. Greenberger, M. Horne, A. Shimony, A. Zeilinger: Am. J. Phys. **58**, 1131 (1990)
12. S.J. Freedman, J.S. Clauser: Phys. Rev. Lett. **28**, 938 (1972)
13. A. Aspect, P. Grangier, G. Roger: Phys. Rev. Lett. **47**, 460 (1981)
14. G. Weihs, T. Jennewein, C. Simon, H. Weinfurter, A. Zeilinger: Phys. Rev. Lett. **81**, 5039 (1998)
15. A. Zeilinger: Phil. Trans. Roy. Soc. **1733**, 2401 (1997)
16. A. Zeilinger: Found. Phys. **29**, 631 (1999)
17. Č. Brukner, M. Zukowski, A. Zeilinger: e-print quant-ph/0106119
18. S.L. Braunstein, A. Mann, M. Revzen: Phys. Rev. Lett. **68**, 3259 (1992)
19. A. Zeilinger, M.A. Horne, D.M. Greenberger: 'Higher-Order Quantum Entanglement.' In: *Proc. Squeezed States and Quantum Uncertainty*, ed. by D. Han, Y.S. Kim, W.W. Zachary, NASA Conf. Publ. 3135 (National Aeronautics and Space Administration 1992)
20. C.H. Benett, S.J. Wiesner: Phys. Rev. Lett. **69**, 2881 (1992)
21. K. Mattle, H. Weinfurter, P.G. Kwiat, A. Zeilinger: Phys. Rev. Lett. **76**, (1996) 4656
22. C.H. Bennett, G. Brassard, C. Crepeau, R. Jozsa, A. Peres, W.K. Wootters: Phys. Rev. Lett. **70**, 1895 (1993)
23. D. Bouwmeester, J.-W. Pan, K. Mattle, M. Eibl, H. Weinfurter, A. Zeilinger: Nature **390**, 575 (1997)
24. S. Stenholm: private communication (2000)
25. Č. Brukner, J.-W. Pan, C. Simon, G. Weihs, A. Zeilinger: quant-ph/0109022
26. M. Zukowski, A. Zeilinger, M. Horne, A. Ekert: Phys. Rev. Lett. **71**, 4287 (1993)
27. J.-W. Pan, D. Bouwmeester, H. Weinfurter, A. Zeilinger: Phys. Rev. Lett. **80**, 3891 (1998)
28. T. Jennewein, G. Weihs, J.-W. Pan, A. Zeilinger: Phys. Rev. Lett. **88**, 017903 (2002)
29. T. Jennewein, C. Simon, G. Weihs, H. Weinfurter, A. Zeilinger: Phys. Rev. Lett. **84**, 4729 (2000)
30. A.K. Ekert: Phys. Rev. Lett. **67**, 661 (1991)
31. J.S. Bell: Phys. World, pp. 33–40 (Aug. 1990)

# Part IV

# Quantum Ideas

# 18 The Geometry of the Quantum Paradoxes

John Conway and Simon Kochen

Text of talk by Simon Kochen

## 18.1 Introduction

At the suggestion of the organizers of this conference, I shall start by discussing the genesis of the result known as the Kochen–Specker Paradox or K–S Theorem and its relation to Bell's Theorem. In the second part of my talk I deal with more recent results obtained with John Conway on the geometry and logic of quantum mechanics, which throw new light on these old results.

The original results were essentially negative in character, showing in different ways that quantum mechanics differs sufficiently radically from classical mechanics so as not to admit hidden variables (at least of a certain kind). The new results are positive, allowing one in effect to reconstruct the quantum mechanical formalism from a minimal amount of physical data. What was remarkable to us was that simple questions about spin systems lead quickly to rather deep geometric problems about symmetric polytopes and root lattices of exceptional Lie algebras. To give a prime example, let us start with the spin projections $P_i \otimes 1, 1 \otimes P_i$, where $i = x, y, z$ and $P_i = \frac{1}{2}(1 + \sigma_i)$, of two spin $\frac{1}{2}$ particles in three *fixed* orthogonal directions $x, y, z$. Using only the operations of orthocomplement (logical negation) and the product of commuting projections (logical conjunction), operations that are in principle physically realizable, we obtain a set of rays (i.e. one-dimensional projections) that are dense in the projective space of all rays of $\mathbb{C}^4$. In fact, we can exactly characterize the set of rays so obtained – it corresponds to the set of all vectors of norm (= squared length) a power of 2 in the Gaussian $E_8$ lattice. We shall see how to construct the usual Hilbert space formalism of density, Hermitian, and unitary operators for states, observables, and symmetries from these results.

The resulting set of projections is so rich that it contains an infinite number of K–S paradoxes. We shall discuss the simplest of these that arises from the $D_4$ lattice (or equivalently the 24–cell polytope) which, in a sense, contains the best features of both the original K–S and Bell theorems.

## 18.2 Old Paradoxes

In 1961 Ernst Specker visited the Mathematics Department of Cornell University, where I was a faculty member, and we began discussing the hidden variables problem, based on an idea in *Specker* [1]. We had read the proof of the non-existence of hidden variables in *von Neumann* [2] and were dissatisfied with its assumption. Von Neumann argues that if hidden variables exist, then at any given time every quantum mechanical observable has a precise (hidden) real value. He assumed that this should define a real–valued function which is linear on the set of Hermitian operators representing the observables. We did not see why such a prediction function should be additive on non-commuting operators, although it was a reasonable condition on commuting operators. This objection to von Neumann's proof had already been raised in *Hermann* [3] in 1935.

We were able by February of 1962 to construct a finite set of operators representing spin observables for which there existed no such prediction function which was additive solely on commuting operators. Let me briefly outline the construction, since I need to refer to it subsequently.

We considered a spin 1 system, such as an atom of orthohelium. The component of spin $S_d$ in direction $d$ has eigenvalues, $0, -1, +1$, so $S_d^2$ is a projection operator corresponding to the property (or 0-1 observable) "$S_d \neq 0$". For any (orthogonal) frame $(x, y, z)$ we have

$$S_x^2 + S_y^2 + S_z^2 = 2 \qquad (\text{with}\, \hbar = 1) \,.$$

Moreover, in the spin 1 case $S_d^2$ commutes with $S_{d'}^2$, if $d \perp d'$. Thus, it is possible to simultaneously measure $S_x^2, S_y^2$ and $S_z^2$ and the resulting values must be one 0 and two 1's, in some order.

If $D$ is any set of directions, then a prediction function for the set $\{S_d^2 | d \in D\}$ yields a function $h : D \to \{0, 1\}$ such that

(1) For any orthogonal frame $(x, y, z)$, $h$ takes the value 0 exactly once.
(2) For any orthogonal pair $(d, d')$, $h$ takes the value 0 at most once.

We now considered a set of eight directions representing the following graph, in the sense that vertices of the graph are connected by an edge if the corresponding directions are orthogonal. The directions given by the following vectors yield the graph, in this sense

$$\begin{aligned} &a_5 = (1,0,0),\ a_1 = (0,1,-1),\ a_3 = (0,1,1) \\ &a_6 = (0,0,1),\ a_2 = (1,-1,0),\ a_4 = (1,1,0) \\ &a_0 = (1,1,1),\ a_7 = (1,-1,1) \end{aligned}$$

Thus, since $|\langle a_0/|a_0|, a_7/|a_7|\rangle|^2 = \frac{1}{9}$, the transition probability from $S_{a_0} = 0$ to $S_{a_7} = 0$ is $\frac{1}{9}$.

On the other hand, if $h(a_0) = h(a_7) = 0$, then $h(a_1) = h(a_2) = h(a_3) = h(a_4) = 1$ by (2), so that $h(a_5) = h(a_6) = 0$ by (1), which contradicts (2).

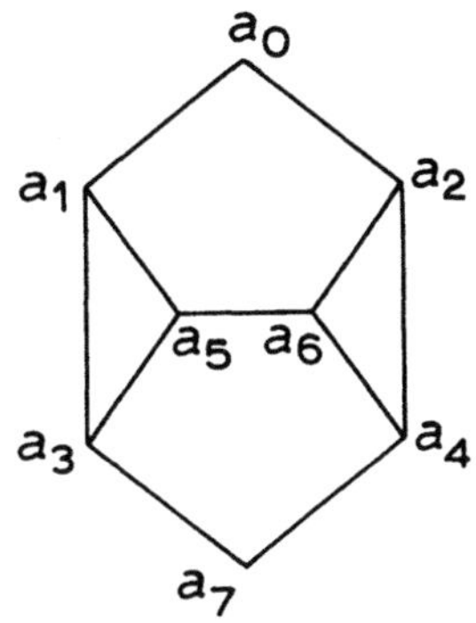


**Fig. 18.1.** Graph 1

This construction thus gave us a contradiction between probabilities assigned to observables by a hidden variable theory and those assigned by quantum mechanics. This result is similar to the one that was to be given by the Bell Theorem. Being logicians we were, however, dissatisfied with this result. We wished to obtain the sharper contradiction that a $0-1$ observable which has probability 1 of being true quantum mechanically has probability 1 to be false in a hidden variable theory. The contradiction obtained would then not be merely statistical in character, but a contradiction of a logical nature regarding truth and falsehood. To achieve this we considered the more complicated set of 117 directions in Graph 2, obtained by suitably rotating the Graph 1 about the axes of frame $(p_0, q_0, r_0)$.

Graph 2 is obtained by identifying the points $p_0$ and $a$, $q_0$ and $b$, and $r_0$ and $c$.

The frame $(p_0, q_0, r_0)$ must have one element, say $p_0$, such that $h(p_0) = 0$, $h(q_0) = h(r_0) = 1$. Hence, from the Graph 1 argument, $0 = h(p_0) = h(p_1) = h(p_2) = h(p_3) = h(p_4) = h(q_0)$, a contradiction.

In this way, we proved the non-existence of hidden variables without the use of probabilities. I should add here that we only became aware of Gleason's Theorem, which implies that there is no such prediction function on the whole unit sphere, after we had completed this proof.

Now what is the relation of this K–S Theorem to Bell's Theorem? We have seen that the conclusion of the K–S Theorem is stronger than Bell's result, since it sharpens the contradiction between quantum mechanics and a hidden variable theory by forcing a probability 1 property to have probability 0. In fact, as we saw, the far smaller Graph 1 of eight spins would suffice to give a conclusion of the weaker sort that Bell obtained.

The hypotheses of the two theorems are also different – Bell's is locality and K–S's is non-contextuality. In the above proof, non-contextuality takes the form of assuming that the hidden value $h(S_z^2)$ is independent of the different orthogonal pairs $(x, z)$, or $(x', z)$ to which the direction $z$ belongs in Graph 2. There is a large literature about these two different assump-

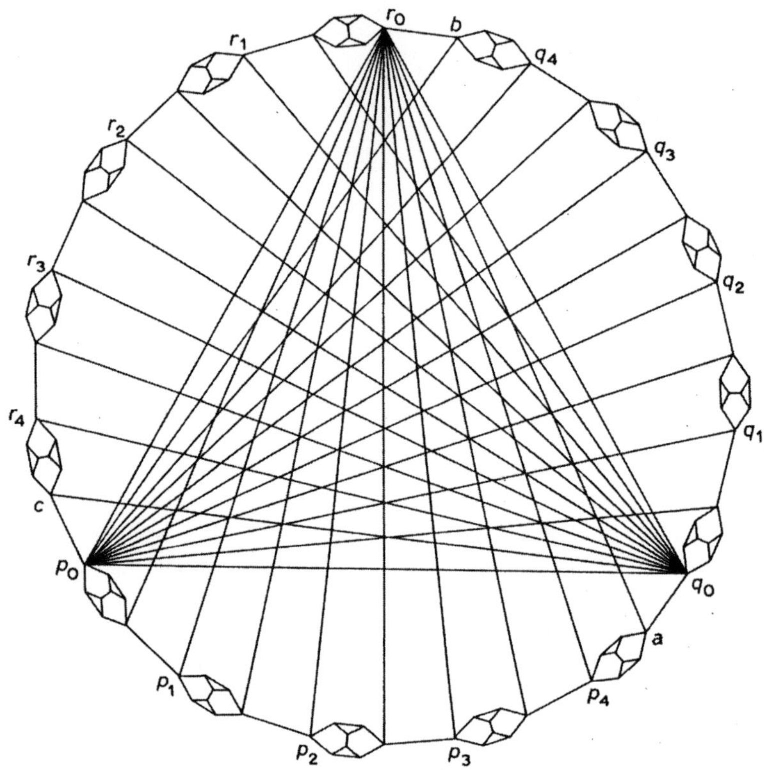


**Fig. 18.2.** Graph 2

tions. However, there is in fact a close connection between locality and non-contextuality, as I noticed when I first saw Bell's paper: Bell's locality implies K–S non-contextuality.

The rough idea is to consider a pair of spin 1 systems prepared so that the combined system is in the singlet state, i.e. a state of spin 0. Then, assuming locality, a value for $S_x$ for one spin 1 system determines the value for $S_x$ for the other without disturbing that system, leading to the K–S non-contextuality condition. This idea seemed rather obvious to me, so I never published it, although I mentioned it in a letter to Abner Shimony (see *Shimony* [4]). In retrospect, given all the subsequent literature which appears to me to miss the point, I rather regret not making the argument more public. In any case, the idea was re–discovered by *Heywood* and *Redhead* [5], where the interested reader can get a detailed version.

Thus, in a formal mathematical sense, the K–S Theorem is stronger than Bell's Theorem, since it deduces a stronger conclusion from a weaker hypothesis. Physically, however, the situation is otherwise. The very simplicity of the Bell result, referring to a small number of states of two spin $\frac{1}{2}$ particles, allows for a simpler experimental verification, culminating in the beautiful experiments of Aspect and Grangier. We shall later discuss a construction which combines the simplicity of Bell's pair of spin $\frac{1}{2}$ particles with the sharper conclusions of the K–S Theorem.

For the purposes of explaining the more recent results, it is necessary to define a formal structure which is the natural setting for the K–S Theorem. A *partial boolean algebra*, or pba, consists of a set $B$ of projections on a Hilbert space $\mathcal{H}$ containing the 0 projection and closed under the orthocomplement $P^\perp(= 1 - P)$ and product $PQ$ of commuting $P$ and $Q$. In particular, the set $B(\mathcal{H})$ of all projections on $\mathcal{H}$ forms a pba.

If $B$ is a pba, then a *homomorphism* $h$ of $B$ into the boolean algebra $\{0,1\}$ is a map $B \xrightarrow{h} \{0,1\}$ such that (i) $h(0) = 0$, (ii) $h(P^\perp) = h(P)'$, and (iii) $h(PQ) = h(P)h(Q)$.

If $S$ is a set of projections, the pba $B(S)$ generated by $S$ is, by definition, the smallest pba containing $S$. The pba $B(S)$ exists since it is the intersection of all pba's containing $S$. If $S$ is finite, we may obtain $B(S)$ constructively by successively adjoining the orthocomplement and product of commuting elements already constructed from $S$. Any function $S \xrightarrow{h} \{0,1\}$ satisfying (i)–(iii) above has an obvious extension to a homomorphism $B(S) \xrightarrow{h} \{0,1\}$. We may restate the K–S Theorem as stating that there is no homomorphism $B(S) \to \{0,1\}$ for $S = \{S_d^2\}$, where $d$ varies over the directions represented by Graph 2.

Projections and closed linear subspaces are in a natural 1–1 correspondence, viz. projection $P$ corresponds to its image space $P(\mathcal{H})$. Moreover, $P^\perp$ corresponds to $P(\mathcal{H})^\perp$, the orthocomplement of $P(\mathcal{H})$, and $PQ$ corresponds to $P(\mathcal{H})Q(\mathcal{H})$, the intersection of $P(\mathcal{H})$ and $Q(\mathcal{H})$, if $[P,Q] = 0$. Thus, we may equally well speak of pba's of closed linear subspaces of $\mathcal{H}$. We shall occasionally, by an abuse of language, refer to a $d$-dimensional projection $P$ as a $d$-dimensional subspace, such as a line or a plane, meaning the image $P(\mathcal{H})$ by this.

There is another way to think of the K–S theorem that relates it to the logic of quantum mechanics. Let $\varphi(x_1, \dots, x_n)$ be a formula in standard propositional logic formed from conjunction $\wedge$ and negation $\neg$. We proved in [8] that a necessary and sufficient condition that there exists no homomorphism of the pba $B$ into $\{0,1\}$ is that there exist a classical tautology $\varphi(x_1, \dots, x_n)$ which is false ($=0$ projection) under a meaningful substitution $\varphi(P_1, \dots, P_n)$ of some elements $P_1, \dots, P_n$ in $B$. The exact definition of meaningful is given in [8], but roughly speaking it means that the intermediate projections corresponding to the sub–formulas out of which $\varphi$ is formed commute when they are combined. For example, if the formula is $(x_1 \wedge x_2) \wedge (\neg x_3 \wedge x_4)$, we require that $[P_1, P_2] = [P_3, P_4] = [P_1P_2, P_3^\perp P_4] = 0$. In the case of the pba $B(S)$ considered earlier, we may take $\varphi$ to be

$$\varphi(x_1, \dots, x_{117}) = \neg \bigwedge_{\substack{(i,j)\in E \\ (k,\ell,m)\in\Delta}} \left[\neg(x_i \wedge x_j) \wedge (x_k \vee x_i \vee x_m)\right] ,$$

where $E$ and $\Delta$ are the sets of edges and triangles in Graph 2. Thus, $\varphi$ encapsulates the whole argument of the proof for the non-existence of a homomorphism of $B(S)$. Later we shall give a far simpler example of such a $\varphi$ in just four variables.

## 18.3 New Paradoxes and Their Geometry

At a conference in 1989 Asher Peres mentioned to me that he had slightly reduced the number of vectors needed for a K–S paradox from 117 to 108. *Jost* [6] had already made a similar reduction. Peres asked me if a significant reduction was possible. John Conway and I studied the problem, and came up with a set of 31 lines with integer co-ordinates. When I showed this result to Peres, he then constructed another set with 33 lines with co-ordinates using $\sqrt{2}$ (see *Peres* [7]). In unpublished work from many years ago, in 1965, Kurt Schütte had reduced the number to 33 in a construction similar to ours. Undoubtedly 31 is still not the smallest number in $\mathbb{C}^3$, although it may be so in $\mathbb{R}^3$. Although from a combinatorial viewpoint, it would be interesting to know the smallest number in $\mathbb{C}^3$, I do not believe the question has much physical significance.

These examples show that $\mathbb{C}^3$, and, even the real space $\mathbb{R}^3$, contains many pba's without homomorphisms into $\{0,1\}$. The pba $B(\mathbb{C}^2)$ of all projections of a two dimensional Hilbert space $\mathbb{C}^2$ has by contrast a plethora of homomorphisms into $\{0,1\}$. In fact, for $\mathbb{C}^2$, essentially the space of a single spin $\frac{1}{2}$ particle, Specker and I defined a non-contextual hidden variable space [8]. This showed incidentally that von Neumann's linearity criterion for hidden variables was unreasonable, since it applied even to $\mathbb{C}^2$. Thus, the first non-classical phenomenon of this type occurs at dimension 3. In 1990, John Conway and I showed that a new and much stronger non-classical phenomenon appears at dimension 4.

In the 1960's, Specker and I had asked the question whether every finitely generated pba is itself finite, i.e. whether $S$ finite implies $B(S)$ finite. In 1970, *Coray* [9] proved that a pba generated by any three projections has at most 256 elements, and is embeddable in a boolean algebra. In 1990, John Conway and I found five projections in $\mathbb{C}^4$ (and hence in all higher dimensional Hilbert spaces) such that the pba generated is infinite. Before describing this construction, we contrast this with the known situation in $\mathbb{C}^3$.

The various finitely generated pba's in $\mathbb{R}^3$ mentioned earlier are, in fact, all finite. The pba generated by the 31 lines that Conway and I constructed has 124 elements, and the one generated by the 33 lines of Peres has 116 elements. We were able to prove that if all the elements of the finite generating set $S$ have rational co-ordinates in $\mathbb{C}^3$, then the pba $B(S)$ must be finite. This has been generalized to the case where the elements of $S$ have algebraic co-ordinates by *Derek Smith* in his thesis [10]. This covers all the existing cases in $\mathbb{C}^3$. The general case of arbitrary co-ordinates in $\mathbb{C}^3$ is still not proved, but my guess is that it is true that $B(S)$ is finite then also. At any rate, the 5 elements in $\mathbb{C}^4$ that generate an infinite pba all have rational co-ordinates, so $\mathbb{C}^4$ exhibits new non-classical features not present in $\mathbb{C}^3$.

Let me now give an indication of why this is so for $\mathbb{C}^4$. To motivate the proof we first look at the more visualizable case of a real Hilbert space $\mathbb{R}^4$. Just as there are five platonic solids in $\mathbb{R}^3$, there are six regular polytopes in $\mathbb{R}^4$. (See e.g. *Coxeter* [11]). The largest of these is the 600–cell $\mathcal{T}$. This polytope has 120 vertices and 600 tetrahedra (or cells) forming the boundary of $\mathcal{T}$. We denote the symmetry (or automorphism) group of $\mathcal{T}$, the group of all orthogonal maps of $\mathbb{R}^4$ which keep $\mathcal{T}$ invariant, by $\operatorname{Aut}\mathcal{T}$.

The 120 vertices of $\mathcal{T}$ lie on a set $T$ of 60 lines through the origin (i.e. diametral lines), since the vectors $\pm v$ lie on one line. We form the pba $B(T)$. Every symmetry of $\mathcal{T}$ extends naturally to a symmetry of $B(T)$. Thus

$$\operatorname{Aut}\mathcal{T} \subset \operatorname{Aut}B(T) \subset O(4)\ .$$

Now, $\operatorname{Aut}\mathcal{T}$ has the interesting property that it is a maximally finite subgroup of the orthogonal group $O(4)$, so that a subgroup of $O(4)$ that is properly larger than $\operatorname{Aut}\mathcal{T}$, is automatically infinite. This is the case for $\operatorname{Aut}B(T)$ since a rotation (through angle $\frac{\pi}{4}$) of the 120 vertices to 120 cell centers is an automorphism of $B(T)$, which clearly does not lie in $\operatorname{Aut}\mathcal{T}$. To see that such cell centers define lines lying in $B(T)$, so that this rotation lies in $\operatorname{Aut}B(T)$, we consider a typical case.

In the standard co-ordinatization of $\mathcal{T}$, the vectors

$$(1,0,0,0),\ (0,0,0,1),\ \frac{1}{2}(1,1,1,1),\ \frac{1}{2}(1,1,-1,-1)$$

are vertices of $\mathcal{T}$. We now compute, using only compatible spaces for each operation,

$$\begin{aligned}
&\big([(1,0,0,0)] \vee [(0,1,0,0)]\big) \wedge \left(\left[\frac{1}{2}(1,1,1,1)\right] \vee \left[\frac{1}{2}(1,1,-1,-1)\right]\right)\\
&\quad = \big([(1,1,0.0)] \vee [(1,-1,0,0)]\big) \wedge \big([(1,1,0,0)] \vee [(0,0,1,1)]\big)\\
&\quad = [(1,1,0,0)]\ ,
\end{aligned} \tag{18.1}$$

the line through the cell center $(1,1,0,0)$[1].

Thus $\operatorname{Aut}B(T)$ is infinite, so that $B(T)$ is itself infinite. A more involved argument shows that $\operatorname{Aut}B(T)$ is dense in $O(4)$ and that $B(T)$ is dense $B(\mathbb{R}^4)$. We may reduce the number of generators of $B(T)$ to 15, since the 60 lines of $T$ can be generated by 15 planes.

---

[1] $[v]$ denotes the line through the vector $v$. Note that the vector $(1,1,0,0) = (1,0,0,0) + (0,1,0,0) = \frac{1}{2}(1,1,1,1) + \frac{1}{2}(1,1,-1,-1)$ has twice the norm (= squared length) of the original vectors. To represent the cell center and later for the lattice computations, it is important not to normalize this vector to a unit vector.

To treat the complex case of $\mathbb{C}^4$, we start with an analogous complex symmetric polytope called the complex Witting polytope $\mathcal{W}$. This polytope has 240 vertices which lie on a set $K$ of 60 diametral lines[2] (since the vectors $\pm v, \pm iv$ lie on one diameter). As before, $\mathrm{Aut}\,\mathcal{W} \subset \mathrm{Aut}\,B(K) \subset U(4)$, the unitary group on $\mathbb{C}^4$, and $\mathrm{Aut}\,\mathcal{W}$ is a maximally finite subgroup of $U(4)$. A similar argument to (1) shows that there is an automorphism of $B(K)$ from the vertices of $\mathcal{W}$ to the cell centers which is not in $\mathrm{Aut}\,\mathcal{W}$. Thus, $\mathrm{Aut}\,B(K)$ and $B(K)$ are infinite. Again, we may extend the argument to show $B(K)$ is dense in $B(\mathbb{C}^4)$.

As in the real case, we may reduce the number of generators, since the 60 diametral lines in $K$ are all generated by 5 planes. What is of particular interest is that these planes have a simple physical meaning. We write $\mathbb{C}^4 = \mathbb{C}^2 \otimes \mathbb{C}^2$, the tensor product of two 2-dimensional spaces. We can then interpret $\mathbb{C}^4$ as the Hilbert space of two spin $\frac{1}{2}$ particles. We choose a fixed orthogonal frame $(x, y, z)$ in space.[3] Let $P_i = \frac{1}{2}(1 + \sigma_i), i = x, y, z$, be the projection that corresponds to the property that the system has spin $\frac{1}{2}$ in direction $i$. Then

$$P_x \otimes 1, P_y \otimes 1, P_z \otimes 1, 1 \otimes P_x, 1 \otimes P_y, 1 \otimes P_z$$

are the spin projections of the two particles in the directions $x, y$, and $z$. We let $S$ be any 5 of these 6 two-dimensional projections. Then one can show that $S$ generates all the 60 rays in $K$. Thus $B(S) = B(K)$ and $B(S)$ is dense in $B(\mathbb{C}^4)$.

The fact that $B(S)$ is dense in $B(\mathbb{C}^4)$ shows that it has a rich structure. It is a surprising fact that this structure has a simple mathematical description. It suffices here to consider the lines in $B(S)$ since the 3–spaces are orthocomplements of lines, and the planes are spans of orthogonal lines. We start with the 240 vectors of the Witting polytope which projectively yield the 60 lines obtained from the original 5 two-dimensional projections. We saw how by the process (1) we obtained cell centers as new vectors lying projectively in $B(S)$. Moreover, the new vectors have norm twice that of the vertex vectors. Continuing this process yields further vectors with norm a power of 2, – but the question is which vectors in $\mathbb{C}^4$ arise this way.

Now a polytope may be the contact polytope of an integer lattice, i.e. the vertices of the polytope are the minimal (i.e. shortest) vectors of the lattice. The Witting polytope is the contact polytope of the Gaussian $E_8$ lattice, $E_8[\mathrm{i}]$. The $E_8$ lattice is an integer lattice in $\mathbb{R}^8$, the root lattice of the exceptional Lie group $E_8$. The space $\mathbb{R}^8$ may be given a complex structure $\mathbb{C}^4 = (\mathbb{R}^8, J)$ by means of the map $J = \left(\begin{smallmatrix} 0 & -I \\ I & 0 \end{smallmatrix}\right)$, where $I$ is the

[2] The projective version of $\mathcal{W}$ is the Klein configuration $60_{15}$ first constructed by Klein over 100 years ago in studying the 16 nodes of Kummer's quartic surface. See, e.g. *Hudson* [12].

[3] We could equally well choose two separate frames $(x, y, z)$ and $(x', y', z')$ for the two particles.

4-dimensional identity. The $E_8$ lattice thereby becomes a Gaussian integer lattice in $\mathbb{C}^4$, the Gaussian $E_8$ lattice, $E_8[\mathrm{i}]$. As stated, the 240 vectors of the Klein configuration are the minimal vectors in $E_8[\mathrm{i}]$. The process of obtaining new vectors always leads to vectors in $E_8[\mathrm{i}]$. Moreover, by the $n^{th}$ stage all the vectors in $E_8[\mathrm{i}]$ of norm $2^n$ are obtained this way. Thus, the set of lines of $B(S)$ consists projectively of precisely the set of vectors in $E_8[\mathrm{i}]$ of norm a power of 2. An immediate consequence is that the infinite discrete set of such vectors in $\mathbb{C}^4$, when normalized to unit vectors, yields a uniformly dense set of points on the unit–sphere of $\mathbb{C}^4$ (reminiscent of Olber's Paradox in astronomy). In fact, the set of unit vectors becomes dense very quickly, since the process which yields the norm–doubled vectors increases the number of new vectors by a factor of 8 at each stage.

It is easily seen that the pba $B(S)$ has no homomorphisms into $\{0,1\}$. In fact, the rich structure of $B(S)$ implies that there are an infinite number of sub–pba's with this property. The simplest of these is obtained by considering the sub–polytope $\mathcal{V}$ of the Witting polytope called the 24–cell. The polytope $\mathcal{V}$ is also a sub–polytope of the 600–cell polytope $\mathcal{T}$ in $\mathbb{R}^4$, with 24 vertices and 24 octahedra forming the boundary. It is one of the six regular polytopes of $\mathbb{R}^4$, and is the contact polytope of the root lattice $D_4$. The 24 vertices lie on a set $V$ of 12 lines. We now consider the pba $B(V)$. The construction in (1) applies also to $\mathcal{V}$ to yield the cell centers of the 24 octahedra, giving 12 more lines in $B(V)$. The rotation through angle $\frac{\pi}{4}$ of the vertices to the cell centers thus, as before, lies in $\operatorname{Aut} B(V)$ but not in $\operatorname{Aut} \mathcal{V}$, so $\operatorname{Aut} B(V)$ is properly larger than $\operatorname{Aut} \mathcal{V}$. However, in this case $\operatorname{Aut} B(V)$ is a finite group, generated by $\operatorname{Aut} \mathcal{V}$ and this rotation. The pba $B(V)$ is itself finite. It has 140 elements, including 24 lines, 90 planes, and 24 3–spaces. The generating set $V$ of 12 lines can itself be generated by a subset $U$ of the previous set $S$ of 5 spin projections, viz.

$$U = \{P_x \otimes 1, P_z \otimes 1, 1 \otimes P_x, 1 \otimes P_z\}\ .$$

We shall illustrate the connection between homomorphisms and tautologies mentioned earlier by showing that $B(U)$ has no homomorphism into $\{0,1\}$ via a classical tautology that is false in $B(U)$. *Kochen* and *Specker* [8] defined abstract partial boolean algebras, of which pba's of projections were one class of examples. In [13] we constructed such a general pba in which the classical tautology

$$[(x \leftrightarrow y) \leftrightarrow (z \leftrightarrow w)] \longleftrightarrow [(x \leftrightarrow z) \leftrightarrow (y \leftrightarrow w)] \tag{18.2}$$

is false. Here the logical equivalence $a \leftrightarrow b$ is defined by $x \cdot y \vee x^{\perp} \cdot y^{\perp}$.

It turns out that this tautology is also false in $B(U)$ under the substitution

$$\begin{aligned} x = P_z \otimes 1 &= [(1,0,0,0),(0,1,0,0)] , \\ y = 1 \otimes P_z &= [(1,0,0,0),(0,0,1,0)] , \\ z = 1 \otimes P_x &= [(1,1,1,1),(1,1,-1,-1)] , \\ w = P_x \otimes 1 &= [(1,1,1,1),(1,-1,1,-1)] . \end{aligned}$$

We have

$$\begin{aligned} P_z \otimes 1 \leftrightarrow 1 \otimes P_z &= [(1,0,0,0),(0,0,0,1)] \\ 1 \otimes P_x \leftrightarrow P_x \otimes 1 &= [(1,1,1,1),(1,-1,-1,1)] \end{aligned}$$

so that

$$(P_z \otimes I \leftrightarrow 1 \otimes P_z) \longleftrightarrow (1 \otimes P_x \leftrightarrow P_x \otimes 1) = [(1,0,0,1),(0,1,-1,0)] .$$

On the other hand,

$$\begin{aligned} P_z \otimes 1 \leftrightarrow 1 \otimes P_x &= [(1,1,0,0),(0,0,1,-1)] , \\ 1 \otimes P_z \leftrightarrow P_x \otimes 1 &= [(1,0,1,0),(0,1,0,-1)] \end{aligned}$$

so that

$$(P_z \otimes 1 \leftrightarrow 1 \otimes P_x) \longleftrightarrow (1 \otimes P_z \leftrightarrow P_x \otimes 1) = [(1,1,1,-1),(-1,1,1,1)] .$$

But

$$[(1,0,0,1),(0,1,-1,0)] = [(1,1,1,-1),(-1,1,1,1)]^{\perp}$$

so

$$[(1,0,0,1),(0,1,-1,0)] \longleftrightarrow [(1,1,1,-1),(-1,1,1,1)] = 0 .$$

This proves that there is no homomorphism of $B(U)$ into $\{0,1\}$. This is the simplest tautology one could expect to be false in a pba, for the result of *Coray* [9] mentioned earlier implies that every tautology in 3 variables in true in every pba.

Without benefit of the geometry described here, Peres and Mermin independently gave a direct proof of a K–S paradox based on the spin operators $\sigma_x \otimes 1, \sigma_z \otimes 1, 1 \otimes \sigma_x, 1 \otimes \sigma_z$ (see *Mermin* [14]), essentially the same example. Incidentally, the original tautology Conway and I constructed was actually

$$[(x+y)+(z+w)] \longleftrightarrow [(x+z)+(y+w)] ,$$

where $a+b$ is the exclusive or $(a \wedge \neg b) \vee (\neg a \wedge b)$. The formula (2) is easier to compare with the Peres–Mermin proof since e.g. $(\sigma_z \otimes 1)(1 \otimes \sigma_x)$ has eigenspace $P_z \otimes 1 \longleftrightarrow 1 \otimes P_x$ with eigenvalue 1, etc.

## 18.4 Reconstructing Quantum Mechanics

We now indicate how to use the theorem that $B(S)$ is dense in $B(\mathcal{H})$ to reconstruct the formalism of quantum mechanics from a small set of physical data. For simplicity, we stay with the case $\mathcal{H} = \mathbb{C}^4$; *Kochen* [15], which this section follows, shows how to generalize the results to high dimensions.

Although our point of view is by no means an operational one, we shall emphasize the strictly constructive nature of this approach by speaking in terms of measurements. We shall assume that if $P$ and $Q$ are two measurable properties which correspond to commuting projections, then there exists an apparatus to measure the product $PQ$. This is certainly true in principle, and it becomes a pragmatic matter to design such an experiment in practice.

Since every stage of the construction from $S$, obtained by adjoining orthocomplements and commutative products of prior stages, realizes only a finite number of projections, we cannot expect to attain every one of the continuum $B(\mathcal{H})$ of projections. Our point of view is that the full continuum of these objects is a mathematical artifact, a model designed to allow for a smooth analysis of the actual objects occurring in nature. It thus suffices to represent a pure state, for instance, by a sequence of constructed one-dimensional projections which approximate as close as we please to the state.

We have used the Hilbert space to define $B(S)$, but we now drop all reference to the Hilbert space, since we wish to reconstruct it. We start with two spin-1/2 particles, and three specified orthogonal directions in physical space and the spins of the particles in these directions. These observables are certainly measurable. At the next level, we consider the conjunction of the two properties specifying the spin of each particle in one of the three directions. This property is measurable since it corresponds to the product $(P_i^{(\perp)} \otimes I)(I \otimes P_j^{(\perp)})$ (where $i, j = x, y, z$ and $P^{(\perp)}$ denotes either $P$ or $P^{\perp}$), and $P_i^{(\perp)} \otimes I$ commutes with $I \otimes P_j^{(\perp)}$. The later stages continue this process which preserves measurability of the properties. It is actually possible, at least in principle, to design the corresponding experiments by using the neutron interferometry experiments described by *Rauch* [16]. Since $B(S)$ is dense in $B(\mathcal{H})$ we may in this manner approximate as closely as desired to any projection on $\mathcal{H}$.

We note that we do not make any assumptions about physical space in our construction. In fact, we can reconstruct physical space itself from this approach along with the Hilbert space. Among the projections we construct are the spin components $P_d = \frac{1}{2}(1 + \sigma_d)$ for each particle for a dense set of directions $d$ in physical space. In this way, we can approximate every direction in space via the spin projections.

So far, although we have indicated how to construct the set of all projections on a Hilbert space, the structure on the set of a partial boolean algebra is apparently a very weak one. It appears to be a far cry from the full formalism of a complex Hilbert space with Hermitian, density, and unitary operators representing the observables, states and symmetries of quantum mechanics. We now briefly show how to recover this structure.

We shall define the observables, states and symmetries of a physical system in terms of its partial boolean algebra $B$ of properties in an intuitively natural way which reduces to the usual concepts in classical physics when $B = B_{\Omega}$, the boolean algebra of Borel sets of the phase space $\Omega$. We shall see that if $B = B(\mathcal{H})$, then these definitions are equivalent to the usual quantum mechanical ones. In conformity to our view of the use of infinite mathematical operations in physics, we shall allow the use of boolean $\sigma$-algebras, with their infinite union, on the understanding that infinite unions may be approximated by finite unions.

The results we give below are more or less familiar in the context of the lattice of projections of Hilbert space. The point we are emphasizing is that the relevant theorems (such as Gleason and Wigner Theorems) are valid for the weaker structure $B(\mathcal{H})$ as a pba.

(i) A *state* is a measure $\mu : B \to [0,1]$, i.e. the restriction of $\mu$ to any $\sigma$-algebra in $B$ is a standard probability measure.
If $B = B_{\Omega}$, then this is the classical concept of a mixed state on $\Omega$. If $B = B(\mathcal{H})$ then Gleason's Theorem (*Varadarajan* [17], Theorem 7.23) implies that the states correspond to the trace–class operators on $\mathcal{H}$ of trace 1 (density operators). The pure states are, by definition, the extreme points of the convex set of states. If $B = B_{\Omega}$, they correspond to the points of $\Omega$. If $B = B(\mathcal{H})$, they correspond to the rays of $\mathcal{H}$.

(ii) An *observable* is a $\sigma$-homomorphism $h : B_{\mathcal{R}} \to B$, where $B_{\mathcal{R}}$ is the $\sigma$-algebra of Borel sets of the reals $\mathcal{R}$.
If $B = B_{\Omega}$, the observables correspond to Borel functions $f : \Omega \to \mathcal{R}$ (*Varadarajan* [17], Theorem 1.4). If $B = B(\mathcal{H})$, the image of $h$ is a $\sigma$-algebra in $B$. The Spectral Theorem shows that the observables correspond to Hermitian operators on $\mathcal{H}$ (*Varadarajan* [17], Theorem 7.11).

(iii) A *symmetry* is an automorphism $\sigma\colon B \to B$.
If $B = B_{\Omega}$, the allowed automorphisms are those that preserve the symplectic structure on $\Omega$. If $B = B(\mathcal{H})$ Wigner's Theorem (*Varadarajan* [17], Theorem 7.29) shows that symmetries correspond to unitary or anti-unitary transformations of $\mathcal{H}$.

Since the Projection Postulate is usually considered as an axiom to be added to the assumptions of quantum mechanics, we show how it fits into our scheme.

(iv) *Conditioning* is a map $(\mu(x), a) \mapsto \mu(x|a)$ with $\mu(x)$ and $\mu(x|a)$ states of $B$ and $a \in B$ such that $\mu(a) \neq 0$, with the property that for every $\sigma$-algebra $\sigma$ of $B$ with $a \in \sigma$ and for every $b \in \sigma$,

$$\mu(b|a) = \frac{\mu(b \cdot a)}{\mu(a)} .$$

If $B = B_\Omega$, $\mu(b|a)$ is the conditioned state given that $a$ is true. If $B = B(\mathcal{H})$, an elementary proof via Gleason's Theorem shows that the state $\mu(x|a)$ corresponds to the density operator

$$\frac{awa}{\mathrm{Tr}\,(wa)}\,,$$

where $w$ is the density operator corresponding to $\mu$. This is Lüder's form of the Projection Postulate of von Neumann.

These results indicate briefly how the construction of the partial boolean algebra $B(\mathcal{H})$ of projections on complex Hilbert space allows one to recover, in a natural way, the complete formalism of quantum mechanics.

## References


1. E. Specker: Dialectica **14**, 239–246 (1960); English translation in: *The Logico-Algebraic Approach to Quantum Mechanics Vol. I* (Reidel, Dordrecht 1975) pp. 135–140
2. J. von Neumann: *Mathematical Foundations of Quantum Mechanics* (Springer, Berlin 1932)
3. G. Hermann: *Die naturphilosophischen Grundlagen der Quantenmechanik*, Abhandlungen der Fries'schen Schule (1935)
4. A. Shimony: Brit. J. Philos. Sci. **35**, 25–45 (1984)
5. A. Heywood, M. Redhead: Found. Phys. **13**, 481–499 (1983)
6. R. Jost: 'Measures on the finite dimensional subspaces of a Hilbert space: Remarks on a theorem by A.M. Gleason'. In: *Studies in Mathematical Physics* (Princeton Univ. Press, Princeton 1976)
7. A. Peres, *Quantum Theory Concepts and Methods* (Kluwer Academic, Dordrecht 1993)
8. S. Kochen, E. Specker: J. Math and Mech.**17**, 59–87 (1967)
9. G. Coray: Commun. Math. Helv. **45**, 49–82 (1970)
10. D. Smith: On finitely–generated quantum logic. PhD Thesis, Princeton University (1999)
11. H.S.M. Coxeter: *Regular Polytopes* (MacMillan, New York 1963)
12. R. Hudson: *Kummer's Quartic Surface* (Cambridge Univ. Press, Cambridge 1990)
13. S. Kochen, E. Specker: *Logical Structures Occurring in Quantum Theory*, Symp. on the Theory of Models, (North Holland, Amsterdam 1965) pp. 177–189
14. N.D. Mermin: Phys. Rev. Lett **65**, 3373–3376 (1990)
15. S. Kochen: 'Construction of quantum mechanics via commutative operations'. In: *Perspectives on Quantum Reality* (Kluwer Academic, Dordrecht 1996) pp. 237–243
16. H. Rauch: 'Quantum measurements in neutron interferometry'. Proc. 2nd Int. Symp. on Foundations of Quantum Mechanics, Tokyo (1986) pp. 3–17
17. V.S. Varadarajan, *Geometry of Quantum Theory*, Vol. I (Van Nostrand, Amsterdam 1968)

# 19 Whose Knowledge?

N. David Mermin

I met John Bell two and a half times. The two times, both memorable for me, were in what turned out to be in the last year of his life. The first was at a summer school in Erice in August 1989, where he delivered what came close to being the most spell-binding lecture I have ever heard. (The only competitors are Richard Feynman's 1965 Messenger lectures at Cornell.) The text was subsequently published in a much quoted article in *Physics World*, "Against Measurement" [1]. The article conveys his brilliance and his wit, but not, of course, the music of his voice.

The other moments I remember from our first meeting are Bell's polite but evidently skeptical response to my telling him that I had long ago written a book on relativity [2] that took the view he advocated several years later in "How to teach special relativity" [3]. Later I was pleased to get a letter saying that he had looked up my book and agreed that I did indeed do it right.

And I remember a reception at which there suddenly rang out over the noise of the crowd a cry of "Don't be a sissy!" It was John Bell, encouraging a younger physicist not to let the scope of his speculative investigations be overly constrained by the wisdom of his elders.

Our second meeting was for a week in Amherst in the summer of 1990. George Greenstein and Arthur Zajonc had invited about a dozen people, among them John and Mary Bell, Kurt Gottfried, Viki Weisskopf, Phillip Pearle, Tony Leggett, Danny Greenberger, Mike Horne, Anton Zeilinger, and Abner Shimony, to spend a week in a fraternity house chatting about the nature of quantum mechanics. No prepared talks, no schedule, no proceedings. Just wonderful conversations.

Shortly before the meeting I had written John – this was at the very end of the epoch in which one communicated at a leisurely pace on paper – to tell him about a simplified version I had constructed of a little known argument of Greenberger, Horne, and Zeilinger. I got back a wonderful reply, which I have searched for in my files unsuccessfully ever since he died. He said that the argument "filled me with admiration", and he wrote down a relevant three-particle Bell inequality, which inspired me to find a stronger one. (So somewhere, lost in my filing cabinet, is proof that the very first Bell inequality for "Bell's Theorem without Inequalities" was Bell's Bell inequality.) There were lots of entertaining discussions and arguments in Amherst. It was far and away the finest conference I have ever been to.

I also have evidence that over a third of a century ago we had a half meeting. The evidence is in the form of a conference photograph in my office taken in Birmingham England in the summer of 1967 at a meeting to celebrate the 60th birthday of Rudi Peierls. Amongst 100 white-shirted necktied gentlemen is a figure in the back row toward the left in a plaid shirt without a tie. He sports one of the only two beards in the photograph and is identified only as Bell; on the right in the rear is a more conventional looking youth identified as Mermin.

This Bell looks like an early version of the John Bell I later knew, and since the even younger John Bell spent a year in Birmingham in 1953–1954 (where I spent two years in 1961–1963), it seems clear that the Bell in the photo was indeed John. So I could have met John Bell almost a quarter of a century before I actually did meet him. It is probable that I was even introduced to him at that time, quite oblivious of the fact that I was meeting the discoverer of what would come to be called Bell's theorem, just as it was quietly tiptoeing onto the intellectual scene.

Our virtual meeting through Peierls is relevant to my contribution to this volume, because what I have to say was inspired by another, more recent, link between Bell and Peierls. After John's Erice lecture appeared in *Physics World*, but too late, alas, for him to read and reply to it, there appeared something between a commentary and a rebuttal by Peierls [4], by then, I calculate, a sprightly 84, which raised for me a puzzle which to this day I have not been able to answer to my satisfaction.[1]

Peierls agreed with Bell that we should have "a clearly formulated presentation of the physical significance of the [quantum] theory without relying on ill-defined concepts" and that "I do not know of any textbook which explains these matters to my satisfaction." On the other hand, he added, "I do not agree with John Bell that these problems are very difficult. I think it is easy to give an acceptable account." According to Peierls all it boils down to is this:

"In my view the most fundamental statement of quantum mechanics is that the wavefunction, or, more generally the density matrix, represents our *knowledge* of the system we are trying to describe".

Interestingly, *knowledge* is not on Bell's now famous list of

"words which, however legitimate and necessary in application, have no place in a *formulation* with any pretension to physical precision".

But "information" is on the proscribed list, the charge against it being:

"*Information? Whose* information? Information about *what?*"

[1] Some progress has been made since I wrote the above. See Todd A. Brun, J. Finkelstein, N. David Mermin: Phys. Rev. A **65**, 032315 (2002) (quant-ph/0109041).

I believe we can take "information" and "knowledge" to be synonymous in this context.

Until quite recently I was entirely on Bell's side on the matter of knowledge-information. But then I fell into bad company. I started hanging out with the quantum computation crowd, for many of whom quantum mechanics is self-evidently and unproblematically all about information. I digress to remark that among the tragedies of John Bell's early death is his missing the quantum computation revolution of the 1990s. It has been widely regretted that death deprived us of Einstein's reaction to Bell's theorem; another great loss is never to know what Bell would have had to say about Peter Shor's factoring algorithm or Lov Grover's search algorithm, and the novel view of quantum mechanics that emerges when you take the theory seriously as a source of impossibly fast algorithms for the processing of knowledge-information.

Partly from my associations with quantum computer scientists and partly from endless debates with constructivist sociologists of science, I have come to feel that "Information about *what*?" is a fundamentally metaphysical question that ought not to distract tough-minded physicists. There is no way to settle a dispute over whether the information is about something objective, or is merely information about other information. Ultimately it is a matter of taste, and, like many matters of taste, capable of arousing strong emotions, but in the end not really very interesting.

On the other hand "*Whose* information?" raises serious practical questions that simply have to be addressed. For while information, in the form of sense perceptions, is all any of us have direct access to, we have developed powerful ways to exchange with each other some of the content of our own private information. It is entirely legitimate and unambiguous to raise the question of what constraints, if any my possession of my knowledge imposes on the knowledge that you can possess, and vice versa.

Peierls was surely aware of Bell's double-barreled shot at information-knowledge and immediately after declaring that it's all about knowledge, he promised to "return later to the question 'whose knowledge'". I infer that like the very recent me, he did not find it necessary to deal with the question "knowledge about what?" But towards the end of his article he dealt briskly and efficiently with "whose knowledge?":

"[Density matrices] may differ, as the nature and amount of knowledge may differ. People may have observed the system by different methods, with more or less accuracy; they may have seen part of the results of another physicist. However, there are limitations to the extent to which their knowledge may differ. This is imposed by the uncertainty principle. For example if one observer has knowledge of $S_z$ of our Stern-Gerlach atom, another may not know $S_x$, since measurement of $S_x$ would have destroyed the other person's knowledge of $S_z$, and vice versa. This limitation can be compactly and conveniently expressed by the condition that the density matrices used by the two observers must commute with each other".

In a chapter in *More Surprises in Theoretical Physics* to which he referred the reader for further details, Peierls added a second condition, which is easy to understand if the density matrices do indeed commute [5]:

"At the same time, the two observers should not contradict each other. This means the product of the two density matrices should not be zero".

Peierls' rationale for his first condition has a highly plausible ring to it. After all, if Alice *knows* that a spin-$\frac{1}{2}$ particle is in an eigenstate of $S_z$, then absolutely no information can be available to anybody about any components of its spin orthogonal to the $z$-direction. A measurement of any spin component in the $x$-$y$ plane is as likely to come out up as down. Mathematically this requires that nobody's density matrix can have off diagonal elements in the $S_z$ representation – i.e. everybody's density matrix must commute with the pure-state density matrix of Alice.

Things become a little less clear if Alice's knowledge is not characterized by a pure-state density matrix, and considerably less clear if we are dealing with more than a two-state system. For some time I struggled to find a general way to understand Peierls' first criterion. I stopped struggling when Chris Fuchs pointed out to me that it was clearly incorrect, even for two-state systems and even when Alice's density matrix described a pure state. I give here a slightly embellished version of Fuchs' simple counterexample, which makes explicit how a disparity in knowledge can arise that leads Alice and Bob to describe one and the same physical system with non-commuting density matrices.

Consider a pair of qubits (spins-$\frac{1}{2}$, if you prefer, 0 being $\uparrow$ along $\boldsymbol{z}$ and **1** being $\downarrow$) in the state

$$|\Psi\rangle = \cos\theta\,|0\rangle\,|0\rangle + \sin\theta\,|1\rangle\,|1\rangle \ , \tag{19.1}$$

with $0 < \theta < \pi/4$. After the state has been prepared the qubits cease to interact. To emphasize the absence of subsequent interaction the qubit on the right (hereafter called the right qubit) can be carried far away from the other (hereafter called the left qubit). Anybody who knows how the two-qubit state was prepared – and, in particular, Alice and Bob, who both know – will agree that although the two-qubit system is in a pure state, each separate qubit, and in particular the left qubit, has no pure state of its own but is characterized, at the most fundamental level, by the mixed-state density matrix

$$\rho = \mathrm{Tr}_r\,|\Psi\rangle\,\langle\Psi| = \cos^2\theta\,|0\rangle\,\langle 0| + \sin^2\theta\,|1\rangle\,\langle 1| \ , \tag{19.2}$$

where $\mathrm{Tr}_r$ is the partial trace over the degrees of freedom of the right qubit.

Suppose Alice now goes to the right qubit and secretly measures it in the computational basis. She does not report to Bob the result of her measurement or even whether she has measured it at all. Since the right qubit is far away and does not interact with the left qubit and since Alice communicates no information whatever to Bob, regardless of what she has or has not done to the right qubit, Bob must surely continue to describe the left qubit with the same density matrix (19.2). Alice, on the other hand, knowing the strong correlations present in the state (19.1), is able after measuring the right qubit to update the density matrix she uses to express her knowledge about the qubit on the left to one of the pure-state density matrices $|0\rangle\langle 0|$ or $|1\rangle\langle 1|$, depending on whether the result of her measurement on the right is 0 or 1. Since either of these commutes with Bob's density matrix (19.2), Peierls' first condition is confirmed.

Suppose, however, that prior to making her measurement Alice secretly applies to the right qubit the unitary Hadamard transformation,

$$|0\rangle \to \frac{1}{\sqrt{2}}(|0\rangle + |1\rangle)\,, \quad |1\rangle \to \frac{1}{\sqrt{2}}(|0\rangle - |1\rangle)\,. \tag{19.3}$$

This transforms the two-qubit state into

$$\begin{aligned} |\Phi\rangle &= \frac{1}{\sqrt{2}}\left[\cos\theta\,|0\rangle\,(|0\rangle + |1\rangle) + \sin\theta\,|1\rangle\,(|0\rangle - |1\rangle)\right] \\ &= \frac{1}{\sqrt{2}}\left[(\cos\theta\,|0\rangle + \sin\theta\,|1\rangle)\,|0\rangle + (\cos\theta\,|0\rangle - \sin\theta\,|1\rangle)\,|1\rangle\right]\,. \end{aligned} \tag{19.4}$$

Now, depending on whether Alice gets 0 or 1 from her measurement of the right qubit, she will assign the left qubit one or the other of the pure-state density matrices

$$\begin{aligned} \rho_a &= \big(\cos\theta\,|0\rangle \pm \sin\theta\,|1\rangle\big)\big(\cos\theta\,\langle 0| \pm \sin\theta\,\langle 1|\big) \\ &= \cos^2\theta\,|0\rangle\,\langle 0| + \sin^2\theta\,|1\rangle\,\langle 1| \pm \cos\theta\sin\theta\big(|0\rangle\,\langle 1| + |1\rangle\,\langle 0|\big)\,. \end{aligned} \tag{19.5}$$

But Bob, who does not know what, if anything Alice did to the right qubit, must continue to use (19.2) for his density matrix $\rho_b$ describing the left qubit. We therefore have

$$[\rho_b, \rho_a] = \pm\frac{1}{4}\sin 4\theta\big(|0\rangle\,\langle 1| - |1\rangle\,\langle 0|\big)\,. \tag{19.6}$$

This commutator is non-zero for $0 < \theta < \pi/4$ regardless of the result of Alice's measurement. Peierls' criterion fails to hold.

Having lived and worked for two years in Peierls' Birmingham department, where I came to admire him immensely, I am adding to my list of reactions I wish death had not snatched from us Peierls' response to Fuchs' counterexample. All I know for sure, is that he would not have attempted to escape by maintaining that Alice's action on the right qubit had somehow altered the objective condition of the left qubit. He would have agreed that whatever she did to and learned about the qubit on the right altered only her *knowledge* of the faraway qubit on the left. Since Alice communicates none of this updated knowledge to Bob, there is no way for him to update his density matrix as a result of her actions. But since she now *knows* that the density matrix for the left qubit is a particular pure state density matrix, she is surely entitled to use it as the appropriate expression of her current knowledge.

Here is what seems to go wrong with Peierls' justification for his first condition. Depending on the result of her measurement on the right qubit, Alice's knowledge of the left qubit is indeed encapsulated in one of the two pure states

$$|\psi\rangle_+ = \cos\theta\,|0\rangle + \sin\theta\,|1\rangle \tag{19.7}$$

or

$$|\psi\rangle_- = \cos\theta\,|0\rangle - \sin\theta\,|1\rangle\ . \tag{19.8}$$

In spin language she therefore either knows that the left qubit is in an eigenstate of the spin along the direction $\boldsymbol{n}_+ = \cos 2\theta\boldsymbol{z} + \sin 2\theta\boldsymbol{x}$ or she knows that it is in an eigenstate of the spin along the direction $\boldsymbol{n}_- = \cos 2\theta\boldsymbol{z} - \sin 2\theta\boldsymbol{x}$. As Peierls noted, it is inconsistent with Alice's knowledge for Bob to know anything about the spin along any direction orthogonal to $\boldsymbol{n}_+$ in the former case, or orthogonal to $\boldsymbol{n}_-$ in the latter case.

But Bob does not know which case is the case. Indeed, he does not even know whether Alice knows which case is the case and therefore (to get slightly metaphysical) whether there *is* a case which is the case. Should Alice happen to know that the state of the left qubit is $|\psi\rangle_+$, then it is not incompatible with Alice's knowledge for Bob to be able to make more than a completely random guess about the result of measuring a spin component in the plane perpendicular to $\boldsymbol{n}_-$, and vice-versa. The only direction along which Bob cannot improve on a random guess, regardless of whether $|\psi\rangle_+$ or $|\psi\rangle_-$ encapsulates Alice's knowledge of the left qubit, is the direction $\boldsymbol{y}$ orthogonal to both $\boldsymbol{n}_+$ and $\boldsymbol{n}_-$. And Bob's density matrix (19.2) does indeed give equal probabilities for his finding either value of $S_y$. That this state of affairs holds whatever unitary transformation Alice applies to the right qubit before she measures it in the computational basis, is shown in the Appendix below. So Bob gets away with violating Peierls' constraint on what he can know because he does not know what knowledge Alice knows he cannot have, and therefore must continue to use a density matrix that embodies knowledge that it *might* be possible for him to have.

I do not believe that this counterexample is contrived or artificial. The initial density matrix (19.2) that both Alice and Bob assign to the left qubit is required by basic principles of quantum mechanics and the fact that what each of them knows about the two-qubit system is given by the pure state (19.1). Alice's subsequent acquisition of information enabling her to refine her density matrix down to one or the other of the two choices of pure-state density matrix (19.5) cannot alter Bob's assignment of density matrix, because the actions she takes to make the refinement are taken far from the left qubit. There is therefore no material alteration of the left qubit that Bob is, in some way, willfully ignoring. Nor can there be any change in Bob's *knowledge* of the left qubit, since neither the nature of Alice's actions nor the outcomes of those actions are communicated to Bob.

So to appropriate a remark John Bell once made about the Einstein–Podolsky–Rosen argument [6], Peierls' first condition "doesn't work. The reasonable thing just doesn't work." We are left with the question of what constraints of mutual consistency, if any, one *can* impose on a collection of density matrices that encapsulate the knowledge available to Alice, Bob, Charles, Doris, Edward, ... about one and the same physical system. I do not know the answer.[2] What I can show, at a minimum, is that Peierls' second condition remains a valid one, even though he justified it only under the assumption that his primary condition held.

Peierls' second condition is the concise mathematical expression of the fact that whatever other mutual consistency conditions one might argue for or against, there is one rock-bottom requirement. The different probabilities their density matrices $\rho_a$ and $\rho_b$ enable Alice and Bob to assign to the outcome of any subsequent measurement made on one and the same individual physical system cannot contradict one another. This is a very weak restriction, since neither of their probability assignments can be refuted by the subsequent occurrence or nonoccurence of any measurement outcome they both assign a probability that is neither 1 nor 0. The knowledge one of them has can contradict the knowledge possessed by the other only if one of them knows that something will certainly happen, while the other knows that it certainly will not. So a minimalist consistency condition is that there can be no outcome of any measurement to which one of them assigns a probability 1 and the other assigns a probability 0. This consistency condition turns out to be equivalent to Peierls' condition

$$\rho_a \rho_b \neq 0 \, . \tag{19.9}$$

To see this note first that if (19.9) fails to hold, so that $\rho_a \rho_b = 0$, then since density matrices are hermitian we also have $\rho_b \rho_a = 0$. So $\rho_a$ and $\rho_b$ commute, and there is a basis of joint eigenstates. Since the vanishing of $\rho_a \rho_b$ requires at least one of $\rho_a$ and $\rho_b$ to have zero eigenvalue for each eigenstate, we can resolve the identity (not necessarily uniquely) into the sum of two projections,

$$1 = P_a + P_b \, , \tag{19.10}$$

where $P_a$ and $P_b$ project onto subspaces of eigenstates of $\rho_a$ and $\rho_b$ with eigenvalue zero. The outcomes of any measurement that discriminates be-

[2] But see footnote 1 above.

tween these two orthogonal subspaces will be assigned probabilities 1 or 0 by Alice, and 0 or 1 by Bob. Whatever Alice says must happen Bob will say cannot happen, and vice versa.

The converse is slightly more subtle. Suppose there is some particular measurement with an outcome that Alice gives probability 0 and Bob gives probability 1. This means there is some hermitian operator[3] $M$ with eigenvalues $m$ in the range

$$0 \leq m \leq 1 \tag{19.11}$$

with

$$\mathrm{Tr}\ \rho_a\ M = 0\ , \quad \mathrm{Tr}\ \rho_b M = 1\ . \tag{19.12}$$

If the eigenstates of $M$ are $|\psi_j\rangle$ with eigenvalues $m_j$ then expanding (19.12) in the basis of those eigenstates gives

$$\sum_j \langle\psi_j|\,\rho_a\,|\psi_j\rangle\, m_j = 0\ , \tag{19.13}$$

and

$$\sum_j \langle\psi_j|\,\rho_b\,|\psi_j\rangle\, m_j = 1\ , \tag{19.14}$$

or, since $\mathrm{Tr}\,\rho_b = 1$,

$$\sum_j \langle\psi_j|\,\rho_b\,|\psi_j\rangle\,(1 - m_j) = 0\ . \tag{19.15}$$

Since the diagonal matrix elements of $\rho_a$ and $\rho_b$ are non-negative and since the $m_j$ satisfy (19.11), we conclude from (19.13) and (19.15) that

$$\langle\psi_j|\,\rho_a\,|\psi_j\rangle = 0\ , \quad m_j \neq 0\ , \tag{19.16}$$
$$\langle\psi_j|\,\rho_b\,|\psi_j\rangle = 0\ , \quad m_j \neq 1\ . \tag{19.17}$$

But if

$$\langle\psi|\,\rho\,|\psi\rangle = 0 \tag{19.18}$$

[3] If you take an old-fashioned view of what constitutes a measurement it suffices to take the hermitian operator to be a projection operator – i.e. every eigenvalue $m$ is either 0 or 1. If you prefer to think of measurements in terms of positive-operator-valued measures (POVMs), there is no reason to make that restriction. The argument that follows works for either reading of "any measurement". It thereby demonstrates that the existence of any such generalized measurement implies the existence of an ordinary von Neumann measurement.

for a density matrix $\rho$, then we must have[4]

$$\rho\,|\psi\rangle = 0\,. \tag{19.19}$$

Therefore (19.16) and (19.17) require every eigenstate $|\psi_k\rangle$ of $M$ to be an eigenstate with zero eigenvalue of either $\rho_a$ or $\rho_b$. It follows that

$$\langle\psi_i|\,\rho_a\rho_b\,|\psi_j\rangle = \sum_k \langle\psi_i|\,\rho_a\,|\psi_k\rangle\,\langle\psi_k|\,\rho_b\,|\psi_j\rangle = 0 \tag{19.20}$$

for arbitrary $i$ and $j$. Since the $|\psi_i\rangle$ are a complete set, we then have

$$\rho_a\rho_b = 0\,. \tag{19.21}$$

The existence of any measurement whatever with an outcome that Alice's density matrix requires and Bob's forbids (or vice versa) means that the product of their density matrices must vanish.

So Bob's knowledge can contradict Alice's if and only if their density matrices violate Peierls' second condition (19.9). This minimalist constraint (19.9) clearly generalizes to any number of density matrices $\rho_a$, $\rho_b$, $\rho_c\ldots$ used by Alice, Bob, Carol, ... to encapsulate what each of them knows about one and the same system. No two of them can assign probabilities of 1 and 0 to the outcome of any measurement, and therefore all possible pairs of density matrices must have a non-zero product.

I do not know if one can impose any conditions beyond requiring all pairs of density matrices to satisfy the minimalist condition (19.9). I have the feeling that if quantum mechanics is really about knowledge and only knowledge, then there ought to be further elementary constraints on the possible density matrices describing one and the same physical system that are stronger than the very weak second condition of Peierls, but not as strong as his overly restrictive first condition. I wish somebody would tell me what they are, or provide me with a convincing argument that there are none.[5]

And I wish it were still possible to talk about these things with John Bell.

## Acknowledgment

For some time I went around like Coleridge's Ancient Mariner, cornering people and asking if they could explain to me Peierls' first criterion. It was Chris Fuchs who finally pointed out to me that the criterion obviously fails if Alice knows the pure state of a qubit but Bob knows only that it is one or the other of two nonorthogonal states. A remark by Ben Schumacher turned my attention to Peierls' second criterion, and I have had a stimulating correspondence about it with Rudiger Schack. This work was supported by the National Science Foundation, Grants No. PHY9722065 and PHY0098429.

[4] For a density matrix $\rho$ has a Hermitian square root, $\sqrt{\rho}$. Since (19.18) requires the norm of $\sqrt{\rho}|\psi\rangle$ to vanish, so must $\sqrt{\rho}|\psi\rangle$ itself. So $\rho|\psi\rangle = \sqrt{\rho}\sqrt{\rho}|\psi\rangle = 0$.

[5] See footnote 1 above.

## Appendix

Every pure spin-$\frac{1}{2}$ state is a spin-up eigenstate along some direction $\boldsymbol{n}$ and the projection operator on such an eigenstate is $\frac{1}{2}(1+\boldsymbol{n}\cdot\sigma)$. So regardless of what observable Alice chooses to measure on the right qubit – i.e. regardless of what unitary transformation she applies before measuring it in the computational basis – the post-measurement mixed-state density matrix for the left qubit can be expressed as a probability-weighted sum of the two possible post-measurement pure state density matrices known to Alice as a result of her measurement of the right qubit:

$$\rho = \frac{p}{2}(1+\boldsymbol{n}\cdot\sigma) + \frac{q}{2}(1+\boldsymbol{m}\cdot\sigma)\,, \tag{19.22}$$

for probabilties $p$ and $q = 1-p$. But the density matrix for the left qubit is unaltered by anything Alice does to the right qubit. It must therefore continue to be the reduced density matrix (19.2), which can be rewritten as

$$\rho = \frac{1}{2}(1+\cos 2\theta\, \boldsymbol{z}\cdot\sigma)\,. \tag{19.23}$$

Thus $p$, $\boldsymbol{m}$, and $\boldsymbol{n}$ are subject to the constraint that

$$\boldsymbol{z} = \sec 2\theta\,(p\boldsymbol{n} + q\boldsymbol{m})\,. \tag{19.24}$$

So the unique direction orthogonal to both $\boldsymbol{n}$ and $\boldsymbol{m}$, along which quantum mechanics forbids Bob to have any knowledge of the spin regardless of the outcome of Alice's measurement, is necessarily orthogonal to $\boldsymbol{z}$. Bob's density matrix (19.2) does indeed afford him no knowledge whatever of the spin along that special direction.

## References

1. J.S. Bell: Phys. World, 33–40 (August 1990)
2. N.D. Mermin: *Space and Time in Special Relativity* (McGraw Hill, New York 1968); reissued by Waveland Press, Prospect Heights, IL (1989)
3. J.S. Bell: Prog. Sci. Culture **1**(2) (1976); reprinted in *Speakable and unspeakable in quantum mechanics* (Cambridge Univ. Press, Cambridge 1987) pp. 67–80
4. R.E. Peierls: Phys. World, 19–21 (January 1991)
5. R.E. Peierls: *More Surprises in Theoretical Physics*, (Princeton Univ. Press, Princeton, NJ 1991) p. 11
6. Quoted by J. Bernstein in *Quantum Profiles* (Princeton Univ. Press, Princeton, NJ 1991) p. 84

# 20 The History of the GHZ Paper

Daniel M. Greenberger

The story I am about to tell, while it is not directly about John Bell, could not have taken place without him. There have been many situations in science where the original work of a great innovator ultimately gets surpassed, but where this cannot detract from the importance and value of the original. I guess Isaac Newton comes to mind first. His equations have been modified, but he is still the giant hovering over the history of mechanics. Another case that comes to mind is that of DNA. James Watson has written about the great competition, in his mind, between his work with Frick, and that of Linus Pauling, over the race to the double helix. But one can hardly discuss this "race" without being aware that the original single helix had been discovered by Pauling. Furthermore the entire field of quantum chemistry was largely set up and defined by Pauling. So all the rules that they were playing by were Pauling's rules. It hardly diminishes his status that these young scientists edged him out of one very important discovery (partly by withholding important data).

And John Bell's status in our field has that same mythic quality. Before him, there was nothing, only the philosophical disputes between famous old men. He showed that the field contained physics, experimental physics, and nothing has been the same since. However far we progress beyond his famous theorem, nothing can extinguish the fact that he was first, with a result that was not only amazing, but also took great courage to pursue, given the climate of the time. We only follow in his footsteps, and that is the context for the story below.

I met Mike Horne and Anton Zeilinger at a neutron conference in Grenoble in 1979. They were collaborating at Cliff Shull's lab at MIT. Al Overhauser and I had written a paper on the COW experiment (named for the authors, Colella, Overhauser, and Werner. The experiment measures the phase shift of a neutron falling in the earth's gravitational field.) Al had been asked to give a talk at Grenoble about it, but he didn't want to go and asked me if I would. There I met most of the people doing coherent neutron optics, mostly with the neutron interferometer, for the first time.

But I had met Cliff Shull before. I had invented a complicated scheme to measure the gravitational phase shift on a neutron, using the hyperfine structure of atoms, and a colleague at CCNY, Ken Rubin, and I had actually received a small grant to explore the possibility. When it seemed too hard for our resources, I started to think about using free neutrons. The neutron interferometer had not been invented yet, and so I took a trip out to Brookhaven to see some of the reactor people there. They said they weren't doing coherent neutron experiments and directed me to Shull. So I went up MIT to see him, and he told me that he thought the experiment was possible, and that he might be able to do it, but his reactor was down, as they were remodeling it, and it would remain so for a couple of years. (I can only imagine that he was thinking about a neutron interferometer, which had not yet been invented.) I was in no hurry, as I could not imagine that anyone else was thinking about the problem. This was in the early 1970s, and I was shocked to read a year or so later in Phys. Rev. Letters a proposal by COW to do the experiment with a neutron interferometer. This by the way was the first time I had heard of a neutron interferometer. A year later they did the experiment. Some time after that I wrote to Overhauser pointing out a problem, and he invited me to give a talk at Purdue, which led to his teaching me how the interferometer worked and to our collaboration, and was the beginning of a wonderful friendship.

At Grenoble, I hit it off with Mike and Anton, and took a sabbatical leave in 1980 at Cliff's lab, collaborating with their group, and after that I came up regularly to MIT to work with them. Cliff was a fantastic physicist, a lovely person, and a man with a very strong sense of scientific integrity. The group he had around him, Mike, Anton, and sometimes Abner Shimony and Herb Bernstein, some wonderful graduate students and occasional visitors, were a superb group of collaborators, and I think this period has to have been the high point of my professional career, in terms of having fun and doing interesting things with great colleagues.

Anton had come to Cliff's lab originally as a postdoc, from Helmut Rauch's laboratory in Vienna. Rauch and Bonse had invented the neutron interferometer, an act of courage in itself. The X-ray interferometer had been invented earlier, but the general consensus was that the reason it could work was due to the fact that part of the X-ray beam is absorbed in the crystal. The neutron beam was totally elastically scattered and everyone thought that one could not make a neutron interferometer, and so no one had tried before.

Once the interferometer had been perfected a host of fundamental experiments were turned out by the Rauch group (such as proving that a neutron has to rotate through 720° before returning to its original state), and Anton had been trained on them. Pretty quickly Sam Werner's group (originally at Ford Motors and subsequently at Missouri) and Cliff Shull at MIT also started doing fundamental experiments as a major activity. (Other places, such as Grenoble, also did some.) But Cliff had also been the grandfather of solid-state structure experiments with neutrons, and it was natural that Anton should have ended up at Cliff's lab.

Mike Horne had started as a student of Abner Shimony at Boston University, turning John Bell's original theorem, which had assumed 100% efficient counters, into a practical test such that actual experiments could be done.

When they heard that John Clauser had been working along the same lines, they invited him to join them, which led to the famous Clauser–Horne, and CHSH (Clauser–Horne–Shimony–Holt) papers.

But in the mid 1980s Cliff started to wind things down at the lab so he could retire, and Anton left and went to Vienna, to the reactor at the Atominstitut where Prof. Rauch, his former mentor, was co-director. In 1986 the lab had cornered a Fulbright Fellowship opening and I had a sabbatical coming, so I ended up at the Institute for a semester, plus the summer. While looking for a problem to work on, I had been thinking about three or four particle correlations, something I had been hoping to do for quite a while, but had never been able to make the time.

My interest in this problem came from a misplaced idea of what one could do with such correlations. Back at a conference at Cesena in Italy, in 1983 or 1984, I remember that John Wheeler was having a discussion at the blackboard with someone. I can't remember who it was or what the subject was, but I do remember that I piped in with the suggestion that they could possibly resolve their difference if they would look at the case of a particle decaying into two particles, and each of these in turn making a secondary decay into two more particles. They both nodded and promptly forgot the suggestion, which is probably all to the good, since I had a serious misconception about what would happen in this case. In particular I thought that the time order of events would be very important, especially if you could measure the results of one of the secondary decays before the other one took place. But of course quantum mechanics doesn't work this way, and the results are all independent of when the decays take place.

But I had resolved to myself to someday explore this case, and no one had ever followed up my suggestion, and so when I got to Vienna I decided that taking a look at trying to create Bell-type theorems for multiparticle states would be a good idea.

It turns out that Mike and Anton had also been thinking about 3-particle states. They had invented the idea of the two-particle interferometer in the early 1980s. At the time the only way they could think of using it was with the two photons coming from electron-positron (positronium) annihilation, a very difficult experiment. In those days, they knew little about quantum optics, or that their idea was soon to be realized in the amazing down-conversion photon experiments of Leonard Mandel and his group at Rochester, and independently by Carroll Alley and H.-W. Shih at Maryland. (Of course, when this happened, Anton was by now starting up his own laboratory, first at Innsbruck, and now at Vienna. His decision was to make it a state-of-the-art quantum optics laboratory, that by now has probably performed more really fundamental experiments on entangled states than any other.) but for them at that time, their interferometer was primarily a wonderful theoretical tool. It allowed one to perform the first Bell-type experiments that had nothing to do with spin, and they had thought of generalizing it to three particles, because they were aware of the alternative three-photon decay of positronium, but they had done nothing about it.

So when I arrived at Vienna, Anton and I decided that the untapped field of 3-particle-correlations was a good project to pursue, although neither of us had any idea of what might turn up! The first task was to find a good system to analyze. After looking at several cases, I realized that it wouldn't be easy to get interesting results. Most simple cases, such as a spin-0 particle decaying into 2 spin-1 particles, in turn decaying into 4 spin-1/2 particles, gave uninteresting correlations between the end products, or one never had a maximum or minimum correlation of $\pm 1$, which meant that one never had a state where one could make a definite prediction of the state of one of the particles, if one knew the states of the others, which was necessary for one to apply the Einstein reality criterion.

Finally, I hit on the case where if one had initially a particle of spin-1, in the $m = 0$ state, decaying into 2 spin-1 particles, which each finally decayed into 2 spin-1/2 particles, the final state would be

$$|J = 1, m = 0\rangle = \frac{1}{\sqrt{2}}(\uparrow\uparrow\downarrow\downarrow - \downarrow\downarrow\uparrow\uparrow) ,$$

where the arrows denote the spin state (up or down) of each of the four final particles. Since this was somewhat close to the two-particle case, it looked worth pursuing. If one measured the spins of the four particles in coincidence, all in the $x - y$ plane, one got

$$E = -\cos(\phi_1 + \phi_2 - \phi_3 - \phi_4) ,$$

for the expectation value of the product of the four $z$-components of the separate spins. This seemed to be a straightforward generalization of the case for two particles, and so looked promising. Also, if

$$\phi_1 + \phi_2 - \phi_3 - \phi_4 = 0 , \quad \text{or} \quad \pi ,$$

then the expectation value would be 1 or $-1$. So, if one were to measure the spin of three of the particles in this case, one would be able to predict the spin of the fourth, and the reality condition would apply.

Thus I was very encouraged to start searching for Bell-type inequalities. After some practice, I found that I was getting better and better inequalities. Each morning I would discuss what was going on with Anton, and we would try to figure out some strategy. But I never did try a systematic approach. I figured that that could wait until later. We were trying to get some physical feeling for the results I was getting. Since the results were complicated, Anton suggested restricting ourselves to perfect correlations for a while, as these should be easier to understand, and there seemed to be lots of inequalities involving only these. This seemed a good idea, as the perfect correlations are the guts of Einstein's arguments, and the starting point in analyzing the simpler 2-particle case.

Many good inequalities followed and I saw lots of ways of improving the results, and each day, I would announce that my new inequality was much more restrictive than that of the day before. Finally, I walked in one day and said, "Anton, I don't understand what is going on in the least. There doesn't seem to be any freedom at all for the particles." I had reached the point where I had such a good inequality, that it ruled out all possible cases! At first this seemed impossible to believe, but then we were able to see that all classical results were self-contradictory, and we finally understood the point of the calculation, and worked out a simple proof. But it had to hit us over the head. Like everyone else, we found our result totally unexpected.

So after mulling over it for a while, I said "let's publish it." Anton thought this was a good idea, but he said that he and Mike also had all sorts of unpublished results, so that maybe we should just write an article on "What we know about Bell-type theorems" and include this result with all the others. I said "fine", but I didn't have the time to write such an article then, and neither did Anton.

The state of the field in those days was such that it didn't bother us at all that this result and all the others we had sat around for several years and we hadn't published them. It wasn't until 1988, when I came back to europe for a semester, to Munich to work at Prof. Herbert Walther's laboratory at the Max Planck Institute for Quantum Optics at Garching, on a Humboldt fellowship, that I had the time and started to write the article that Anton had suggested. So I started writing, and pretty soon I called up Anton and said, "I've barely begun and already I have over 70 typed pages. This is turning into a monograph, and I don't want to write a monograph." So I stopped working on it, although this time I started giving talks about the theorem at various conferences. I remember one such talk I gave was as a guest lecturer at a very good workshop on quantum problems at Erice, at a meeting organized by Arthur Miller, a historian of physics. But nobody asked me to write it up, so I didn't.

Suddenly, as I was about to leave europe, we received a preprint of a paper by Michael Redhead's group, where they talked about our result and claimed to have a much more rigorous derivation of it. Now we were in the situation where people were starting to quote our result, and we had never even published it. So I told Anton we had better publish something, or we will get lost in the static. The opportunity came when I got home and was asked by Michael Kafatos to participate in a daylong symposium on Bell's theorem and other problems in modern physics at Fairfax University in Virginia. So I gave a talk there and wrote up a short four-page paper for their proceedings, in which I gave no details of the proof, although I did note that one of the particles acted as a spectator, and the proof would hold for a three particle system. It was published rather quickly by Kluwer, but my paper was so short that, although they published it, they never bothered to send me a copy of the book, so that I have actually never seen a copy of the article in print.

They did send me ten reprints of the article though. (Physically, it looks awful! It was written on my HP home computer, which in those days had about 32 K of memory, and the dot-matrix printer had one ugly font.)

Although Mike Horne hadn't taken a direct role in the day-to-day development of the theorem in Vienna, there was never any question of whether he would be a co-author, as everything we had learned about Bell's theorem over the years we had done in collaboration with him, and we were in constant touch with him during this period, and he was at that time the world's leading authority on the subtleties of the Bell inequalities.

The next problem was writing up a proper paper on the subject. Since the philosophers who were worried about the problem were good mathematicians and experts in measure theory, we felt that we needed a collaborator who could fit those qualifications. The obvious choice was Abner Shimony, who had been collaborating with us off and on since the days in Cliff Shull's lab, and whom we were all very fond of. The paper proved to be very hard to write. The four of us spent two full weeks together, sometimes in person, sometimes by phone, fighting over every detail (the way good friends fight). Then it went back and forth through several revisions, and was altogether a very exhausting experience, although it actually was greatly improved in the end by the input of each of us. We published it in the American Journal of Physics, because we felt that they had an audience that was concerned about such problems, and had a long and distinguished history of publishing papers on the interpretation of quantum mechanics by such people as Eugene Wigner.

Some improvements in the paper were suggested by David Mermin, who also developed a very pretty model of three spin 1/2 particles. It was also he who realized that one didn't need the kind of realistic decay we were talking about, but that any of the eight three-particle maximally entangled states would do equally as well. He also did us the favor of mentioning it in his point of view column in Physics Today, which brought the theorem to public attention.

But it should be obvious that everything we did was inspired by and was an extension of John Bell's work, and his spirit hovered over the results. We were all at that wonderful workshop at Amherst in the summer of 1990 with John, and we didn't at all suspect that we were about to lose him.

Both this paper and the work described were supported in part by the National Science Foundation, grant PHY-97-22614, and its precursors. It was written while I was on sabbatical leave at Anton's laboratory at the University of Vienna during the spring of 2001, and I would like to thank him and his students for their hospitality.

# 21 John Stewart Bell and the Dynamical Reduction Program

GianCarlo Ghirardi

It is a great pleasure to have the opportunity of expressing once more my deep feelings for J.S. Bell, a unique individual, a great scientist, and a deep thinker whose memory we are honoring. He was always fascinated by the unsuspected depth of quantum mechanics (in its orthodox formulation) and at the same time worried by its puzzling aspects, and he was able to grasp more deeply than any other scientist the subtle features of the theory. Moreover, with his unparalleled contributions to the subject, he has imposed on all of us a radical change of attitude about "the quantum world".

He was constantly, lucidly and passionately involved in his search for "rigour", "exactness" and "real understanding" in science. In this he played a fundamental role in stimulating the "revival" of interest in and the flourishing of investigations on the foundational problems of quantum mechanics (QM) which characterize the period following the derivation of his celebrated inequality. He also compelled the scientific community to reconsider with a much more "critical" attitude the various *easy solutions* – to use his words – to the basic problems of the theory which had been put forward by eminent scientists.

John also had an incredibly pleasant personality, a subtle sense of humour and an extremely rich "human side". Having had the opportunity of knowing him has been, for all of us who had the luck to do so, an unforgettable and extremely enriching experience.

Concerning the subject of this chapter, let me begin by recalling that in the last five years of his life John devoted a lot of attention to the so-called dynamical reduction program (DRP) and, in particular, to the proposal [1, 2], usually referred to as the GRW theory, that *Rimini*, *Weber* and myself put forward in 1985. As John put it, our attitude in developing such a program corresponds to having chosen the latter of two [3] "*inescapable conclusions that either the wavefunction as given by the Schrödinger equation is not everything or it is not right.*"

In fact, the DRP contemplates the possibility of modifying standard quantum dynamics by adding stochastic and non-linear terms in such a way as to: [4] "*have little impact on small systems but nevertheless suppress macroscopic superpositions*". Actually, our work has shown that such a goal can be consistently reached, by exhibiting an explicit model theory, a rival theory of QM, which respects all its well-established predictions for microsystems and implies, on the basis of a unique and universal dynamics, the objectification of macroscopic properties. It is useful to remark that, since the theory differs from quantum mechanics, it can suggest crucial tests about the possible breakdown of linearity (even though at present no one has been able to identify practically feasible tests of this type). Finally, the theory has shed some light [5–9] on the line one should follow to account for reduction processes in a relativistic context, a quite old and well-debated problem [10–13].

Before proceeding I would like to make perfectly clear my position with regard to the DRP. I believe that the real interest of the theory does not stem from its precise formal structure – in fact it is a fundamentally phenomenological theory with two (or one if one considers the variant involving gravity [14] we have proposed subsequently) phenomenological parameters – but from the fact that it has shown that a line of thought which has always been considered as not viable by the scientific community turns out to be perfectly workable and leads to a completely consistent and fully self-contained theoretical scheme. Note that even *Einstein* [15] considered such an approach as not practicable: "*the 'macroscopic' and the 'microscopic' are so inter-related that it appears impracticable to give up this program* [scientific realism in the classical sense, requiring that all systems possess objective properties] *in the microscopic alone.*"

## 21.1 John Stewart Bell and His Role

I am firmly convinced that the specific attention that John paid to our work has played an extremely important role in raising the interest of the scientific community in it. As a consequence, we are deeply indebted to him. But, besides this, his stimulating writings and his cute remarks have certainly played an extremely important role in the investigations of dynamical reduction, *before*, *during* and *after* the elaboration of our theory.

We consider it appropriate to call attention to the fact that he was always extremely sharp in stressing the very reasons which are at the basis of the DRP, i.e., that the assumptions of the unrestricted validity of the superposition principle and of the completeness of the standard theory imply the *indefiniteness, the waviness of the micro world*, as opposed to the particularity of the world of our experience. With reference to this point he put great emphasis [16] on the fact that the superposition of two states, e.g., those corresponding to different spatial locations of an object,

$$|\psi\rangle = \frac{1}{\sqrt{2}}\left[|\mathrm{Here}\rangle + |\mathrm{There}\rangle\right] \tag{21.1}$$

must be associated to a "state of affairs" that cannot be in any consistent way reconcilied with the logical alternative of "*either* Here *or* There". Finally he contributed in a remarkable way to compelling the scientific community to confront itself with the entanglement of composite quantum systems, a state of affairs which implies the loss of individuality of the constituents, the strict correlations of outcomes and, at the same time, the logical impossibility of entertaining the idea that the constituents themselves have, in general, definite properties prior to any act aimed at ascertaining them. Finally, he never ceased to call attention to the fact that the superposition principle clashes with the definiteness, the particularity of the world of our experience, and with the idea that macroscopic objects "possess objectively" definite macroscopic properties. He clearly identified the logical inconsistency of the reduction postulate with the assumption of an unlimited validity of the standard theory, and he stressed the fundamental ambiguity of most proposed solutions to the macro-objectification problem, since they are all characterized by a fundamental vagueness. The borderline between micro and macro, reversible and irreversible and quantum and classical is fundamentally shifty, since the theory does not contain any element which can qualify a system as "a measurer". As a consequence he has always criticized the 'complacency' of the orthodox interpretation of the theory and of the many *easy solutions* to its problems.

The remarkable and deep papers in which he declared for years his fundamental dissatisfaction with the theory in its present form can be legitimately considered to have been the most important stimulus leading to the elaboration of theories, such as the dynamical reduction models, which are genuine rival theories to the standard one and which solve the macro-objectification problem.

## 21.2 His Role Before 1985

To make more precise my previous assertions, let me focus on precise arguments and statements by J.S. Bell. His dissatisfaction with standard QM derived mainly from the measurement problem, or better, from the one which is now referred to as the macro-objectification problem.

As already stated, it was clear to him from the very beginning that the solution of this problem required *either the completion or the modification of the theory* in such a way to be able to locate precisely the split between the processes governed by the linear nature of the theory and those for which the superposition principle cannot hold. In 1966 he wrote [17]:

- "*Although we would not wish to cast doubt on the practical adequacy of* [the reduction postulate] *it is clear that if we leave it un-analyzed the theory can at best be described as a phenomenological makeshift. The fact already stressed that observation implies a dynamical interference, together with the belief that instruments after all are no more than large assemblies of atoms, and that they interact with the rest of the world largely through the well-known electromagnetic interaction, seems to make this a distinctly uncomfortable level at which to replace analysis by axioms.*"
- "*... if interference is destroyed, then the Schrödinger equation is incorrect.*"

- "... *if it is not destroyed the quantum mechanical description is revealed as not wrong but certainly incomplete.*"

An invaluable contribution to the debate came from his deep critical analysis of the hidden-variable models and in particular the pilot wave theory, which led him to put forward many illuminating remarks [4] focusing on the basic features that any serious theory should share and on the correct way to found it:

"*The concept of 'measurement' becomes so fuzzy on reflection that it is quite surprising to have it appearing in physical theory at the most fundamental level. ... Does not any analysis of measurement require concepts more fundamental than measurement? And should not the fundamental theory be about these more fundamental concepts?*"

Such a position can be appropriately and concisely summarized by the clear requests he put forward [18] a few years later for any sensible theory:

"*In your theory you must identify some things as being really there, as distinct from the many mathematical concepts that you can easily devise – like the projection of the side of a triangle to infinity and so on. We must decide that some things are really there and that you are going to take them seriously. These are the beables. ... Another good word is kinematics. ... I would want the kinematics of your theory tell me what it is you are talking about before you tell me what about it.*"

Moreover his deep critical reconsideration of the unavoidable contextual features of the de Broglie–Bohm theory, as well as of any hidden-variable theory, led him to stress [19] the relevance of the non-contextual variables and the fact that the contextual ones cannot be endowed with a real status:

"*In physics the only observations we must consider are position observations, if only the positions of measurement pointers. It is a great merit of the de Broglie–Bohm picture to force us to consider this fact. If you make axioms, rather than definitions and theorems, about the measurement of anything else, then you commit redundancy and risk inconsistency.*"

This fact, among other things, makes it clear that, within such a context, one has to make a choice about *what is real in his theory* and that one cannot pretend to attribute a status of reality analogous to the one of classical physics to too many variables. This fact, too, represents a deep lesson for the DRP. In fact, the above situation is paralleled, within the DRP, by the circumstance that when one tries to build a genuine Hilbert space and complete theory inducing reductions, one must choose the *preferred basis*, i.e. the observables that the modified dynamics strives to make objective. They are the strict analogue of the non-contextual variables of hidden-variable theories. Before analyzing J.S. Bell's role during the elaboration of the DRP let us briefly recall its development.

## 21.3 The Birth of Dynamical Reduction Theories

Various investigations during the 1970s can be considered as preliminary steps for the subsequent developments. Let us briefly mention them.

### 21.3.1 Stochastic Equations for the Description of Decay Processes

In the years 1970–1973 *Fonda*, *Rimini*, *Weber* and myself were seriously concerned with quantum decay processes and in particular with the possibility of deriving, within a quantum context, the exponential decay law [20, 21]. Some features of our approach are extremely relevant for the DRP; they are the following:

- One deals with individual physical systems.
- The state vector is supposed to suffer random processes occurring at random times, leading to appropriate sudden changes of it:

  $$|\Psi\rangle \Rightarrow P_u\,|\Psi\rangle \,/\, \|P_u\,|\Psi\rangle\| \ .$$

  When $P_u$ is identified with the projection operator on the unstable state manifold, one obtains the desired result.
- To make the treatment quite general (the apparatus does not know which kind of unstable system it is testing), we identify the random processes with localization processes of the relative coordinates of the decay fragments. Such an assumption, combined with the peculiar "resonant dynamics" of an unstable system, yields, completely in general, the desired result. The "relative position basis" is the preferred basis of this theory.
- We have also applied our ideas to measurement processes [22].
- The final equation for the evolution at the ensemble level is of the quantum-dynamical semigroup type and has a structure extremely similar to the final one of the GRW theory.

### 21.3.2 Stochastic Differential Equations

In the same years (actually starting from 1973) Pearle, Gisin and others had entertained the idea of accounting for the reduction process in terms of a stochastic differential equation. However they had no idea and have presented no suggestion about how to identify the states to which the dynamical equation should lead. Without any clear indication on this point there was no way to identify a mechanism whose effectiveness could have been negligible for microsystems but extremely relevant for the macroscopic ones. John, who always paid a lot of attention to all proposals aiming to solve the macro-objectification problem, studied such papers (incidentally P. Pearle had spent a sabbatical year at the University of Geneva in 1973), and he realized that the lack of the identification of the preferred basis, i.e. of what "is out there", was the main obstacle to the success of the program.

It is interesting to note that, with precise reference to the proposals of resorting to stochastic differential equations, in 1981, he wrote [4],

"*None of these modifications has the property required here, of having little impact for small systems but nevertheless suppressing macroscopic superpositions. It would be good to know how this could be done, ...*"

and that, about one year after we had presented (at Heidelberg, in October 1984) a preliminary version of our paper [1], and at the time as the so-called GRW model [2]appeared, with precise reference to the above-mentioned attempts he wrote [23],

"*The necessary theoretical development involves introducing what is called nonlinearity and perhaps what is called stochasticity into the basic Schrödinger equation. There have been interesting pioneer efforts in this direction but not yet a breakthrough.*"

It is important to remark that *Gisin* subsequently published an extremely interesting proof [24] that non-linear modifications of the standard equation by themselves are unacceptable since they lead to faster-than-light effects. In the same period *Grassi* and I [25] showed that stochastic modifications without non-linearity can at most induce ensemble and not individual reductions.

### 21.3.3 The Breakthrough – Quantum Mechanics with Spontaneous Localizations

We will present only a very rough description of the original version [2] of the first consistent and satisfactory model (QMSL) of dynamical reduction, the one on which all subsequent attempts are based.

The key assumption of the model is that each elementary constituent of any system is subjected, at random times according to a Poisson distribution with mean frequency $\lambda$, to random and spontaneous localization processes around appropriate positions. For a system of $N$ distinguishable particles the random processes (hits) – assuming that it is the $i$th particle which suffers one of them – are described in the following way:

$$\Psi(\boldsymbol{q}_1, \boldsymbol{q}_2, \dots, \boldsymbol{q}_N) \Rightarrow \Psi_x(\boldsymbol{q}_1, \boldsymbol{q}_2, \dots, \boldsymbol{q}_N) = \frac{\Phi_x(\boldsymbol{q}_1, \boldsymbol{q}_2, \dots, \boldsymbol{q}_N)}{\|\Phi_x(\boldsymbol{q}_1, \boldsymbol{q}_2, \dots, \boldsymbol{q}_N)\|^2}, \tag{21.2}$$

$$\Phi_x(\boldsymbol{q}_1, \boldsymbol{q}_2, \dots, \boldsymbol{q}_N) = \left(\frac{\alpha}{\pi}\right)^{\frac{3}{4}} \exp\left(-\frac{\alpha}{2}(\boldsymbol{q}_i - \boldsymbol{x})^2\right) \Psi(\boldsymbol{q}_1, \boldsymbol{q}_2, \dots, \boldsymbol{q}_N). \tag{21.3}$$

One assumes that the probability density, $P(\boldsymbol{x})$, of its taking place at point $\boldsymbol{x}$ is given by:

$$P(\boldsymbol{x}) = \|\Phi_x(\boldsymbol{q}_1, \boldsymbol{q}_2, \dots, \boldsymbol{q}_N)\|^2, \tag{21.4}$$

so that hits occur with a higher probability at those places where, in the standard quantum description, there is a higher probability of finding the particle. Note that the above prescription introduces non-linear and stochastic elements into the dynamics.

It is extremely easy to convince oneself that the model has embodied in it the so-called *trigger mechanism*, i.e., that reductions become more and more frequent with an increasing number of particles in the system under consideration. As an intuitive example, consider the superposition, $|\Psi\rangle$, with equal weights, of two macroscopic pointer states, $|H\rangle$ and $|T\rangle$, corresponding to a pointer pointing at two different positions, $H$ and $T$, respectively:

$$|\Psi\rangle = K\left(|H\rangle + |T\rangle\right) \equiv K\left(|1 \cong h\rangle \ldots |N \cong h\rangle + |1 \cong t\rangle \ldots |N \cong t\rangle\right) \tag{21.5}$$

$[h(t)$ being an arbitrary point near to $H(T)]$. If any of the particles suffers a hitting process, e.g. near the point $h$, the second term is (practically) suppressed. The hitting frequency is therefore effectively amplified proportionally to the number of constituents.

With these premises, one easily understands that one can choose the parameters $\lambda$ and $1/\sqrt{\alpha}$ in such a way that the model [4],

"*has the property required here, of having little impact for small systems but nevertheless suppressing macroscopic superpositions. [It would be good to know how this could be done.]*"

In fact, the choice

$$\lambda \cong 10^{-16}\,\mathrm{s}^{-1} \quad \text{and} \quad 1/\sqrt{\alpha} \cong 10^{-5}\,\mathrm{cm} \tag{21.6}$$

implies that a microscopic system suffers a localization, on average, every $10^8$ yr, while for a macroscopic one it occurs every $10^{-7}$ s. In short, within QMSL [3]:

"*The cat is not both dead and alive for more than a split second.*"

## 21.4 His Role Immediately After the Publication of Our Paper

John Bell was immediately extremely interested by the proposal. On February 12, 1986, he wrote a letter to us, to which I will return below, which is quite indicative of his kind-heartedness and sensitivity. Then a sort of escalation of interest and comments followed, in the form of talks and writings describing our proposal. This is a simple list:

- The talk [3] at the Schrödinger's Centenary Conference held at Imperial College, London, UK, in 1987: "*Are there quantum jumps?*"
- The Bruno Touschek lectures in Rome.

- The talk [17] at the Erice Conference in 1989: "*Against Measurement*".
- The talk [19] at Schwinger's 70th birthday celebration: "*Towards an Exact Quantum Mechanics*".
- The talk at Trieste for the XXV Anniversary of the ICTP: "*First and second type difficulties of quantum mechanics*".

Meanwhile the exchange of letters continued.

All these contributions have played an important role for the program. In the following, I will analyze in greater detail the first letter he addressed to us and the paper: "Are there quantum jumps?", while I will simply mention some points from the remaining items on the list.

### 21.4.1 The Letter

He expresses his deep interest in our work, as one can see from the opening sentence:

**CERN** 1986 Feb 12

Dear Professors Ghirardi, Rimini, and Weber,

I read with very great interest and admiration your paper IC/84/240 'A model for a unified description of macroscopic and microscopic systems'.

I think one can easily imagine what this meant to us. It is then interesting to recall the last sentence of his letter, which is absolutely exquisite and indicative of his exceptionally amiable attitude:

> Of course I will be happy to receive any reply from you. But I hold the right not to reply to letters to be the most fundamental of human freedoms.
>
> With best wishes
>
> John Bell

The letter also contained an interesting physical comment. He complained about our having resorted to the statistical operator formalism:

> There is a point in your presentation that makes me fear I have not fully understood you. It is your emphasis on the reduced density matrix $\rho_Q$. It seems to me that this only obscures the argument — in so far as I have grasped it at all!

In this he was perfectly correct. In fact, in spite of the fact that our paper [2] contained clear-cut sentences such as,

*"We remark that in the case under discussion there is a macroscopic number of particles which are located in macroscopically different positions when the state is the first or the second, respectively. If a spontaneous localization takes place for one of such particles, this particle is found to be either in the spatial region corresponding to the first or the second wavefunction, ..."*

in most of it we had used, for computational purposes, the language of the statistical operator. This might have been misleading due to the fact that, being that the correspondence between statistical ensembles and statistical operators is infinitely many to one, it is essential to make a clear distinction between theories yielding only ensemble reductions (diagonalization of the statistical operator) and those giving individual reductions (each individual statevector ends up in the desired linear manifold).

### 21.4.2 The Paper: Are There Quantum Jumps?

At the Schrödinger Centenary Conference in London, John Bell chose to deliver a talk [3] in which he essentially described what we have achieved. However in his presentation he added a remark, a theorem and an interesting suggestion for a possible interpretation of the theory:

- *The remark*: The model does not respect the symmetry requirements for identical constituents.
- *The theorem*: He presented a very smart indirect proof that the model violates Bell's locality by violating only outcome independence. He resorted to the consideration of a two-times Schrödinger formulation of the process and he showed that the probability of occurrence of hits on the constituents of a part of a composite system is independent of the occurrence of hits on far-apart parts of the same system. This fact led him to claim that the GRW Theory, "*is as Lorentz invariant as it could be in the nonrelativistic version. It takes away the grounds of my fear that any exact formulation of quantum mechanics must conflict with fundamental Lorentz invariance.*"

Concerning the proof he gave, it is useful to mention that, subsequently, *Butterfield*, *Fleming*, *Grassi* and I [26, 27] proved (along completely different lines) a result which is conceptually equivalent to what he had obtained, i.e. that the GRW theory violates Bell's locality requirement because it violates the outcome independence requirement. This is a quite relevant result. In fact, *Grassi* and I [28] have been able to prove that any theory violating Bell's locality requirement by violating parameter independence cannot admit a genuine relativistic generalization (i.e. one without a preferred reference frame).

- *The suggestion*: A clear hint about a possible interpretation: "*There is nothing in the theory but the wavefunction. It is in the wavefunction that we must find an image of the physical world, and in particular of the arrangement of things in ordinary three-dimensional space. But the wavefunction* *as a whole lives in a much bigger space, of 3N-dimensions. It makes no sense to ask for the amplitude or phase or whatever of the wavefunction at a point in ordinary space. It has neither amplitude nor phase nor anything else until a multitude of points in ordinary three-dimensional space are specified. However, the GRW jumps(which are part of the wavefunction, not something else) are well localized in ordinary space. Indeed each is centred on a particular space–time point, $(x, t)$. So, we can propose these events as the basis of the "local beables" of the theory. These are the mathematical counterparts in the theory to real events at definite places and times in the real world (as distinct from the many purely mathematical constructions that occur in the working out of physical theories, as distinct from things which may be real but not localized, and as distinct from "observables" of other formulations of quantum mechanics, for which there is no use here). A piece of matter then is a galaxy of such events. As a schematic psychophysical parallelism we can suppose that our experience is more or less directly of events in particular pieces of matter, our brains, which events are in turn correlated with events in our bodies as a whole, and they in turn with events in the outer world."*

Finally, during a dinner at the conference, he discussed with us a possible generalization for dealing correctly with identical particles. The proposal involved the introduction of a further phenomenological parameter. When later I met him in Padua he told me: "If you do not write the paper on identical constituents I will do it, I do not care about adding a parameter if this makes the theory more general."

### 21.4.3 The Relativistic Issue

As already stated John Bell understood immediately that the QMSL model had some appealing features also concerning the relativistic aspects. However it did not exhibit relativistic invariance, and one could easily guess that in trying to make it relativistic, one would probably meet various difficulties. A few months later, John was invited to lecture at Rome at a meeting in honor of Bruno Touschek. He delivered three lectures.

- The first one stressed once more the problems of standard quantum mechanics:

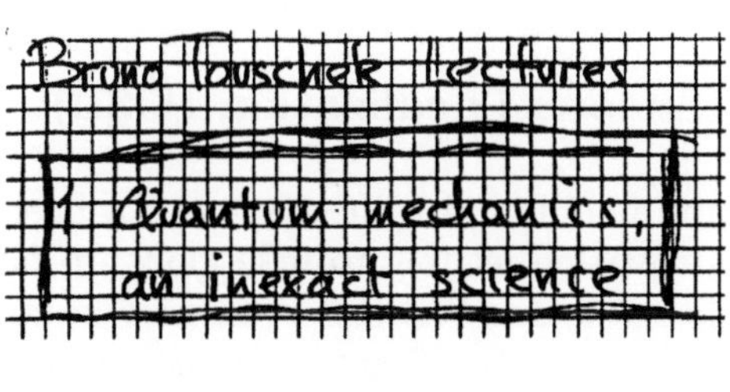

- The third one was exclusively devoted to the GRW theory:

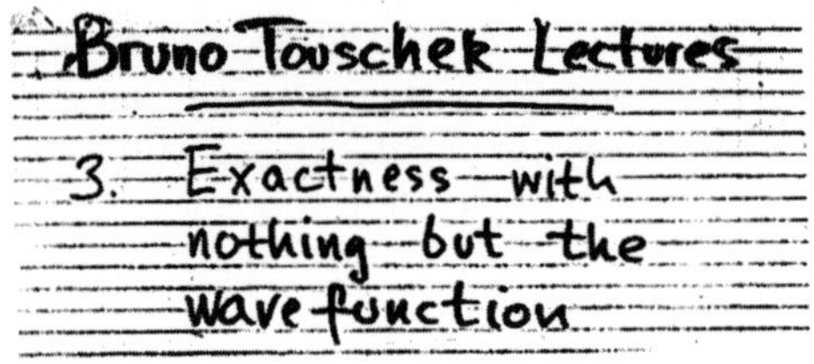


- Finally, the concluding sentence of the lectures launched the relativistic challenge to the DRP (PQM: practical quantum mechanics):

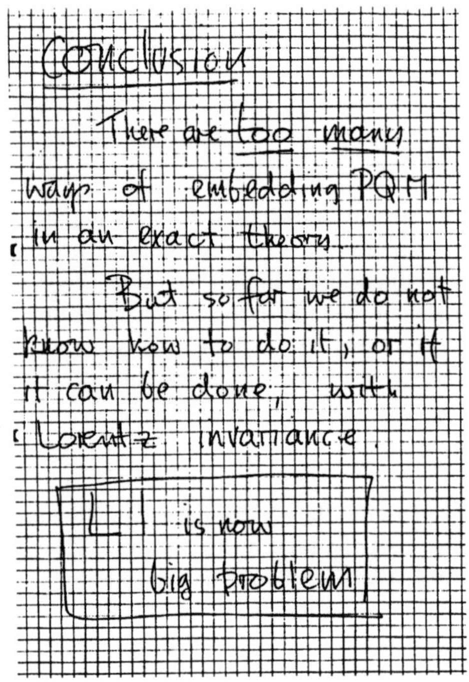


## 21.5 From QMSL to Continuous Spontaneous Localization (CSL)

As already stated, John Bell was quite familiar with the attempts of Pearle to work out a stochastic differential equation leading to reductions. In fact he wrote [19]: "*Philip Pearle kept alive this idea that it is some modification of the Schrödinger equation that is required.*" However, as we have already remarked, he also made it clear that, up to 1984: "*There have been interesting pioneer efforts in this direction but not yet a breakthrough.*" In 1989 Pearle took a sabbatical leave and wrote to John to investigate whether he could spend part of it at CERN. John, who was already a convinced supporter of our model suggested him to come to Trieste and wrote the following letter to us:

ORGANISATION EUROPÉENNE POUR LA RECHERCHE NUCLÉAIRE
EUROPEAN ORGANIZATION FOR NUCLEAR RESEARCH

CERN CH-1211 GENÈVE 23 SUISSE/SWITZERLAND
Téléphone: GENÈVE (022) 83 61 11 Telex: 419000 CER CH - Télégramme: CERNLAB-GENÈVE

SIÈGE: GENÈVE, SUISSE

13 November 1986

Professor G.C. Ghirardi
Dipartimento di Fisica Teorica
Università di Trieste
Strada Costiera 11
I-34014 - TRIESTE

Professor A. Rimini
Dipartimento di Fisica Nucleare
e Teorica
Università di Pavia
Via A. Bassi 6
I-27100 - PAVIA

Dear Colleagues,

I enclose a copy of a letter from Philip Pearle. As you will see he has a sabbatical year coming, and is considering where to go. He is interested in the problem of quantum mechanics, so CERN would not be a good place. There would be only me to talk to, and most of my time goes on efforts more appropriate for this laboratory. He would feel isolated and frustrated here. It would be a bit frustrating for me too, for even without anyone to talk to here I have difficulty keeping away from the quantum question.

So, as he himself suggests, I pass his letter on to you, in case you might consider inviting him. I have a very good impression of Philip Pearle. Before your own work, his work on spontaneous wave packet reduction seemed to me the most serious in the literature. You could get an idea of his work from Phys.Rev. D33 (1986) 2240, and the references there.

Thank you for the paper of Benatti et al., which I enjoyed reading.

With best wishes,

Yours sincerely,

John Bell

J.S. Bell

As a consequence Pearle came to Trieste and Pavia and, after a long period of work together with Rimini and I trying to obtain relativistic generalizations of QMSL, he saw how to fit the GRW ideas within his stochastic formalism. Thus CSL was born [29, 30]. It is a continuous spontaneous localization model in which the hits take place continuously. It can be summarized in the following terms:

The state vector evolves according to a linear and stochastic evolution equation,

$$\mathrm{d}\,|\Psi_B(t)\rangle = \left[\left(-\mathrm{i}H - \frac{\gamma}{2}\sum_i A_i^2\right)\mathrm{d}t + \sum_i A_i\,\mathrm{d}B_i\right]|\psi_B(t)\rangle\ , \tag{21.7}$$

in which $H$ is the Hamiltonian of the system, $\{B_i\}$ is a Wiener process characterized by

$$\langle\langle \mathrm{d}B_i\rangle\rangle = 0\ , \quad \langle\langle \mathrm{d}B_i\,\mathrm{d}B_j\rangle\rangle = \gamma\delta_{ij}\,\mathrm{d}t\ , \tag{21.8}$$

and $\{A_i\}$ is a set of self-adjoint commuting operators. Note that the equation does not preserve the norm of the state vector but only its stochastic average.

One introduces non-linearity by stating that the probability density for the stochastic process is not the natural one, $P_{\mathrm{raw}}[B_i]$, but the evolution equation must be supplemented by the assumption that the probability under examination is given by

$$P_{\mathrm{cooked}}[B_i] = P_{\mathrm{raw}}[B_i]\|\,|\psi_B(t)\rangle\,\|^2\ . \tag{21.9}$$

The formalism is much more elegant than that of QMSL and, if one makes continuous the index $i$ and one chooses

$$A_i \equiv N(\boldsymbol{x}) = \left(\frac{\alpha}{2\pi}\right)^{\frac{3}{2}} \int \mathrm{d}\boldsymbol{q}\exp\left(-\frac{\alpha}{2}(\boldsymbol{q}-\boldsymbol{x})^2\right)a^\dagger(\boldsymbol{q})a(\boldsymbol{q})\ , \tag{21.10}$$

with $a^\dagger(\boldsymbol{q})$ and $a(\boldsymbol{q})$ being the creation and annihilation operators for a particle at point $\boldsymbol{q}$ (we disregard for simplicity the spin indices), one obtains, in the case of one particle, exactly the GRW theory. Moreover, in general, the theory deals correctly with systems with identical constituents and has essentially the same physical implications as the GRW theory for many-particle systems.

Some comments seem appropriate. It has been suggested that the CSL model is definitely better than the original GRW theory, in particular, due to the fact that it is based on a linear stochastic equation. However, it has to be remarked that two alternative formulations of CSL can be considered:

- The linear stochastic equation plus the cooking procedure we have adopted here.
- A physically equivalent non-linear stochastic equation.

Concerning these two alternatives we would like to mention that in [26, 27] it has been shown that the first choice leads to a theory which has only outcome dependence, while the second involves parameter dependence. So the first alternative (an equation plus an extra prescription) is definitively better for the reasons we have discussed previously concerning the possibility of genuine relativistic generalizations of the scheme. If this is taken into account we can claim that CSL is based on a linear equation plus an independent stochastic prescription, just as QMSL is. The only difference derives from the fact that the stochastic processes take place continuously.

Moreover, in [30] we have proved that adding a further parameter to a hitting process within a GRW model and taking an appropriate limit one obtains precisely the corresponding CSL model. In short, for any given CSL model there are discrete models which are as physically close to it as one wants with concern for the predictions about all observables. Actually to prove this result, we took advantage of the exchange of views we had at Imperial College with John Bell. John who, as already mentioned, was trying to identify the precise positions and times of the localizations with the beables of the theory, told me on various occasions that he preferred the discrete to the continuous version of the theory, precisely for the interpretation he had in mind.

## 21.6 The Subsequent Role of J.S. Bell in the Dynamical Reduction Program

As already mentioned, John S. Bell, from the beginning of 1987, devoted many papers and delivered many talks dealing with the dynamical reduction theories. He gave invaluable indications about several aspects of them, in particular, concerning issues of interpretation and of the difficulties one might meet in finding genuine relativistic formulations of such theories.

### 21.6.1 Problems of Interpretation

I have already mentioned that, in his first paper on the GRW model, J.S. Bell proposed an interpretation making reference to the positions (in real 3-dimensional space) and times of the spontaneous localizations as the *beables* of the theory.

At the Erice Conference he proposed [16] another interpretation:

"*The GRW-type theories have nothing in their kinematics but the wavefunction. It gives the density (in a multidimensional configuration space!) of stuff. To account for the narrowness of that stuff in macroscopic dimensions, the linear Schrödinger equation has to be modified, in the GRW picture, by a mathematically prescribed spontaneous collapse mechanism.*"

I repeatedly discussed this proposal with him. Let me exhibit one of the letters he wrote to me:

> As regards $\Psi$ and the density of stuff, I think it is important that this density is in the 3-N dimensional configuration space. So I have not thought of relating it to ordinary matter or charge density in 3-space. Even for one particle I think one would have problems with the latter. So I am inclined to the view that you mention after "... as it is sufficient for an objective interpretation ..." And it has to be stressed that the "stuff" is in 3-N space — or whatever corresponds in field theory!

Reflecting on his remarks and reconsidering the idea of macroscopic similarity, we have proposed a new interpretation of the theory [31]. It is based on the idea that what the theory speaks of, at the appropriate level, is the averaged (on an appropriate volume of $10^{-15}$ cm$^3$) mass-density distribution of the whole universe. We have made it perfectly plausible that this approach (disregarding the completely unphysical debate about the so-called tails problem) allows one to take a consistent macrorealistic position with regard to natural processes which agrees with our definite perceptions.

### 21.6.2 The Relativistic Program

As discussed before, John faced the relativistic aspects of the theory from the very beginning (at the Schrödinger Centenary Conference, in the Touschek lectures, at Erice and so on). Let us recall some of his remarks about this problem:

- "*The GRW theory takes away the grounds of my fear that any exact formulation of quantum mechanics must conflict with fundamental Lorentz invariance.*"
- "*There are too many ways of embedding PQM in an exact theory. But so far we do not know how to do it, or if it can be done, with Lorentz invariance. LI is now the big problem.*"
- "*The big question, in my opinion, is which, if either, of these two precise pictures* [Bohm theory and GRW] *can be redeveloped in a Lorentz invariant way.*"

He raised some basic questions, gave some important hints and pointed out, on various occasions, some serious difficulties in reaching this goal [19]:

- "*Novel renormalization problem may arise.*"
- "*The jumping process could somehow be formulated within the Schwinger and Tomonaga and Stuckelberg way of presenting quantum field theory.*"
- "*One might meet immense difficulties in replacing particle positions by field variables since all the local quantities in ordinary quantum field theory fluctuate infinitely. Thus in trying to diminish this infinite fluctuation you are going to create things in the vacuum and you are going to do severe damage to the internal economy of the system.*"

All his predictions came true; his indications were followed by those who tried to work out a relativistic DRP, in the first place *Pearle* [5] and then him, *Grassi* and I [6, 7].

It is appropriate to stress two facts: the problem is still an open and a quite stimulating and difficult one. However there seems to be some possibility of carrying it on consistently. In particular we have learned some relevant lessons about possible relativistic reduction mechanisms. I have discussed this matter in great detail in various papers [9, 32, 33], to which I refer the reader for details.

## 21.7 Concluding Remarks

I hope I have been able to convince the reader that John Stewart Bell showed an extraordinary interest in the dynamical reduction program and that he played an important role in making it known to the scientific community and in contributing to the clarification of some subtle points which deserved particular care. I believe that the attention he paid to a phenomenological but consistent attempt to work out what he considered an exact theory (entirely formulated in wave language) represents a clear proof that his interest in the foundational problems was an extremely deep one. He was really worried by the orthodox complacency and the lack of precision in many of the formulations and interpretations of the theory. I consider this attitude towards the foundational problems of 'our best theory' his most important legacy for all those who are interested in the real meaning of quantum mechanics and in the conceptual foundations of modern science.

## References


1. G.C. Ghirardi, A. Rimini, T. Weber: In *Quantum Probability and Applications*, ed. by L. Accardi, W. von Waldenfels (Springer, Berlin, Heidelberg 1985)
2. G.C. Ghirardi, A. Rimini, T. Weber: Phys. Rev. D **34**, 470 (1986); ibid. D **36**, 3287 (1987)
3. J.S. Bell: In *Schrödinger – Centenary Celebration of a Polymath*, ed. by C.W. Kilmister (Cambridge Univ. Press, Cambridge 1987)
4. J.S. Bell: In *Quantum Gravity 2*, ed. by C. Isham, R. Penrose, D. Sciama (Clarendon, Oxford 1981)
5. P. Pearle: In *Sixty-Two Years of Uncertainty*, ed. by A. Miller (Plenum, New York 1990)
6. G.C. Ghirardi, R. Grassi, P. Pearle: Found. Phys. **20**, 1271 (1990)
7. G.C. Ghirardi, R. Grassi, P. Pearle: In *Symposium on the Foundations of Modern Physics 1990*, ed. by P. Lahti, P. Mittelstaedt (World Scientific, Singapore 1990)
8. G.C. Ghirardi, R. Grassi: Stud. Hist. Phil. Sci. **25**, 397 (1994)
9. G.C. Ghirardi: Found. Phys. Lett. **9**, 313 (1996)
10. L. Landau, R. Peierls: Z. Phys. **69**, 56 (1931)
11. K.-E. Hellwig, K. Kraus: Phys. Rev. D **1**, 566 (1970)
12. Y. Aharonov, D.Z. Albert: Phys. Rev. D **24**, 359 (1981)
13. Y. Aharonov, D.Z. Albert: Phys. Rev. D **29**, 228 (1984)
14. G.C. Ghirardi, R. Grassi, A. Rimini: Phys. Rev. A **42**, 1057 (1990)
15. A. Einstein: In *Albert Einstein: Philosopher-Scientist*, ed.by P.A. Schilpp (Tudor, New York 1949)
16. J.S. Bell: In *Sixty-Two Years of Uncertainty*, ed. by A. Miller (Plenum, New York 1990)
17. J.S. Bell, M. Nauemberg: In *Preludes in Theoretical Physics*, ed. by A. De Shalit, H. Feshbach, L. Van Hove (North Holland, Amsterdam 1966)
18. J.S. Bell: In *Themes in Contemporary Physics II, Essays in Honour of Julian Schwinger's 70th birthday*, ed. by S. Deser, R.J. Finkelstein (World Scientific, Singapore 1989)
19. J.S. Bell: Found. Phys. **12**, 989 (1982)
20. L. Fonda, G.C. Ghirardi, A. Rimini, T. Weber: Nuovo Cimento A **15**, 689 (1973)
21. L. Fonda, G.C. Ghirardi, A. Rimini: Rep. Prog. Phys. **41**, 587 (1978)
22. L. Fonda, G.C. Ghirardi, A. Rimini: Nuovo Cimento B **18**, 1 (1973)
23. J.S. Bell: In *Proceedings of the Nobel Symposium 65: Possible Worlds in Art and Sciences*, Stockholm, 1986 (de Gruyter, Berlin 1989)
24. N. Gisin: Helv. Phys. Acta **62**, 363 (1989)
25. G.C. Ghirardi, R. Grassi: In *Nuovi Problemi della Logica e della Filosofia della Scienza*, ed. by D. Costantini et al. (Clueb, Bologna 1991)
26. G.C. Ghirardi, R. Grassi, J. Butterfield, G.N. Fleming: Found. Phys. **23**, 341 (1993)

27. J. Butterfield, G.N. Fleming, G.C. Ghirardi, R. Grassi: Int. J. Theor. Phys. **32**, 2287 (1993)
28. G.C. Ghirardi, R. Grassi: In *Bohmian Mechanics and Quantum Theory: An Appraisal*, ed. by J. Cushing et al. (Kluwer, Dordrecht 1996)
29. P. Pearle: Phys. Rev. A **39**, 2277 (1989)
30. G.C. Ghirardi, P. Pearle, A. Rimini: Phys. Rev. A **42**, 78 (1990)
31. G.C. Ghirardi, R. Grassi, F. Benatti: Found. Phys. **25**, 5 (1995)
32. G.C. Ghirardi: In *Open Systems and Measurement in Relativistic Quantum Theory*, ed. by H-P. Breuer, F. Petruccione, Lecture Notes in Phys. (Springer, Berlin, Heidelberg 1999)
33. G.C. Ghirardi: Found. Phys. **30**, 1337 (2000)

# 22 How Does God Play Dice? (Pre-)Determinism at the Planck Scale

Gerard 't Hooft

In deterministic theories, one can start from a set of ontological states to formulate the dynamical laws, but these may not be directly observable. Observable are only equivalence classes of states, and these will span a basis of "beables", to be promoted to an orthonormal basis of Hilbert Space. After transforming this basis to a more conventional basis, a theory may result that is fundamentally quantum mechanical. It is conjectured that the quantum laws of the real world may be understood from exactly such a procedure.

## 22.1 Dice

Quantum Mechanics works. It first served as a successful doctrine for describing the dynamical laws for electrons inside atoms. Subsequently, it was realized that quantum mechanics must be more universally valid, so it should also apply to the atomic nuclei and to the elementary particles out of which these nuclei are composed.

Including the requirements of special relativity has been a major challenge. At first sight, it seemed that quantum mechanics was to undergo substantial revisions in order to meet this challenge. The words "second quantization" underlined the fact that Schrödinger's equation for single particles was to be replaced by the equations for quantized fields.

But the fundamental starting points of quantum mechanics were not seriously altered by the advent of quantum field theory. Rather, they were generalized and deepened. Non-Abelian gauge theories – essential building blocks of the Standard Model – added a further complication: the emergence of ghost particles, required for achieving unambiguously renormalized amplitudes, but again, one has to conclude that the fundamental dogmas of quantum mechanics not only survived, but also served as a backbone for these theories. In accordance with these rules, the quantized field theory must provide for the scattering matrix ($S$-matrix) amplitudes, and the scattering probabilities should follow from these. Should we depart from the usual "Copenhagen" rules for interpreting the amplitudes, then we would not even know how to start formulating consistent theories.

For the last couple of decades, theoreticians have been considering the challenge to include the one remaining theoretical requirement: general relativity, i.e., the laws of gravity. The problem appears to be a beautiful one, since now the dynamics of the curvature in space and time must be submitted to the laws of quantum mechanics. The virtual contributions to the amplitudes due to space-time curvature diverge at the small-distance end, just like Fermi's earliest models for the weak force used to do. In the latter case, the divergences were successfully tamed by the introduction of the Standard Model. Naturally, one expects similar solutions to the problem at hand, and indeed, string theories and their successors are claimed to come close to bringing just such a solution.

But, they have not done so yet, and upon closer inspection one finds that there really are reasons to be skeptical. Not only are there numerous technical difficulties – the identification of the ground state, our lack of understanding the supersymmetry breaking mechanism, the smallness of the cosmological constant –, there are also more fundamental ones, which are difficulties that have little to do with the fact that we are dealing with strings, $D$-branes, or what not. Rather, they have to do with the fact that we are attempting to apply quantum mechanics to dynamical laws that should be the ultimate driving forces of the entire Universe; one will be forced to consider statistical ensembles of universes, and such notions will be much more questionable than the notion of an ensemble of experiments inside one universe. It should be kept in mind that one will never be able to do experiments in more than one universe, and that the 'averaged value' of a quantity measured in different universes cannot be checked against any theory.

Quantum fluctuations of the space-time metric at distances at or below the Planck scale diverge, and the implications of this may well be deeper than the need to re-sum some divergent perturbation series. Under these circumstances, causality and unitarity may become impossible to keep under control, regarding the difficulties in establishing time ordering. It should be clear that, at this point, the most serious difficulties may be associated to our insistence to keep the quantum rules unchanged.

It is illustrative, in this respect, to study a beautiful toy model for a universe: point particles gravitating in a $2+1$ dimensional universe [1], while obeying Einstein's equations. The classical equations are exactly soluble in $2+1$ dimensional space-times, during finite time intervals, but they become chaotic near a big bang or a big crunch. Poisson brackets can be written down for this system, but replacing these by commutators fails, exactly because the notion of a Hilbert space becomes problematic when it refers to entire universes.

It is here that I would like to advocate a different approach. We should not forget that quantum mechanics does not really describe what kind of dynamical phenomena are actually going on, but rather gives us probabilistic results. To me, it seems extremely plausible that *any* reasonable theory for the dynamics at the Planck scale would lead to processes that are so complicated to describe, that one should expect *apparently stochastic* fluctuations in any approximation theory describing the effects of all of this at much larger scales. It seems quite reasonable first to try a classical, deterministic theory for the Planck domain. One might speculate then that what we call quantum mechanics today, may be nothing else than an ingenious technique to handle this dynamics statistically.

Of course I am aware of the numerous studies regarding the difficulties in devising hidden variable theories for quantum mechanics. Deterministic theories appear to lead to the famous Bell inequalities [2], and the Einstein–Rosen–Podolsky paradox [3]. There are various possible reasons nevertheless to continue along this avenue. Generally speaking, we could take the attitude that every "no-go theorem" comes with some small-print, that is, certain conditions and assumptions that are considered totally natural and reasonable by the authors, but which may be violated in the real world. Certainly, physics at the Planck scale will be quite alien to us, and therefore, expecting some or several of the "natural" looking conditions to be violated is not so objectionable. More specifically, one might try to identify some of such conditions. One example might be the following: turning some apparatus at will, in order to measure either the $x$- or the $y$-component of an electron's spin, requires invariance of the device under rotations, allowing the detector to rotate independently of the rest of the scenery. This is unlikely to be totally admissible in terms of Planck scale variables.

Another attitude could be to assume that deterministic approaches might only succeed partially but not completely. Locally deterministic approaches might become useful for the identification of local dynamical rules, but maybe some quantum mechanical constraint will have to be added at a later stage, thus restoring the quantum mechanical nature of the entire description. In some of the calculations shown below, it can be seen that such constraints can exist, since already at an early stage we do turn to Hilbert space techniques, and there, one may well introduce fully quantum mechanical projection operators that only slightly modify the dynamics.

Let us simply keep an open mind, and just see where a deterministic approach leads us. Whether or not to call my most recent results promising, the reader may decide for him- or herself.

The most logical domain of physics where one may expect quantum mechanics to become replaceable by a more deterministic scenario is the Planck scale. One reason for this expectation was already mentioned: the unwieldy strangeness of the world of stringlike phenomena which might invalidate some of the assumptions made, be it consciously or tacitly, when arguments using the Bell inequalities are applied to prove hidden variables to be impossible or at least non local. Another reason for taking the Planck regime is the emergence of black holes there. On the one hand, a conventional quantum mechanical description of black holes appears to contradict General Relativity; on the other hand, a classical (read: deterministic) description of black holes implies the feature of information loss. Information loss may indeed be admitted in deterministic theories, and, as we shall argue further, information loss may be the key ingredient that could turn a local, deterministic theory into a quantum theory.

Nevertheless, one could observe that reference to the gravitational force is not truly necessary for our considerations. One might hope that the search for a theory behind quantum mechanics could provide for further costraints on the apparently arbitrary constants of nature in the Standard Model, long before the Planck scale is reached, but I am not counting on such a miracle.

My starting point is basically very simple. The dividing line between quantum physics and classical physics is more subtle than usually advertised. Hilbert space techniques, as an aid to discuss statistical features, can be introduced for deterministic systems [4–7] just like in notoriously "quantum mechanical" theories. A deterministic system is characterized by a set of evolution equations that tell unambiguously how it evolves in time, given its initial configuration. The definition of time does not need to be very strict. One might have a continuous time variable or discretized time, or time might be defined in terms of Cauchy surfaces in a general relativistic setting. There is one important condition that must be met by the time variable, however: it must be on a real line (possibly with a beginning and/or an end to it). Time is not allowed to be cyclic. If closed time-like trajectories would exist, this would lead to clashes and our theory would no longer be unambiguous. Closed time-like loops, popular in some versions of gravity theories, will be excluded.

First, we assume that the evolution is time-reversible (an assumption that may later be relaxed). We attach a basis element of a Hilbert space to every possible configuration of the Universe at a given time $t$, so the ensemble of all possible configurations spans a Hilbert space. If now time is discrete, then one step in time is associated to a permutation operator in the space of states, and it corresponds to a unitary evolution operator $U(\Delta t)$ in Hilbert space, where $\Delta t$ is the time quantum. In each row and on each column of $U$ there is one entry equal to 1 and the other entries are 0.

Subsequently, one may define an operator $H$ such that

$$U(\Delta t) = \exp(-\mathrm{i}H\Delta t) . \tag{22.1}$$

This operator is not unambiguously defined; one may freely add or subtract multiples of $2\pi/\Delta t$ to its eigenvalues.

If time is continuous, we have equations in a space of coordinates $\{q^i\}$, of the form

$$\frac{\mathrm{d}}{\mathrm{d}t}q^i = f^i(\boldsymbol{q}) , \tag{22.2}$$

where $f^i$ may be any kind of functions of the coordinates $q^i$. We limit ourselves to first-order differential equations because, of course, higher order equations can be reduced to first order ones by enlarging the space of independent variables $q^i$. In our Hilbert space, this equation is the Schrödinger equation corresponding to the Hamiltonian

$$H = \sum_i p^i f^i(\boldsymbol{q}) + g^i(\boldsymbol{q}) \ , \tag{22.3}$$

where

$$p_i \stackrel{\text{def}}{=} -\mathrm{i}\frac{\partial}{\partial q^i} \ , \tag{22.4}$$

and $g^i$ is an arbitrary function. Note that $p_i$ are the true momentum operators in the quantum mechanical sense, even though we are discussing deterministic theories. The point here is, that the Hamiltonian (22.3) is linear in the momenta, as opposed to the usual quadratic expressions in 'real' quantum mechanics.

Using either the Hamiltonian defined in (22.1) or the one in (22.3), one may use all conventional prescriptions from quantum mechanics and calculate the evolution of any 'wave function'. The absolute squares of its coefficients may always be interpreted as probabilities. It is important to observe that in these systems the 'collapse of the wave function' means absolutely nothing; there are probability distributions, and as soon as one measures something, our knowledge concerning the state collapses towards only one of various alternative possibilities, and that is all there is to it.

But of course our systems are not truly quantum mechanical. With our choices of the Hamiltonian, wave functions do not 'spread'; there seem to be no interference effects, and it seems to be difficult to create 'coherent states'. The question is, can one make a link between the deterministic systems described above and theories that look more like genuine quantum mechanical systems?

Let us first introduce some definitions. The basis elements of our Hilbert space refer to states the 'universe can really be in'. We call these states 'ontological states'. They may be realized or not realized, and, at least in our deterministic models, there is no possibility in between, although in our analysis we shall often take our refuge into probability concepts.

An operator that measures which of these ontological states we are in, will be called a *beable*, à la John Bell. Beables form a set of operators that are defined at all times, and they all commute with one another. An operator that replaces an ontological state by another ontological state will be called a *changeable*. Changeables may also add any kind of phase factors to the states. All operators presently known, and used in the Standard Model, must be changeables. Changeables do in general not commute, but they can sometimes be diagonalised. Physicists living in a deterministic world may be unable to distinguish beables from changeables.

At first sight, our deterministic models may seem to be trivial, and one may suspect that there is a wide gap between these and truly quantum mechanical models.

Just like in 'true quantum mechanics', we have a Hilbert space and a Hamiltonian. Where exactly is the gap between deterministic and 'truly quantum mechanical' theories? A very important distinction appears to be that the Hamiltonian (22.3) is not bounded from below. There is no ground state. In ordinary quantum mechanics we do have a ground state. But here also one might have to look at the gravitational force. When gravitational collapse is admitted, and protons are allowed to decay, states usually considered to be ground states are not truly ground states. There may be an alley here. Conversely, one might consider the Hamiltonian (22.3) and impose constraints. In some models, this Hamiltonian can be seen to be the difference between two commuting, positive operators [6]:

$$H = H_1 - H_2 \; ; \quad [H_1, H_2] = 0 \; . \tag{22.5}$$

Imposing the constraint

$$H_2 \, |\psi\rangle = 0 \; , \tag{22.6}$$

on our states $|\psi\rangle$, leaves us the Hamiltonian $H_1$, which may be bounded from below. Are there other manipulations that can be performed to turn our system into one where the Hamiltonian does have a ground state?

If time is discrete, the Hamiltonian is bounded, so of course there is a ground state. Here, however, the eigenvalues are well-defined apart from multiples of $2\pi/\Delta t$, so here the problem is that the choice of the ground state appears to be arbitrary. If our world could be related to a deterministic scenario at the Planck scale, then why does the vacuum state of our world appear to be unique, whereas it is either arbitrary or ill-defined in our deterministic models?

The answer to this may be that one possible refinement has not yet been applied. In the beginning of our formalism, we assumed that, in the deterministic model, the evolution law is time-reversible. If time is discrete, this needs not be the case. Two different configurations might evolve into the same final state. We call this 'information loss', or 'dissipation'. At first sight, the introduction of information loss appears to be a disaster. We now no longer have a unitary evolution matrix, so we cannot define a Hamiltonian at all. However, the damage can be repaired. In principle, we could decide to 'remove' from consideration all those states that now cannot be reached at all from any configuration in the distant past. Then time reversibility might be restored, and we are back in the previous situation. In practice, however, this is infeasible. It may be too difficult to identify the states without a past ("Gardens of Eden"). Another avenue may be the following. Instead of single states, we now use the concept of 'equivalence classes of states'. Two states are equivalent iff within some finite time period they both evolve into the same final state. A basis element of Hilbert space is defined to be such an equivalence class. The time evolution of equivalence classes is time-reversible by construction.

How can the introduction of equivalence classes affect the ground state problem? The Hamiltonian is still the one defined in (22.1), and its ground state is still ill-defined. However, the notion of *locality* is severely affected. Locally, one cannot distinguish the equivalence classes, so it may appear to be impossible to define local operators. Notice, that this resembles the situation in gauge theories, where gauge-invariance causes complications in defining local observables (local operators have a high dimensionality). In string theories, local observables are even more difficult to define.

Our ontological states may still evolve according to a completely local law, but the equivalence classes are not locally well-defined. Therefore, our theory may have the property that information cannot spread faster than a certain velocity, say the velocity of light, but its quantum states cannot be characterized locally. We do expect that the Hamiltonian $H$ of such a theory can be expressed as the integral over a Hamiltonian density $\mathcal{H}(\boldsymbol{x})$, with

$$[\mathcal{H}(\boldsymbol{x}),\mathcal{H}(\boldsymbol{x}')]=0 \quad \text{if} \quad \boldsymbol{x}\neq\boldsymbol{x}' \; ; \quad H \stackrel{\text{def}}{=} \int \mathcal{H}(\boldsymbol{x})\,\mathrm{d}\boldsymbol{x} \; . \tag{22.7}$$

The effects of introducing information loss and the equivalence classes may be drastic. The classes may be very large, so that the dimensionality of Hilbert space may be greatly reduced. For instance, in the case of black holes, the total number of quantum states is known to grow exponentially, but only with the area of the horizon, and not its volume – the well-known holographic principle. Clearly, quantum states are related to ontological states after applying sequences of projection operators. It is these projection operators that may be responsible for curtailing the Hamiltonian as well. There was arbitrariness in the definition of $H$, but insisting on equations such as (22.7), we might find that $\mathcal{H}(\boldsymbol{x})$ does have a natural lower bound, so that this $H$ has one as well. We suspect that this is how our Hamiltonian receives a ground state.

In this light, one may regard our newest publication [7]. In there, we show that free bosons are in fact deterministic, but only if one restricts oneself to those observables which are invariant under a specially chosen group of gauge transformations. One may either consider these gauge transformations as an indelible ingredient of our theory, or one might say that these are representing transitions about which information gets lost. It is then the equivalence classes that possess this gauge invariance, not the ontological degrees of freedom.

This is an example where we can see our proposal work. Besides our deterministic model leading to second quantized free bosons of any mass, another deterministic model could be constructed whose Hamiltonian turns out to be that of non-interacting massless fermions. These fermions appear to behave as flat sheets rather than particles. Here also, one may suspect that these sheets represent equivalence classes rather than ontological variables. Both for the fermions and the bosons, the Hamiltonian could be made bounded from below. As yet the models we could produce do not seem to be physically very realistic, let alone impressive, but they are to be taken seriously. The search for more models is under way. It is comforting to realize that superstring theory makes full use of freely propagating bosons and fermions along the string world sheet. Since the construction of deterministic quantum theories works particularly well for free bosons and massless fermions, these theories are ideally suited for being incorporated into the picture advocated here.

We think the above gives some idea about the approach that I wish to advertise. Quantum mechanical features can be obtained from completely deterministic scenarios, in such a way that the Born interpretation of the wave function as a probability amplitude is completely respected. Of course, we expect the reader to be skeptical. A number of issues should be confronted in order to become more convincing. Most of these have to do with 'quantum weirdness'. Quantum mechanics has given us so many surprising and counter intuitive features of the submicroscopic world that many have come to conclude that it will be forever impossible to construct deterministic, or 'ontological' theories. Let me begin with the relatively easy issues.

## 22.2 Quantum Interference

The prototype experiment is the electron two-slit experiment. If an electron is allowed to go only through one of the two slits, a relatively smooth pattern of electrons is detected on the screen behind the slits. If the electron is given the choice between two slits, an interference pattern arises. There may be constructive or destructive interference.

In theories of the type proposed here, this is not at all a forbidden phenomenon. The operators that characterize the presence of an electron apparently cannot be beables, because the operator that detects an electron at one of the slits does not commute with the operator that detects an electron on the screen behind the slits. Electrons are not the ontological variables of the deterministic system. The fields $\psi$ introduced by physicists to describe electrons are changeables, to be constructed from the primordial variables of the theory.

The same holds, in a more obvious way, for the spin of a particle. Spin is the eigenvalue of a rotation operator. Surely, rotations replace ontological states by different states. The operators measuring spins are changeables.

## 22.3 Quantum Coherence and Entanglement

I am often asked how, in deterministic theories, coherent and/or entangled quantum states can arise. How they can arise out of inconspicuous initial states, I do not know, but it should be stressed that the notion of states is here as in genuine quantum mechanics. All quantum states may be considered. We may prepare any initial state we like, and ask how it evolves with time. It is important to emphasize that our description of *states* is exactly as in genuine quantum mechanics. We allow the use of operators with non-trivial commutation rules. These states merely describe our choice of the probablities of the various possible configurations at any given time. So we definitely allow these probabilities to be as in a state with 'quantum coherence'. Where our theories differ from conventional quantum mechanics is our choice of the *dynamics*. The dynamical rules are deterministic, but only after one has been able to identify what the basis of Hilbert space is that corresponds to the ontological states, or the equivalence classes (which are also ontological). The description of this basis may be deeply hidden in the Planckian regime.

## 22.4 The Einstein–Rosen–Podolsky Paradox and the Violation of the Bell Inequalities

This is surely the most difficult aspect to be addressed, and a completely satisfactory response has not yet been given. Part of an answer can perhaps be found in the observations made above: we do admit non-commuting operators and states in which some of these are diagonalized. But this does not explain why and how experiments can be performed that defy any description in terms of local, ontological observables. We must conclude that the observables commonly used are not ontological. They are not beables. This however leads to the question: how can it be that, at any given time, an experimenter can choose to measure either the $x$- or the $y$-, or the $z$-component of a spin, as if these *were* beables?

Before attempting to answer this question, we should decide about a related topic: *is a registration of an experimental result, fixed in a classical way, a beable?* If the motion of planets is partly determined by haphazard events with a quantum mechanical origin (particularly relevant if the motion is a chaotic one, such as most Newtonian systems), are the coordinates of planets, or at least their first 20 or 30 decimal places, beables?

The answer here is most probably yes. So the previous question remains: how can changeables turn into beables in the course of a measurement?

My suspicion is that the relation between changeables and beables at the atomic scale is extremely complex. It is already fairly complicated to express the $x$ variable and the $p$ variable of a harmonic oscillator in terms of its beables, as described in [7]. In more realistic theories that include interactions, it might involve the changeables and beables of the experimenter himself, when the decision is made which component of the spin to measure. This is what Bell called *predeterminism*. The experimenter's decisions also follow from Standard Model interactions in his brain. Predeterminism is here defined by the assumption that the experimenter's 'free will' in deciding what to measure (such as his choice to measure the $x$- or the $y$-component of an electron's spin), is in fact limited by deterministic laws, hence not free at all, so that it must be included in a deterministic interpretation of quantum mechanics. Most physicists dismiss predeterminism as a likely resolution of the EPR paradox. I am not so certain about this. The entire combination, experimenter/experiment, should be understood by performing a renormalization group transformation from the Planck scale to the atomic scale. But exactly how all of this should be worked out is far from understood.

On the other hand, it also remains to be seen whether the EPR paradox and Bell's inequalities are real contradictions at all, or whether they can simply be ignored. If our models work, they explain quantum mechanics, and it could be that the apparent violation of Bell's inequalities will be seen as a natural consequence. Although in any case, more understanding of this situation is desired, it is legitimate to postpone answering these questions until more sophisticated models have been found. The existence of unanswered questions of this sort does not invalidate our approach, so we continue research in this direction.

## References

1. A. Staruszkiewicz: Acta Phys. Pol. **24**, 734 (1963)
   S. Giddings, J. Abbott, K. Kuchar: Gen. Rel. Grav. **16**, 751 (1984)
   S. Deser, R. Jackiw, G. 't Hooft: Ann. Phys. **152**, 220 (1984)
   S. Deser, R. Jackiw: Commun. Math. Phys. **118**, 495 (1988)
   G. 't Hooft: Class. Quant. Grav. **10**, 1023 (1993)
2. J.S. Bell: Physica **1**, 195 (1964)
3. A. Einstein, B. Podolsky, N. Rosen: Phys. Rev. **47**, 777 (1935)
4. G. 't Hooft: Found. Phys. Lett. **10**, 105 (1997); quant-ph/9612018
5. G. 't Hooft: Class. Quant. Grav. **16**, 3263 (1999); SPIN-1999/07, gr-qc/9903084
6. G. 't Hooft: 'Determinism and dissipation in quantum gravity'. In: *Basics and Highlights in Fundamental Physics*, Erice, August 1999, The Subnuclear Series, Vol. 37, ed. by A. Zichichi (World Scientific, Singapore 2001) p. 397, SPIN-2000/07, hep-th/0003005
7. G. 't Hooft: Determinism in free bosons, SPIN-2001/07, hep-th/0104080

# Part V

# Quantum Specials

# 23 John Bell, State Reduction, and Quanglement

Roger Penrose

## 23.1 Encounters with John Bell

The First time that I met John Bell was at a conference on quantum gravity, held in Oxford in April 1980, that Chris Isham, Dennis Sciama and I had organized. In our planning of the meeting, we had thought it appropriate to address the issue of the foundations of quantum mechanics, since we considered it not improbable that a proper understanding of what is to be meant by "quantum gravity" might involve a reappraisment of the very structure of quantum mechanics. Each of the three of us had, in our different ways, been prepared to entertain the possibility that the rules of quantum mechanics might need alteration if its intended marriage with Einstein's general relativity were to be successful. We were aware that the quantum violation Bell's famous inequalities provided one of the most powerful arguments in support of conventional quantum mechanics, and we had in mind that the strength of the conventional case ought to be presented as forcefully as possible, so that the constraints on any alteration of quantum laws should be made clear. Thus, in our ignorance of John Bell's actual views, we thought that we could not do better than to invite him to speak, so that we should be forced to appreciate the strength of this conventional case.

Bell accepted our invitation, and we were delighted with the insights that he provided us with, but he took us quite by surprise by presenting a viewpoint that was much more sympathetic to our own ways of thinking than we had expected. We did not get the bolstering of the conventional perspective that we had foolishly expected, but instead we got something far more valuable, namely a perspective whereby one can maintain an entirely *objective* view of quantum reality if one is prepared to accept a very non-local picture of the world. Bell, in his talk [1], made a particular comparison between the *many-worlds* view, which he realized was a popular one amongst quantum cosmologists, and the de Broglie–Bohm theory, this having been his favourite at the time.

When the Ghirardi–Rimini–Weber scheme arose, in the mid-1980s, Bell took the view that this provided a better alternative. I believe that I first heard of the details of this scheme from Bell himself, in a talk that he gave at Schrödinger's centenary conference in Vienna. In my opinion, John Bell performed an extremely important role then, and also later, in generally supporting – thereby making *respectable* – the apparently "fringe" activities of such people as Károlyházy, Bohm, Pearle, Ghirardi, and many others (including myself) in suggesting schemes that go beyond standard quantum mechanics, in the intended direction of *realism.* No physicist could doubt the scientific credentials of John Bell. The fact that he was prepared to go out of his way to support research of this kind gave it a previously unaccustomed status.

I had later met John Bell on a number of occasions. I remember, in particular, his extreme generosity with his time, when my wife and I visited CERN in 1988(?). With his characteristic charm and good nature, he showed us the accelerator tunnels and many other things, without the hint of a suggestion that he might have other work to do. The last occasion on which we met was at a conference organized by Martin Rees, held at Cambridge in August 1990, only just over a month before John died. In his talk there, John described some exciting new ideas in which *gravity* might be the key to quantum state reduction. I only wish that I had been more fully attentive to its details. I think it likely that the ideas that he was describing then, were ones that he had been developing from *Diósi's* scheme, introduced in the late 1980s.

At that time, the gravitation-based scheme that I had myself been espousing involved what I called "the one-graviton criterion", according to which I tried to give a mass scale, based on gravity, above which quantum superpositions would not occur (as described in [2]). I had envisaged that one ought to calculate the expectation value of the number of gravitons in the difference between two states under quantum superposition. The superposition was supposed to reduce to one or to the other of the superposed states when this number becomes of order unity. In Diósi's [3, 4] scheme, however, there is a gravitational *energy* criterion, and the reciprocal of this energy provides a measure of the *rate* of quantum state reduction. This provided an important distinction.

In 1992, by means of an independent line of thinking, I was led to a scheme of my own, very similar to that of Diósi, but which did not involve the specific stochastic dynamics that Diósi had proposed. I presented this at the International Congress on General relativity and Gravitation in Cordoba Argentina [5]. Later, in 1996, I was to provide a completely different (and more direct) justification for the same type of proposal [6].

When I was writing *Shadows of the Mind* [7], in around 1993, I wrote to Mary Bell, asking her if she had any record of the contents of John's 1990 Cambridge lecture. She very kindly sent me a Xerox copy of the transparencies that John had probably used in that lecture, but I was not able to ascertain how far he had been able to develop his ideas at the time, and to what extent he had been stimulated by Diósi. I recall that at the Cambridge meeting, over lunch, John had asked me various questions relating to the Bekenstein–Hawking black-hole entropy, etc. (probably because I had been basing my own considerations on such matters, at the time). It was clear that he was beginning to take seriously the issues raised by Einstein's general relativity (which, on my present view, are vitally relevant). But I do not suppose that I shall ever properly know what he really had in mind.

## 23.2 Gravitationally Induced State Reduction: A Minimalist View

My own ideas have developed somewhat since 1996, and I shall try to make them a little more specific here. In my proposal, I make no attempt to provide a (stochastic) dynamics, of the nature of that pioneered by Philip Pearle [8], and developed later by others such as Ghirardi and his collaborators, Diósi, Percival, and others. My scheme is a rather "minimalist" one which merely purports to tell us a *lifetime* $\tau$ for a quantum superposition of two states, each of which would be stationary on its own, to reduce spontaneously to one or the other of these two stationary states. By analogy with an unstable nucleus, the quantity $\tau$ is to be thought of as something analogous to the "half-life" of the superposition, after which time, on the average, it would decay into one or other of the two constituent stationary states. To calculate $\tau$ we first work out the gravitational self-energy $E_{\mathrm{G}}$ of the *difference* between the mass distributions of the two states in superposition. In accordance with Heisenberg's principle, $E_{\mathrm{G}}$ is to be thought of as a fundamental *uncertainty* in the energy of the superposition, leading us to expect a lifetime of the order

$$\tau \approx \hbar / E_{\mathrm{G}} .$$

This energy uncertainty $E_{\mathrm{G}}$ comes about through a basic conflict between the principles of quantum mechanics and those of general relativity. Let us try to be fairly specific and consider a situation in which a lump of material is at rest on a horizontal table. This is then put into a quantum superposition of (a) being slowly moved to a neighbouring location A on the table and (b) being left in its initial location B (a Schrödinger's cat! [Fig. 23.1]). In standard quantum mechanics, the Schrödinger equation provides the time-evolution of each state individually and also of the quantum-superposed state quantum state. This evolution is represented by the operator $\partial/\partial t$.

Here, we wish to take into account the effects of gravity, properly according to the principles of general relativity. First, in order to take into account the space–time curvature effects of the Earth's gravitational field, we should replace "$\partial/\partial t$" by the appropriate *Killing vector* $\boldsymbol{T}$, which generates the time-translation symmetry that represents the stationary nature of the background space–time. Assuming that the gravitational field of the lump itself can be ignored, its *stationarity*, in each of its two alternative locations $A$ and $B$, is expressed as:

$$\mathrm{i}\hbar \boldsymbol{T} \left|\psi_A\right\rangle = E \left|\psi_A\right\rangle \quad \text{and} \quad \mathrm{i}\hbar \boldsymbol{T} \left|\psi_B\right\rangle = E \left|\psi_B\right\rangle ,$$

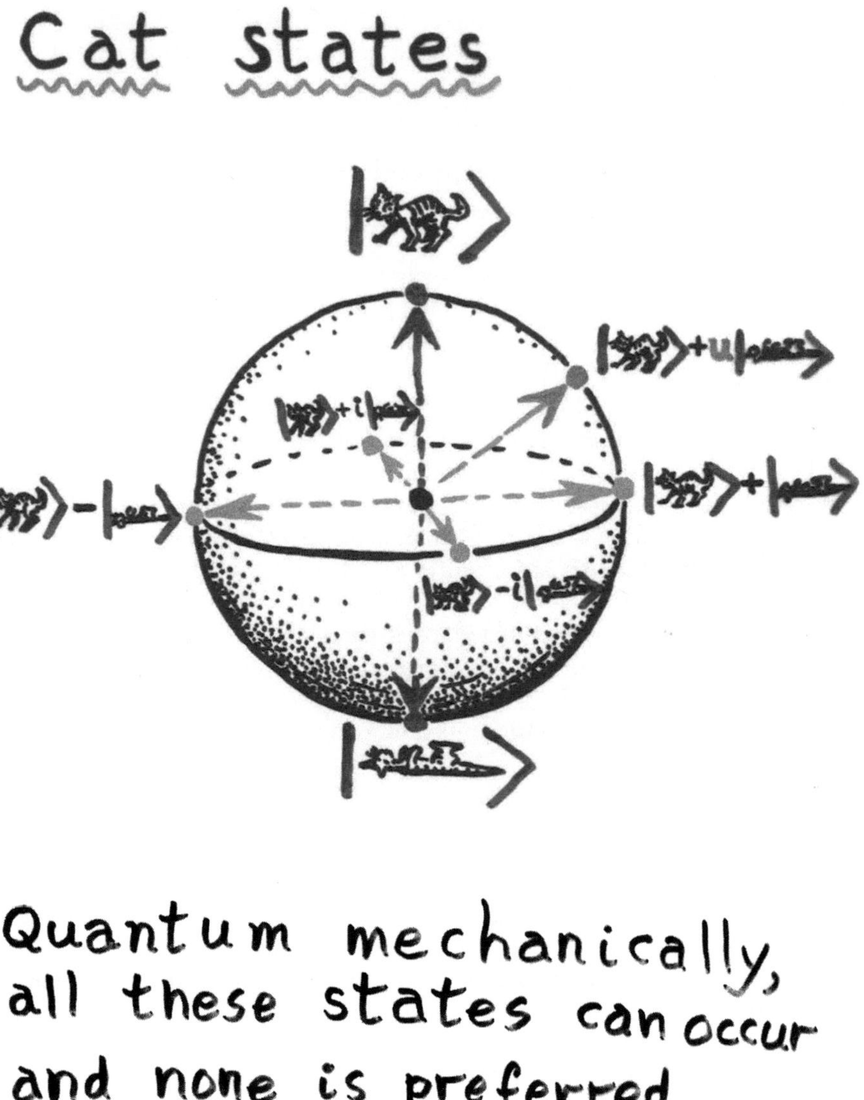


**Fig. 23.1.**

where the respective lump quantum states are $|\psi_A\rangle$ and $|\psi_B\rangle$, and where the energy eigenvalue $E$ is taken to be the same for each. Each superposition $z|\psi_A\rangle + w|\psi_B\rangle$, where $z$ and $w$ are constant complex amplitudes, is clearly just as stationary as are the constituent states $|\psi_A\rangle$, $|\psi_B\rangle$:

$$\mathrm{i}\hbar \boldsymbol{T}\{z\,|\psi_A\rangle + w\,|\psi_B\rangle\} = E\{z\,|\psi_A\rangle + w\,|\psi_B\rangle\}\,,$$

the energy eigenvalue $E$ being the same as before.

This is just standard elementary quantum mechanics. But now suppose that we take the lump's *own* gravitational field into account. I shall assume that the lump is large enough that its gravitational field can be treated, to a good approximation, as a classical solution of the Einstein equation. Then for a single lump location, say $A$, we must modify the (still stationary) space–time so as to incorporate the lump's field as well as that of the Earth. Now we have a Killing vector $\boldsymbol{T}_A$ to describe the stationarity of this combined gravitational field, and our Schrödinger equation now provides us with

$$\mathrm{i}\hbar \boldsymbol{T}_A \left|\psi_A\right\rangle = E \left|\psi_A\right\rangle \ .$$

Likewise, for the original lump location $B$ we have a Killing vector $\boldsymbol{T}_B$, giving

$$\mathrm{i}\hbar \boldsymbol{T}_B \left|\psi_B\right\rangle = E \left|\psi_B\right\rangle \ .$$

However, we cannot now directly apply this to obtain stationarity for the superposition $z|\psi_A\rangle + w|\psi_B\rangle$, because the time derivative operators $\boldsymbol{T}_A$ and $\boldsymbol{T}_B$ are *different.*

In fact, they are different in a *profound* sense which is tied up with the very foundational principles of Einstein's general relativity. It is not just that the Killing vector fields $\boldsymbol{T}_A$ and $\boldsymbol{T}_B$ are different vector fields; they are vector fields on two quite different *spaces*! According to the principle of general covariance (which is a natural consequence of Einstein's foundational principle of equivalence) there is no physically natural way of labelling the points of space–time and, accordingly, there is no physically natural pointwise identification of one space–time with another (in general). If $\boldsymbol{T}_A$ and $\boldsymbol{T}_B$ had been merely different vector fields on one and the same manifold, then one could envisage averaging the two or of obtaining some appropriate "correction" to each, so that a common form of the Schrödinger equation could be found for each stationary component of the state and therefore for the quantum superposition of the two components. But the principle of general covariance seems to be telling us that this would be as *in*appropriate thing to try to do.

Instead, I propose that we do try to make an *approximate* pointwise identification of the two space–times with their respective Killing fields, and try to estimate the *error* in doing so. In order to be explicit, it is convenient to take the limit where the massive ingredients of the system have small velocities, and the gravitational potentials are small, so that Newtonian theory is a good approximation. One might think that the problems would now go away, because there is a universal time $t$ and a flat background space. However, our problems do *not* really go away.

The universality of $t$ is not the point, of course, because the definition of "$\partial/\partial t$" requires our knowing the "$x, y, z$" which are to be held *constant.* It is these spatial variables which are ambiguous with regard to our paying due respect to the principle of general covariance. One could take the view that the spaces of constant $t$ are Euclidean spaces, so the "$(x, y, z)$-freedom" could be regarded as limited, and determined by the mass centre of the entire system (including the Earth, which must be brought in to determine that the mass centre is not moved in the displacement of the lump of material). This is essentially a viewpoint suggested by Christian [9, 10]. In my view, this viewpoint involves a non-locality that is not quite in the spirit of general relativity (although the situation is certainly arguable) and I shall follow a different line.

I propose that we adopt the different perspective whereby Nature would "prefer" that the notions of *free-fall* (i.e. inertial motion, in the Einsteinian view) are to be identified in the two space–times involved in the superposition. This, however, cannot be carried out consistently, so we must try to estimate the *error* involved in *failing* to make the free falls identical, when we try to identify the two space–times. For this, we integrate over all space (at one time $t$) the squared length $|\nabla\Phi_A - \nabla\Phi_B|^2$ of the *difference* between the Newtonian acceleration (spatial) vectors, $\Phi_A$ and $\Phi_B$ being the Newtonian potentials for the two gravitational fields involved. Integrating by parts and introducing Poisson's equation, we finally obtain the quantity $E_G$, which is the *gravitational self-energy of the difference between the two mass distributions involved.* The quantity $E_G$ is taken as a fundamental uncertainty in the "$\partial/\partial t$" of Schrödinger's equation, which I take, by analogy with the application of Heisenberg's time/energy uncertainty to an unstable nucleus, to be related to a *lifetime* of the superposition of the order of $\tau \approx \hbar/E_G$, as was claimed above.

One might think that a gravitational energy of this nature would be so small as to be totally ignorable, for objects that might be involved in a quantum experiment. But $\hbar$ is also very small, in everyday terms, so one has to calculate the expected magnitude of this effect to see whether it has importance. In fact, it is possible to envisage experiments to test the effect [11]. These would be difficult, but not impossible to perform within the limitations of present technology. One such possibility, partly based on a suggestion by Johannes Dapprich, would be to put a tiny crystal (of the nature of what is called a *Mössbauer* crystal, which responds to an impulse as a whole, rather than internal vibrations being set up) into a quantum superposition of two slightly different locations by hitting it with a high-energy photon (or other type of quantum projectile), the photon having been previously beam-split, so that the crystal's state becomes a superposition of having been displaced and of not having been displaced. The part of the photon's state that hits the crystal, as it emerges from the beam splitter, has to be reflected cleanly from the crystal; and it, and the other part of the photon's state, have to be kept coherent for the time $\tau$ or longer until, as this period draws to a close, the whole procedure has to be put into reverse. If quantum coherence is maintained throughout this entire period, in accordance with standard quantum mechanics, then the two beams recombine and the photon re-enters the source. But if my proposal is valid (or else if some decoherence of a standard kind takes place, due to some experimental imperfection in the system) then there will be a 50% probability for a detector, placed in the alternative beam emerging from the beam splitter, to detect the photon [Fig. 23.2].

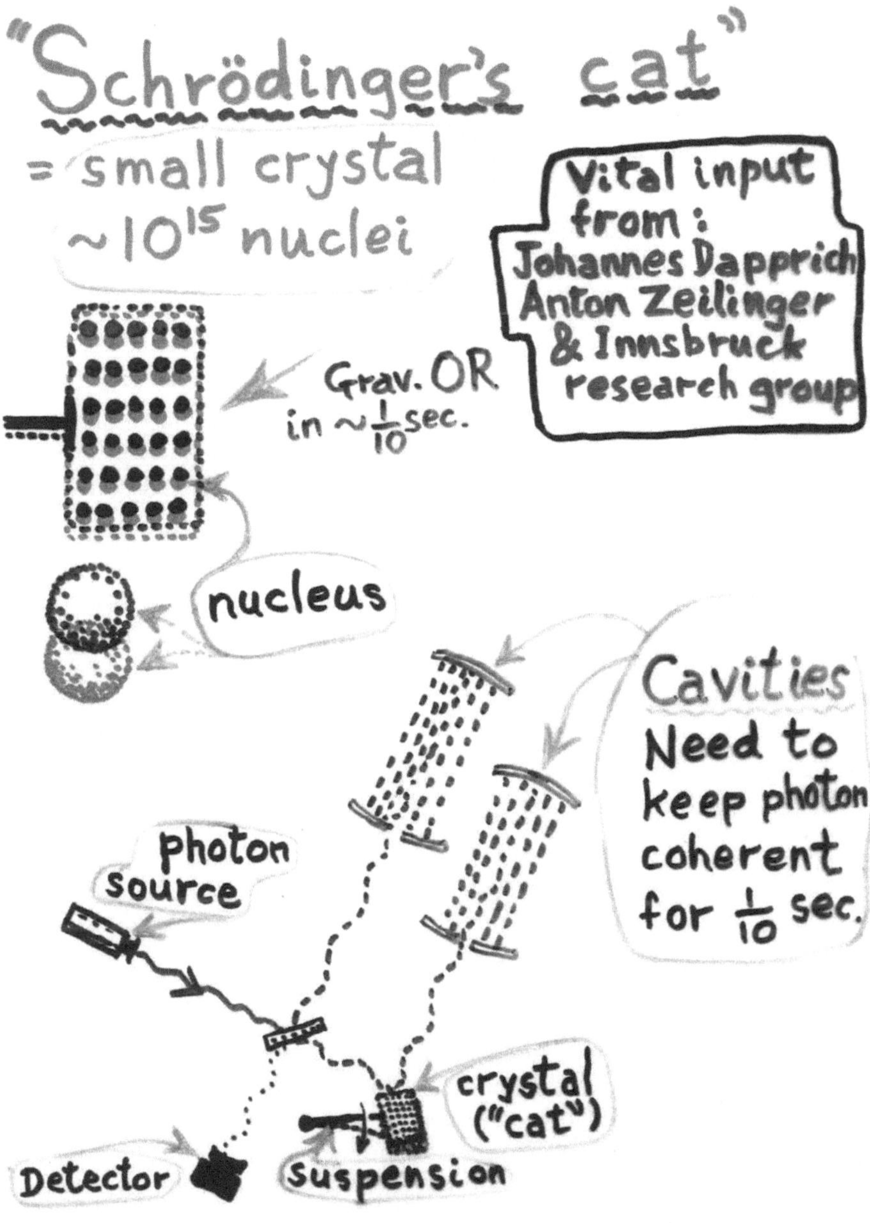


Fig. 23.2.

An important difficulty with this experiment is that for a crystal large enough for its superposed state to reduce to one or the other in a reasonably short time, the photon seems to need to be an *X-ray* photon in order to give the crystal enough of an impulse (according to the scheme being suggested here). A reduction time of $\tau \approx 10^{-1}$ s, might be achieved for a crystal of some $10^{15}$ nuclei. It would be hard to preserve coherence for the photon over this period, but one possibility for achieving this would be to perform the whole experiment in space, where the photon (in each of its superposed components) would be sent from one space platform to another, from which it would be reflected back, the two platforms being perhaps an Earth diameter apart, the round trip for this taking about $10^{-1}$ s. This is the FELIX proposed experiment: Free-Orbit Experiment with Laser-Inerferometry X-rays; [11]. There are other possible experiments for testing the aforementioned suggested gravitational reduction scheme, and it is quite possible that one of these will ultimately prove to be a better prospect than FELIX.

## 23.3 Quanglement

The proposal outlined in the previous section is, at best, a suggestion of a "minimalist" nature, where no dynamics is proposed which actually underlies the state reduction. Moreover, clear predictions as to the reduction time are provided only for superpositions between two states, each of which would be stationary on its own. One should perhaps also insist that the two amplitudes $w$ and $z$ are of the same order of size. It may be imagined that various kinds of improvements could be made to this scheme, in order to make it applicable in more general situations, so that this could lead to a genuine theory with a broad range of application. In the explicit cases under consideration, where an experimental test of these ideas is being proposed, the velocity of the *massive* components of the system is taken to be extremely small, in comparison with the speed of light $c$, and the changes in the gravitational potential that take place during the experiment are likewise very small.

The last two conditions allow a Newtonian-type approximation to be used, and this has the additional advantage of avoiding one of the biggest difficulties confronting any suggested scheme for an *objective* reduction of the quantum state (**OR**), namely the necessary fundamental *non-locality* of any realistic such scheme. When reduction takes place in an entangled state (such as in an EPR experiment to test quantum mechanics, where violations of Bell's inequalities are expected) the reduction takes place across the entire state, which might involve widely separated components. The measurement of one such component would seem to have to cause the other to reduce its state at the very moment that the measurement is made. Somehow the requirements of relativity seem violated, with the "quantum information" describing the connections between the two widely separated components of the state seeming to be propagated instantaneously. With the Newtonian approximation under consideration here, the seeming violation of causality is of no severe consequence, since there is no bar to instantaneous propagation of information in a Newtonian-type theory.

This can be only a temporary respite, however, since any serious theory of the kind that is being put forward here should have the capability of generalization in order that genuinely relativistic situations can be treated. I do not yet have a clear proposal for a relativistic generalization of the above scheme. Moreover, I know that John Bell himself might well have been happier to abandon relativistic invariance than to give up objectivity (i.e. **OR**) in a quantum reduction scheme. Nevertheless, I am of the opinion that a *relativistically invariant* proposal for **OR** *ought* to be sought.

How can this possibly be achieved, in the face of the above-mentioned fundamental difficulty? Many physicists do not regard the seeming causality violation in EPR phenomena as a real problem, because they are not unhappy with the abandonment of an objective picture of the quantum world, and they may point out that the probabilities come out the same no matter in which order it is considered that distant measurements are made. The difficulty lies with the seeming clash between relativity and the *objective* nature of **OR**. I am proposing that a relativistically invariant **OR** scheme may well be found in the future, but this will require some fundamentally novel ingredients. As a step towards such a future understanding, I suggest that we must seriously take on board the notion of *quanglement*.

What is "quanglement"? I am proposing that we use this term for the concept that underlies what is frequently referred to as "quantum information". My problem with the latter term is that it contains the word "information". This word has the apparent connotation that it is something that is possible to disseminate freely; yet we know from the "no-cloning" theorem [12] that the "information" contained in a quantum state cannot be copied, so it is unlike the normal idea of information. Moreover, "information" should be something constrained by the requirements of *relativity theory*, according to which no information-carrying signal can propagate outside the future null cones, whereas "quantum information" seems to violate this in EPR situations. I believe that some of the extremely puzzling features of quantum mechanics – those which lead many physicists to despair of ever finding an objective view of the quantum world – indeed come about from the enigmatic acausal nature of this "quantum information".

How can the rechristening of this notion as "quanglement" help? It can, essentially, because it allows us to abandon those preconceived ideas that the word "information" invokes. To see how this might work, it is useful to consider the so-called *quantum teleportation* experiment [13]. Recall that this concerns two individuals "Alice" and "Bob", separated by a large distance. A friend of Alice hands Alice a quantum state – say, the angular momentum state of an atom of spin 1/2, where the atom is suspended inside a sealed box. Alice now has to transmit to Bob the precise spin state that she has been handed, merely by sending Bob a few bits of actual classical information. Just like this, the task would be impossible, but at some time in the past a single quantum state of spin zero was split into two atoms, each of spin 1/2, and these were put into two suitcases, one of which has been handed to Alice and the other to Bob. We assume that the design of each suitcase is such that the spins of the two atoms remain undisturbed, for an unlimited time, until that suitcase is opened. When Alice receives her friend's sealed box, she opens it, combining its contents with the contents of her suitcase and performs a particular type of spin measurement on the combined spin states. The eigenvalues are 0, 1, 2, and 3, these distinguishing the four alternative orthogonal "Bell states" (up-up±down-down, up-down±down-up). She now transmits one of the four possible values (0, 1, 2, 3) to Bob, by an ordinary classical signal. Upon receiving this message, Bob effects a rotation of about either the $x$-axis, the $y$-axis, or the $z$-axis, or else no rotation at all, according to the message that he receives. Because of the details of the situation, it turns out that the contents of Bob's suitcase now has the same spin state as did the contents of Alice's friend's box, originally!

What is particularly remarkable about this situation is that whereas only *two bits* of actual information has passed from Alice to Bob, there has been, conveyed from Alice to Bob, something requiring an entire (two-dimensional) continuum for its description – namely a point on the Riemann sphere describing all possible states of spin 1/2. How can this be? An examination of the space–time description of the entire situation (Fig. 23.3) makes clear where the route of "continuous information" from Alice to Bob has to be found. The route can be traced from the contents of Alice's friend's box to the contents of Alice's suitcase, then backwards in time to the spin-0 origin of the EPR pair, then forwards in time again, within Bob's suitcase, to his reception of the signal from Alice and then onwards, still within Bob's suitcase after the rotation is performed. What "travels" this route is "quantum information", which is "information" in the *mathematical* sense of the parameters needed to specify the situation completely. Mathematically, it is not an inconsistency that part of this route involves a segment that is oriented in a backwards-time direction. Our troubles with the notion come from our physical understandings that information should not be permitted to propagated into the past. These troubles are eased somewhat as soon as we realize that this "quantum information" is *not* information in the ordinary sense, but something quite different – which is why I prefer the term *quanglement*.

Alice cannot send Bob an actual *message* using quanglement alone. There is no measurement that he can perform upon the atom in his suitcase that will ascertain the actual state that Alice's friend presented her with. Quantum mechanics is like that. But quanglement can be used in *conjunction* with ordinary information, so that tasks can be performed that could not be if the quanglement were absent. Although quanglement is not information, it provides a physical ingredient of an objective nature, provided that we are

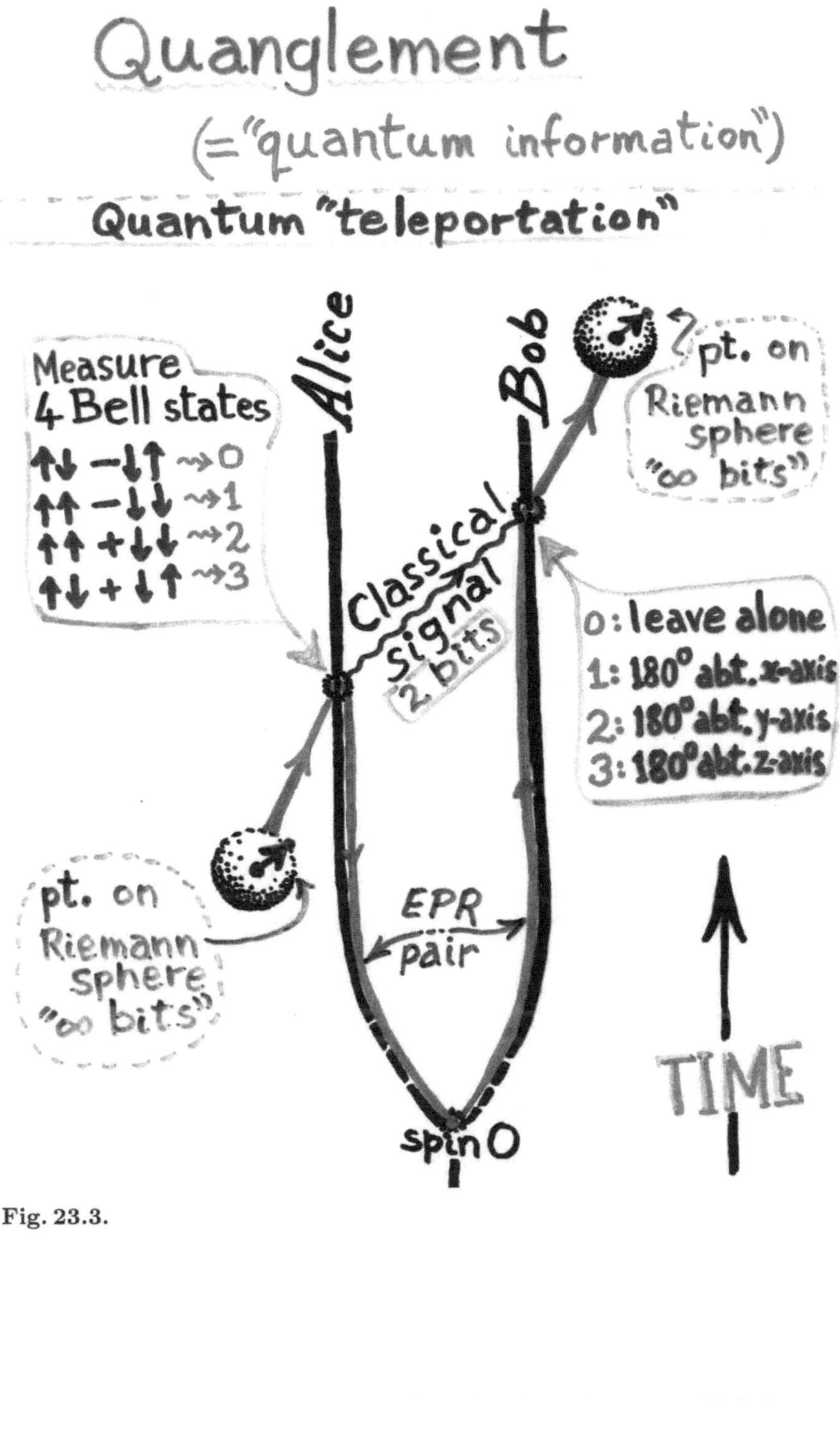
Quanglement
(="quantum information")
Quantum "teleportation"
Alice
Bob
Measure
4 Bell states
↑↓ − ↓↑ ⇝ 0
↑↑ − ↓↓ ⇝ 1
↑↑ + ↓↓ ⇝ 2
↑↓ + ↓↑ ⇝ 3
Classical signal
2 bits
pt. on Riemann sphere "∞ bits"
0: leave alone
1: 180° abt. x-axis
2: 180° abt. y-axis
3: 180° abt. z-axis
EPR pair
spin 0
TIME
pt. on Riemann sphere "∞ bits"

**Fig. 23.3.**

not unhappy with this ingredient having past-pointing as well as future-pointing ingredients. In fact, we can equally read the whole quanglement connection in the reverse sense, with the roles of future-directed and past-directed connections reversed. I prefer to talk in terms of "quanglement links" between events. (In fact, when there is a temporal reversal in a quanglement connection, this is accompanied by an operation of complex conjugation, quanglement being a complex-number quantity; see [14]).

A remark on the nature of the "objectivity" that is relevant here. Alice's friend may well have deliberately arranged that the state in the original box was something very specific, say having its spin direction oriented towards the star Betelgeuse. This actual information has not been transmitted to Bob, but if he ultimately hears, by normal classical means, that this was indeed the state that Alice's friend had placed in the box, then Bob has the potential to "check up" on this contention. He can orient his Stern–Gerlach apparatus in the direction of Betelgeuse, and be *guaranteed* that the measurement of the atom in his suitcase indeed has that as its spin direction. Of course, the answer "yes" does not tell him that the spin direction was in fact that, but for no other direction can the guarantee of a "yes" answer be given. It seems to me that this is certainly something *objective*, even though the spin state is not "information" in the ordinary state. It is this requirement of objectivity that seems to be requiring us to take the notion of quanglement seriously.

Is it possible to construct a theory of quanglement which gives us an objective picture of the physical world, and in which all these strange requirements fit together? I believe that this may be so. My own viewpoint would be to take the *basic* quanglement links to be *null* (i.e. lightlike) rather than timelike, where a timelike quanglement path would be built up of null zig-zags on a tiny scale. This leads us into the theory of *twistors* [15] which has a character of this nature, and in which the very notion of space–time points takes a secondary role. Twistors are (to a first approximation) closely associated with light rays, and these may be related to such basic quanglement links. (This idea will be explored somewhat further in [16].) Twistor theory is an approach to the combining of quantum theory with general relativity, but the theory still remains a very long way from its goals.

## Acknowledgements

I am grateful to many colleagues for discussions, to NSF for assistance under PHY 93-96246, and to the Leverhulme Foundation for an Emeritus Fellowship.

## References

1. J.S. Bell: 'Quantum mechanics for cosmologists'. In: *Quantum Gravity 2: A Second Oxford Symposium*, ed. by D.W. Sciama, R. Penrose, C.J. Isham (Oxford Univ. Press, Oxford 1981) pp. 611–637

2. R. Penrose: *The Emperor's New Mind: Concerning Computers, Minds, and the Laws of Physics* (Oxford Univ. Press, Oxford 1989)
3. L. Diósi: Phys. Lett. A **120**, 377 (1987)
4. L. Diósi: Phys. Rev. A **40**, 1165 (1989)
5. R. Penrose: 'Gravity and quantum mechanics'. In: *General Relativity and Gravitation 13, Part 1: Plenary Lectures 1992*, Proc. Thirteenth Int. Conf. General Relativity and Gravitation, Cordoba, Argentina, 28 June – 4 July 1992, ed. by R.J. Gleiser, C.N. Kozameh, O.M. Moreschi (Inst. Phys. Publ., Bristol 1993) pp.179–189
6. R. Penrose: Gen. Rel. Grav. **28**, 581 (1996)
7. R. Penrose: *Shadows of the Mind; An approach to the missing science of consciousness* (Oxford Univ. Press, Oxford 1994)
8. P. Pearle: 'Models for reduction'. In: *Quantum Concepts in Space and Time*, ed. by C.J. Isham, R. Penrose (Oxford Univ. Press, Oxford 1985)
9. J. Christian: Phys. Rev. D **56**, 4844 (1997)
10. J. Christian: 'Why the quantum must yield to gravity'. In: *Physics Meets Philosophy at the Planck Scale*, ed. by C. Callender, N. Huggett (Cambridge Univ. Press, Cambridge 2000)
11. R. Penrose: 'Wavefunction collapse as a real gravitational effect'. In: *Mathematical Physics 2000*, ed. by A. Fokas, T.W.B. Kibble, A. Grigouriou, B. Zegarlinski (Imperial College Press, London 2000)
12. W.K. Wootters, W.H. Zurek: Nature **299**, 802 (1982)
13. C.H. Bennett, G. Brassard, C. Crepeau, R.O. Jozsa, A. Peres, W.K. Wootters: Phys. Rev. Lett. **70**, 1895 (1993)
14. R. Penrose: Philos. Trans. R. Soc. (London) A **356**, 1927 (1998)
15. R. Penrose: Chaos, Solitons & Fractals **10**, 581 (1999)
16. R. Penrose: *The Road to Reality: The Mathematics and Physics of the Universe* (Vintage, London 2002) (in press)

# 24 Interferometry with Macromolecules: Quantum Paradigms Tested in the Mesoscopic World

Markus Arndt, Olaf Nairz, Anton Zeilinger

## 24.1 A Prototype Quantum Experiment

When thinking about quantum non-locality a modern physicist would probably think of two-particle entanglement first. And many beautiful experimental demonstrations have been published on quantum systems which violate Bell's inequalities [1].

However non-locality as well as almost all other quantum 'mysteries' can be found in even simpler experiments as we will discuss in the context of single-particle interferometry of large molecules. This is why *Feynman* et. al called the double-slit experiment the heart of quantum mechanics [2]. And this is also the reason why the exploration of interference effects with more massive and complex, mesoscopic systems will also help us elucidate important features of physics along the quantum–classical boundary.

Imagine first a classical soccer player aiming with a ball at a wall with a single opening, as sketched in Fig. 24.1a. If we further assume an infinitely thin wall with a slit which is just wide enough to let the ball pass then we would expect only a single point in the goal to be hit by a few balls. All the other points in the goal will be masked by the wall. If we open the gap in the wall and allow for wall thickness then we would expect different balls to hit the goal at different places, both because the player may kick different balls differently and because there may be scattering at the wall edges. Classically we would therefore expect a certain position distribution.

In contrast to that, what would be the result in a quantum version of the game? There we can use an argument based on the *Heisenberg* uncertainty relation [3]. The passage through the hole in the wall improves the knowledge of the position state of the ball. Given the complementarity between position and momentum we conclude that the momentum uncertainty after the wall must be larger than before. In the plane of the goal this translates into a spread of the possible ball positions. Narrowing the wall opening will therefore narrow the position distribution for classical balls but widen the corresponding quantum distribution. The same description obviously holds also for the case where we open the right door in the wall instead of the left one, as sketched in Fig. 24.1b.

If we now open both doors a classical physicist will simply expect the two probability distributions for two single openings to add up to a big hump.

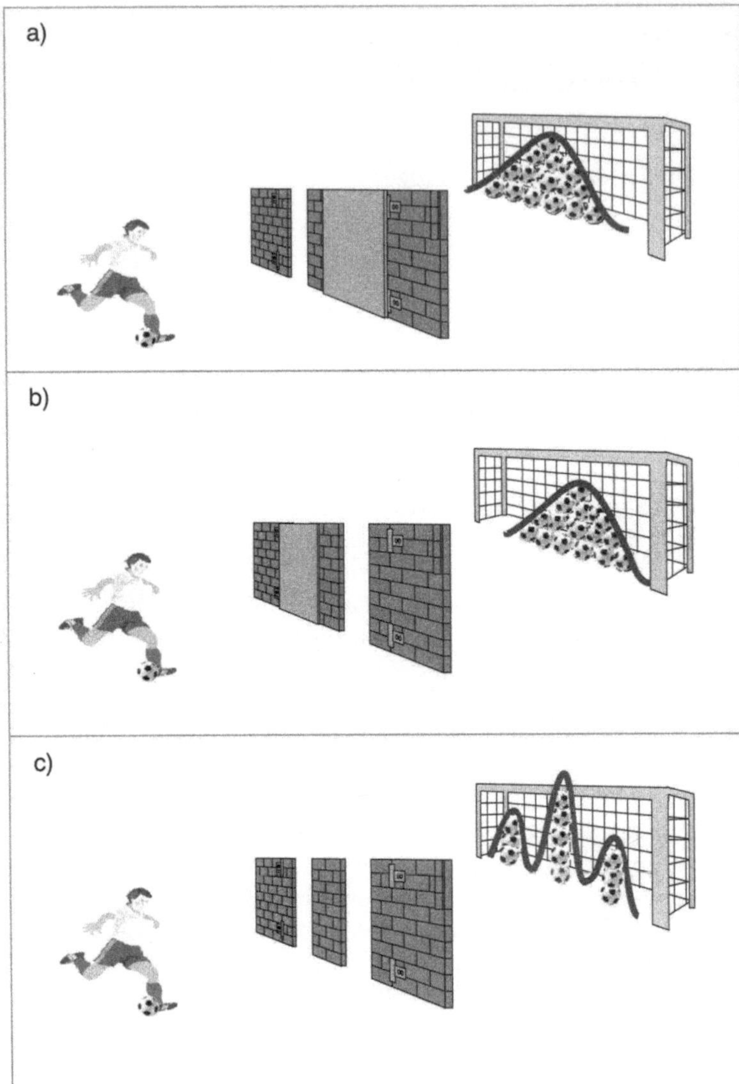


**Fig. 24.1.** Wave particle duality: a quantum paradigm confronted with our everyday experience

She will expect that at any point in the goal the number of balls can only grow when we open the second door. This is the experience made when using real soccer balls in real life.

However, under appropriate quantum circumstances the observer can actually see a pattern which differs dramatically from this expectation. *Opening* the second door *can dramatically reduce* the number of balls hitting the goal in specific places. In particular one can find well-defined places which will be completely avoided by all balls, as sketched in Fig. 24.1c.

A classical model which helps in understanding a part of the phenomenon is the wave picture: If we regard each ball as a wave illuminating the two openings simultaneously then we actually expect to see interference of the wavelets if they take *indistinguishable paths* through the slits in the wall.

While we usually accept this explanation for water waves, we run into trouble when we shoot one ball at a time: How can a commonly *well-localized object*, i.e. a *particle* like a ball, acquire *non-local* information about widely separated wall openings, as we accept the situation to be for a *wave*? In addition, if the ball is a non-local object with *position uncertainty* at the wall openings, how can it decide upon its final well-localized position in the goal? What actually is the famous *collapse* of the wave function? All experiments so far are consistent with the view that each particle chooses its final location on the screen *randomly*, although the total ensemble will finally generate an interference pattern which is completely *deterministic*. How can this be understood?

In summary, we find at least the following quantum notions in the context of double-slit interferometry: complementarity, uncertainty, non-locality, superposition of classically exclusive states, wave–particle duality, objective randomness and the collapse of the wave function. All of these 'mysteries' may be subsumed by the notion that quantum mechanics describes potentiality instead of reality. It is important to note that potentiality is often the only objective reality.

Finally, another key ingredient of quantum physics has to be mentioned: entanglement, the inseparable correlation of at least two quantum systems – which can also be part of the same physical object, as in the entanglement of internal and external degrees of freedom. This property will become relevant in the context of decoherence, as discussed further below. In the double-slit experiment the degree of entanglement between system and environment is actually important for the possibility of observing either the particle or the wave nature.

All these quantum effects have been known for more than 70 years, but even though we have learnt to accept their weirdness, there remain many open questions. Two of them are at the origin of the present work: Firstly, how can we understand the transition from the quantum to the classical world? In our classical world we see neither objective randomness nor spontaneous collapses nor wave effects of macroscopic objects nor quantum correlations between space-like-separated objects. Under which circumstances and at which scale do these properties disappear? Secondly, can we arrange experiments such that we observe and possibly exploit quantum phenomena on the mesoscopic or even macroscopic scale?

## 24.2 Interference of Fullerenes

While the discussion of classical soccer balls in the introduction appears somewhat academic, since we cannot see quantum effects for them, we can construct an experiment in which we may gradually approach the transition from the quantum to the classical case. Our experiment is based on the world's smallest soccer balls, composed of 60 carbon atoms.

Most quantum interference experiments so far have dealt with microscopic objects, such as electrons [4], neutrons [5, 6] and atoms [7] or, more recently, with coherent ensembles of ultra-cold atoms [8]. Complementary to that, our studies investigate strongly bound hot particles which form a mesoscopic single quantum object. Our experiments are in a parameter regime unexplored hitherto and may allow the investigation of the quantum–classical transition in complex systems.

Fullerenes, in particular $C_{60}$ and $C_{70}$, turned out to be very good candidates for studies on the route towards the mesoscopic world. They represent a qualitative step forward with respect to previous experiments, since these carbon cages resemble in many aspects rather classical lumps of carbon than simple quantum systems. They have more than 100 different vibrational modes, leading to broad optical lines, highly excited rotational states and both non-radiative and radiative transitions between most of them. Because of the rapid thermalization it is possible to assign a temperature to the internal energy distribution. Moreover, as for macro-objects, one experimentally finds both line radiation, from vibrational transitions, and a continuous blackbody spectrum as well as thermal electron emission [10, 11]. It is this internal complexity and energy which is new and which may give rise to new phenomena. Also, from a practical point of view, fullerenes are the macromolecules of choice, since they are available in a sufficient amount, can be sublimated to produce a thermal beam and have the required symmetry and stability to survive their time in the vacuum machine without significant fragmentation.

The setup of the fullerene interference experiments is as sketched in Fig. 24.2 and has been described in more detail in [9, 12]. Fullerenes are sublimated and collimated. Depending on the performed experiment, they are either diffracted at the second slit or at a nanofabricated grating immediately behind this slit. The Fraunhofer far-field interference pattern is subsequently scanned with well-focused intense visible laser light, which thermally ionizes the fullerenes. In the following figures the ion count rate is plotted as a function of the detection-laser position.

Since the experiment is based on a thermal source, i.e. a classical mixture of external and internal molecule states, one may wonder why interference is possible at all. In the elementary introduction to double-slit interferometry we emphasized that the indistinguishability of all possible paths is an essential requirement for interference. It is important to note that each molecule is very likely to be in a state completely different to that of any other particle in the

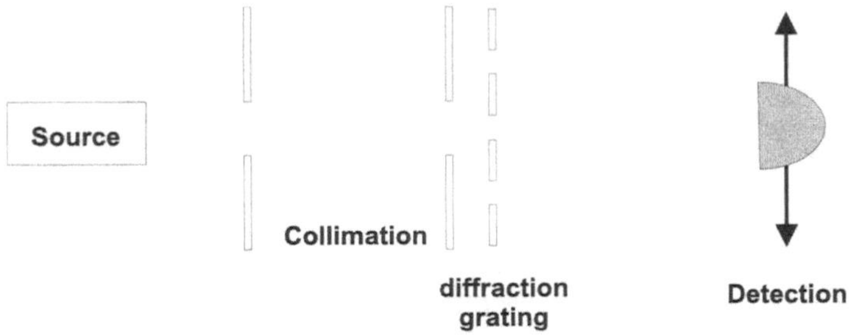


**Fig. 24.2.** Simplified setup of the fullerene interference experiment [9]. Fullerenes are sublimated and collimated. They are either diffracted at the second slit or at a nanofabricated grating immediately behind this slit. The Fraunhofer far-field interference pattern is subsequently scanned with a scanning laser. The laser thermally ionizes the fullerenes. In the following figures this ion count rate is plotted as a function of the detection-laser position. For the single-slit interference of Fig. 24.3 the grating is taken out and the width of slit 2 can be varied between 70 and 20 000 nm. For the diffraction-grating experiments the second slit is fixed at 5 μm and the grating is shifted into the beam

thermal ensemble. We can therefore expect to observe only the interference of each molecule with itself. In order to obtain good contrast one has therefore to restrict the initial momentum and position distribution. This then provides the required spectral and spatial coherence of the interfering waves.

A wide *velocity distribution* generally results in a poor *spectral coherence.* Molecules at different speeds have different diffraction angles. Since their interference patterns add incoherently on the screen, they tend to wash out.

The *spatial coherence* is the second important aspect in the preparation of the beam. Within the source the fullerenes can be regarded as very well localized, in general, to a width corresponding to their thermal de Broglie wavelength [14]. This is of the order of a few picometers and thus more than 100 times smaller than the diameter of $C_{60}$. This is why the notion of a particle is well justified while the particle is in the oven. However, as the molecules fly freely towards the grating, their wave function is spreading out. The transverse coherence width at the grating can then be regarded as being roughly the width of the zeroth-order peak due to diffraction at the first collimation slit [15]. In the grating interference experiments this transverse coherence width amounts to about 1 μm.

That *single-slit diffraction* actually takes place even for the hot and heavy molecule $C_{70}$ is demonstrated in Fig. 24.3. For this figure the width of the second collimation slit was gradually closed from 18.6 μm in Fig. 24.3a via 1.4 μm in Fig. 24.3b down to 70 nm in Fig. 24.3c, and the far field molecular beam width was recorded. While the beam width is actually composed of several contributions – including the detector resolution and the classical collimation – it is evident that for small slit widths the Heisenberg uncertainty relation between position and momentum must become relevant. This

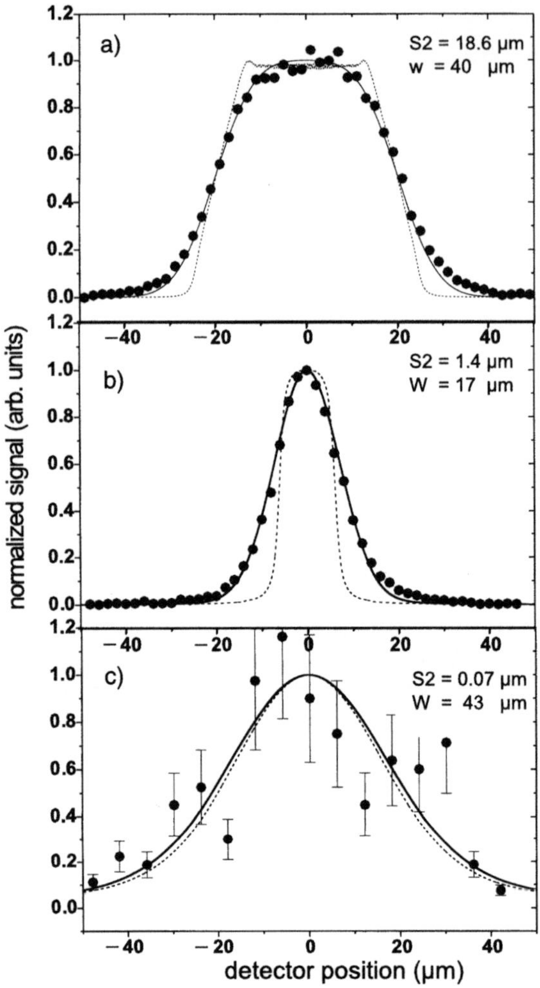


**Fig. 24.3.** Single-slit diffraction for $C_{70}$. (**a–c**) A decreasing width of the second slit. The far-field molecular beam pattern reflects the broadening of the momentum distribution in agreement with Heisenberg's uncertainty relation. See also [13]

qualitative evolution can be seen in Fig. 24.3a–c where the observed molecular beam changes its width (FWHM) from $\Delta X_{\mathrm{mol}} = 40\,\mu\mathrm{m}$ via $\Delta X_{\mathrm{mol}} = 17\,\mu\mathrm{m}$ to $W = 43\,\mu\mathrm{m}$. It can clearly be seen that, when the collimation is improved, the molecular beam first becomes narrower and then spreads out again. In [13] we have shown that this effect is not only in qualitative but also in good quantitative agreement with expectations from a wave calculation for the Heisenberg uncertainty relation.

Based on this experiment we know that diffraction at a 70-nm-wide gap leads to an angular spread of the beam of more than $\sim 33\,\mu\mathrm{rad}$. We now

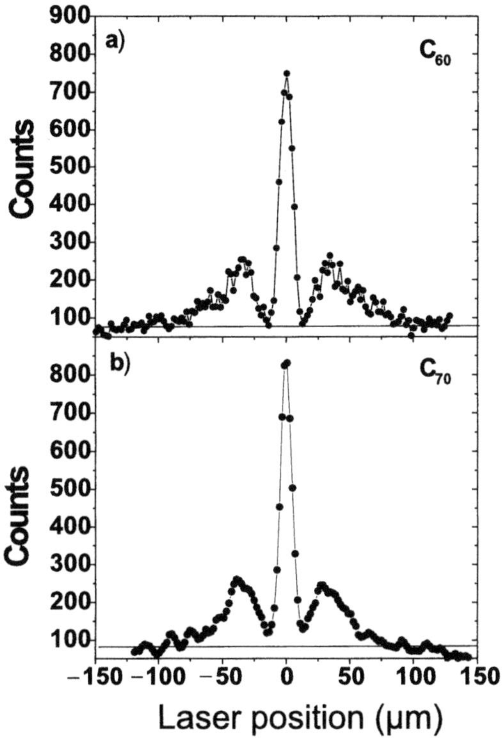


**Fig. 24.4.** High-contrast interference for $C_{60}$ and $C_{70}$. For this picture the collimation was set to its optimum value, limited by diffraction at the second collimation slit [18]. Minima are a signature of destructive interference. In the curves presented here the minimum goes to the background level, indicating almost 100% interference contrast, in spite of the high internal excitation and a non-vanishing probability for internal state changes during the passage from the grating to the detector

ask if this spread is sufficiently coherent for high-contrast interference fringes after diffraction at a grating. In order to test this we look for the key feature of grating interference, namely the complete extinction of the probability amplitude at certain positions on the screen due to destructive interference. The setup is as above, but complemented by a nanofabricated SiN structure close behind the second collimation slit. Many slits are each etched $\sim$ 50 nm wide and into the 200-nm-thick SiN membrane[1] to form a grating with a period of 100 nm. About 20–30 such slits are illuminated by the molecular beam. The coherence width covers about 10 such slits. Figure 24.4a,b shows interference patterns for thermal beams of $C_{60}$ and $C_{70}$ respectively. In both cases one can clearly recognize the first-order interference maxima and minima. The absence of higher orders of diffraction is due to the broad spectral width of

[1] Gratings were provided by H. Smith and T. Savas, MIT, Cambridge, Mass. The manufacturing technology is described in [16].

the molecule source as can also be found in a numerical model.[2] In spite of this limitation, the interference minimum approaches the background count rate. This high interference contrast is a clear proof of nearly perfect coherence.

## 24.3 Decoherence

The de Broglie wavelength is in principle a well-defined concept for any object of arbitrary mass, size and complexity. One could therefore be tempted to say that the above experiments do not show anything surprising. However, in spite of being well defined, quantum interference of a chair or table is never observed.

There may be several reasons for this. On the one hand, a widespread and valid argument is the smallness of the effect. The mass of a chair is so big that at a speed of 1 m/s, i.e. the velocity of a typical pedestrian, the de Broglie wavelength of the chair is close to the Planck length of $10^{-35}$ m. It is clear that a superposition pattern of such wavelengths will be experimentally inaccessible. However, often quantum effects disappear well before any technological resolution limit. While several models have been put forward to complement or modify quantum physics [19, 20], we shall focus here on decoherence theory as developed in much more detail in [14, 21, 22]. This model describes the loss of correlations within a quantum system as being the result of the interaction of the system with the environment.

Any interaction of the quantum system with the outside world – be it the interaction with a conscious observer using a measuring device (a meter) or only an accidental coupling to the environment – has the effect of entangling the system with this environment. The *von Neumann* measurement of the quantum state $|\Phi_n\rangle$ turns the initial meter pointer $|\Phi_0\rangle$ into a position correlated with the quantum state $|\Phi_n\rangle$:

$$|\phi_n\rangle\,|\Phi_0\rangle \overset{\text{measure}}{\longrightarrow} |\phi_n\rangle\,|\Phi_n\rangle \ . \tag{24.1}$$

A reliable meter is necessarily based on this correlating interaction.[3] The situation becomes puzzling when the quantum system is in a superposition state. By linearity of quantum mechanics, a macroscopic superposition of meter states (pointer positions) should then also exist,

$$\left(\sum_n c_n\,|n\rangle\right)|\Phi_0\rangle \overset{\text{measure}}{\longrightarrow} \sum_n c_n\,|n\rangle\,|\Phi_n\rangle \ . \tag{24.2}$$

However this phenomenon has never been observed. As has been pointed out in [21, 22], the conflict with reality can be reduced if one considers that

[2] Experiments with improved longitudinal coherence length show higher-order interference fringes, as shown in [17].

[3] The accidental interaction with the environment does not in general act as an ideal meter. The correlation can be much weaker than indicated in (24.1).

the *entanglement* (non-separability) of the system and the meter may de facto be unobservable due to the large number of degrees of freedom of the meter. Therefore, if the experiment measures the system alone and ignores the environment, the correct answer to any experimental question will be based on a reduced density matrix which has been *traced over the meter states.* If all pointer states are sufficiently distinct, i.e. $\langle \Phi_n | \Phi_m \rangle \simeq 0$, it can be shown that the quantum system is forced into a classical mixed state, lacking all coherence and ability to interfere.

This formalism may be interpreted in a different way, too. If the corresponding environmental states are completely distinct from each other, the information about the quantum systems can be known with absolute certainty after its coupling to the environment. This rephrases, confirms and mathematically justifies Bohr's rule of complementarity that interference must disappear if we obtain definite which-path information [23].

How would this apply to our fullerene experiment? Suppose we had a UV laser with a wavelength of below $\sim 100\,\text{nm}$. Then scattering of a laser photon by a fullerene would give us sufficient information about which way the molecule took through the diffraction grating. Therefore the environmental states would be orthogonal and the interference pattern should disappear. As a matter of fact, one can formulate even a third rule for the limit of interference: The coupling to the environment will in most cases go along with a random momentum transfer. In the described example this random recoil and corresponding phase shift of the molecular wave function is due to the scattering of the UV photons. The different phase shifts acquired by the different molecules will finally wash out the interference pattern. Purist decoherence theorists would claim that the three processes – entanglement with the environment, which-path information and phase shifts due to momentum recoil – are completely different and should not be mixed. One could be tempted to argue that a very massive object, such as a large building, could be entangled with the town and localized by a radio-wave photon without getting any measurable recoil. However, if we follow Bohr's rule that we should not talk about things that we cannot measure, then we can safely claim that all three arguments are faces of the same physics in pure de Broglie interferometers[4]; for example, in order to experimentally prove that there is coherence in the center of mass wave function of the building, one would have to repeat an interference experiment with it. Since the de Broglie wavelength and the corresponding diffraction angles after the grating would be microscopically tiny, the detection screen would have to be placed at an astronomical distance in order to resolve the diffraction orders. It turns out formally that a photon that would be sufficient to locate the building in any

[4] The information storage and entanglement can be very different in interferometers which also exploit internal states of the quantum system. Examples of this phenomenon, emphasizing the aspect of complementarity as opposed to simple phase diffusion, are discussed in [24, 25].

given grating slit – a condition required for producing orthogonal states in the environment – would also give the random recoil required to wash out the interference pattern.

Localization and entanglement via a single strong scattering event, as discussed so far, is not the only way of producing decoherence. One can also gradually gain which-path information by many weak coupling processes. This is actually the more interesting case both from a theoretical point of view [14] and with respect to thermal decoherence. For experiments dealing with spatial superposition states, one can show that the off-diagonal matrix elements, representing the coherences, decay exponentially in time, in the squared separation of the position wave packets and in the decoherence parameter:

$$\rho(x,x',t) = \rho(x,x',0)\exp\left[-\Lambda t(x-x')^2\right] . \tag{24.3}$$

Here the decoherence parameter is given by $\Lambda = \Gamma \cdot k^2$ [22] and is a measure for the interaction rate, $\Gamma$, and the square of the interaction strength, as defined by the wave number, $k$, of the particles which are scattered or emitted into the environment. Many weak scattering processes can yield sufficient entanglement, which-path information and momentum diffusion to destroy the interference pattern.

Experiments on decoherence in atom interferometers have already been performed by different groups [26–28] using the interaction of the atoms with external radiation. In contrast to that, experiments with the fullerene molecules may provide a way to look at naturally occurring *thermal decoherence* processes. Which-path information and entanglement could be produced by the fullerenes in scattering or emission processes. But also correlations of interferometric phase shifts with external fields may become important. Several of these interactions shall be discussed in the following.

### 24.3.1 Vibrational Transitions

At $T \sim 900\,\mathrm{K}$, as in our experiment, each $C_{60}$ molecule has on average a total vibrational energy of $E_v \sim 5\,\mathrm{eV}$ [29] stored in 174 vibrational modes at 46 different energies. *Non-radiative transitions* between many of them are in principle possible. However they are *per definitionem* decoupled from the environment. Any possible transition takes place in a superposition of all position states. One could even say that – since both the energy and entropy remain conserved – all such 'transitions' are actually part of a unitary and coherence-preserving transformation. The center of mass motion, the environment and the available position information are unaffected. The interference pattern must therefore remain unperturbed.

*Radiative transitions* are possible between electronically excited and ground states as well as among vibrational modes. In the case of $C_{60}$ the latter are limited to four threefold degenerate vibrational modes. They emit at $\lambda_{vib} = 7.11$, 8.57, 17.54 and 18.98 μm [30]. A single photon at these wavelengths is certainly insufficient – by about a factor of one hundred in energy – to yield unambiguous which-path information in our grating experiment.

While the wavelengths are precisely known, the values for the line strengths have been a matter of debate in the literature. The most recent values for the Einstein coefficients are of the order of $A_k \sim 2 - 30\,/\mathrm{s}$ [31]. Other authors estimate that, even if one includes combination lines and overtones for $C_{60}$ heated to 3500 K, the total emitted vibrational energy will be below 16 eV/s, which corresponds to one or two IR photons during the 10 ms time of flight through our vacuum can [32].

The situation is slightly different for the less symmetric fullerene, $C_{70}$. It shows 31 different IR emission lines – out of 204 vibrational modes. Again the emission wavelengths are situated between 10 and 20 μm and thus much too long for single-photon decoherence. Experiments in our lab have shown [33] that even at the increased number of allowed transitions the number of photons is still insufficient to induce a sizable loss of contrast.

It is interesting to ask how internal properties will affect the interference of larger objects. It may be assumed that with increasing complexity also the number of allowed transitions will increase and therefore the ability to diffuse quantum coherence into the environment. However, one should also consider that a minimum energy of $\sim 0.1$ eV is generally required for the excitation of a vibrational level. Substantial cooling of the internal degrees of freedom seems to be within the reach of current technologies even for macromolecules using jet expansion or buffer gas techniques [34]. Exploitation of these methods should render coherence experiments feasible.

### 24.3.2 Emission of Blackbody Radiation

Laser-heated $C_{60}$ has been observed [10, 35] to emit blackbody radiation with a continuous spectrum in functional agreement with Planck's radiation law corrected by the factor $8\pi R/\lambda$. This factor takes into account that the radius, $R$, of a fullerene is much smaller than the emitted wavelength, $\lambda$ [36]. The 'black' body therefore radiates rather in bright grey. A wavelength-integrated emissivity of $\varepsilon \sim (4.5 \pm 2.0) \times 10^{-5}$ has been published in [29]. From this value we can derive an upper bound on the energy that can be lost while the buckyball is flying from the grating to the detector. For a typical value of $T \sim 900$ K the average energy emitted during the time of flight can be estimated to be $E_{bb} \sim 0.1$ eV. This corresponds to the energy of a single photon at $\lambda \sim 10$ μm. A natural question arises as to whether it is possible to increase the temperature to see a gradual increase in radiation and decoherence. Using our detection light, derived from an argon ion laser we can easily increase the molecular energy up to 40 eV. The maximal energy that can be stored in the molecule is limited by the onset of fragmentation or ionization on the 10 – 100 μs time scale for temperatures higher than 3000 K (corresponding to $E_{int} = 37$ eV [10]). At a thus threefold-increased temperature we would expect an 81-fold increase in black body-emission intensity. This should already give rise to a measurable phase diffusion. Since many macromolecular parameters are very difficult to access experimentally and because of the sizeable uncertainty for many values published in the literature, we have implemented laser heating for the fullerenes and searched for a broadening of the molecular beam or a decrease in the contrast of the interference pattern. In all preliminary experiments performed so far the contrast in the presence of laser heating was always comparable to the case when we used a thermal beam source alone. We therefore conclude that blackbody emission is still too weak in our present experiments in spite of the high internal temperatures.

However, we expect that for much larger particles than the fullerenes this will change significantly. The emitted intensity scales with the third power of the particle size, $R^3$, both because of the increasing surface ($R^2$) and because of the antenna effect ($R/\lambda$) mentioned above. For particles in the '30 nm' region, corresponding to the size of a small virus or a medium nanocrystal, it will therefore be necessary to lower the internal temperature to below $T \sim 30\,\mathrm{K}$. The generation of sufficiently intense beams of particles having this size and temperature is certainly non-trivial. But we believe that it is still within the reach of the emerging technologies.

### 24.3.3 Absorption of Blackbody Radiation

In our fullerene experiments absorption of blackbody radiation has a smaller influence on decoherence than any emission, since the environment is at a much lower temperature than the fullerene molecule. This changes when we consider larger objects. A billiard ball will certainly absorb enough blackbody photons to decohere more rapidly than we can perform a room temperature interference experiment with it. However, up to at least the size of a small virus we expect that the detrimental effect of blackbody absorption can be suppressed under moderate cryogenic conditions.

### 24.3.4 Rayleigh Scattering of Thermal Radiation

Here we treat the case of a particle of dielectric constant $\epsilon$ and radius $a$ which is bombarded by thermal radiation at temperature $T$. For this particular situation *Joos* et al. have calculated the decoherence parameter of (24.3) due to Rayleigh scattering to be [22]:

$$\Lambda[\mathrm{m}^{-2}\mathrm{s}^{-1}] \sim 10^{36} \cdot a^6 \cdot T^9 \cdot \left(\frac{\varepsilon - 1}{\varepsilon + 1}\right)^2 . \tag{24.4}$$

To obtain a rough estimate for the importance of this effect, let us assume that the typical time of flight through our interferometer is 10 ms for the currently used buckyballs and 0.9 s for a virus beam of mass $10^7$ amu at a velocity of 4.4 m/s. The latter speed is chosen to yield a vertical parabola of 1 m height. It turns out that Rayleigh scattering of 300 K blackbody radiation limits the maximal coherent wave packet separation of the buckyball to 1 m! The dielectric constants of most macromolecules are unknown. For an order of magnitude estimate we therefore use the bulk dielectric constant of water, $\epsilon = 80$, and the radius of a small virus, say 15 nm. For such a water cluster the largest possible coherent separation would already be limited to $\Delta x_{\mathrm{coh}} < 2.3\,\mu\mathrm{m}$, which is small but still an order of magnitude larger than the beam splitting in a reasonable near-field interferometer. For objects with a higher dielectric constant or a larger diameter, the dramatic size dependence in (24.4) can in principle be counteracted by a significant reduction of the temperature of the vacuum machine.

### 24.3.5 Fragmentation and Ionization

The emission of either $C_2$ subunits or electrons, is a possible process in fullerene interferometry at high temperatures. However, it only becomes relevant at temperatures around 3000 K. Moreover both would rather contribute to a loss of molecules than to decoherence, since the imparted recoil would be sufficient to kick the molecules beyond the detection region. Most biological macromolecules are much more thermo-labile than $C_{60}$. Even at moderate temperatures configuration changes and fragmentation are likely. But anyway a realistic interference experiment would have to be performed at low molecule temperatures for the reasons given above.

### 24.3.6 Influence of Collisions with the Residual Gas

Collisions with gas molecules are in principle very effective in determining positions and destroying coherence. This is due to the very small de Broglie wavelength of the fast nitrogen or oxygen molecules. Several authors [22] therefore claim that even at the best possible laboratory vacuum, decoherence must be very fast. We argue in this section that the perturbative approach to decoherence theory does actually not apply to our experiment, although the computed decoherence parameters are correct. As described earlier, the decoherence parameter, $\Lambda$, of (24.3) is the product of a scattering rate, $\Gamma$, and the square of the wave number, $k^2$, of the incident molecule. While the first is generally small the momentum transfer is generally much too strong to allow the application of perturbation theory.

As a matter of fact the influence of collisions on the observability of the interference fringes depends on the setup of the experiment. Since small angle collisions have a much higher scattering cross-section than wide-angle collisions, far-field diffraction patterns with widely separated beams are generally more robust against collisional decoherence than near-field interferometers. Cross-sections for scattering processes between large molecules and background gas are not very well known. So we can only extrapolate from our experiments with fullerenes, where we still see a measurable influence of the background pressure on the collimation of the beam down to $p \sim 10^{-5}$ Pa. Deflections of the order of more than 30 nrad would already be detrimental in near-field interferometers for molecules or clusters around $m \sim 10^7$ amu. This shows the need for sophisticated vacuum techniques. Luckily, it has been demonstrated in positron trapping experiments [37] that it is possible to obtain a base pressure down to $10^{-15}$ Pa. This leaves enough room for the hope that collisions can be sufficiently suppressed – although at the expense of a substantial cryogenic effort.

### 24.3.7 Quasi-Static Interactions

In the preceding sections we have discussed the interaction with the environment through radiation. But also quasi-static interactions could betray the particle's position. Constant static field gradients shift the interference pattern by a constant amount. The magnitude of the field is thus encoded by and entangled with the position of the individual interference pattern on the screen. Decoherence would be induced by slowly and randomly varying field gradients which would shift the molecular phases in an uncontrolled way and therefore wash out the ensemble-averaged interference structure. The loss of information in this case is due to the inability of the experimentalist to correlate the instantaneous field values with the observed screen patterns. If he did this, the interference pattern could actually be completely restored.

This shows a general trend in the quantum physics of large systems: The number of degrees of freedom generally becomes so huge and the coupling to the environment so efficient that even a very enthusiastic and hard-working physicist will soon lose the details of all the available information. In the picture of decoherence, quantum mechanics is always valid but less and less visible, since the phase information somehow diffuses into the environment.

External fields could interact, for example, with the magnetic susceptibility or dipole moment, the electric polarizability or dipole moment or also with the mass of the molecule.

### 24.3.8 Magnetic Interactions

Isotopically pure $^{12}C_{60}$ in the electronic ground state is totally diamagnetic and has a small magnetic susceptibility. It has neither nuclear spin nor orbital angular momentum nor electronic spin. The interaction with magnetic fields in the lab can therefore be completely neglected. However, if we take into account the natural abundance of carbon (99% $^{12}$C, 1% $^{13}$C) we find [30] that in almost 50% of all fullerenes there is at least one $^{13}$C which possesses one extra nuclear magneton. While now there is at least a handle for the interaction between the molecule and an external magnetic field, this handle and the separation of the arms of the interferometer is still too small to permit a significant phase shift in our experiment, even in the presence of very large magnetic field gradients. For larger objects with, for instance, a ferro-magnetic core this effect could become decisive.

### 24.3.9 Electric Interactions

The fullerenes have no permanent electric dipole moment but a sizable static polarizability of $\alpha_{\text{stat}} \sim 4\pi\epsilon_0 \times 80\,\text{Å}^3$. Again, a significant phase difference between two paths in our experiment would require a setup especially designed for this purpose.[5] Ordinary laboratory electric fields are too small by far to induce decoherence in buckyballs. Even for objects which possess a permanent electric dipole moment, such as large biomolecules[6], we estimate that it should still be possible to control external stray fields well enough to avoid electrically induced decoherence in an interferometer.

### 24.3.10 Inertial Forces

Molecular de Broglie interferometers are very sensitive to inertial forces. The *gravitational phase shift* between two arms of an interferometer is due to the different energies of the molecule at different heights in the earth potential, while the rotation of an interferometer results in the quantum version of the Coriolis force, which is the *Sagnac effect.*[7] In the diffraction experiment described above, both the gravitational and the rotational phase difference from the source to the screen are proportional to the mass of the quantum object. In a near-field interferometer – which is the more realistic device for coherence experiments with very large quantum objects – the sensitivity to inertial forces even increases quadratically with the mass. It turns out that for an earth-bound stationary experiment the gravitational phase shift is generally much more important than the Sagnac effect.

In principle, both effects can be zeroed for an appropriate choice of the orientation of the interferometer: The gravitational shift disappears if the surface vector of the interferometer is in line with the direction of the gravitational acceleration. Similarly, the rotational shift will disappear in an interferometer where the surface vector is orthogonal to the rotation axis. In three-dimensional space both directions can be adjusted independently. The earth's rotation and its gravitational field are sufficiently constant to allow the observation of high-contrast fringes even if the interferometer is somewhat misaligned. A constant phase shift would only displace the interference pattern. However, it is important to note that the phases introduced by the intertial forces are dispersive: different velocity classes will acquire different phase shifts, and therefore even a small misalignment may have a detrimental

[5] One way to realize this is to implement an optical grating, i.e. a standing-wavelight field, which – due to its small wavelength – can generate important electric field gradients. This setup has actually be used to demonstrate interference at an optical grating for the first time for large molecules [38].

[6] For hemoglobin the dipole moment amounts to 300 Debye (1 Debye = $3.33 \times 10^{-30}$ Cm), roughly 200 times the value of water.

[7] Theses two effects have been discussed by various authors. A good introduction can be found in [6].

effect. While in the case of our fullerenes an angle control to within a few milliradians is sufficient to maintain a good fringe visibility, the requirements on angular alignment and stability for an object the size of a virus is of the order of nanoradians. This is certainly at the extreme edge of currently available technologies.

### 24.3.11 Which-Way Information in Internal Clocks

As an alternative to entanglement with the environment, we may ask whether which-way information could be encoded in an internal clock.[8] Suppose a molecule actually has a marker sitting on its surface and is prepared in a well-defined rotational state. One would then say that the marker will rotate with a given angular frequency and the number of turns as well as its phase can be taken as a measure of time. If now a molecule in a superposition of spatial wave packets passes two neighboring slits and proceeds to the left-hand first-order interference maximum, then – classically – one would expect the internal clock to show slightly different times. The wave passing the left-hand slit will arrive somewhat earlier than that taking the right-hand track. A constant time slip in the center of mass coordinates would simply shift the interference pattern. However, in the case described here the wave function is a product of the translational center of mass motion and the rotational evolution: $|\psi\rangle = |\psi_{\mathrm{cm}}\rangle \otimes |\psi_{\mathrm{rot}}\rangle$. The time scales for their evolution can be rather different. It may happen that the phase difference, $\Delta\phi_{\mathrm{cm}} = k_{\mathrm{dB}}\Delta L$, between the two interfering amplitudes of the center of mass motion is set to constructive interference, while the phase difference in the internal part, $\Delta\phi_{\mathrm{rot}} = \omega_{\mathrm{rot}}\tau$, is shifted and destructive. Actually the situation is even a bit more complicated since the molecular beam is in a thermal mixture of highly excited rotational states with an average angular momentum of $J \sim 150$, or an average angular frequency of about $\omega \sim 6 \times 10^{11}\,\mathrm{s}^{-1}$. But the question then is – even if we are neither able to prepare nor to read a well-defined initial or final rotational state: Which phase difference will be accumulated for the average rotational frequency due to the time difference related to the path-length difference, $\Delta L$, towards the first-order interference maximum? If we take $\Delta L = \lambda_{\mathrm{dB}} = 2\,\mathrm{pm}$ and $v = 200\,\mathrm{m/s}$ the time difference amounts to only $\tau \sim 10^{-14}\,\mathrm{s}$, yielding a phase shift of only $\Delta\phi_{\mathrm{rot}} \sim 0.006\,\mathrm{rad}$. Such a tiny phase difference is indeed very far from producing an orthogonal state. The internal clock is therefore unable to induce decoherence.

The relevance of internal clocks becomes even weaker for larger objects, since particles of larger mass and size possess a greater moment of inertia. Already at the same temperature they rotate at a lower frequency than their smaller counterparts. Interferometers working with even a reduced molecular temperature should therefore not be affected by internal clocks.

[8] The question of internal clocks was brought to our attention by Chris Westbrook, IOTA, Orsay.

## 24.4 Conclusion

The sections above show that internal and external degrees of freedom are essentially decoupled and explain the persistence of interference in our fullerene experiments so far. A variety of novel decoherence experiments will be possible in a future extension of the experiment towards a large-area interferometer. A three-grating Talbot–Lau interferometer is being set up and will have a molecular beam separation of up to 1 μm. In this case the which-path information through a few thermal photons may become relevant.

Based on all the preceding arguments one can summarize that there is still a long way to go if one ever wants to find a fundamental limit for earth-based de Broglie interference experiments. But in order to actually see de Broglie interference with molecules in the range of $m \sim 1000 - 100\,000$ amu, the development of many novel and difficult techniques will be needed: The generation of intense, cold and neutral molecular beams is a big challenge, as is the detection with high spatial resolution of single or few macromolecules. Also, various coherent manipulation methods still have to be tested for such heavy particles.

However, the severest constraint in physics may actually be the limited power of our human imagination. Stretching the experimental limits of quantum mechanics is therefore always at the same time an effort to shift the limits of common belief. Thus, any such experiment is also a tribute to John Bell who – through his famous inequalities – gave us a criterion to test our classical prejudices against the reality of nature.

## Acknowledgements

This work has been supported by the European TMR network, contract no. ERBFMRXCT960002, and by the Austrian Science Foundation (FWF), within the project F1505. O.N. acknowledges a grant by the Austrian Academy of Sciences. We acknowledge help in the setup of the experiment by Julia Petschinka.

## References

1. For an overview and references, see the many other articles in this book
2. R.P. Feynman, R.B. Leighton, M.L. Sands: *The Feynman Lectures on physics, Vol. III: Quantum Mechanics* (Addison-Wesley, Reading, MA 1965)
3. W. Heisenberg: Z. Phys. **43**, 172 (1927)
4. C.J. Davisson, L.H. Germer: Nature **119**, 558 (1927);
   G. Möllenstedt, C. Jönsson: Z. Phys. **155**, 472 (1959)
5. H. v. Halban Jr, P. Preiswerk: C. R. Acad. Sci. Paris **203**, 73 (1936)
6. H. Rauch, A. Werner: *Neutron interferometry: lessons in experimental quantum mechanics* (Oxford Univ. Press, London 2000)

7. I. Estermann, O. Stern: Z. Phys. **61**, 95 (1930)
8. M.H. Anderson, J.R. Ensher, M.R. Matthews, C.E. Wieman, E.A. Cornell: Science **269**, 198 (1995)
9. M. Arndt, O. Nairz, J. Voss-Andreae, C. Keller, G. van der Zouw, A. Zeilinger: Nature **401**, 680 (1999)
10. R. Mitzner, E.E.B. Campbell: J. Chem. Phys. **103**, 2445 (1995)
11. K. Hansen, E.E.B. Campbell: Phys. Rev. E **58**, 5477 (1998)
12. O. Nairz, M. Arndt, A. Zeilinger: J. Mod. Opt. **47**, 2811 (2000)
13. O. Nairz, M. Arndt, A. Zeilinger: Phys. Rev. A 032109 (2002)
14. E. Joos: In: *Decoherence and the Appearance of the Classical World in Quantum Theory*, ed. by D. Giulini et al. (Springer, Berlin, Heidelberg 1996)
15. M. Born, E. Wolf: *Principles of Optics*, 6th edn. (Pergamon Press, Oxford 1980) p. 510 & 514
16. T.A. Savas et al.: J. Vac. Sci. Technol. B **13**, 2732 (1995)
17. O. Nairz, M. Arndt, A. Zeilinger: Am. J. Phys. (in preparation)
18. M. Arndt, O. Nairz, J. Petschinka, A. Zeilinger: C. R. Acad. Sci. Paris **2**, série IV, 581 (2001)
19. R. Penrose: Gen. Rel. Grav. **28**, 581 (1996)
20. G.C. Ghirardi, A. Rimini, T. Weber: Nuovo Cimento **27**, 293 (1980)
21. W.H. Zurek: Phys. Today October, 36 (1991); W.H. Zurek: Philos. Trans. R. Soc. **356**, 1793 (1998)
22. E. Joos, H.D. Zeh: Z. Phys. D. **59**, 223 (1985)
23. N. Bohr (1927): In: *Albert Einstein, Philosopher-Scientist*, ed. by P. Schilpp (Tudor, New York 1949)
24. M.O. Scully, B.G. Englert, H. Walther: Nature **351**, 111 (1991)
25. S. Dürr, T. Nonn, G. Rempe: Nature **395**, 33 (1998)
26. J.F. Clauser, S. Li: Phys. Rev. A **50**, 2430 (1994)
27. T. Pfau, S. Spälter, C. Kurtsiefer, C.R. Ekstrom, J. Mlynek: Phys. Rev. Lett. **73**, 1223 (1994)
28. M.S. Chapman, T.D. Hammond, A. Lenef, J. Schmiedmayer, R.A. Rubenstein, E. Smith, D.E. Pritchard: Phys. Rev. Lett. **75**, 3783 (1995)
29. E. Kolodney, A. Budrevich, B. Tsipinyuk: Phys. Rev. Lett. **74**, 510 (1995)
30. M.S. Dresselhaus, G. Dresselhaus, P.C. Eklund: *Science of Fullerenes and Carbon Nanotubes* (Academic, New York 1996)
31. J.U. Andersen, E. Bonderup: Eur. Phys. J. D **11**, 435 (2000)
32. W.A. Chupka, C.E. Klots: Int. J. Mass. Spectrom. Ion. Proc. **167/168**, 595 (1997)
33. O. Nairz, M. Arndt, A. Zeilinger: unpublished
34. G. Scoles (ed.): *Atomic and Molecular Beam Methods* (Oxford Univ. Press, Oxford 1988)
35. P. Heszler, J.O. Carlsson, P. Demirev: J. Chem. Phys. **107**, 10440 (1997)
36. C.E. Bohren, D.R. Huffman: *Absorption and Scattering of Light by Small Particles* (Wiley, New York 1983)
37. G. Gabrielse, X. Fei, L.A. Orozco, R.L. Tjoelker, J. Haas, H. Kalinowsky, T.A. Trainor, W. Kells: Phys. Rev. Lett. **65**, 1317 (1990)
38. O. Nairz, B. Brezger, M. Arndt, A. Zeilinger: Phys. Rev. Lett. **87**, 160401 (2001)

# 25 Towards More Quantum Complete Neutron Experiments

Helmut Rauch

Neutrons are proper tools for testing quantum mechanics because they are massive, they couple to electromagnetic fields due to their magnetic moment and they are subject to all basic interactions and they are sensitive to topological effects as well. Related experiments will be discussed. Deterministic and stochastic partial absorption experiments can be described by Bell-type inequalities. Recent neutron interferometry experiments based on postselection methods renewed the discussion about quantum nonlocality and the quantum measuring process. It has been shown that interference phenomena can be revived even when the overall interference pattern has lost its contrast. This indicates persisting coupling in phase space even in cases of spatially separated Schrödinger-cat-like situations. These states are extremely fragile and sensitive to any kind of fluctuation or other decoherence processes. More complete quantum experiments also show that a complete retrieval of quantum states behind an interaction region becomes impossible in principle.

## 25.1 Introduction – Basic Relations

Different kinds of neutron interferometers based on wavefront and amplitude division have been tested in the past [1–4]. The perfect crystal interferometer – first tested in 1974 at our 250 kW TRIGA reactor – provides the highest intensity and became the most frequently used neutron interferometer due to its wide beam separation and its universal availability for fundamental, nuclear and solid-state physics [5]. It represents a macroscopic quantum device with characteristic dimensions of several centimeters (Fig. 25.1).

The basis for this kind of neutron interferometry is provided by the undisturbed arrangement of atoms in a monolithic perfect silicon crystal [2, 6]. An incident beam is split coherently at the first crystal plate, reflected at the middle plate and coherently superposed at the third plate. From general symmetry considerations it follows immediately that the wave functions in both beam paths, which compose the beam in the forward direction behind the interferometer, are equal ($\psi_0^{\mathrm{I}} = \psi_0^{\mathrm{II}}$), because they are transmitted-reflected-reflected (TRR) and reflected-reflected-transmitted (RRT), respectively. The theoretical treatment of the diffraction process from the perfect crystal is described by the dynamical diffraction theory [7, 8]. To preserve the interference properties over the length of the interferometer, the dimensions of

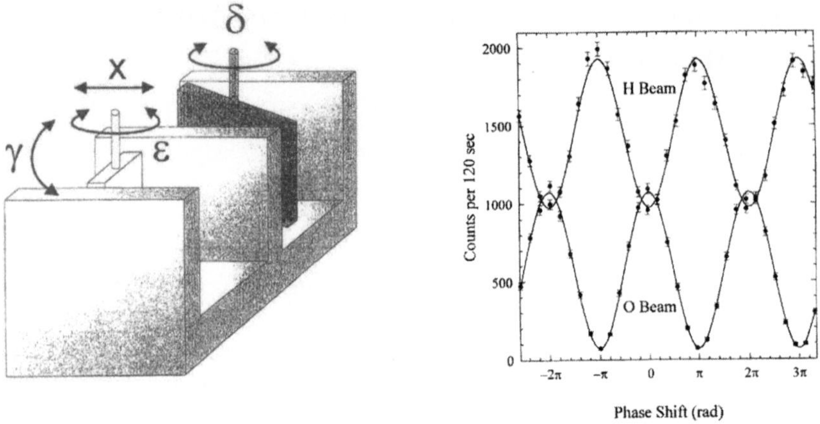


**Fig. 25.1.** Sketch of a symmetric perfect crystal neutron interferometer and a typical interference pattern

the monolithic system have to be accurate on a scale comparable to the so-called Pendellösung length ($\sim 50\,\mu\mathrm{m}$). The whole interferometer crystal has to be placed on a stable goniometer table under conditions avoiding temperature gradients and vibrations. A phase shift between the two coherent beams can be produced by nuclear, magnetic or gravitational interactions. In the first case, the phase shift is most easily calculated using the index of refraction [9, 10]:

$$n = \frac{\boldsymbol{k}}{k_0} = 1 - \frac{\lambda^2 N}{2\pi}\sqrt{b_\mathrm{c}^2 - \left(\frac{\sigma_\mathrm{r}}{2\lambda}\right)^2} + \mathrm{i}\frac{\sigma_\mathrm{r} N \lambda}{4\pi} \cong 1 - \lambda^2 \frac{N b_\mathrm{c}}{2\pi}\,, \tag{25.1}$$

where $b_\mathrm{c}$ is the coherent scattering length, $\sigma_\mathrm{r}$ the reaction cross-section, and $N$ is the particle density of the phase-shifting material. The different $\boldsymbol{k}$ vector inside the phase shifter causes a spatial shift, $\boldsymbol{\Delta}$, of the wave packet which depends on the orientation of the sample surface, $\hat{s}$, and which is related to the scalar phase shift, $\chi$, by

$$\psi \to \psi_0\,\mathrm{e}^{\mathrm{i}\boldsymbol{\Delta}\cdot\boldsymbol{k}} = \psi_0\,\mathrm{e}^{-\mathrm{i}N b_\mathrm{c}\lambda D} = \psi_0\,\mathrm{e}^{\mathrm{i}\chi}\,, \tag{25.2}$$

where $\chi$ can be written as a path integral of the canonical momentum, $\boldsymbol{k}_\mathrm{c}$, along the beam paths, $\chi = \oint \boldsymbol{k}_\mathrm{c}\,\mathrm{d}\boldsymbol{s}$ [11]. Therefore, the intensity behind the interferometer becomes

$$I_0 \propto \left|\psi_0^\mathrm{I} + \psi_0^\mathrm{II}\right|^2 \propto (1 + \cos\chi)\,. \tag{25.3}$$

The intensity of the beam in the deviated direction, $I_\mathrm{H}$, follows from particle conservation, $I_0 + I_\mathrm{H} = \mathrm{const}$. Thus, the intensities behind the interferometer vary as a function of the thickness, $D$, of the phase shifter, the particle density, $N$, or the neutron wavelength, $\lambda$.

Neutron optics is a part of quantum optics, and many phenomena can be described properly in that terminology, where the coherence function plays an important role [12, 13],

$$\Gamma(\boldsymbol{\Delta}) = \langle \psi(0)\psi(\boldsymbol{\Delta}) \rangle \ , \tag{25.4}$$

which is the autocorrelation function of the wave function. Using a wave packet description for the wave functions [amplitudes $a(\boldsymbol{k})$],

$$\psi(\boldsymbol{x}) \propto \int a(\boldsymbol{k})\,\mathrm{e}^{\mathrm{i}\boldsymbol{k}\boldsymbol{x}}\,\mathrm{d}k \ , \tag{25.5}$$

one obtains

$$I_0(\boldsymbol{\Delta}_0) \propto \left|\psi_0^{\mathrm{I}} + \psi_0^{\mathrm{II}}\right|^2 \propto 1 + |\Gamma(\boldsymbol{\Delta}_0)| \cos\chi_0 = 1 + |\Gamma(\boldsymbol{\Delta}_0)| \cos(\boldsymbol{\Delta}_0.\boldsymbol{k}_0) \ , \tag{25.6}$$

where $\boldsymbol{\Delta}_0$ and $\chi_0$ denote the phase shifts related to the mean momentum, $\boldsymbol{k}_0$. This gives

$$|\Gamma(\boldsymbol{\Delta})| \propto \left| \int g(\boldsymbol{k})\,\mathrm{e}^{\mathrm{i}\boldsymbol{k}\cdot\boldsymbol{\Delta}}\,\mathrm{d}^3\boldsymbol{k} \right| \ . \tag{25.7}$$

Thus the absolute value of the coherence function can be obtained from the fringe visibility, $|\Gamma(\boldsymbol{\Delta})| = (I_{\mathrm{Max}} - I_{\mathrm{Min}})/(I_{\mathrm{Max}} + I_{\mathrm{Min}})$, or as the Fourier transform of the momentum distribution, $g(\boldsymbol{k}) = |a(\boldsymbol{k})|^2$.

The mean square distance related to $|\Gamma(\boldsymbol{\Delta})|$ defines the coherence length, $\Delta_i^{\mathrm{c}}$, which is for Gaussian distribution functions directly related to the minimum uncertainty relation ($\Delta_i^{\mathrm{c}}\delta k_i = 1/2$). Similar relations can be obtained for time-dependent phenomena, where the spectral distribution, $g(\omega)$, and the temporal coherence function come into play.

Any experimental device deviates from the idealized situations: the perfect crystal can have slight deviations from its perfectness, and its dimensions may vary slightly; the phase shifter contributes to such deviations by variations in its thickness and inhomogeneities; and even the neutron beam itself contributes to a deviation from the idealized situation because of its momentum spread, $\delta k$. Therefore, the experimental interference patterns have to be descibed by a generalized relation,

$$I \propto A + B\,|\Gamma(\boldsymbol{\Delta})| \cos(\chi + \Phi_0) \ , \tag{25.8}$$

where $A$, $B$ and $\Phi_0$ are characteristic parameters of a certain set-up. It should be mentioned, however, that the idealized behaviour described by (25.3) can nearly be approached by a well-balanced set-up (Fig. 25.1). Phase shifts can be applied in the longitudinal, transverse and vertical directions, and the related coherence properties can be measured [14]. In the transverse direction the phase shift becomes wavelength independent ($\chi_{\mathrm{T}} = -2d_{hkl}Nb_{\mathrm{c}}D_0$; $d_{hkl}$: lattice distance), which implies a much larger coherence length in that direction.

All the results of interferometric measurements obtained up until now can be explained well in terms of the wave picture of quantum mechanics and the complementarity principle of standard quantum mechanics. Nevertheless, one should bear in mind the fact that the neutron also carries well-defined particle properties, which have to be transferred through the interferometer. These properties are summarized in Table 25.1 together with a formulation in the wave picture. Both particle and wave properties are well established, and, therefore, neutrons seem to be a proper tool for testing quantum mechanics with massive particles, where the wave-particle dualism becomes very obvious. In connection to this the author remembers a conversation with John Bell during a meeting in Grand Montana in 1983, where John explained that a neutron even it is massive should be considered as a much smaller object than a photon or an electron.

All neutron interferometric experiments pertain to the case of self-interference, where during a certain time interval, only one neutron is inside the interferometer, if at all. Usually, at that time the next neutron has not yet been born and is still contained in the uranium nuclei of the reactor fuel. Although there is no interaction between different neutrons, they have a certain common history within predetermined limits which are defined, e.g., by the neutron moderation process, by their movement along the neutron guide tubes, by the monochromator crystal and by the special interferometer set-up. Therefore, any interferometer pattern contains single-particle and ensemble properties together.

**Table 25.1.** Properties of neutrons

| Particle Properties | Connection | Wave Properties |
|---|---|---|
| $m = 1.6748220(25) \times 10^{-27}$ kg | | $\lambda_{\mathrm{C}} = \frac{h}{mc}$ |
| $s = \frac{1}{2}\hbar$ | de Broglie | $= 1.319695(20) \times 10^{-15}$ m |
| $\mu = -9.6491783(18) \times 10^{-27}$ J/T | $\lambda_{\mathrm{B}} = \frac{h}{mv}$ | For thermal neutrons: |
| $\tau = 882.6(2.7)$ s | Schrödinger | $\lambda = 1.8\,\mathrm{A}, v = 2200\,\mathrm{m/s}$ |
| $R = 0.7$ fm | $H\psi(\boldsymbol{r},t) = \mathrm{i}\hbar\frac{\delta\psi(\boldsymbol{r},t)}{\delta t}$ | $\lambda_{\mathrm{B}} = \frac{h}{mv} = 1.8 \times 10^{-10}$ m |
| $\alpha = 12.0(2.5) \times 10^{-4}\,\mathrm{fm}^3$ | and | $\Delta_{\mathrm{c}} = \frac{1}{2\delta k} \cong 10^{-8}$ m |
| u - d - d quark structure | boundary conditions | $\Delta_{\mathrm{p}} = v\Delta t \cong 10^{-2}$ m |
| | | $\Delta_{\mathrm{d}} = v\tau = 1.942(5) \times 10^{6}$ m |
| | | $0 \leq \chi \leq 2\pi(4\pi)$ |
| $m$: mass; $s$: spin; $\mu$: magnetic | | $\lambda_{\mathrm{C}}$: Compton wavelength; |
| moment; $\tau$: $\beta$-decay lifetime; | | $\lambda_{\mathrm{B}}$: de Broglie wavelength; |
| $R$: (magnetic) confinement radius; | $-\mu$B ——— | $\Delta_{\mathrm{c}}$: coherence length; |
| $\alpha$: electric polarizability; all | Two-level | $\Delta_{\mathrm{p}}$: packet length; |
| other measured quantities like | system | $\delta k$: momentum width; |
| electric charge magnetic monopole | $\mu$B ——— | $\Delta t$: chopper opening time; |
| and magnetic dipole moment are | | $v$: group velocity; |
| indistinguishable from zero | | $\chi$: phase |

## 25.2 Classic Neutron Interference Experiments

### 25.2.1 Gravity Experiments

The gravitational interaction of neutrons for usual laboratory conditions is comparable to the mean nuclear and magnetic interaction, and, therefore, a measurable interference signal is to be expected. The interaction Hamiltonian in this case reads

$$H_{\mathrm{g}} = m\boldsymbol{g}\boldsymbol{r} - \boldsymbol{\omega}\hbar(\boldsymbol{r} \times \boldsymbol{k}) , \tag{25.9}$$

where $\boldsymbol{g}$ is the gravitational acceleration directed towards the center of the earth and $\omega$ the angular rotation frequency of the earth.

The phase shift within the interferometer is calculated by using the path integral with the canonical momentum as mentioned earlier. In this way, after several intermediate steps, one obtains the *gravitational phase shift*:

$$\beta = \frac{m^2 g \lambda A \sin \Phi}{2\pi\hbar^2} + \frac{2m\omega A \sin \phi_{\mathrm{L}} \sin \varepsilon}{\hbar} , \tag{25.10}$$

where $A$ is the area enclosed by the coherent beam trajectories in the interferometer, $\Phi$ is the angle at which the interferometer is turned out of the horizontal plane, $\Phi_{\mathrm{L}}$ is the latitude of the point at which the experiment takes place, and $\varepsilon$ is the angle of rotation around the vertical axis. The first expression in the above equation describes the familiar gravitational term and was proven by *Colella, Overhauser* and *Werner* [15] by rotating the interferometer around a horizontal axis (COW experiment). This phase shift can be understood as the difference in gravitational potentials of the two coherent beams, as one travels higher than the other.

The *Coriolis* or *Sagnac term* in (25.9) was observed experimentally for the first time by *Werner* et al. [16, 17], by directing a neutron beam vertically upwards and by turning a perfect crystal interferometer around this vertical

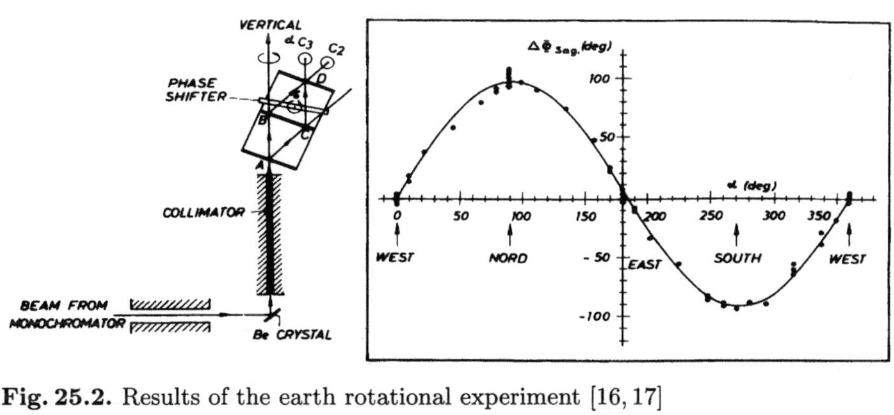


**Fig. 25.2.** Results of the earth rotational experiment [16, 17]

axis (Fig. 25.2). The result gives an impression of how sensitive the interferometric measuring method actually is. The easiest way to visualize this effect is by imagining the area encompassed by the two coherent beams as a differently oriented flag on the rotating earth. A more detailed discussion can be found in *Greenberger* [18].

A complimentary investigation to the gravitational measurements was performed by *Bonse* and *Wroblewski* [19], who brought the interferometer into a slightly oscillatory motion and, in so doing, also observed a phase shift, this being proportional to the respective acceleration of the interferometer plates. In summary, this proved the validity of the classical transformation laws for *noninertial frames of reference* in the quantum limit.

### 25.2.2 Neutron Fizeau Effect

An additional phase shift arises when there is a relative motion between the beam and the phase shifter. The calculation of this effect can be done on the basis of Galileian transformation because the velocity of the neutron ($\boldsymbol{v}$) is much smaller than the velocity of light. Therefore, the momentum of a particle, $\boldsymbol{K}'$, inside a material which moves with the velocity $\boldsymbol{w}$ is given as

$$\boldsymbol{K}' = \boldsymbol{K} - \frac{m}{\hbar}\boldsymbol{w} \,. \tag{25.11}$$

The *Fizeau phase shift* arises from the different phase shifts of a static and moving phase shifter [20]:

$$\Delta\chi_{\mathrm{F}} = (1-n')k'D - (1-n)kD = -\frac{VDm}{\hbar^2 k}\frac{w}{v-w} \,. \tag{25.12}$$

Quantum mechanics predicts that a Fizeau phase shift occurs only when the boundary is moving relative to the neutron beam, which is quite a difference to the optical Fizeau effect, which depends on the motional state of the phase shifter material. The first observation of this effect was achieved by *Klein* et al. [21] with a double-slit interferometer and a moving quartz phase shifter.

### 25.2.3 $4\pi$ Spinor Symmetry

This is probably one of the most discussed neutron interference experiments. Based on elementary principles of quantum mechanics, the propagation of a wave function can be described by a unitary transformation, given by the relevant Hamiltonian. For magnetic interaction, $H_{\mathrm{m}} = -\boldsymbol{\mu B}$, the propagation of the two-component *spinor wave function*, which describes the neutron as a fermion, can be represented as follows:

$$\psi(t) = \psi(0)\,\mathrm{e}^{\mathrm{i}Ht/\hbar} = \psi(0)\,\mathrm{e}^{-\mathrm{i}\boldsymbol{\mu B}/\hbar} = \psi(0)\,\mathrm{e}^{-\mathrm{i}\boldsymbol{\sigma\alpha}/2} = \psi(\alpha) \,, \tag{25.13}$$

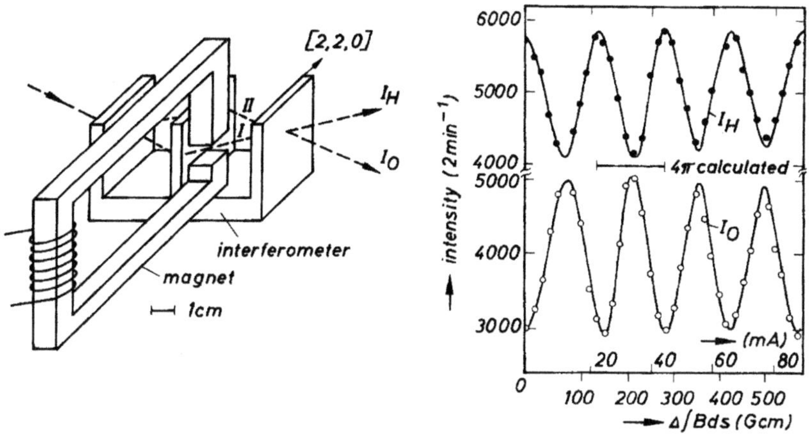


**Fig. 25.3.** First verification of the $4\pi$ symmetry of a spinor wave function [22]

where $\alpha$ is the *Larmor precession angle*,

$$|\alpha| = \frac{2\mu}{\hbar}\int B\,\mathrm{d}t = \frac{2\mu}{\hbar v}\int B\,\mathrm{d}s\,. \tag{25.14}$$

When inserting the Pauli spin operators, one can easily show that $\psi(\alpha)$ has $4\pi$ *symmetry*, and not the $2\pi$ symmetry we are used to with respect to expectation values and within the scope of classical physics,

$$\begin{aligned}\psi(2\pi) &= -\psi(0)\,,\\ \psi(4\pi) &= \psi(0)\,.\end{aligned} \tag{25.15}$$

These facts, which were not previously regarded as verifiable, can be elucidated very easily with neutron interferometry by observing the intensity modulations, while one of the coherent beams passes through a magnetic field,

$$I_0 = |\psi_0(0) + \psi_0(\alpha)|^2 \propto 1 + \cos\frac{\alpha}{2}\,. \tag{25.16}$$

The above relation is valid for polarized and unpolarized neutrons, which points to the inner symmetry properties of fermions. From (25.14) and (25.15) one recognizes that only for $\alpha = 4\pi$ is the original state reproduced. This was verified, nearly simultaneously, in measurements by *Rauch* et al. [22] and by *Werner* et al. [23] (Fig. 25.3). Afterwards, this effect was also proven through several other methods and for a series of other fermion systems. A distinction between dynamical and topological phases will be discussed in Sect. 25.5.

### 25.2.4 Spin Superposition

Spin superposition is an often used principle of quantum mechanics. Its curiosity value has been stressed by *Wigner* [24]. In the experiment the wave function of both coherent beams are originally polarized in the $|z\rangle$ direction. One beam is then inverted to a polarization in the $|-z\rangle$ direction, whereas the other remains unchanged. Both beams are then superimposed. This spin flip can be produced, for example, by Larmor precession around a magnetic field perpendicular to the $z$ direction. The result for superposition of these two beams, thus prepared, can be obtained by applying the rotation operator (25.13) to the spin-flipped beam for a rotation of 180° in the $y$ direction. If we also allow for a nuclear phase shift, one finds

$$\psi(\chi,\pi) = \mathrm{e}^{\mathrm{i}\chi}\,\mathrm{e}^{-\mathrm{i}\sigma_y\pi/2}\,|+z\rangle = -\mathrm{i}\pi\sigma_y\,\mathrm{e}^{\mathrm{i}\chi}\,|+z\rangle = \mathrm{e}^{\mathrm{i}\chi}\,|-z\rangle\ . \tag{25.17}$$

The total wave function, $\psi = |+z\rangle + \mathrm{e}^{\mathrm{i}\chi}|-z\rangle$, leads to the following polarization of the outgoing beam:

$$\boldsymbol{p} = \frac{\langle\psi^*\boldsymbol{\sigma}\psi\rangle}{\langle\psi^*\psi\rangle} = \begin{pmatrix}\cos\chi\\ \sin\chi\\ 0\end{pmatrix}\ . \tag{25.18}$$

Consequently, this polarization lies in the $x, y$ plane, and is perpendicular to the polarizations of the two superimposed, coherent beams. This implies that a pure quantum state in the $|z\rangle$ direction, e.g. for $\chi = 0$, has been transformed into a quantum state in the $|x\rangle$ direction, and, in the sense of self-interference, which definitively applies here, it seems that each neutron has information about the physical situation in both of the widely separated coherent beams. The experiment by *Summhammer* et al. [25] has fully confirmed this process. Intensity modulations appear only when the polarization analysis is done in the $x, y$ plane.

This above-mentioned experiment was repeated with a *Rabi resonance flipper*, which is also routinely used as a spin-flipping device in polarized neutron physics. Along with the spin flip, a simultaneous energy exchange is taking place between the resonator system and the neutrons ($\Delta E = \hbar\omega_\mathrm{r} = 2\mu B_0$, where $B_0$ is the strength of the guide field). Here one has to use the time-dependent Schrödinger equation and to take into consideration the change in the total energy of the neutron system,

$$\psi(\chi,\omega_\mathrm{r}) = \mathrm{e}^{\mathrm{i}\chi}\,\mathrm{e}^{-\mathrm{i}(\omega-\omega_\mathrm{r})t}\,|-z\rangle\ , \tag{25.19}$$

which leads to the following polarization, when the other unchanged coherent beam $|z\rangle$ is superimposed:

$$\boldsymbol{p} = \begin{pmatrix}\cos(\chi-\omega_\mathrm{r}t)\\ \sin(\chi-\omega_\mathrm{r}t)\\ 0\end{pmatrix}\ . \tag{25.20}$$

The polarization is also in the $x, y$ plane; however, it rotates within this plane synchronously with the resonance flip-field. It was possible to demonstrate this effect with a stroboscopic measurement, where the polarization in a given direction was measured synchronously with the phase of the flip-field [26].

In connection with these results, the obvious question arises as to whether the measurement of the energy transfer makes a determination of the beam path possible. One can, however, show that this is impossible, because interference vanishes in the presence of a measurable energy shift (i.e. larger than the energy width of the beam), and because the measurement of the energy change of the flipfield is impossible due to the photon-number–phase-uncertainty relationship ($\Delta\phi\Delta N > 1$).

These energy-exchange measurements have been extended by *Summhammer* et al. [27] to multiphoton exchange experiments. In this case an oscillating magnetic field with a frequency of 7.534 kHz was inserted into one beam and up to five photon emission and absorption processes were identified from the time-resolved interference pattern.

### 25.2.5 Neutron Josephson Effect

A double coil arrangement has been used for the observation of a new quantum beat effect, which is the magnetic analog to the well-known superconducting electric Josephson effect. If the frequencies of the two coils are chosen to be slightly different, the energy transfer becomes different too [$\Delta E = \hbar(\omega_{r1} - \omega_{r2})$]. The flipping efficiencies for both coils are always very close to unity (better than 99%). Now, the wave functions change accordingly to

$$\psi \to e^{i(\omega-\omega_{r1})t} \left|\downarrow\right\rangle + e^{i\chi} e^{i(\omega-\omega_{r2})t} \left|\downarrow\right\rangle . \tag{25.21}$$

Therefore, the intensity behind the interferometer exhibits a typical quantum beat effect, given by

$$I \propto 1 + \cos[\chi + (\omega_{r1} - \omega_{r2})t] . \tag{25.22}$$

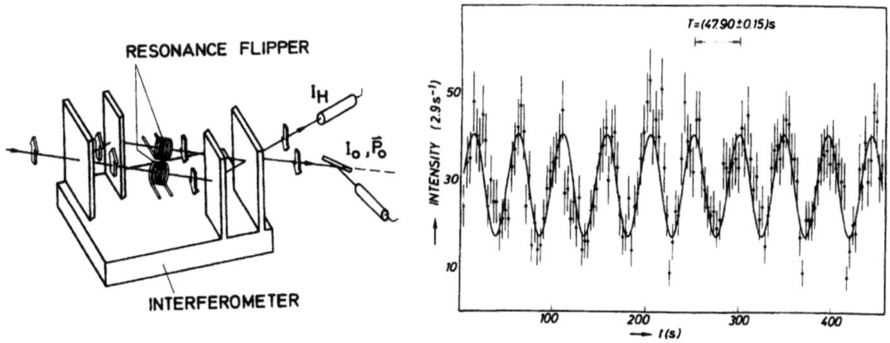


**Fig. 25.4.** Quantum beat effect observed when the frequencies of the two flipper coils differ by about 0.02 Hz around 71.89979 kHz [28]

Thus, the intensity behind the interferometer oscillates between the forward and deviated beam without any apparent change inside the interferometer [28]. The time constant of this modulation can reach a macroscopic scale which is correlated to the uncertainty relation, $\Delta E \Delta t \leq \hbar/2$.

Figure 25.4 shows the result of an experiment where the periodicity of the intensity modulation, $T = 2\pi/(\omega_{r1} - \omega_{r2})$, amounts to $T = 47.90 \pm 0.15\,\mathrm{s}$ caused by a frequency difference of about 0.02 Hz. This corresponds to a mean energy transfer difference $\Delta E$ between the two beams of $\Delta E = 8.6 \times 10^{-17}\,\mathrm{eV}$ and to an energy sensitivity of $2.7 \times 10^{-19}\,\mathrm{eV}$, which is many orders of magnitudes larger than that of other advanced spectroscopic methods. This high resolution is strongly decoupled from the monochromaticity of the neutron beam, which was $\Delta E_{\mathrm{B}} = 5.5 \times 10^{-4}\,\mathrm{eV}$ around a mean energy of the beam of $E_{\mathrm{B}} = 0.023\,\mathrm{eV}$ in this case. The quantum beat effect can also be interpreted as the magnetic Josephson effect analog where the phase difference, $\Delta(t)$, is driven by the magnetic energy, whereas in the well-known Josephson effect in superconducting tunnel junctions [29] the phase of the Cooper pairs in both superconductors is driven by the electrical energy.

### 25.2.6 Stochastic Versus Deterministic Beam-Path Detection

A certain amount of beam attenuation can be achieved either by a semi-transparent material or by a proper chopper or slit system. The transmission probability in the first case is defined by the attenuation cross-section, $\sigma_{\mathrm{a}}$, of the material $[t = I/I_0 = \exp(-\sigma_{\mathrm{a}} N D)]$. The change in the wave function is obtained directly from the complex index of refraction (25.1):

$$\psi \rightarrow \psi_0\, \mathrm{e}^{\mathrm{i}(n-1)kD} = \psi_0\, \mathrm{e}^{\mathrm{i}\chi}\, \mathrm{e}^{-\sigma_{\mathrm{a}} N D/2} = \mathrm{e}^{\mathrm{i}\chi} \sqrt{\underline{t}}\,\psi\,. \tag{25.23}$$

Therefore, the beam modulation behind the interferometer is obtained in the following form:

$$I_0 \propto \left|\psi_0^{\mathrm{I}} + \psi_0^{\mathrm{II}}\right|^2 \propto \left[(1+\underline{t}) + 2\sqrt{\underline{t}}\cos\chi\right]\,. \tag{25.24}$$

On the other hand, the transmission probability of a chopper wheel or another shutter system is given by the open-to-closed ratio, $\underline{t} = t_{\mathrm{open}}/(t_{\mathrm{open}} + t_{\mathrm{closed}})$, and one obtains after straightforward calculations

$$I \propto \left[(1-\underline{t})\left|\psi_0^{\mathrm{II}}\right|^2 + \underline{t}\left|\psi_0^{\mathrm{I}} + \psi_0^{\mathrm{II}}\right|^2\right] \propto \left[(1+\underline{t}) + 2\underline{t}\cos\chi\right]\,, \tag{25.25}$$

i.e. the contrast of the interference pattern is proportional to $\sqrt{\underline{t}}$ in the first case and proportional to $t$ in the second case, although the same number of neutrons are absorbed in both cases. The absorption represents a measuring process in both cases, i.e. a beam-path detection, because compound nuclei are produced with an excitation energy of several MeV, which are usually deexcited by capture gamma rays. The measured contrast lies along

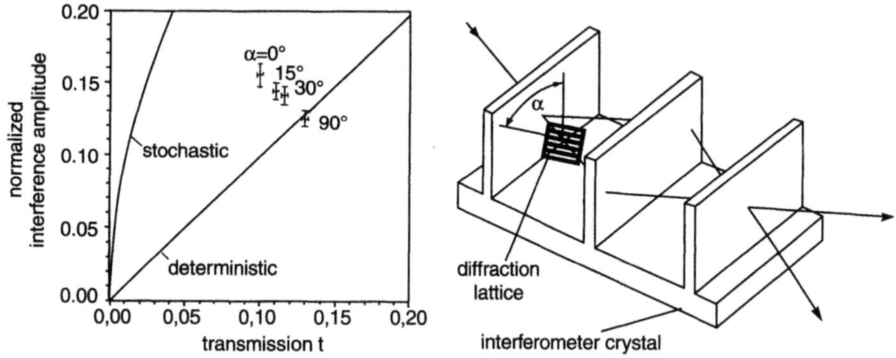


**Fig. 25.5.** Lattice absorber in the interferometer approaching the classical limit when the slits are oriented horizontally and the quantum limit when they are oriented vertically [34]

the lines "stochastic" and "deterministic" in Fig. 25.5 [30, 31]. The different contrast becomes especially obvious for low transmission probabilities. The discrepancy diverges for $\underline{t} \to 0$, but it has been shown that in this regime the variations of the transmission due to variations of the thickness or of the density of the absorber plate have to be taken into account, which shifts the points below the $\sqrt{\underline{t}}$ ("stochastic") curve [32].

The region between the linear and the square-root behaviour can be reached by using very narrow chopper slits or narrow transmission lattices, where one starts to lose information regarding through which individual slit the neutron went. This is exactly the region which shows the transition between a deterministic and a stochastic situation, and, therefore, it can be formulated using a Bell-like inequality ($\sqrt{\underline{t}} > x > \underline{t}$ [33]).

The stochastic limit corresponds to the quantum limit when one does not know anymore through which individual slit the neutron went. Which situation exists depends on how the slit width, $l$, compares with the coherence lengths in the related direction. When the slit widths become comparable to the coherence lengths, the wave function behind the slits show distinct diffraction peaks which correspond to new quantum states ($n \neq 0$). The creation of the new quantum states means that those labeled neutrons carry information about the chosen beam path and, therefore, do not contribute to the interference amplitude [34] (Fig. 25.5). A related experiment has been carried out by rotating an absorption lattice around the beam axis, where one changes from $l \lesssim \Delta_x^c$ (vertical slits) to $l \gg \Delta_y^c$ (horizontal slits). Thus, the attenuation factor, $\underline{t}$, has to be generalized to include not only nuclear absorption and scattering processes but also lattice diffraction effects if they remove neutrons from the original phase space.

The partial absorption and coherence experiments are closely connected to the quantum duality principle, which states that the observation of an interference pattern and the acquisition of which-way information are mutually exclusive. Various inequalities have been formulated to describe this mutual exclusion principle [35, 36]. The most concise formulation reads

$$V^2 + P^2 =\leq 1 \, , \tag{25.26}$$

where $V$ denotes the fringe visibility (25.5) and $P$ is the predictability of the way through the interferometer, which is a quantitative measure of the a priori which-way knowledge.

## 25.3 Postselection Experiments

Various post selection measurements in neutron interferometry have shown that interference fringes can be restored by proper filtering methods even in cases when the overall beam does not exhibit any interference fringes due to spatial phase shifts larger than the coherence lengths of the interfering beams [37, 38]. Postselection procedures can be applied to various parameters of an experiment:

- spatial postselection;
- momentum postselection;
- counting statistic postselection;
- phase postselection;
- topology postselection.

Figure 25.6 shows some of them schematically. Here we discuss momentum postselection and phase-echo experiments and refer the reader for other methods to the literature [39–41].

### 25.3.1 Postselection of Momentum States

The experimental arrangement with an indication of the wave packets at different parts of the interference experiment is shown in Fig. 25.6. An additional monochromatization is applied behind the interferometer by means of a single crystal adjusted to a Bragg position or by time-of-flight systems. Adding the momentum-dependent intensity to the plane-wave interference pattern (25.3) one obtains for Gaussian momentum distributions

$$I_0(k) = \exp\left[-(k-k_0)^2/2\delta k^2\right]\left[1+\cos\left(\chi_0\frac{k_0}{k}\right)\right] \, . \tag{25.27}$$

The spatial phase-shift-dependent intensity is given by (25.5). The formula shows that the overall interference fringes disappear for spatial phase shifts much larger than the coherence lengths $[\Delta_i \geq \Delta_i^c = 1/(2\delta k_i)]$. The surprising feature is that $I_0(k)$ becomes oscillatory for large phase shifts where

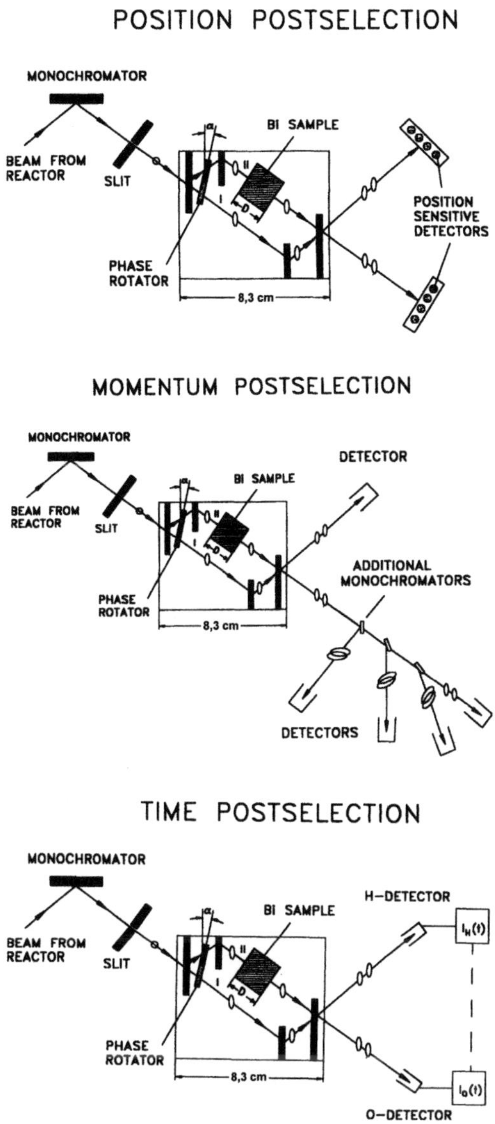


**Fig. 25.6.** Sketch of various postselection methods

the interference fringes described by (25.5) disappear ([37], see Fig. 25.7). This indicates that interference in phase space has to be considered [42].

The amplitude function of the packets arising from beam paths I and II determines the spatial shape of the packets behind the interferometer,

$$I_0(x) = |\psi(x) + \psi(x + \Delta)|^2 \,, \tag{25.28}$$

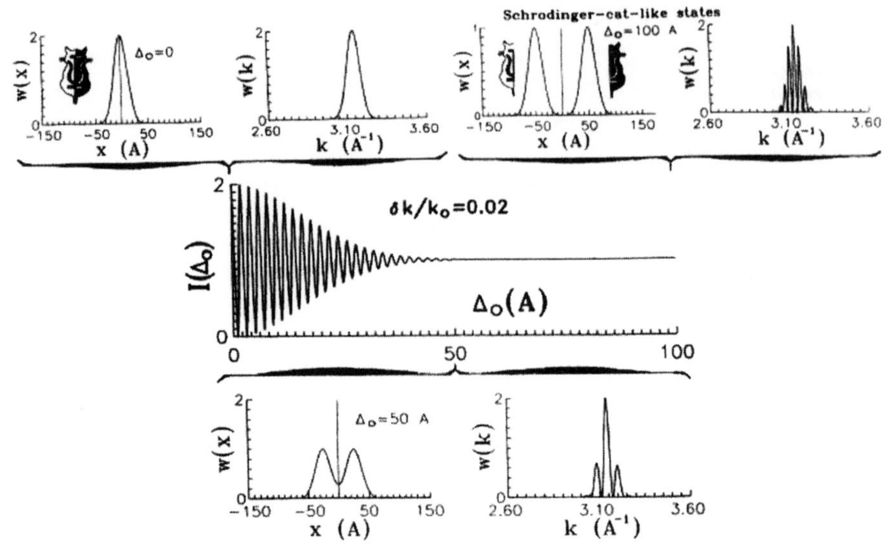


**Fig. 25.7.** Interference pattern as a function of the relative phase shift (middle) and related wave packets and momentum spectra behind the interferometer for different values of the phase shift [37]

which separates for large phase shifts into two peaks (Fig. 25.7). For an appropriately large displacement ($\Delta \gg \Delta^c$), the related state can be interpreted as a superposition state of two macroscopically distinguishable states, that is a stationary Schrödinger cat-like state [43, 44], but here first for massive particles. These states – separated in ordinary space and oscillating in momentum space – seem to be notoriously fragile and sensitive to dephasing effects [44–48]. The duality equation (25.26) is still valid because distinguishable states are produced ($P \mp 1$).

Measurements of the wavelength spectrum were made with a silicon crystal with a rather narrow mosaic spread, which reflects in the parallel position a rather narrow band of neutrons only ($\delta k'/k_0 \approx 0.0003$), causing a restored visibility of the interference pattern at large phase shifts [38] (Fig. 25.8). This feature shows that an interference pattern can be revived even behind the interferometer by means of a proper postselection procedure. In this case the overall beam does not show interference fringes anymore, and the wave packets originating from the two different beam paths do not overlap. The momentum distribution was measured by scanning the analyzer crystal through the Bragg position. These results clearly demonstrate that the predicted spectral modulation (25.27) appears when the interference fringes of the overall beam disappear. The modulation is somehow smeared out due to averaging processes across the beam due to various imperfections, which are unavoidable in any experimental arrangement. The contrast of the empty interferometer was 60%.

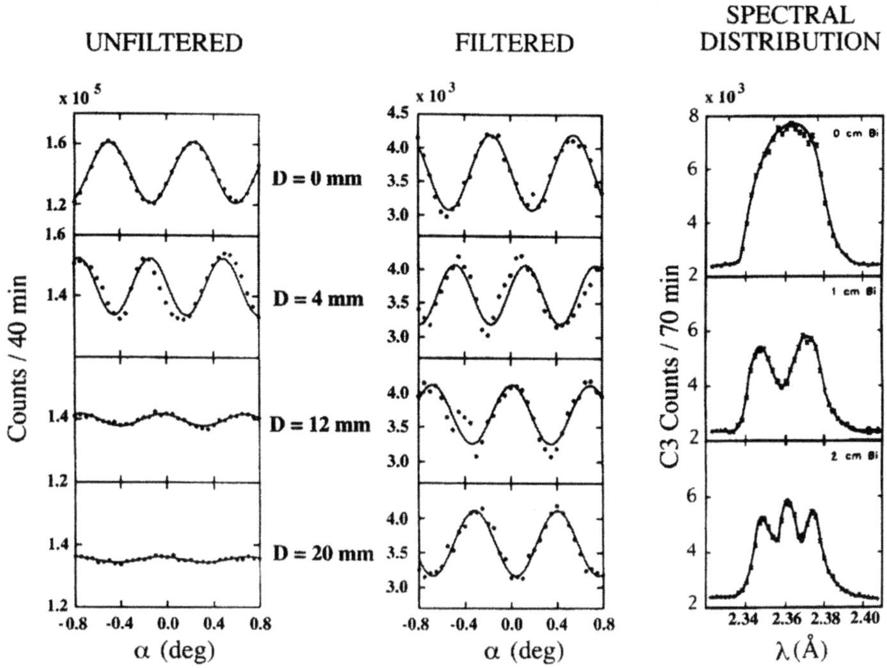


**Fig. 25.8.** Interference pattern of the unfiltered overall beam ($\delta k/k_0 = 0.012$, middle) and the filtered beam reflected from a nearly perfect crystal analyzer in the antiparallel position ($\delta k'/k_0 = 0.0003$, left) and the observed spectral modulation (right) of the outgoing beam for different phase-shifter thicknesses [38]

The new quantum states created behind the interferometer can be analyzed with regard to their uncertainty properties. Analogies between a coherent state behavior and a free but coherently coupled particle motion inside the interferometer have been addressed [31]. In such cases, the dynamical conjugate variables $x$ and $p$ minimize the uncertainty product with identical uncertainties $(\Delta x)^2 = (\Delta k)^2 = 1/2$ (in dimensionless units). Simple calculations show that for $(\Delta k)^2$ a value below the coherent state value can be achieved, which in quantum optic terminology means squeezing [49–52]. One emphasizes that a single coherent state does not exhibit squeezing, but a state created by superposition of two coherent states can exhibit a considerable amount of squeezing. Thus highly nonclassical states are made by the power of the quantum-mechanical superposition principle.

It should be mentioned that momentum post selection in typical Bell experiments with entangled photons may also give a less mystic view of these experiments [37].

### 25.3.2 Contrast Retrieval by Phase Echo

A large phase shift ($\Delta > \Delta^c$) can be applied in one arm of the interferometer, which can be compensated by a negative phase shift acting in the same arm or by the same phase shift applied to the second beam path [53]. Because the phase shift is additive, the coherence function depends on the net phase shift only. Thus, the interference pattern can be restored, as is shown in the form of an experimental example in Fig. 25.9. The phase-echo method can also be applied behind the interferometer loop when multiplate interferometers are used [41]. In this case, the situation becomes even more alike to the situation discussed in the previous section. The experimental results completely confirmed that behaviour. Phase echo is a similar technique to spin echo [3], which is routinely used in neutron spectroscopy and which represents an interference experiment as well.

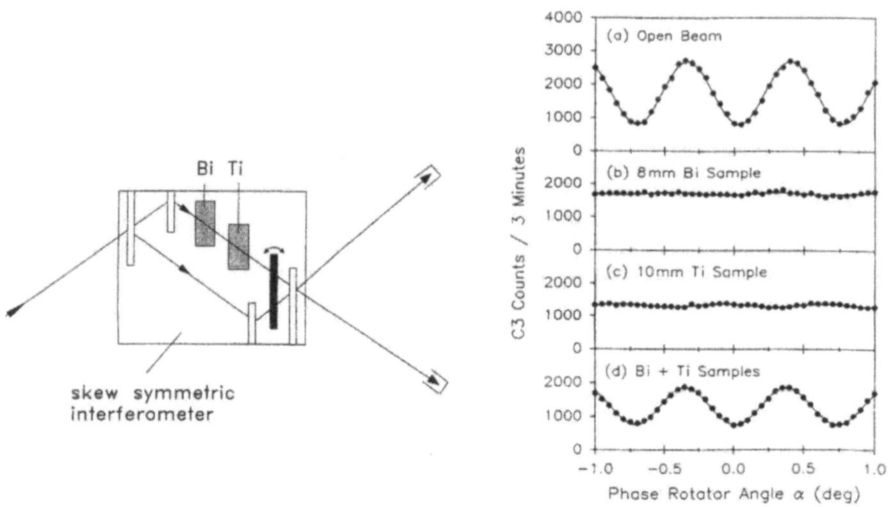


**Fig. 25.9.** Loss of contrast at high interference and its retrieval by an opposite phase shifter inserted into the same beam [53]

## 25.4 Phase-Space Coupling

In the previous sections one noticed that coherence phenomena can be exchanged between various parts of the phase space. The modulation of the momentum distribution that appears when the spatial interference pattern disappears may be the most direct evidence of this phenomena (Fig. 25.8) [38]. In quantum optics, many phenomena are visualized by Wigner quasi-distribution

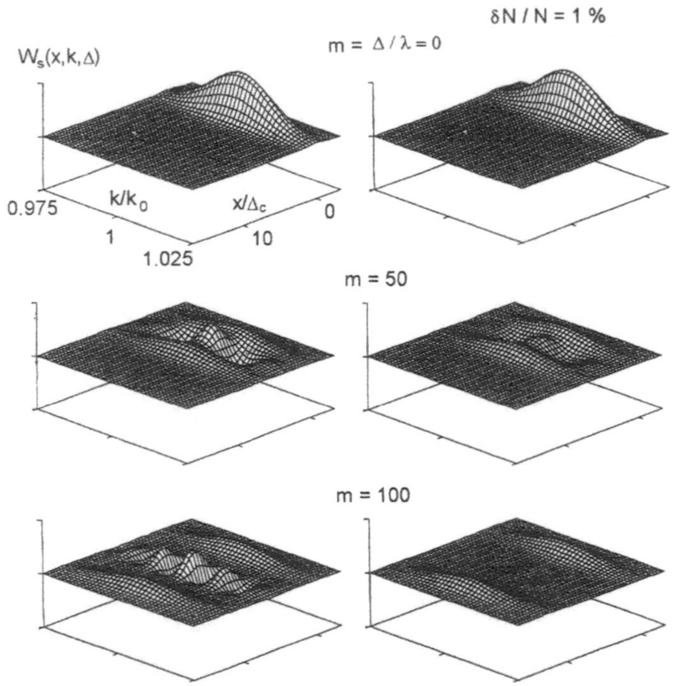


**Fig. 25.10.** Wigner functions for various phase shifts without (*left*) and with (*right*) fluctuations in the phase shifter [55]

functions, which are defined as [13, 54]

$$W_S(k,x) = \frac{1}{\pi}\int_{-\infty}^{+\infty} \mathrm{e}^{\mathrm{i}kx'}\psi^*\left(x+\frac{x'}{2}\right)\psi\left(x-\frac{x'}{2}\right)\mathrm{d}x' , \tag{25.29}$$

where in our case,

$$\psi(x) = \psi^{\mathrm{I}} + \psi^{\mathrm{II}} = \psi(x) + \psi(x+\varDelta) , \tag{25.30}$$

we arrive at

$$\begin{aligned} &W_S(x,k,\varDelta) \\ &\quad = W(x,k) + W(x+\varDelta,k) + 2W\left(x+\frac{\varDelta}{2},k\right)\cos\left(\varDelta\cdot k\right) \\ &\quad \propto \exp[-(k-k_0)^2/2\delta k^2]\left\{\exp(-x^2/2\delta x^2)\right. \\ &\qquad \left. +\exp[-(x+\varDelta)^2/2\delta x^2] + 2\exp\left[-\left(x+\frac{\varDelta}{2}\right)^2/2\delta x^2\right]\cos(\varDelta\cdot k)\right\} . \end{aligned} \tag{25.31}$$

Integration over the momentum variable gives the spatial distribution (25.5), and integration over the spatial variable gives the momentum distribution (25.27). Typical results are shown in Fig. 25.10 [55].

When fluctuations of the phase shifter ($\delta N$ or $\delta D_0$) are included, one notices that the wiggle structure between the separated peaks is more sensitive at high interference order than at low order. This causes a decrease in the coherence and a transition from a pure quantum state to a mixture. As a result, upper limits for the separation of massive (Schrödinger-cat) systems due to unavoidable zero-point fluctuations can be derived. It also indicates why the retrieval of a quantum state from one phase space to another one becomes intrinsically more difficult the larger the separations in one phase space happen. Since the spatial distribution, the momentum distribution and the offdiagonal traces of the Wigner function can be measured, the Wigner function can be reconstructed, which means a quantum-state reconstruction [56–58]. Related neutron optical experiments are in progress [59].

## 25.5 Topological Effects

Topological and geometrical effects appear in the solution of the Schrödinger equation due to special geometric forms of the interaction [60–62]. Thus they are part of quantum mechanics, but they are easily overlooked by a pure intensity experiment. It also shows that a wave function often carries more information than those extracted in a standard experiment. A typical example is the spin superposition experiment discussed in Sect. 25.2.4, where the exact result also depends on around which axis the spin has been rotated into

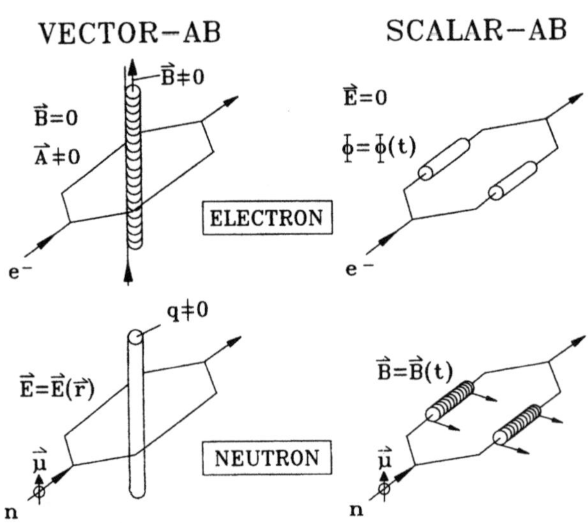


**Fig. 25.11.** Comparison of various Aharonov–Bohm phenomena for electrons and neutrons

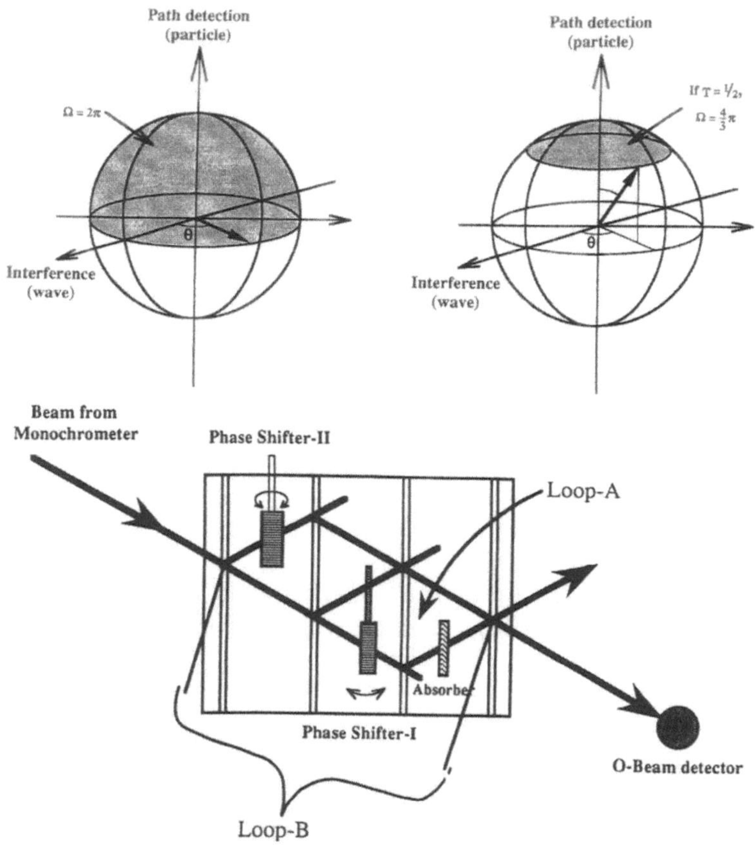


**Fig. 25.12.** Poincaré presentation of the absorption topological phase and the experimental arrangement of the related experiment [66]

the opposite direction. In this respect the action of a Hamiltonian can be separated into its strength (dynamical) and its geometry, which results from the sum of state changes along the excursion in phase space:

$$\begin{aligned}\phi &= -\frac{1}{\hbar}\int_0^T \langle\psi(t)|\,H\,|\psi(t)\rangle\,\mathrm{d}t + \mathrm{i}\int_0^T \left\langle \phi(t)\frac{\mathrm{d}}{\mathrm{d}t}\middle|\phi(t)\right\rangle \mathrm{d}t \\ &= \alpha + \phi_{\mathrm{g}}\,. \end{aligned} \tag{25.32}$$

*Wagh* et al. [63] recently performed a related experiment and showed clearly the existence of the topological phase. In a similar sense the scalar and the vector Aharonov–Bohm effects of neutrons have been verified by neutron interferometric methods [64, 65]. The geometric nature can be seen in Fig. 25.11 in comparison to the well-known electron Aharonov–Bohm effects.

The concept of topological phases can be extended to the description of absorption phenomena as they are discussed in Sect. 25.2.6. Increasing absorption can be attributed to a more particle-like behaviour [see (25.26)], and the state moves away from the equator to a latitude circle of the Poincaré sphere with solid angle $\Omega$ related to the geometrical phase, $\phi_g$:

$$\Omega(C) = 4\pi \frac{\underline{t}}{1+\underline{t}} = 2\phi_g \,. \tag{25.33}$$

This can be detected by a four-plate, double-loop interferometer as done by Hasegawa et al. [66] (Fig. 25.12). Good agreement between the theoretical predictions and the experimental results has been achieved.

## 25.6 Discussion

It has been shown in the previous sections that more information about a quantum system can be extracted when more experimentally accessible parameters are measured. It becomes obvious that a system remains coupled in phase space even when it becomes separated in any parameter space. Thus, interference properties can be shifted from one parameter space to another one and back again. Related bands of plane-wave components which compose the wave packets may be considered as a factor responsible for the understanding of the coupling and nonlocality phenomena in quantum mechanics. It looks like that these plane-wave components of the wave packets, i.e. narrow bands, interact over much larger distances than the size of the packets. This interaction guides neutrons of certain momentum bands to the O or H beam. These phenomena throw a new light on the discussion about locality and nonlocality in quantum mechanics [67, 68]. This may be considered as a contribution to the speakable and unspeakable aspects of quantum mechanics [69]. Spatially separated packets remain entangled in phase space and nonlocality appears as a result of this entanglement. The analogy with optical experiments performed in the time–frequency domain is striking [70]. Since entanglement exists not only between objects but also between different degrees of freedom [71], Bell-type experiments can also be done in single-particle experiments [72].

The summaries drawn for the different experimental situations discussed in this chapter are followed by statements to the effect that the retrieval of the interference properties by several post selection procedures become increasingly more difficult the wider the separation of the quantum system before. This is also demonstrated by means of the Wigner distributions (Sect. 25.4, Fig. 25.10), where it has been shown that the transition from a quantum state to a mixture is always related to some statistical features of the interaction acting on the quantum system. Such fluctuations are, in principle, unavoidable due to residual quantum fluctuations inherent in any physical system.

Unavoidable fluctuations (even zero-point fluctuations) cause an irreversibility effect, which becomes more influential for widely separated Schrödinger-cat-like states. All these effects can be described by an increasing entropy inherently associated with any kind of interaction. This also supports the idea that irreversibility is a fundamental property of nature and reversibility, an approximation only, as stated by several authors [73–76].

All the results of the neutron interferometric experiments are described well by the formalism of quantum mechanics. According to the complementarity principle of the Copenhagen interpretation, the wave picture has to be used to describe the observed phenomena. The question of how the well-defined particle properties of the neutron are transferred through the interferometer is not a meaningful one within this interpretation, but from the physical point of view it should be an allowed one.

More complete quantum experiments show that a complete retrieval of all wave components behind an interaction the quantum system experienced becomes impossible, in principle. This implies on a high accuracy level a basic noncommutivity of operators, $A.B|\psi\rangle \neq B.A|\psi\rangle$.

## Acknowledgements

Most of the experimental work has been supported by projects of the Austrian Science Foundation (FWF), most recently by the project SFB1513.

## References

1. H. Maier-Leibnitz, T. Springer: Z. Phys. **167** (1962)
2. H. Rauch, W. Treimer, U. Bonse: Phys. Lett. A **47**, 369 (1974)
3. F. Mezei: Z. Phys. **25**, 146 (1972)
4. A.I. Ioffe, V.S. Zabiyankan, G.M. Drabkin: Phys. Lett. **111**, 373 (1985)
5. H. Rauch, S.A. Werner (eds.): *Neutron Interferometry* (Clarendon, Oxford 2000)
6. U. Bonse, M. Hart: Appl. Phys. Lett. **6**, 155 (1965)
7. W. Bauspiess, U. Bonse, W. Graeff: J. Appl. Cryst. **9**, 68 (1976)
8. H. Rauch, D. Petrascheck: *Neutron Diffraction*, ed. by H. Dachs (Springer, Berlin, Heidelberg 1978) Chap. 9
9. M.L. Goldberger, F. Seitz: Phys. Rev. **71**, 294 (1947)
10. V.F. Sears: *Neutron Optics* (Oxford Univ. Press, Oxford 1989)
11. R.P. Feynman, R.B. Leighton, M. Sands: *The Feynman Lectures on Physics*, Vol. III (Addison-Wesley, New York 1965)
12. R.J. Glauber: Phys. Rev. **130**, 2529; ibid. **131**, 2766 (1963)
13. L. Mandel, E. Wolf: *Optical Coherence and Quantum Optics* (Cambridge Univ. Press, Cambridge 1995)
14. H. Rauch, H. Wölwitsch, H. Kaiser, R. Clothier, S.A. Werner: Phys. Rev. A **53**, 902 (1996)
15. R. Colella, A.W. Overhauser, S.A. Werner: Phys. Rev. Lett. **34**, 1472 (1975)
16. S.A. Werner, J.L. Staudenmann, R. Colella: Phys. Rev. Lett. **42**, 1103 (1979)
17. J.L. Staudenmann, S.A. Werner, R. Colella, A.W. Overhauser: Phys. Rev. A **21**, 1419 (1980)
18. D.M. Greenberger: Rev. Mod. Phys. **55**, 875 (1983)

19. U. Bonse, T. Wroblewski: Phys. Rev. Lett. **51**, 1401 (1983)
20. M.A. Horne, A. Zeilinger, A.G. Klein, G.I. Opat: Phys. Rev. A **28**, 1 (1983)
21. A.G. Klein, G.I. Opat, A. Cimmino, A. Zeilinger, W. Treimer, R. Gähler: Phys. Rev. Lett. **46**, 1551 (1981)
22. H. Rauch, A. Zeilinger, G. Badurek, A. Wilfing, W. Bauspiess, U. Bonse: Phys. Lett. A **54**, 425 (1975)
23. S.A. Werner, R. Colella, A.W. Overhauser, C.F. Eagen: Phys. Rev. Lett. **35**, 1053 (1975)
24. E.P. Wigner: Am. J. Phys. **31**, 6 (1963)
25. J. Summhammer, G. Badurek, H. Rauch, U. Kischko, A. Zeilinger: Phys. Rev. A **27**, 2523 (1983)
26. G. Badurek, H. Rauch, J. Summhammer: Phys. Rev. Lett. **51**, 1015 (1983)
27. J. Summhammer, K.A. Hamacher, H. Kaiser, H. Weinfurter, D.L. Jacobson, S.A. Werner: Phys. Rev. Lett. **75**, 3206 (1995)
28. G. Badurek, H. Rauch, D. Tuppinger: Phys. Rev. A **34**, 2600 (1986)
29. B.D. Josephson: Rev. Mod. Phys. **46**, 251 (1974)
30. J. Summhammer, H. Rauch, D. Tuppinger: Phys. Rev. A **36**, 4447 (1987)
31. H. Rauch, J. Summhammer, M. Zawisky, E. Jericha: Phys. Rev. A **42**, 3726 (1990)
32. M. Namiki, S. Pascazio: Phys. Lett. A **147**, 430 (1990)
33. J.S. Bell: Phys. **1**, 195 (1965)
34. H. Rauch, J. Summhammer: Phys. Rev. **46**, 7284 (1992)
35. W.K. Wooters, W.H. Zurek: Phys. Rev. D **19**, 473 (1979)
36. B.-G. Englert: Phys. Rev. Lett. **77**, 2154 (1996)
37. H. Rauch: Phys. Lett. A **173**, 240 (1993)
38. D.L. Jacobson, S.A. Werner, H. Rauch: Phys. Rev. A **49**, 3196 (1994)
39. W. Bauspiess, U. Bonse, H. Rauch: Nucl. Instrum. Meth. **157**, 495 (1978)
40. M. Zawisky, H. Rauch, Y. Hasegawa: Phys. Rev. A **50**, 5000 (1994)
41. M. Heinrich, D. Petrascheck, H. Rauch: Z. Physik B **72**, 357 (1988)
42. W. Schleich, J.A. Wheeler: Nature **326**, 574 (1987)
43. A. Legett: Proc. *Foundations of Quantum Mechanics*, ed. by S. Kamefuchi (Phys. Soc. Japan, Tokyo 1984) p. 74
44. W. Schleich, M. Pernigo, F. Le Kien: Phys. Rev. A **44**, 2172 (1991)
45. D.F. Walls, G.J. Milburn: Phys. Rev. A **31**, 2403 (1985)
46. H. Zurek: Physics Today, 36 (October 1991)
47. P. Blanchard, A. Jadczyk: Phys. Lett. A **175**, 157 (1993)
48. V. Buzek, C.H. Keitel, P.L. Knight: Phys. Rev. A **51**, 2594 (1995)
49. D.F. Walls: Nature **306**, 141 (1983)
50. R. Loudon, P.L. Knight: J. Mod. Opt. **34**, 709 (1987)
51. J. Janski, A.V. Vinogradov: Phys. Rev. Lett. **64**, 2771 (1990)
52. E. Wolf: Phys. Rev. Lett. **63**, 2220 (1989)
53. R. Clothier, H. Kaiser, S.A. Werner, H. Rauch, H. Wölwitsch: Phys. Rev. A **44**, 5357 (1991)
54. E.P. Wigner: Phys. Rev. **40**, 749 (1932)
55. H. Rauch, M. Suda: Appl. Phys. B **60**, 181 (1995)
56. M. Freyberger, S.H. Kienle, V.P. Yakovlev: Phys. Rev. A **56**, 195 (1997)
57. U. Leonhardt: *Measuring the Quantum State of Light* (Cambridge Univ. Press, Cambridge 1997)
58. P.W. Schleich: *Quantum Optics in Phase Space* (Wiley-VCH, Berlin 2001)

59. M. Baron, H. Rauch, M. Suda: Appl. Phys. A (2001), in press
60. S. Pancharatram: Proc. Ind. Acad. Sci. A **44**, 247 (1956)
61. M.V. Berry: Proc. R. Soc. London A **392**, 45 (1984)
62. Y. Aharonov, J. Anandan: Phys. Rev. Lett. **58**, 1593 (1987)
63. A.G. Wagh, V.C. Rakhecha, J. Summhammer, G. Badurek, H. Weinfurter, B.M. Allman, H. Kaiser, K. Hamacher, D.L. Jacobson, S.A. Werner: Phys. Rev. Lett. **78**, 755 (1997)
64. B.E. Allman, A. Cimmino, A.G. Klein, G.I. Opat, H. Kaiser, S.A. Werner: Phys. Rev. Lett. **68**, 2409 (1992)
65. A. Cimmino, G.I. Opat, A.G. Klein, H. Kaiser, S.A. Werner, M. Arif, R. Clothier: Phys. Rev. Lett. **68**, 380 (1989)
66. Y. Hasegawa, M. Zawisky, H. Rauch, A.I. Ioffe: Phys. Rev. A **53**, 2406 (1996)
67. A. Einstein, B. Podolsky, N. Rosen: Phys. Rev. **47**, 777 (1935)
68. D.M. Greenberger, M.A. Horne, A. Zeilinger: In: *Bell's Theorem, Quantum Theory and Conceptions of the Universe*, ed. by M. Kafatos (Kluwer, Dordrecht 1989) p. 69
69. J.S. Bell: *Speakable and Unspeakable in Quantum Mechanics* (Cambridge Univ. Press, Cambridge 1987)
70. X.Y. Zou, T.P. Grayson, L. Mandel: Phys. Rev. Lett. **69**, 3041 (1992)
71. B.-G. Englert: Z. Naturforsch. A **54**, 11 (1999)
72. S. Basu, S. Bandyopadhyay, G. Kar, D. Home: Phys. Lett. A **279**, 281 (2001)
73. H.A. Lorentz: *Theorie der Strahlung* (Akad. Verlagsges., Leipzig 1927)
74. I. Prigogine: *Proc. Ecol. Phys. Chem.* (Elsevier, Amsterdam 1991) p. 8
75. F. Haag: Commun. Math. Phys. **123**, 245 (1990)
76. N. Komo, K. Machida, M. Namiki, S. Pascazio: Phys. Rev. A **54**, 1064 (1996)

# Part VI

# Particles and Fields

# 26 John Bell's Observations on the Chiral Anomaly and Some Properties of Its Descendants

Roman Jackiw

John Bell's emphasis of the essential ambiguities in anomaly calculations is recalled. Some descendants of the anomaly are reviewed.

## 26.1 John Bell and the Chiral Anomaly

I expect that most everyone in this audience knows that John Bell codiscovered the mechanism of anomalous symmetry breaking in quantum field theory. Indeed, our paper on this subject [1] is his (and my) most-cited work. The symmetry breaking in question is a quantum phenomenon that violates the correspondence principle; it arises from the necessary infinities of quantum field theory. Over the years it has become evident that theoretical/mathematical physicists are not the only ones to acknowledge this effect. Nature makes fundamental use of the anomaly in at least two ways: the neutral pion's decay into two photons is controlled by the anomaly [1, 2] and elementary fermions (quarks and leptons) arrange themselves in patterns such that the anomaly cancels in those channels to which gauge bosons – photon, W, Z – couple [3]. (There are also phenomenological applications of the anomaly to collective, as opposed to fundamental, physics – for example, to edge states in the quantum Hall effect.) Beyond physics, in mathematics one finds closely related structures, such as Atiyah–Singer index theory, zero modes of Dirac operators, Chern–Pontryagin characteristics of gauge fields, Chern–Simons secondary characteristics. The mathematical ideas were developed at nearly the same time as the physical ones, and this unexpected conjunction between physics and mathematics seeded joint activity that flourishes to this day.

Once it was appreciated that the anomaly is not merely an obscure pathology of quantum field theory, but reflects an as-yet-to-be-understood wrinkle in the field-theoretic description of Nature, many people wrote many papers providing various and alternative derivations of the result. But not John Bell. It seems that all he had to say on the subject was contained in our first joint paper. He did follow the subsequent developments and elaborations, and he commented on them to me whenever we met in Geneva or Cambridge.

In his characteristically diffident manner, he did not always support the various fancy elaborations, and he remained skeptical about their value. On the contrary, he insisted on one statement, which already appeared in our original paper, but which perhaps had not been forcefully enunciated, so I want to call attention to it here.

Our original analysis concerns the correlation function for three currents – one axial vector and two vector currents – which is given in lowest order by the fermionic triangle graph. With massless fermions, this correlation function should be transverse in all three channels, the axial vector and both vectors, as a consequence of various symmetries in the theory. The anomaly manifests itself in that any evaluation of the relevant diagram is ambiguous up to a local term, owing to the underlying infinities of the quantum field theory. Moreover, no matter how one fixes the ambiguity, the transversality conditions fail – the calculated amplitude is not transverse in all three channels. Thus the full extent of symmetry, which would ensure full transversality, is broken. While specific choices for resolving the ambiguity allow transversality in some (but not all three) channels, John Bell always insisted that there is no intrinsic way to select a "correct" result. And this opinion informed his criticism of alternative derivations of the anomaly, which usually preserve vector transversality at the expense of axial vector transversality. Whenever we discussed yet another new approach to the anomaly, he always wanted to verify that there remained the possibility of obtaining a variety of results. If this freedom were not present, he would dismiss the rederivation as too restrictive.

One alternative viewpoint did appeal to John Bell. The lack of transversality in a particular channel can be ascribed to an anomalous nonconservation of the appropriate "symmetry" current, with the nonvanishing anomalous current divergence determined by the gauge fields with which the fermions interact. But from another perspective, one could state that transversality fails because the commutators between the relevant current operators differ from the corresponding Poisson brackets of their classical antecedents – the commutators contain additional, quantal (anomalous) contributions. (Indeed, this is the point of view one must adopt in the interaction picture, where operator equations do not see the interaction.) The form for the anomalous commutators is uniquely determined by the triangle graph, yet the transversality conditions retain their ambiguity, because they continue to reflect the arbitrariness in the local contribution to the graph. This gives an appealing algebraic characterization of the anomaly [4]. Either point of view signals absence of the expected symmetries and exposes a breakdown of the correspondence principle.

John Bell's insistence that the triangle graph supports a variety of anomalies is not merely a pedantic nicety. In fact it has a physical realization in Gerard 't Hooft's calculation of fermion-number nonconservation in the standard model [5], where the vectorial fermion number current in the triangle graph carries the anomaly, and the chiral current is anomaly free [6].

## References

1. J.S. Bell, R. Jackiw: Nuovo Cimento A **60**, 47 (1969)
2. S.L. Adler: Phys. Rev. **182**, 1459 (1969)
3. D.J. Gross, R. Jackiw: Phys. Rev. D **6**, 477 (1972); C. Bouchiart, J. Iliopoulos, Ph. Meyer: Phys. Lett. 38 **B**, 519 (1972); G. 't Hooft: 'Naturalness, chiral symmetry, and spontaneous chiral symmetry breaking'. In: *Recent Development in Gauge Theories*, ed. by G. 't Hooft et al. (Plenum, New York 1980), reprinted in *Dynamical Gauge Symmetry Breaking*, ed. by E. Farhi, R. Jackiw (World Scientific, Singapore 1982)
4. R. Jackiw, K. Johnson: Phys. Rev. **182**, 1459 (1969); L.D. Faddeev: Phys. Lett. B **145**, 81 (1984); J. Mickelsson: Comm. Math. Phys. **97**, 361 (1985)
5. G. 't Hooft: Phys. Rev. Lett. **37**, 8 (1976)
6. For further reading, see S. Treiman, R. Jackiw, B. Zumino, E. Witten: *Current Algebra and Anomalies* (Princeton Univ. Press/World Scientific, Princeton, NJ/Singapore 1985); R. Bertlmann: *Anomalies in Quantum Field Theory* (Oxford Univ. Press, Oxford 2000)

## 26.2 Descendants of the Anomaly

The axial anomaly, that is, the departure of transversality of the 3-fermion current correlation function, involves $^*FF$, an expression constructed from the gauge fields to which the fermions couple. Specifically, in the Abelian case one encounters

$$^*F^{\mu\nu}F_{\mu\nu} = \frac{1}{2}\varepsilon^{\mu\nu\alpha\beta}F_{\mu\nu}F_{\alpha\beta} = -4\boldsymbol{E}\cdot\boldsymbol{B}\ , \tag{26.1}$$

where $F_{\mu\nu}$ is the covariant electromagnetic tensor

$$F_{\mu\nu} = \partial_\mu A_\nu - \partial_\mu A_\nu \tag{26.2a}$$

while $\boldsymbol{E}$ and $\boldsymbol{B}$ are the electric and magnetic fields

$$E^i = F^{io}\ , \quad B^i = -\frac{1}{2}\varepsilon^{ijk}F_{jk}\ . \tag{26.2b}$$

The non-Abelian generalization reads

$$^*F^{\mu\nu a}F^a_{\mu\nu} = \frac{1}{2}\varepsilon^{\mu\nu\alpha\beta}F^a_{\mu\nu}\ F^a_{\alpha\beta}\ , \tag{26.3}$$

where $F^a_{\mu\nu}$ is Yang-Mills gauge field strength

$$F^a_{\mu\nu} = \partial_\mu A^a_\nu - \partial_\nu A^a_\mu + f^{abc}A^b_\mu A^c_\nu \tag{26.4}$$

and $a$ labels the components of the gauge group, whose structure constants are $f^{abc}$.

The quantity $^*FF$ is topologically interesting. Its integral over 4-space is quantized, and measures the topological class (labeled by integers) to which the vector potential $A$ belongs. Consequently, the 4-volume integral of $^*FF$ is a topological invariant and we expect that, as befits a topological invariant, it should be possible to present $^*FF$ as a total derivative, so that its 4-volume integral becomes converted by Gauss' law into a surface integral, sensitive only to long distance, global properties of the gauge fields. That the total derivative form for $^*FF$ holds is seen when $F_{\mu\nu}$ is expressed in terms of potentials. In the Abelian case, we use (26.2a) and find immediately

$$\frac{1}{2}{}^*F^{\mu\nu}F_{\mu\nu} = \partial_\mu\left(\varepsilon^{\mu\alpha\beta\gamma}A_\alpha\partial_\beta A_\gamma\right) . \tag{26.5}$$

For the non-Abelian fields, (26.4) establishes the desired result:

$$\frac{1}{2}{}^*F^{\mu\nu a}F^a_{\mu\nu} = \partial_\mu\varepsilon^{\mu\alpha\beta\gamma}\left(A^a_\alpha\partial_\beta A^a_\gamma + \frac{1}{3}f^{abc}A^a_\alpha\, A^b_\beta A^c_\gamma\right) . \tag{26.6}$$

The quantities whose divergence gives $^*FF$ are called Chern–Simons terms. By suppressing one dimension they become naturally defined on a 3-dimensional manifold (they are 3-forms), and we are thus led to consider the Chern–Simons terms in their own right [1]:

$$\mathrm{CS}(A) = \varepsilon^{ijk}A_i\partial_j A_k \qquad \text{(Abelian)} \tag{26.7}$$

$$\mathrm{CS}(A) = \varepsilon^{ijk}\left(A^a_i\partial_j A^a_k + \frac{1}{3}f^{abc}A^a_i\, A^b_j A^c_k\right) \qquad \text{(non-Abelian).} \tag{26.8}$$

The 3-dimensional integral of these quantities is again topologically interesting. When the non-Abelian Chern–Simons term is evaluated on a pure gauge, non-Abelian vector potential

$$A_i = g^{-1}\partial_i g \tag{26.9}$$

the 3-dimensional volume integral of $\mathrm{CS}(g^{-1}\partial g)$ measures the topological class (labeled by integers) to which $g$ belongs. The integral in the Abelian case – the case of electrodynamics – is called the magnetic helicity $\int \mathrm{d}^3r\boldsymbol{A}\cdot\boldsymbol{B}$, $\boldsymbol{B} = \boldsymbol{\nabla}\times\boldsymbol{A}$, and measures the linkage of magnetic flux lines. An analogous quantity arises in fluid mechanics, with the local fluid velocity $\boldsymbol{v}$ replacing $\boldsymbol{A}$, and the vorticity $\boldsymbol{\omega} = \boldsymbol{\nabla}\times\boldsymbol{v}$ replacing $\boldsymbol{B}$. Then the integral $\int \mathrm{d}^3r\boldsymbol{v}\cdot\boldsymbol{\omega}$ is called kinetic helicity [2].

I shall not review here the many uses to which the Chern–Simons terms, Abelian and non-Abelian, introduced in [1], have been put. The applications range from the mathematical characterization of knots to the physical description of electrons in the quantum Hall effect [3], vivid evidence for the deep significance of the Chern–Simons structure and of its antecedent, the chiral anomaly.

Instead I pose the following question: Can one write the Chern–Simons term as a total derivative, so that (as befits a topological quantity) the spatial volume integral becomes a surface integral. An argument that this should be possible is the following: The Chern–Simons term is a 3-form on 3-space, hence it is maximal and its exterior derivative vanishes because there are no 4-forms on 3-space. This establishes that on 3-space the Chern–Simons term is closed, so one can expect that it is also exact, at least locally, that is, it can be written as a total derivative. Of course, such a representation for the Chern–Simons term requires expressing the potentials in terms of "pre-potentials", since the formulas (26.7, 26.8) show no evidence of derivative structure. [Recall that the total derivative formulas (26.5, 26.6) for the axial anomaly also require using potentials to express $F$.]

There is a physical, practical reason for wanting the Abelian Chern–Simons term to be a total derivative. It is known in fluid mechanics that there exists an obstruction to constructing a Lagrangian for Euler's fluid equations, and this obstruction is just the kinetic helicity $\int \mathrm{d}^3 r \boldsymbol{v} \cdot \boldsymbol{\omega}$, that is, the volume integral of the Abelian Chern–Simons term, constructed from the velocity 3-vector $\boldsymbol{v}$. This obstruction is removed when the integrand is a total derivative, because then the kinetic helicity volume integral is converted to a surface integral by Gauss' theorem. When the integral obtains contributions only from a surface, the obstruction disappears from the 3-volume, where the fluid equation acts [4].

It is easy to show that the Abelian Chern–Simons term can be presented as a total derivative. We use the Clebsch parameterization for a 3-vector [5]

$$\boldsymbol{A} = \boldsymbol{\nabla}\theta + \alpha \boldsymbol{\nabla}\beta \, . \tag{26.10}$$

This nineteenth-century parameterization of the a 3-vector $\boldsymbol{A}$ in terms of the prepotentials ($\theta$, $\alpha$, $\beta$) is an alternative to the usual transverse/longitudinal parameterization. In modern language it is a statement of Darboux's theorem that the 1-form $A_i \, \mathrm{d}r^i$ can be written as $\mathrm{d}\theta + \alpha \, \mathrm{d}\beta$ [6]. With this parameterization for $\boldsymbol{A}$, one sees that the Abelian Chern–Simons term is indeed a total derivative:

$$\begin{aligned} \mathrm{CS}(A) &= \varepsilon^{ijk} A_i \partial_j A_k \\ &= \varepsilon^{ijk} \partial_i \theta \partial_j \alpha \partial_k \beta \\ &= \partial_i \left( \varepsilon^{ijk} \theta \partial_j \alpha \partial_k \beta \right) \, . \end{aligned} \tag{26.11}$$

When the Clebsch parameterization is employed for $\boldsymbol{v}$ in the fluid dynamical context, the situation is analogous to the force law in electrodynamics. While the Lorentz equation is written in terms of field strengths, a Lagrangian formulation needs potentials from which the field strengths are reconstructed. Similarly, Euler's equation involves the velocity vector $\boldsymbol{v}$, but in a Lagrangian for this equation the velocity must be parameterized in terms of the prepotentials $\theta$, $\alpha$, and $\beta$.

In a natural generalization of the above, one asks whether a non-Abelian vector potential can also be parameterized in such a way that the non-Abelian Chern–Simons term (26.8) becomes a total derivative. We have answered this question affirmatively and we have found appropriate prepotentials that do the job [4, 7, 8], but the details of the construction are too technical to be presented here. We hope that our non-Abelian generalization of the Clebsch parameterization will be as interesting and useful as the Abelian one.

## References

1. S. Deser, R. Jackiw, S. Templeton: Ann. Phys. (NY) **149**, 372 (1982); Ann. Phys. (E) **185**, 406 (1985)
2. Fluid mechanics and magnetohydrodynamics were the contexts in which the Abelian Chern–Simons term made its first appearance: L. Woltier: Proc. Nat. Acad. Sci. **44**, 489 (1958)
3. S. Deser: *IHES Publications in Mathematics*, 1847 (1998); Faddeev Festschrift, Proc. V.A. Steklov Inst. Math. **226**, 180 (1999)
4. For a review, see R. Jackiw: *Lectures on Fluid Dynamics* (Springer, Berlin 2002)
5. For a review, see H. Lamb: *Hydrodynamics* (Cambridge Univ. Press, Cambridge 1932) p. 248 and [4]
6. A constructive discussion of the Darboux theorem is by R. Jackiw: '(Constrained) Quantization without Tears'. In: *Constraint Theory and Quantization Methods*, ed. by F. Colomo, L. Lusann, G. Marmo (World Scientific, Singapore 1994), reprinted in R. Jackiw: *Diverse Topics in Theoretical and Mathematical Physics* (World Scientific, Singapore 1996)
7. R. Jackiw, S.-Y. Pi: Phys. Rev. D **61**, 105015 (2000)
8. R. Jackiw, V.P. Nair, S.-Y. Pi: Phys. Rev. D **62**, 085018 (2000)

# 27 Fractional Charge

R. Rajaraman

The origin and quantum status of Fractional Charge are reviewed, along with reminiscences of collaboration with John Bell on the subject.

## 27.1 Introduction

I cannot claim the privilege of having known Professor John Bell as intimately or for as long as some of the other participants in this conference to honour his memory. But there was a period of about 3 years from 1982 to 1985 when we interacted a fair amount and conducted (mostly at long distance) a collaboration trying to understand the then recently discovered phenomenon of fractional charge. During this period we had the opportunity of hosting him at my Institute in Bangalore. He came home for dinner and met my family. I in turn went to CERN, first to write up our work and later for a long visit, when I had the pleasure of meeting his wife Dr. Mary Bell and was the beneficiary of their kindness in many ways.

During this interaction I developed, like so many people before me, great respect for Professor Bell, not just for his great originality and prowess as a theoretical physicist, but equally for the precision of thought and language that he demanded of himself and of those who were fortunate enough to work with him. John Bell was more than an intellectual force in physics – his was a moral presence. One can ill-afford the loss of such a moral presence in today's world, and it is a great tragedy that he was taken away from us when we continue to need him so much.

Long before I met John, I had of course heard of him and his many important contributions. In fact the very first time I had to study his work was in the early 'sixties, when I was a foot-soldier in Professor Hans Bethe's army attacking the nuclear matter problem. All of you know that John had worked on a wide range of subjects, but not all might know that one of his early interests was nuclear matter, on which he had co-authored a long review article with E.J. Squires way back in 1961. A few years after that as I gravitated towards particle physics and quantum field theory I had to study in great detail his landmark discovery with Roman Jackiw of the axial current anomaly.

Despite such overlap of interests, I met him only much later in life when he spent a couple of weeks at the Indian Institute of Science in Bangalore where I was then working. A colleague introduced us and he made the usual polite enquiry – "What are you working on these days?" I thought that rather than inflict on him the different problems I was working on at that time, I would instead get his "take" on the very interesting but puzzling claims of fractionally charged states that had recently emerged.

As it turned out, he had not heard about these developments. "You mean fractionally charged like quarks? Isn't that by now old and well established?", says John, a slight note of wariness creeping into his voice. "No, no", I hasten to add, "I don't mean quarks. By fractionally charged I refer to states with fractional eigenvalues of the number operator. These have been discovered not just theoretically in soliton states of field theory models, but are claimed to be present in down to earth systems like polymer chains. What I don't understand is how a real polymer made of some finite number of electrons could, no matter how it twists itself into a soliton configuration, carry a fractional number of electrons." This produced a flicker of a smile from him and I knew I had caught his interest. "Would you like to tell me about it?", he asked. So I ushered him into my office and closed the door. This was a lovely opportunity, having trapped the distinguished John Bell in one's office with all avenues of escape sealed off, to summarize for him *ab initio* the curious phenomenon of fractional charge and get his opinion.

Let me begin the next section with a pedagogical expansion of what I described to John in the fading light of that winter afternoon in Bangalore.

## 27.2 Fractional Charge in Field Theory

The phenomenon of fractional charge was first discovered by *Jackiw* and *Rebbi* in their pioneering work in 1976 in the context of soliton states in quantum field theory [1]. Three years after the Jackiw–Rebbi work, but quite independently and in an entirely different context, *Su*, *Schrieffer* and *Heeger* [2], argued that a similar phenomenon can occur for soliton states in the long chain molecule *trans*-Polyacetylene. See also the work by *Rice* [3]. Subsequently *Jackiw* and *Schrieffer* [4] wrote a paper drawing attention to the similarity between these results discovered in entirely different areas of physics.

Let us begin with the Jackiw–Rebbi work. They discussed the phenomenon in both one-dimensional and three-dimensional models. Let me for simplicity discuss just the one-dimensional example, which contains all the essential ideas. Consider in (1+1) dimensions a Fermi field $\Psi(x,t)$ coupled to a scalar field $\Phi(x,t)$ through a Lagrangian density

$$\mathcal{L} = \mathcal{L}_B + \mathcal{L}_F \tag{27.1}$$

where,

$$\mathcal{L}_B = \frac{1}{2g^2}\left[\left(\frac{\partial\Phi}{\partial t}\right)^2 - \left(\frac{\partial\Phi}{\partial x}\right)^2 - (1/2)(\Phi^2-1)^2\right] \tag{27.2}$$

and

$$\mathcal{L}_F = \bar{\Psi}\left[\mathrm{i}\partial_\mu \gamma^\mu - m\Phi(x,t)\right]\Psi \ . \tag{27.3}$$

This example corresponds to a quartic double well potential in $\Phi$. It is straighforward to generalise the results to other potentials $U(\Phi)$ with symmetric double wells. The phenomenon of fractional charge occurs in the soliton sector of this system. To a wider readership it may be helpful to recall what is meant by the soliton sector and how fermi fields are treated in such sectors [5].

The bosonic sub-system $\mathcal{L}_B$, in the absence of the Fermi field, has as its field equation

$$\left(\frac{\partial^2}{\partial t^2} - \nabla^2\right)\phi - \phi + \phi^3 = 0 \ . \tag{27.4}$$

This has the following two lowest energy classical solutions:

$$\phi(x,t) = \pm 1 \ . \tag{27.5}$$

As is well known the presence of these two degenerate classical solutions indicates spontaneous breaking of the $\Phi \leftrightarrow -\Phi$ symmetry of the Lagrangian. Around each of these classical solutions a separate vacuum state and a whole tower of Fock states can be built corresponding to the two phases this system permits. We will call these the vacuum sectors.

In addition this system also has two other static (time independent) solutions. Those are the topological soliton solutions – the so called kink solution and its reflected partner, the anti-kink, given respectively by

$$\phi_S(x) = \pm \tanh\left(x/\sqrt{2}\right) \ . \tag{27.6}$$

As per the general theory of semi classical quantization of quantum fields, one can build two other separate towers of states, one around each of these soliton solutions. The topological index $n = \phi(\infty) - \phi(-\infty)$ becomes a superselection quantum number upon quantization and prevents any states from these soliton sectors from decaying into the vacuum sector. In short, we have four sectors of states for this bosonic system: two are the vacuum sectors built around $\phi = \pm 1$ and the other two are the soliton and antisoliton sectors built around the kink and antikink solutions in (27.6). For more details on soliton sectors of states and their properties see [5].

Now consider the full system in (27.1) including the Fermi field.

### 27.2.1 Vacuum Sector

To leading order in $\hbar$ one can replace in each sector the Bose field operator $\Phi$ occurring in the fermi field Lagrangian (27.3) by the corresponding classical solution. Thus in the vacuum sector built around $\phi = 1$, the Fermi Lagrangian (27.3) reduces in leading order to

$$\mathcal{L}_F = \bar{\Psi}\,(\mathrm{i}\partial_\mu \gamma^\mu - m)\,\Psi\ . \tag{27.7}$$

This is just the free Dirac system with mass $m$ discussed in textbooks. In order to contrast with what later happens in the soliton sector, let us recall why the number (total charge) operator has only integral eigenvalues for the free Dirac system. Let us denote by $u_k(x)$ and $\tilde{u}_k(x)$ the positive and negative energy spinorial solutions of the Dirac equation

$$\begin{aligned}(-\mathrm{i}\alpha\partial_x + \beta m)u_k(x) &= E_k u_k(x)\ ,\\ (-\mathrm{i}\alpha\partial_x + \beta m)\tilde{u}_k(x) &= -E_k \tilde{u}_k(x)\ ,\end{aligned} \tag{27.8}$$

where $E_k = +\sqrt{k^2+m^2}$ and spinor indices have been suppressed. The Dirac matrices for this 1+1 dimensional system can be taken to be $\alpha = \sigma_2$ and $\beta = \sigma_1$. The Dirac field is expanded in terms of these solutions and the destruction operators $b_k$ and $d_k$ obeying the usual anticommutation rules

$$\Psi(x,t) = \sum_k \left(b_k u_k\,\mathrm{e}^{-\mathrm{i}E_k t} + d_k^\dagger \tilde{u}_k\,\mathrm{e}^{\mathrm{i}E_k t}\right)\ . \tag{27.9}$$

The vacuum state in the $\phi = 1$ sector is given by the conditions

$$b_k\,|\mathrm{vac}\rangle = d_k\,|\mathrm{vac}\rangle = 0 \tag{27.10}$$

with all the bosonic oscillators being in the ground state.

Note that the third Pauli matrix $\sigma_3$ acts as the charge conjugation matrix. It anti commutes with the Dirac hamiltonian in (27.8) and generates for every positive energy solution $u_k(x)$ of energy $E_k$ the corresponding negative energy solution of energy $-E_k$:

$$\sigma_3 u_k(x) = \tilde{u}_k(x)\ . \tag{27.11}$$

Hence all modes of the expansion (27.9) come in pairs with positive and negative energy. There are no zero energy solutions in the free massive Dirac equation.

Finally consider the charge density operator (which is really the number density operator)

$$\rho(x,t) = \frac{1}{2}\left[\Psi^\dagger(x,t), \Psi(x,t)\right]\ . \tag{27.12}$$

This commutator form of $\rho$ is standard; it is designed to be regularised and odd under charge conjugation. Inserting the mode expansion (27.9) and using the orthonormality of the Dirac solutions the total charge becomes

$$\begin{aligned} Q &\equiv \int \mathrm{d}x \rho(x,t) \\ &= \frac{1}{2} \sum_k \left( \left[ b_k^\dagger, b_k \right] + \left[ d_k, d_k^\dagger \right] \right) \\ &= \sum_k \left[ \left( b_k^\dagger b_k - 1/2 \right) - \left( d_k^\dagger d_k - 1/2 \right) \right] \\ &= \sum_k \left( b_k^\dagger b_k - d_k^\dagger d_k \right) . \end{aligned} \tag{27.13}$$

Notice that the half-integers cancel term by term because of the existence of paired positive and energy modes. Since

$$\left( b_k^\dagger b_k \right)^2 = b_k^\dagger b_k \, , \quad \left( d_k^\dagger d_k \right)^2 = d_k^\dagger d_k \tag{27.14}$$

these have eigenvalues of only 0 or 1. Hence the familiar result in the vacuum sector that the charge operator has only integer eigenvalues.

### 27.2.2 Soliton Sector

Let us repeat exactly the same steps in the soliton sector. Now we have to substitute for the Bose field operator in the Dirac Lagrangian the classical kink function in (27.6). The corresponding Dirac equation now becomes

$$\left( -\mathrm{i}\alpha \partial_x + \beta m \tanh \frac{x}{\sqrt{2}} \right) \psi(x) = E \psi(x) \, . \tag{27.15}$$

For the two components $\psi_{1,2}$ of the Dirac spinor, this yields the coupled equations

$$\begin{aligned} \left( -\partial_x + m \tanh \frac{x}{\sqrt{2}} \right) \psi_2 &= E \psi_1 \, , \\ \left( \partial_x + m \tanh \frac{x}{\sqrt{2}} \right) \psi_1 &= E \psi_2 \, . \end{aligned} \tag{27.16}$$

This equation will also have a set of positive energy solutions $\eta_k(x)$ with some associated energy $E_k$. We need not find these solutions explicitly. But since the charge conjugation matrix $\sigma_3$ again anticommutes with the Dirac hamiltonian, we know that for every positive energy solution $\eta_k(x)$ there will exist a negative energy solution $\tilde{\eta}_k(x)$ with energy $-E_k$. But now there is also an unpaired zero-energy solution

$$\eta_0 = \begin{pmatrix} A \exp \left[ -m \int^x \mathrm{d}y \tanh \left( y/\sqrt{2} \right) \right] \\ 0 \end{pmatrix} . \tag{27.17}$$

Such a normalisable solution to (16) exists because the soliton function $\tanh x/\sqrt{2}$ which forms the background potential for the Dirac spinor tends to opposing limits $\pm 1$ as $x \to \pm\infty$. In infinite spatial volume this solution has no partner. It is self charge conjugate: $\sigma_3\eta_0 = \eta_0$.

The mode expansion of the Fermi field operator now becomes

$$\Psi(x,t) = \sum_{k\neq 0}\left[b_k\eta_k(x)\,\mathrm{e}^{-\mathrm{i}E_kt} + d_k^\dagger\tilde{\eta}_k(x)\,\mathrm{e}^{\mathrm{i}E_kt}\right] + a\eta_0(x)\,, \tag{27.18}$$

where a is the destruction operator for the zero-mode.

Unlike the vacuum sector built around $\phi = 1$ which had a unique ground state ("the vacuum" in that sector), in the soliton sector there are two degenerate ground states because of the existence of the fermionic zero-mode. They are $|\mathrm{sol}\rangle$ and $|\hat{\mathrm{sol}}\rangle$ obeying

$$a\,|\mathrm{sol}\rangle = b_k\,|\mathrm{sol}\rangle = d_k\,|\mathrm{sol}\rangle = 0 \tag{27.19}$$

and

$$\left|\hat{\mathrm{sol}}\right\rangle \equiv a^\dagger\,|\mathrm{sol}\rangle\ ; \quad a\left|\hat{\mathrm{sol}}\right\rangle = |\mathrm{sol}\rangle\ . \tag{27.20}$$

These are the two basic quantum soliton states of this system. They are energetically degenerate, but are distinguishable by their Charge.

$$\begin{aligned}
Q &\equiv \frac{1}{2}\int \mathrm{d}x\,\left[\Psi^\dagger(x,t),\Psi(x,t)\right] \\
&= \frac{1}{2}\sum_k\left(\left[b_k^\dagger,b_k\right] + \left[d_k,d_k^\dagger\right]\right) + 1/2[a^\dagger,a] \\
&= \sum_k\left[\left(b_k^\dagger b_k - 1/2\right) - \left(d_k^\dagger d_k - 1/2\right)\right] + (a^\dagger a - 1/2) \\
&= \sum_k\left(b_k^\dagger b_k - d_k^\dagger d_k\right) + a^\dagger a - 1/2\,.
\end{aligned} \tag{27.21}$$

Notice that the piece $(-1/2)$ coming from the zero-mode commutator remains uncancelled because it does not have a charge conjugate partner. Since the operators $b_k^\dagger b_k, d_k^\dagger d_k$ and $a^\dagger a$ all have eigenvalues of 0 or 1, it then follows that the total charge (number) operator $Q$ has half-integral eigenvalues. It should be emphasized that (27.21) is an operator equation for $Q$. The half-integer appearing in it will be reflected in its *eigenvalues* and not just its expectation values. In particular, the two degenerate soliton states have eigenvalues of $\pm 1/2$, respectively, for the total number operator $Q$:

$$Q\,|\mathrm{sol}\rangle = -(1/2)\,|\mathrm{sol}\rangle\ ; \quad Q\left|\hat{\mathrm{sol}}\right\rangle = (1/2)\left|\hat{\mathrm{sol}}\right\rangle\ . \tag{27.22}$$

This, in its barest form, is the prototype example of the Jackiw–Rebbi discovery that in some field theories there can be states carrying fractional eigenvalues of the Number operator. For a fuller discussion including similar results in three-dimensional models see their original paper [1].

## 27.3 Polyacetylene

This discovery by Jackiw and Rebbi was clearly very remarkable. But it is not easy to "understand" the result physically in terms of our familiar intuition with quantum field theory. How can the number operator which is widely used to count the number of particles in QED and so many other theories, and which even in the Higgs model used by Jackiw and Rebbi has only integer eigenvalues in the vacuum sector, yield fractional values in another sector? Yet the proof given is so simple and transparent that, stare at it as we may, we have no choice but to accept the result within the parameters of its derivation. In my own attempts to make peace with the result (before embarking on the more detailed study with John), I loosely attributed it to the vagaries of the infinite degrees of freedom of continuum field theory. (We know from high school that formally summing divergent series of integers can yield fractions.) But clearly some further clarification of this phenomenon was called for.

The need for clarification became more compelling when a similar result was derived not in a model field theory, but in a down-to-earth experimentally accessible polymer system by *Su*, *Schrieffer* and *Heeger* [2]. Three years after the Jackiw-Rebbi work, but quite independently they showed that the same phenomenon of fermion number 1/2 can occur for soliton states in the long chain molecule *trans*-polyacetylene.

That the same phenomenon occurs in polyacetylene (we will henceforth drop the prefix "trans") as in the Jackiw–Rebbi field theory is not accidental. As pointed out by *Jackiw* and *Schrieffer* [4], the former system has the same structure in the continuum limit as the field theory model. Polyacetylene is the molecule $(\mathrm{CH})_\mathrm{n}$ with large $n$, where the carbon ions form a long chain with the H atoms sticking out transverse to the chain. If we consider the flow of electrons along the chain, the system can be viewed as consisting of electrons and bosons (the phonons of lattice vibration of the carbon ions along the chain) in one space dimension just as in the field theory model.

The Hamiltonian for the electron–lattice system can be taken, for each of the two spin states of the electron and for small ion displacements as [4]:

$$H = \sum_n \left\{ \left[ \frac{p_n^2}{2\mu} + K/2(u_{n+1} - u_n)^2 \right] \right.$$
$$\left. + \left( D_{n+1}^\dagger D_n + \mathrm{h.c.} \right) \left[ u_{n+1} - u_n - 1/(2a) \right] \right\} , \qquad (27.23)$$

where $n$ labels lattice sites, $p_n$ and $u_n$ are ion momenta and displacements, respectively, $D_n$ is electron destruction operator and $K$ and $a$ are constants. All other constants have been absorbed into definitions for simplicity.

Our primary interest is in the fermions. As far as the lattice vibrations go, let us just state the *Su* et al. result [2] that the lattice system undergoes dimerisation doubling its spatial period (a Peierls transition). There are two degenerate ground states. In one, the mean value of the displacement $u_n$

instead of being zero takes the staggered value of

$$\phi_n \equiv 4(-1)^n u_n = 1 \tag{27.24}$$

for all $n$. In the other

$$\phi_n = -1\,. \tag{27.25}$$

The two degenerate ground states further provide the possibility of domain wall (soliton) configurations connecting the two phases, i.e. $\phi_n \to \pm 1 (or \mp 1)$, as $n \to \pm\infty$, respectively. Thus the boson subsystem has four sectors of states – two of them being soliton sectors – just as in the field theory discussed in the previous section.

Furthermore, the boson coordinates $u_n$ again act as the background potential for the electrons. In fact the electronic part of the Hamiltonian (27.23) becomes in the continuum approximation exactly the same as that of the Dirac field theory. John and I, as part of our work on these systems [9], offered a simple derivation showing that the polyacetylene hamiltonian is just a realisation of the *Kogut–Susskind* lattice regularisation of the Dirac system [6]. Let me sketch that derivation. The electronic part of (27.23) is

$$H_{\text{elec}} = \sum_n \left\{ D^\dagger_{n+1} D_n \left[u_{n+1} - u_n - 1/(2a)\right]\right\} + \text{h.c.} \tag{27.26}$$

Define staggered variables

$$B_{2r-1} \equiv (-1)^r D_{2r-1}\,; \quad C_{2r} \equiv (-1)^r D_{2r} \tag{27.27}$$

with $\phi_n \equiv 4(-1)^n u_n$ for all $n$ as already defined above. Then,

$$\begin{aligned} H_{\text{elec}} &= \sum_r \Big\{ \Big[ D^\dagger_{2r+1} D_{2r}(u_{2r+1} - u_{2r} - 1/2a) \\ &\quad + D^\dagger_{2r} D_{2r-1}(u_{2r} - u_{2r-1} - 1/2a)\Big] + \text{h.c.}\Big\} \\ &= (1/4)\sum_r \Big[ B^\dagger_{2r+1} C_{2r}(\phi_{2r+1} + \phi_{2r} + 2/a) \\ &\quad + C^\dagger_{2r} B_{2r-1}(\phi_{2r} + \phi_{2r-1} - 2/a)\Big] + \text{h.c.} \\ &= \sum_r \left(\frac{B^\dagger_{2r+1} - B^\dagger_{2r-1}}{2a}\right) C_{2r} + (1/4) B^\dagger_{2r+1} C_{2r}(\phi_{2r+1} + \phi_{2r}) \\ &\quad + (1/4) B^\dagger_{2r-1} C_{2r}(\phi_{2r} + \phi_{2r-1}) + \text{h.c.} \end{aligned} \tag{27.28}$$

In the continuum limit, as $a \to 0$ and $\sum_r \to \int dx$, this reduces to

$$\begin{aligned} H_{\text{elec}} &= \int dx \left(\frac{\partial B^\dagger}{\partial x} C + C^\dagger \frac{\partial B}{\partial x}\right) + \phi(x)(B^\dagger C + C^\dagger B) \\ &= \int dx \left(-B^\dagger \frac{\partial C}{\partial x} + C^\dagger \frac{\partial B}{\partial x}\right) + \phi(x)(B^\dagger C + C^\dagger B)\,. \end{aligned} \tag{27.29}$$

This is just the Dirac Hamiltonian with Yukawa coupling:

$$H_F = \Psi^\dagger \left[ -\mathrm{i}\sigma_2 \frac{\partial}{\mathrm{d}x} + \sigma_1 \phi(x) \right] \Psi \tag{27.30}$$

with the upper and lower components of the 2-spinor being identified respectively with the odd-and even-site electron operators:

$$\Psi(x) \equiv \begin{pmatrix} B(x) \\ C(x) \end{pmatrix} \, . \tag{27.31}$$

The charge operator operator at each site also has the same commutator form as in the continuum theory's charge density:

$$\rho_n = D_n^\dagger D_n - (1/2) = (1/2) \left[ D_n^\dagger, D_n \right] \, , \tag{27.32}$$

where the $1/2$ subtracted at each site can be attributed to the neutralizing background (ionic) charge *per spin state.*

Given this mapping from the polyacetylene system to the field theory model used by Jackiw and Rebbi, it is not surprising that fractional charged states arise in the former system too.

But in the context of a real polymer the result becomes still more mysterious. A molecule of polyacetylene has after all some finite number of electrons and it is hard to imagine how, even if it twists itself into some topological soliton state, it could have a fractional number of electrons. The need to understand this better became all the more compelling because *Su* et al. referred to experimental signals supporting the result. See for instance [7]. [Because of the presence of the two spin degrees of freedom in actual polyacetylene, as distinct from a truly 1+1 dimensional model, the fractional charge gets doubled and becomes integral; but a signature of the effect can still be observed in the form of an unfamiliar charge–spin combination of excitations. There will be neutral excitations with spin and charged excitation which are spinless.]

It should be emphasised that an *expectation value* of one-half would cause no surprise. Obviously that can easily arise for an operator with integral eigenvalues. For instance if we take any particle in the ground state of a double-well potential and look for it in one of the wells, you will find it there half the times and not find it there the other half, giving an expectation value of $1/2$. While in the Su et al. paper, fractional values were not claimed as eigenvalues, in the prototype field theoretic case we have explicitly seen in Sect. 27.2 that the eigenvalues themselves are half-integers. This is what puzzled me and I was fortunate in having John Bell join me in worrying about it further.

## 27.4 Eigenvalue or Expectation Value?

Now let me summarize what John and I did in our attempt to understand better what these half integral values of the number operator mean [8, 9].

[We later learnt that *Kivelson* and *Schrieffer* had also been independently investigating the eigenvalue status of fractional charge around the same time with essentially a similar resolution of the issues [10]. See also the work of *Jackiw* et al. which provided further clarification [11]. I will describe here only the work done with John Bell.] Our strategy was to treat the infrared and ultraviolet limits of the problem more carefully. In one paper [8], we looked at the continuum field theory used by Jackiw and Rebbi, but in a finite volume, which could be later taken to infinity. In a subsequent paper we considered what would happen if we put an ultraviolet cut-off as well, by going to the finite polyacetylene chain [9].

Let me begin with the field theory example of Jackiw and Rebbi which was discussed in Sect. 27.2. The result derived there that the total charge operator has half-integral eigenvalues cannot be disputed in the infinite spatial volume case ($L \to \infty$ in the one dimensional field theory model considered there). However suppose we kept the spatial volume $2L$ finite to start with, and later go the $L \to \infty$ limit [8]. In particular, in the soliton sector, we go back and solve the same Dirac equation

$$\begin{aligned} \left(-\partial_x + m \tanh \frac{x}{\sqrt{2}}\right) \psi_2 &= E\psi_1 \ , \\ \left(\partial_x + m \tanh \frac{x}{\sqrt{2}}\right) \psi_1 &= E\psi_2 \end{aligned} \tag{27.33}$$

but in a finite volume ($-L \leq x \leq L$) with appropriate boundary conditions on the spinor $\psi_\alpha$ at $x = \pm L$. The natural boundary conditions which leave the Dirac Hamiltonian hermitian are $\psi_{1,2}(-L) = \psi_{1,2}(L)$. Once again every positive energy solution $\eta_k$ will have a negative energy partner $\tilde{\eta}_k$. We do not need their details. But now there are *two* zero energy solutions:

$$\eta_0(x) = \begin{pmatrix} A \exp\left[-m \int_0^x \mathrm{d}y \tanh\left(y/\sqrt{2}\right)\right] \\ 0 \end{pmatrix} \tag{27.34}$$

and

$$\tilde{\eta}_0(x) = \begin{pmatrix} 0 \\ A \exp\left[-mL + m \int_0^x \mathrm{d}y \tanh\left(y/\sqrt{2}\right)\right] \end{pmatrix} . \tag{27.35}$$

While the solution (27.34) is the same one as before, localised around the origin, the second solution (27.35) has support near the edges $\pm L$. In Sect. 27.2 where we started with an infinite volume problem, this second solution could not be entertained. But for any finite $L$ however large this second solution, finite everywhere and normalisable, is certainly present. Each of these two zero-modes is charge self-conjugate and both will appear in the expansion of the Dirac field operator, along with the non-zero energy modes:

$$\Psi(x,t) = \sum_{k \neq 0} \left[b_k \eta_k(x) \mathrm{e}^{-\mathrm{i}E_k t} + d_k^\dagger \tilde{\eta}_k(x) \mathrm{e}^{\mathrm{i}E_k t}\right] + a\eta_0(x) + c^\dagger \tilde{\eta}_0(x) \ . \tag{27.36}$$

Correspondingly there will now be four degenerate ground states in the soliton sector, namely, $|\mathrm{sol}\rangle$ defined by

$$b_k\,|\mathrm{sol}\rangle = d_k\,|\mathrm{sol}\rangle = a\,|\mathrm{sol}\rangle = c\,|\mathrm{sol}\rangle = 0 \tag{27.37}$$

and $\widetilde{|\mathrm{sol}\rangle} = a^\dagger|\mathrm{sol}\rangle, \overline{|\mathrm{sol}\rangle} = c^\dagger|\mathrm{sol}\rangle$, and $\widetilde{|\mathrm{sol}'\rangle} = a^\dagger c^\dagger|\mathrm{sol}\rangle$.

The expansion (27.36), when inserted into the definition of the charge operator

$$Q \equiv \frac{1}{2}\int_{-L}^{L} \mathrm{d}x\, \left[\Psi^\dagger(x,t), \Psi(x,t)\right] \tag{27.38}$$

will yield

$$Q = \sum_k \left[(b_k^\dagger b_k - 1/2) - (d_k^\dagger d_k - 1/2)\right] + (a^\dagger a - 1/2) + (c^\dagger c - 1/2)\,. \tag{27.39}$$

We see that there are now no unpaired terms of 1/2 and the total charge operator has only integral eigenvalues. One can now let the volume $2L$ become larger and larger and at each stage both zero modes will be present and the total charge will have only integral eigenvalues.

That one can restore integral valued charges by this strategy of starting with a finite volume and then letting $2L \to \infty$ can also be understood in a different way. Note that if we had closed our line $(-L \leq x \geq L)$ by identifying the points $x = \pm L$, that is equivalent to putting the system on a circle with a soliton at $x = 0$ *and* an antisoliton at $x = L$. This in turn amounts to working in the *vacuum sector*, where we saw that the charge is integral.

While integral charge has been thus restored, if we were to stop here we would have lost all the important physics unearthed by Jackiw and Rebbi. Notice that while there are now two zero modes, both localised with width of order $1/m$, one of them, $\tilde{\eta}_0(x)$ given in (27.35) is stuck at the edges of the sample. But the other zero mode $\eta_0(x)$ in (27.34), is in the middle and can move about as an excitation, carrying charge 1/2 in some sense that needs to be made more precise.

To pick just the charge carried by this middle zero-mode, we can try and define a "partial charge operator":

$$P_l \equiv \int_{-l}^{l} \mathrm{d}x \rho(x)\,, \tag{27.40}$$

where $l, L \to \infty$ in the order $L >> l >> 1/m$. In this limit the partial charge $P_l$ clearly includes the central zero mode while excluding the second zero mode at the edges $\pm L$. However, the soliton ground state will not be an eigenstate of $P_l$, but only of the total charge operator $Q$ in (27.39). The charge density operator $\rho(x)$ given in (27.12) and partial charge operators such as $P_l$ will excite particle-hole pairs when acting on the soliton state $|\mathrm{sol}\rangle$. But consider expectation values in this soliton state.

$$\begin{aligned}\langle\mathrm{sol}|\,\rho(x)\,|\mathrm{sol}\rangle &= (-1/2)\sum_{k\neq 0}\left[\eta_k^*(x)\eta_k(x) - \tilde{\eta}_k^*(x)\tilde{\eta}_k(x)\right] \\ &\quad +(1/2)\left[\tilde{\eta}_0(x)\tilde{\eta}_0(x) - \eta_0(x)\eta_0(x)\right] \\ &= (1/2)\left[\tilde{\eta}_0(x)\tilde{\eta}_0(x) - \eta_0(x)\eta_0(x)\right] , \end{aligned} \tag{27.41}$$

where in the last line, contributions from non-zero modes cancel for each $k$ by charge conjugation. Hence

$$\begin{aligned}\langle\mathrm{sol}|\,P_l\,|\mathrm{sol}\rangle &= \int_{-l}^{l} \mathrm{d}x(1/2)\left[\tilde{\eta}_0^2(x) - \eta_0^2(x)\right] \\ &= -(1/2)\int_{-l}^{l} \mathrm{d}x\eta_0^2(x) = -(1/2) . \end{aligned} \tag{27.42}$$

Because this is not an eigenstate of $P_l$ there will be fluctuations. Since

$$(P_l - < P_l >)\,|\mathrm{sol}\rangle = \sum_{k,k'}\int_{-l}^{l} \mathrm{d}x\left[\eta_k(x)\tilde{\eta}_{k'}(x)\right]|k,k'\rangle , \tag{27.43}$$

we have

$$\langle\mathrm{sol}|\,(P_l - < P_l >)^2\,|\mathrm{sol}\rangle = \sum_{k,k'}\left|\int_{-l}^{l} \mathrm{d}x\eta_k(x)\tilde{\eta}_{k'}(x)\right|^2 . \tag{27.44}$$

The integrand is positive and hence the fluctuations are non-zero. In fact they diverge logarithmically.

Such large fluctuations in partial charges are not special to soliton states. They occur even in familiar vacuum states and are because of the sharp boundary of the defining domain. One can kill the fluctuations while still retaining a value of 1/2 by defining a more sophisticated partial charge operator, one with fuzzy edges. Let

$$\tilde{P}_{l,d} \equiv \frac{1}{d}\int_{l}^{l+d} \mathrm{d}l' P_{l'} . \tag{27.45}$$

When $L, l, d \to \infty$ with $L >> l, d$ then once again,

$$\langle\mathrm{sol}|\,\tilde{P}_{l,d}\,|\mathrm{sol}\rangle = -(1/2) . \tag{27.46}$$

To obtain the fluctuations of this operator note that $[\rho(x) - < \rho(x) >]$ connects the state $|\mathrm{sol}\rangle$ to the particle-hole states $|k, k'\rangle$. We therefore have

$$\begin{aligned}\langle k, k'|\, P_l - < P_l > |\mathrm{sol}\rangle &= \langle k, k'| \int_{-l}^{l} \mathrm{d}x(\rho - < \rho >)\, |\mathrm{sol}\rangle \\ &= \frac{1}{\mathrm{i}(E_k + E_{k'})} \langle k, k'| \int_{-l}^{l} \mathrm{d}x \frac{\partial \rho}{\mathrm{d}t}\, |\mathrm{sol}\rangle \\ &= \frac{1}{\mathrm{i}(E_k + E_{k'})} \langle k, k'|\, (j(l) - j(-l))\, |\mathrm{sol}\rangle \end{aligned} \quad (27.47)$$

by current conservation. Note that the current $j$ has matrix elements $\langle k, k'|j(x)|\mathrm{sol}\rangle = \eta_k^*(x)\mathrm{i}\alpha\tilde{\eta}_{k'}(x)$. Hence,

$$\langle k, k'|\, \tilde{P}_{l,d} - < \tilde{P}_{l,d} > |\mathrm{sol}\rangle = \frac{1}{d} \int_{l}^{l+d} \mathrm{d}l' \frac{1}{\mathrm{i}(E_k + E_{k'})} [\eta_k^* \mathrm{i}\alpha\tilde{\eta}_{k'}]_{-l'}^{l'} \,. \quad (27.48)$$

To obtain the exact result for this one must know analytic expressions for all the non-zero energy modes $\eta_k(x)$ and $\tilde{\eta}_k(x)$. We had in fact obtained these solutions for the simplified case where the soliton function is taken as a step function $\phi(x) = \Theta(x)$ instead of $\tanh x$. This simplification does not affect the issues of interest to us. But even quite generally one can see that for large $l$ far away from the soliton center ($l \gg 1/m$) the solutions $\eta_k(l)$ will be trigonometric functions of $kl$. Since $E_k \approx k$ for large k, the matrix element in (27.47) will behave as $1/k$ for large k and give rise to the logaritmically divergent fluctuations in (27.44). But the charge operator with fuzzy edges $\tilde{P}_{l,d}$ defined in (27.45) has an additional integral over $dl'$ which will bring down an additional convergence factor of $1/k$. There is also the extra factor $1/d$ in the definition which goes to zero. Altogether

$$\begin{aligned}\langle \mathrm{sol}|\, (\tilde{P}_{l,d} - < \tilde{P}_{l,d} >)^2\, |\mathrm{sol}\rangle &= \sum_{k,k'} \left| \langle k, k'|\, \tilde{P}_{l,d} - < \tilde{P}_{l,d} > |\mathrm{sol}\rangle \right|^2 \\ &\approx \frac{1}{d^2} \sum_{k,k'} f(k, k') \,, \end{aligned} \quad (27.49)$$

where the function $f(k, k'))$ has a convergent sum over $k, k'$. Clearly this vanishes as $d \to \infty$. In other words, one can define a suitable partial charge operator with fuzzy edges for which the soliton state has a value of $-1/2$ with *no* fluctuations! This fractional charge then *is* an eigenvalue.

It is very possible that if one were to measure the charge in the neighborhood of the soliton's center, one is using some such partial charge operator with fuzzy edges. In that case the measured value of $1/2$ can be promoted to the status of an eigenvalue. Further, since these fractions are eigenvalues of complicated partial charge operators and not of the total charge (number) operator Q, there is no conceptual problem reconciling the result with our intuition about the latter.

### 27.4.1 Polyacetylene Re-Visited

Let us describe briefly what a corresponding analysis [9] yields for polyacetylene which is a lattice chain of some finite number $N$ of sites amongst which the electrons hop. I will emphasize just the additional insights that emerge here due to the fact that it is an even better regulated system than the field theory in finite volume that we just studied. It has only a finite number of degrees of freedom and hence no ultraviolet or infrared problems. The lattice Dirac Hamiltonian in (27.28) yields the following coupled equations for the electron wave functions $b_n$ and $c_n$ at odd and even sites, respectively:

$$\begin{aligned} E\, b_n &= (1/4)c_{n+1}(\phi_{n+1} + \phi_n - 2/a) + (1/4)c_{n-1}(\phi_n + \phi_{n-1} + 2/a)\ , \\ E\, c_n &= (1/4)b_{n+1}(\phi_{n+1} + \phi_n + 2/a) + (1/4)b_{n-1}(\phi_n + \phi_{n-1} - 2/a)\ . \end{aligned} \tag{27.50}$$

Clearly, given a solution for some positive energy $E$ we can get another solution of negative energy $(-E)$ by replacing $b_n \to -b_n, c_n \to c_n, E \to -E$ in the above equations. Thus positive and negative energy solutions again come in pairs. But suppose the lattice had altogether only an *odd* number $N$ of sites. Then the Dirac operator can have only the same odd number of independent solutions. Therefore there must be one or more zero-energy solutions.

Notice that this argument for the existence of zero modes has nothing to do with whether the background boson field is in the topological soliton sector or the normal vacuum sector. All you need is an odd number of lattice sites! One can explicitly verify this result. Take $N = 2M + 1$ with, say, $M$ odd. A zero energy solution of the lattice Dirac equation (27.50) clearly exists in which the even site variable $c_n = 0$ for all $n$ while the odd site variables $b_n$ satisfy

$$\frac{b_{n+1}}{b_{n-1}} = \frac{2 - a(\phi_n + \phi_{n-1})}{2 + a(\phi_n + \phi_{n+1})}\ . \tag{27.51}$$

This will be true for any background Bose field $\phi_n$. In the uniform phase (the "vacuum sector" here) $\phi_n = 1$ for all $n$. Hence the solution is

$$b_n = b_0 \exp(-Kan)\ ; \quad c_n = 0\ , \tag{27.52}$$

where the constant K obeys $\tanh(Ka) = a$. If $\phi_n$ is a soliton, say having the simple form $\phi_n = n/|n|$ then the fermion zero mode is

$$b_n = b_0 \exp(-Ka|n|)\ ; \quad c_n = 0\ . \tag{27.53}$$

Given the single zero mode in both sectors, one can show that the ground states in both sectors will have total charge of $(-1/2)$ as eigenvalues. By simply adding $+1/2$ to the definition of charge one can trivially make eigenvalues integral in both sectors as was the case in the finite-volume field theory.

Similarly, if the number of lattice points $N$ had been even, there would have been either no zero modes or an even number of them in both sectors, leading to integral total charge.

Thus the important lesson we learn is that as far as the *total* charge is concerned, there is no half integral eigenvalue in the soliton sector *as compared* to the uniform phase. This is true in finite chains even if there is a single isolated zero mode in the soliton sector, because a zero mode will then exist in the vacuum sector too.

Of course one can also define for lattice chains partial charges that pick up the zero mode in the soliton sector but not the one in the vacuum sector. [Recall in the example mentioned that the zero mode in the soliton sector given in (27.53) is located near the center whereas in the vacuum sector even though a zero mode (27.52) was present it was near one of the edges.] As happened in the field theory case discussed earlier in this section, such partial charge operators would have a half-integral value in the soliton state. This value would get promoted to the status of an eigenvalue with no fluctuations if, in the limit of a very long chain, the operator covered a region with suitably fuzzy edges (see [9] for more details).

## 27.5 Conclusion

I have tried to outline in the last section the work that John Bell and I had done on Fractional charge, preceded in earlier sections by some background on where matters stood when we began our work. I believe our work had thrown light on which operators do have eigenvalues of 1/2 (as distinct from expectation values ) and which don't. In the process we had also laid to rest the concern that had motivated us to study this problem – which was that the total number operator of a physical system should not have fractional eigenvalues. That would have made no sense. We showed that the total charge operator defined to have integral eigenvalues in the absence of solitons will, in a well regulated theory, continue to have integer eigenvalues even in the presence of solitons. But suitable localised partial charge operators can have fractional eigenvalues and fractional charge should be interpreted as corresponding such an operator. Indeed these may well be what charge measurements in experiments employ. As mentioned already, *Kivelson* and *Schrieffer* [10] had come to the same conclusion independently.

The work discussed above is nearly 20 years old. This article is primarily devoted to John Bell and his work, so this is not the place to discuss in any detail other developments on this topic around that time or since then. However, for the sake of completeness let me summarize in a few sentences some of these developments.

Soon after charge 1/2 states were discovered in field theory and polyacetylene, the possibilities of charge at other fractions were unearthed both in polymer physics and in model field theories (see for example [12] and [13]).

A major addition to the list of fractionally charged objects was the theoretical discovery by *Laughlin* in an entirely new arena. He showed that quasi-particles in fractional quantum Hall (FQH) systems carry fractional charge [14]. Subsequently these have also been experimentally observed in "shot-noise" experiments [15]. To the best of my knowledge the quantum mechanical status of this quasi-particle charge in FQH has not been analyzed as carefully as in the case of polyacetylene or field theory models. The former have been identified through ingenious but indirect arguments involving plasma analogies and Aharanov–Bohm/Berry phases rather than by an explicit analysis of the charge operators and their eigenstates. The FQH quasi-particles are more difficult states to study. They correspond to genuinely correlated many-body states as distinct from the polyacetylene and Jackiw–Rebbi soliton states considered above, which could be analysed in terms of single particle solutions of the Dirac equation. But broadly speaking, the status of the fractional charge in FQH is believed to be similar to those discussed above. It is localized near the center of the quasi-particle wavefunction and the complementary missing fraction is believed to be near the edge of the Hall sample.

Finally, very recently, it has been claimed by *Maris* [16] that some "bubbles" formed in liquid helium which contain a trapped electron can then fission into daughter bubbles each of which carries a fragment of the original electron. However, in an analysis of this phenomenon, the pioneers of fractional charge *Jackiw*, *Rebbi* and *Schrieffer* have argued that these fractions in helium bubbles are just expectation values of the sort familiar in local charge measurements in any double well quantum system [17].

## Acknowledgements

I would like to thank the organizers of the conference Professors R. Bertlmann and A. Zeilinger, the University of Vienna and the Erwin Schroedinger Institute for their hospitality and travel support.

## References

1. R. Jackiw, C. Rebbi: Phys. Rev. D **13**, 3398 (1976)
2. W.P. Su, J.R. Schrieffer, A.J. Heeger: Phys. Rev. Lett **42**, 1698 (1979); Phys. Rev. B **22**, 2099 (1980)
3. M.J. Rice: Phys. Lett. A **71**, 152 (1979)
4. R. Jackiw, J.R. Schrieffer: Nucl. Phys. **190** [FS3], 253 (1981)
5. R. Rajaraman: *Solitons, Instantons* (North Holland, Amsterdam 1982)
6. J. Kogut, L. Susskind: Phys. Rev. D **11**, 395 (1975)
7. I.B. Goldberg, H.R. Crowe, P.R. Newman, A.J. Heeger, A.G. MacDiarmid: J. Chem. Phys. **70**, 1132 (1979)
8. R. Rajaraman, J.S. Bell: Phys. Lett. B **116**, 151 (1982)
9. J.S. Bell, R. Rajaraman: Nucl. Phys. B **220** [FS 8], 1 (1983)

10. S. Kivelson, J.R. Schrieffer: Phys. Rev. B **25**, 6447 (1982)
11. R. Jackiw, A. Kerman, I. Klebanov, G. Semenoff: Nucl. Phys. B **225** [FS225], 233 (1983)
12. W.P. Su, J.R. Schrieffer: Phys. Rev. Lett. **46**, 741 (1981)
13. J. Goldstone, F. Wilczek: Phys. Rev. Lett. **47**, 986 (1981)
14. R.B. Laughlin: Phys. Rev. Lett. **50**, 1395 (1983)
15. R. de-Picciotto, M. Reznikov, M. Heiblum, V. Umansky, G. Bunin, D. Mahalu: Nature **389**, 162 (1997)
16. H. Maris: J. Low Temp. Phys. **120**, 173 (2000)
17. R. Jackiw, C. Rebbi, J.R. Schrieffer: J. Low Temp. Phys. **122**, March (2001)

# 28 Thermal Excitations of Accelerated Electrons

Jon Magne Leinaas

For a uniformly accelerated thermometer Minkowski vacuum appears hot, with a temperature proportional to the acceleration. This is a quantum effect; the fluctuations of the quantum fields excite the accelerated system. However, an extremely robust detector is needed to register the temperature, since even very low temperatures correspond to extremely large accelerations. Several years ago John Bell and I suggested that electrons could be viewed as such robust detectors, with the spin degree of freedom used to measure the temperature. We also showed that for electrons in a storage ring, a known spin depolarization effect is closely related to the thermal effect of linearly accelerated electrons. I give here a brief discussion of how accelerated electrons are excited by the vacuum fluctuations and in particular focus attention on the importance of the coupling between spin and orbital motion for electrons in circular motion.

## 28.1 Electrons as a Thermometer

As a Fellow at CERN in 1981–82 I had the good fortune to get to know John Bell and to work with him. Now, many years later, it is with pleasure I recollect this period, with our many interesting physics discussions related to the work we were doing. Our collaboration lasted for a period after I had moved to Stavanger, and I regularly went to CERN. John also came to the annual meeting of the Norwegian Physical Society in 1987, and in particular I recall a pleasant walk we had together in the hillside outside the city.

My acquaintance with John started with a discussion with him about a subject that had caught my interest, namely the so-called Unruh effect. It had been noticed by Unruh some years earlier that a quantum detector that has a uniform, linear acceleration through Minkowski space will get excited by interactions with vacuum fluctuations of the quantum fields [1] (see also the earlier paper by Paul Davis [2]). The excitation spectrum will have a thermal form, as if the detector were immersed in a heat bath, with a temperature determined by the acceleration. The effect was by Unruh related to the temperature effect of black holes, namely that a black hole will emit thermal radiation, with a temperature determined by the mass of the black hole [3]. An accelerated detector in Minkowski space will register a temperature in a similar way as a stationary detector located in the gravitational field outside the black hole.

Our discussion was the start of a collaboration that resulted in three publications, one of them with Richard Hughes, then a Fellow at CERN, now a researcher at Los Alamos [4–6]. The physics discussions we had I enjoyed much, and I appreciated John's combination of challenging views, sharp analysis and good physics understanding. In addition I felt, like many others, his good will in all respects. There were plesant breaks in the cafeteria, with his wife Mary and their small lunch group, and with herbal tea (and my coffee) in the afternoon. There was with John also this feeling of energy and strong engagement under the usually peaceful surface. To do physics was a challenge of basic importance, and there were matters that could at rare occasions increase his temper to a point that was felt like a small eruption.

The background for our first discussion was the relation between the Unruh temperature $T_{\mathrm{U}}$ and acceleration $a$ that is given by the formula

$$kT_{\mathrm{U}} = \frac{a\hbar}{2\pi c} \,. \tag{28.1}$$

With numbers inserted for Boltzmann's constant $k$, Planck's constant $\hbar$ and the speed of light $c$ one finds that an extremely large acceleration is needed to produce even a feeble temperature. For example, a temperature of 1 K corresponds to an acceleration of $2.5 \times 10^{20}\,\mathrm{m/s^2}$. Obviously the formula (28.1) can apply only to an idealized thermometer. It would seem that a real thermometer would be strongly deformed, even destroyed long before any (Unruh) temperature could be registered. Only an extremely robust system could have a chance to resist the effects of such a strong acceleration, and an atom for example would not satisfy this condition. Elementary particles would be robust, but could they be used as a thermometer?

In our discussion John showed me a paper he had received from *Jackson* on a spin polarization effect of electrons circulating in a storage ring [7]. He suggested that we should look at this and consider whether there was a connection to the Unruh effect. The idea we discussed was whether electrons in a magnetic field could be considered as defining a thermometer, in the sense that the degree of occupation of the upper spin energy level would determine the temperature. Electrons in a storage ring would be accelerated, and by using the Unruh formula (28.1) we could find a value for the electron polarization and compare with the value calculated in the more conventional approach. This was soon done, and we found a well-defined, small depolarization, but smaller than the one found in the conventional way. The polarization effect could, it seemed, be considered as a kind of Unruh effect, but a naive application of the acceleration-temperature formula (28.1) would be a bit too simple. There were obvious points to examine further: What difference does it make that the electrons are circularly rather than linearly accelerated? Is the orbital motion and not only the spin motion of importance? These questions were discussed in some detail in our papers. Our conclusion was that there was indeed a close connection between the polarization effect for circular motion and the Unruh effect for linear motion. But the coupling between spin and orbital motion is important, and Thomas precession complicates things. I would like here to give a brief account of my present understanding of these matters. (For an earlier review see [8].)

## 28.2 Linear Acceleration and the Unruh Effect

Let me first consider the case where electrons are linearly accelerated in an external electric field (in the $x$-direction). If such an electron is treated as a classical pointlike object it will follow a path of constant proper acceleration $a$, which is a hyperbolic path in space-time,

$$x = \frac{c^2}{a}\cosh\left(\frac{a\tau}{c}\right), \quad t = \frac{c}{a}\sinh\left(\frac{a\tau}{c}\right), \quad y = y', \quad z = z', \tag{28.2}$$

where $\tau$ is the proper time of the particle and $c$ the speed of light. If there is an additional magnetic field in the same direction, there will be a splitting of the spin energy levels,

$$H_{\text{spin}} = \frac{1}{2}\hbar\boldsymbol{\omega}\cdot\boldsymbol{\sigma}, \tag{28.3}$$

with $\boldsymbol{\sigma}$ as the Pauli spin matrices and $\boldsymbol{\omega}$ as the spin precession vector determined by the magnetic field,

$$\boldsymbol{\omega} = -\frac{e}{2mc}g\boldsymbol{B}, \tag{28.4}$$

where $e$ is the charge and $m$ is the mass of the electron and $g$ is the $g$-factor, which is very close to 2. In this description the spin is treated as a quantum system, while the motion of the electron itself is treated classically. Obviously this is a crude approximation, but for the purpose of illustrating the Unruh effect it will do.

The magnetic field $\boldsymbol{B}$ includes the external field which gives rise to an energy splitting of the two spin levels, but it also includes the (quantum) radiation field. This is affected by quantum fluctuations of the vacuum state and gives rise to spin interactions which cause transitions between the two spin levels.

Transition probabilities per unit time can be found by use of first order time-dependent perturbation theory. By summing over all possible final states of the radiation field the transition probabilities for transitions up and down in spin (along the direction of the external field) can be written as

$$\Gamma_{\pm} = \left(\frac{eg}{4mc}\right)^2 \int_{-\infty}^{\infty} \exp(\mp \mathrm{i}\omega\tau)C(\tau - \mathrm{i}\epsilon)\,\mathrm{d}\tau, \tag{28.5}$$

where $\omega$ here means the spin precession frequency as determined by the external magnetic field and $\pm$ refers to transitions up/down in spin. $C(\tau)$

is the vacuum correlation function of the magnetic field along the electron path,

$$\begin{aligned} C_B(\tau) &= \langle B_\pm(\tau/2)B_\mp(-\tau/2)\rangle \\ &= \frac{\hbar a^4}{2\pi c^7}\left[\sinh\left(\frac{a}{2c}\tau\right)\right]^{-4}, \quad B_\pm = B_x \pm \mathrm{i}B_y\,, \end{aligned} \tag{28.6}$$

with $B_\pm$ meaning (components of) the radiation part of the magnetic field as measured in the instantaneous rest frame.

The correlation function $C(\tau)$ has a periodicity property with respect to shifts in the imaginary time direction,

$$C_B\left(\tau + \frac{4\pi c}{a}\mathrm{i}\right) = C_B(\tau)\,, \tag{28.7}$$

which makes it easy to solve the integral (28.5) by closing the integration contour along the shifted path $\tau + 4\pi\mathrm{i}\,c/a$. For the closed contour we find [4]

$$\left[1 - \exp\left(\pm 2\pi c\frac{\omega}{a}\right)\right]\Gamma_\pm = \mp\frac{e^2g^2\hbar}{6m^2c^5}\omega\left(\omega^2 + \frac{a^2}{c^2}\right), \tag{28.8}$$

and the ratio between the probabilities for transitions up and down is

$$R = \frac{\Gamma_+}{\Gamma_-} = \exp\left(-2\pi c\frac{\omega}{a}\right). \tag{28.9}$$

In an equilibrium situation the ratio between the probabilities for transitions up and down (for one particle) defines the relative occupation probabilities of the two spin levels. The dependence of the energy shows that it has the form of a Boltzmann factor corresponding to a temperature

$$kT_{\mathrm{U}} = \frac{a\hbar}{2\pi c}. \tag{28.10}$$

This is the Unruh temperature, which is generally associated with an accelerated system. As mentioned there is a connection between the Unruh effect and Hawking radiation. This is clearly indicated by the temperature formula for a black hole,

$$kT_{\mathrm{BH}} = \frac{\kappa\hbar}{2\pi c}, \tag{28.11}$$

which has the same form as (28.10), whith the acceleration $a$ replaced by the the surface gravity $\kappa$ of the black hole [3].

The spin effect discussed above is a special realization of the Unruh effect. The thermal distribution has been derived with reference to a particular correlation function of the vacuum fields. However, in reality only general features of the vacuum state and special properties of the accelerated trajectory, which lead to the symmetry properties (28.7), are important. This is the interesting point; the effect has a universal character, it does not depend on details of the accelerated system or on its coupling to the radiation field.

In principle a beam of linearly accelerated electrons could be used to demonstrate the Unruh effect in the way discussed above. Thus, the spin polarization would depend on the spin precession frequency as

$$P(\omega) = \frac{1-R}{1+R} = \tanh\left(\pi c \frac{\omega}{a}\right) , \tag{28.12}$$

and by varying the strength of the external magnetic field the functional form of $P$ could be demonstrated. Unfortunately, this is not a realistic situation for real particle accelerators. If we use a value for the electric field $E = 10\,\mathrm{MV/m}$ we find a rather low corresponding temperature $T = 0.7 \times 10^{-3}\,\mathrm{K}$. However, the main complication is the long time for reaching equilibrium. From the expression for the transition probabilities we find a typical time of $\tau = 3 \times 10^{18}\,\mathrm{s}$. In the lab frame this would be enhanced even further by the time dilatation effect. So this is not very promising. If much larger accelerations can be obtained the thermalization time $\tau$ would be strongly reduced (it varies with $a$ as $a^{-3}$), but the limited time available in a linear accelerator is nevertheless a serious problem.

For circular accelerators the situation is different. Here much higher values for the proper acceleration are obtained, mainly due to the relativistic $\gamma$ factor for the transformation from the lab frame magnetic field to the rest frame electric field. A typical acceleration is $a = 3 \times 10^{23}\,\mathrm{m/s^2}$ corresponding to a temperature of $T = 1200\,\mathrm{K}$. And more importantly, for electrons in a storage ring the time needed to reach equilibrium is available, typically of the order of minutes to hours. It is indeed well-known, since the original calculation by by *Sokolov* and *Ternov* [9–11] that in the magnetic field of a circular accelerator, the electrons will gradually build up a transverse polarization due to spin flip radiation. The question addressed in our discussion was whether this can be seen as a "circular Unruh effect".

A simple (naive) comparison between the Sokolov–Ternov effect and Unruh effect, is to use the polarization formula (which is derived for linear acceleration), insert values for $\omega$ and $a$ valid for circular motion and find the polarization expressed as a function of the electron $g$-factor. This function can be compared with the conventional one derived in [11] for the Sokolov–Ternov effect.

In Fig. 28.1 these two functions are plotted and compared. There is, as pointed out in our first paper [4], a basic similarity between the two curves, but there is also an important difference in the relative shift along the $g$-axis. The interesting point is to understand the basis for this similarity and for the difference. In order to do so, the spin excitations of the circulating electrons have to be considered more seriously. Just as for linear acceleration, the circulating electron can be treated as an accelerated quantum system coupled to the vacuum fields. That I will do, but first I will consider briefly what are

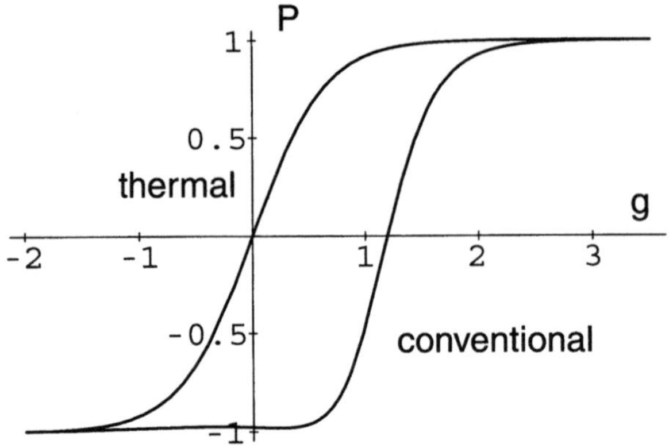


**Fig. 28.1.** The equilibrium polarization $P$ as a function of $g$. The curve referred to as "conventional" represents the conventional result, whereas the curve referred to as "thermal" is based on the use of the relation between temperature and acceleration derived in the case of linear acceleration. The electrons are considered to be ultra-relativistic, with $v \approx c$

the types of motion where an accelerated system will settle in a stationary state with a well defined probability distribution over the energy levels.

## 28.3 Stationary World Lines

A special feature of the hyperbolic world line is that all points along the path are (geometrically) equivalent. For an accelerated system following the world line this means that the situation is stationary. Therefore, when the system interacts with the vacuum fluctuations, it will (eventually) settle in a stationary state. There are other world lines with the same property, in particular the one corresponding to circular motion with constant speed [13]. In general a stationary world line is generated by an infinitesimal, time independent Poincaré transformation. We can think of this as a transformation between inertial rest frames associated with neighbouring points of the world line. By using the freedom to choose directions of axes, the time evolution operator $H'$, which performs this continuous jump between inertial frames can be written in the form

$$H' = H + \frac{a}{c} K_x - \omega_z J_z - \omega_x J_x \,, \tag{28.13}$$

with $H$ as the generator of time evolution in the laboratory frame, $K_x$ as the boost operator in the $x$ direction, $J_z$ as the generator of rotations about the $z$-axis and $J_x$ about the $x$-axis. Thus, in general $H'$, and therefore the corresponding stationary world line, is characterized by three parameters, $a$, $\omega_z$ and $\omega_x$. The physical interpretation of the parameters is that they correspond to acceleration and angular velocity of an accelerated frame which at each time coincides with the inertial frame generated by the Poincaré transformation. They also have a geometric interpretation as curvature, torsion and hypertorsion of the world line [13].

Uniform linear acceleration corresponds to $\omega_x = \omega_z = 0$ and the circular motion with constant speed corresponds to $\omega_x = 0, \omega_z = a/v$ with $v$ as the speed of the particle. It is interesting to note that a continuous interpolation can be made between the two cases by changing $\omega_z$ while $\omega_x = 0$. For $\omega_z > a/c$ the motion is circular (in a correctly chosen frame) while for $\omega_z < a/c$ the motion is non-periodic. For any of these world lines an accelerated quantum system interacting with the vacuum fluctuations will settle in a stationary state with a certain probability distribution over the energy levels. However, only for linear acceleration ($\omega_z = 0$) it will have a truly thermal form, while for $\omega_z \neq 0$ there will be deviations. The situation is somewhat similar to that seen by a stationary observer in the geometry of a rotating black hole. With no rotation the observer will measure a well-defined local (Hawking) temperature, but with rotation the situation is not so simple.

However, even though there is no exact temperature associated with the motion when $\omega_z \neq 0$, the notion of an effective temperature is meaningful. The vacuum state, in some approximate meaning seems hot in the accelerated frame, and in addition there is a relative rotation between the stationary detector and the vacuum. I will next consider the case of electrons circulating in a storage ring to discuss in what sense this is a correct picture.

## 28.4 Electrons in a Storage Ring

When we consider electrons in a storage ring under ideal conditions, where they move in a rotationally symmetric magnetic field and with the correction due to radiation loss neglected, they can be described by a time-independent Hamiltonian $H'$ of the form previously discussed,

$$H' = H + \frac{a}{c}K_x - \frac{a}{v}J_z \,. \tag{28.14}$$

This Hamiltonian refers to an accelerated coordinate system which rotates with the frequency of a reference particle moving on a classical trajectory in the magnetic field; $a$ is the proper acceleration and $v$ the velocity of this reference particle. The operators $H$, $K_x$ and $J_z$ are Dirac operators in the inertial rest frame of the reference trajectory at time $\tau = 0$. The electrons described by the Hamiltonian are not restricted to the classical trajectory. Also quantum fluctuations in the orbital motion are included in the description. However, we may assume deviation from this orbit both in position and velocity to be small, which makes it possible to linearize in the deviation and to consider a non-relativistic approximation. In this approximation (and with $v \approx c$) the electron Hamiltonian gets the form

$$H_{\mathrm{e}} = \frac{p^2}{2m} + \frac{1}{2}m\Omega^2 z^2 + \frac{1}{2}\omega\hbar\sigma_z - \frac{1}{mc}\omega\hbar p\sigma_y - \frac{\hbar}{4c}g\Omega^2 z\sigma_x \,, \tag{28.15}$$

where $p$ is the momentum in the vertical ($z$-)direction and $\omega = \frac{a}{2c}(g-2)$ is the spin precession frequency. A confining harmonic oscillator potential has been introduced for the vertical motion. In a weak focussing machine this is produced by a gradient in the magnetic field, $\Omega^2 = -\frac{e}{m}\frac{\partial B}{\partial x}$. One should note that to lowest order only the vertical motion is coupled to the spin.

The coupling to the radiation field gives an additional term,

$$H_1 = -\frac{e\hbar}{4mc} g\boldsymbol{B}\cdot\boldsymbol{\sigma} - ezE_z \ , \tag{28.16}$$

which we treat as a perturbation. The fields $\boldsymbol{B}$ and $E_z$ are rest frame fields.

Thus, the transformation to the accelerated frame gives us a fairly simple and straight forward way to calculate the equilibrium properties of the electron beam. We first determine the eigenvectors and eigenvalues of $H_{\mathrm{e}}$, which is a two-level system coupled to a harmonic oscillator, and we then find the occupation of the levels by calculating the transition probabilities induced by $H_1$.

The coupling terms between spin and orbital motion in (28.15) are normally quite small. If they are treated perturbatively we find to first order the following expressions for the eigenvectors of $H_{\mathrm{e}}$

$$\begin{aligned} |\chi_{n,\pm}\rangle = |n,\pm\rangle - \mathrm{i}\sqrt{\frac{\hbar\omega}{32mc^2}} \Bigg[ (2\omega \mp g\Omega)\frac{\sqrt{n+1}}{\omega \mp \Omega} |n+1,\mp\rangle \\ + (2\omega \pm g\Omega)\frac{\sqrt{n}}{\omega \pm \Omega} |n-1,\mp\rangle \Bigg] \ . \end{aligned} \tag{28.17}$$

$|n,\pm\rangle$ is the eigenvector of the uncoupled system, with $n$ referring to the harmonic oscillator and $\pm$ to the spin levels. Note the resonance between spin and orbital motion for $\Omega = \omega$. To higher order in perturbation theory also higher order resonances, for $\Omega = n\omega$, will be present[1].

Let us first consider the transition matrix elements for non-spin flip transitions. It is clear from the form of $H_1$ that the small terms of (28.17) will only give small contributions to the matrix elements. If they are neglected the result is

$$\langle \chi_{n+1,\pm} | H_1 | \chi_{n,\pm} \rangle = \mathrm{i}e\sqrt{\frac{\hbar\,(n+1)}{2m\Omega}} E_z \ . \tag{28.18}$$

For spin transitions the situation is different. Even if the coupling terms are small they give significant contributions since they are influenced by the coupling of the charge to the radiation field, which is much stronger than the spin coupling. The result for spin flips is

$$\langle \chi_{n,\pm} | H_1 | \chi_{n,\mp} \rangle = -\frac{e\hbar}{4mc}\left\{ gB_{\mp} + \left[ 2 - (g-2)\frac{\Omega^2}{\omega^2 - \Omega^2} \right] E_z \right\} \ . \tag{28.19}$$

---

[1] At resonance the energy levels are degenerate, and improved expressions should be obtained by use of degenerate perturbation theory.

Note that due to the $g-2$ factor the resonance term is of importance only close to resonance, and it will be neglected in the following.

The equilibrium populations $p_{n,\pm}$ of the energy levels can be found be assuming *detailed balance*. Thus, the transition probabilities can be expressed in terms of correlation functions for the electromagnetic fields, and the relative population of two levels can be determined as the ratio between the probabilities for transition between the levels, up and down in energy. For fixed spin the ratio is determined by the matrix element (28.18), and thereby by the correlation function of the (rest frame) electric field,

$$R(\Omega) = \frac{p_{n+1,\pm}}{p_{n,\pm}} = \frac{C_E(\Omega)}{C_E(-\Omega)} , \tag{28.20}$$

$$C_E(\Omega) = \int d\tau \exp(-i\Omega\tau) \langle E_z(\tau) E_z(0) \rangle . \tag{28.21}$$

Note that the ratio $R(\Omega)$ is independent of the state $(n, \pm)$. This implies that the excitation spectrum has a thermal form,

$$p_{n,\pm} = N_\pm \exp(n \ln R) , \tag{28.22}$$

with an effective temperature [6]

$$kT_{\text{eff}} = -\Omega\hbar / \ln R . \tag{28.23}$$

This temperature is frequency dependent, but the dependence on frequency is not very strong [12]. It is mainly somewhat higher than the Unruh temperature associated with the same acceleration. It is interesting to note that in this way the vertical fluctuations, at least in principle, could be used to demonstrate the temperature effect of accelerated motion. However, in practice that may not be so simple, and the polarization, which I consider next, is more easily available for measurements.

The relative population of spin up and down for the same $n$ is determined by ratios between spin flip transitions. It has a similar form as for non-flip transitions,

$$R'(\omega) = \frac{p_{n,+}}{p_{n,-}} = \frac{D(\omega)}{D(-\omega)} , \tag{28.24}$$

but now with a composite correlation function

$$D(\omega) = g^2 C_B(\omega) + 2g C_{EB}(\omega) + 4 C_E(\omega) , \tag{28.25}$$

where $C_B$ is the correlation function of the magnetic fields $B_+$ and $B_-$ and $C_{EB}$ is the mixed correlation function of $B_\pm$ and $E_z$.

With the resonance term of (28.19) neglected, the expression (28.24) for the relative populations of the spin levels defines a polarization curve that agrees with the standard result earlier shown in Fig. 28.1. This is so even if the metod used is somewhat different from the conventional one.

It is now interesting to see how the result for the polarization is built up from contributions from the direct coupling between the spin and magnetic field and the indirect one transmitted through fluctuations in the orbit. The effect of the fluctuations is demonstrated by not including the two terms $C_{EB}$ and $C_E$ in (28.25). The fluctuations in the orbit are then effectively suppressed and only the coupling to the magnetic moment of the electron is retained. The result is a changed curve $P(g)$, denoted "direct" in Fig. 28.2. It indeed has a form very similar to the one derived from the temperature formula, denoted "thermal" in Fig. 28.1. The effective temperature indicated by the curve is however somewhat higher than the Unruh temperature $T_{\mathrm{U}}$ for the same acceleration.

It is instructive also to consider the curve obtained if the contributions to the spin transitions caused by the direct coupling is suppressed and only the indirect one, due to fluctuations in the orbit, are retained. (That means including only $C_E$ in (28.25).) The result is referred to as "indirect" in Fig. 28.2. It is very similar to the one obtained from the direct coupling to the radiation field, although differing by a shift of 2 units along the $g$-axis.

The relative shift of the curves obtained from the direct and indirect coupling of the spin to the radiation field is in fact not so difficult to understand. The direct coupling of the radiation field to the spin is rotationally invariant and the corresponding occupation probabilities have (approximately) a thermal form in the frame which is non-rotational (with respect to the vacuum). This is the Fermi-Walker transported frame, where the direct coupling term is proportional to $g$. The vertical fluctuations, on the other hand, are insensitive to rotations in the $x, y$-plane, but the coupling between the orbital motion and the spin is time independent in the stationary frame, which is rotating relative

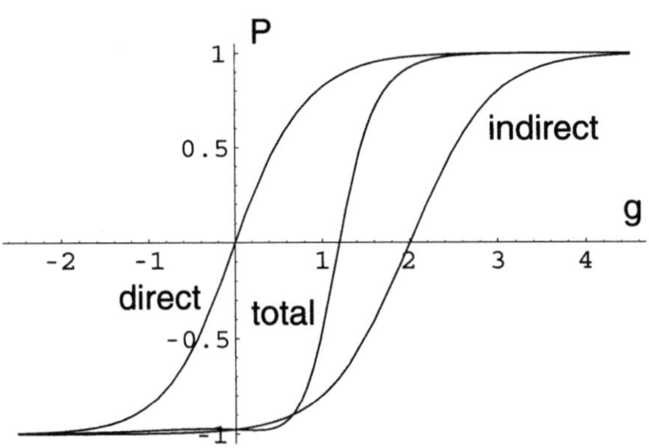


**Fig. 28.2.** Evaluated polarization curves, based on (28.24) in the text. *Total* refers to the full expression, and agrees with the conventional result displayed in Fig. 28.1. *Direct* refers to the curve where fluctuations in the orbit are suppressed, and *indirect* refers to the curve where only the spin excitations due to fluctuations in the orbit are retained

to the Fermi-Walker transported frame (mainly due to the Thomas precession effect). This gives rise to occupation probabilities which are (approximately) thermal in the stationary frame. The relative rotation between these two frames is represented by the shift along the $g$-axis.

Based on this understanding we may conclude that much of the difference between the two curves in Fig. 28.1 is due to the way the electron acts as a detector. The coupling between spin and vertical oscillations is important, and since this coupling is not rotationally invariant in the accelerated frame the polarization will be shifted and changed relative to the naive thermal polarization curve. This is so, even if the spin effect and the vertical oscillations, viewed separately, give rise to effects that are rather well described in terms of an effective temperature.

## 28.5 Concluding Remarks

The Unruh effect – like the related Hawking effect – is a quantum vacuum effect which usually is assumed to be detectable only under extreme situations. Therefore it is highly interesting that the effect is closely related to measurable polarization effects of electrons in a storage ring. The natural way to see this connection is to describe the electron beam in the accelerated frame of an orbiting reference particle. The electron can be described as a simple quantum mechanical system, which includes the vertical motion coupled to the spin, with transitions between the states of this system induced by the radiation field. Since the Minkowski vacuum state appears excited in this frame, transitions both up and down in energy are induced, and equilibrium is produced as a balance between these two processes.

Due to the coupling between spin and orbital motion, the electron acts as a non-trivial detector. Thus, the system is not rotationally invariant, and this complicates the detection of the vacuum effects. The vacuum can be seen as being hot in the accelerated frame, not in the stationary frame of the detector, but rather the non-rotational (Fermi–Walker transported) frame. Separately, the spin excitations and the vertical oscillations are well described in terms of an effective temperature that depends linearly on the acceleration. However, the combined effect is complicated by the fact that the direct coupling to the spin and the indirect one, due to vertical fluctuations, appear approximately thermal in two different frames which involve a relative shift in terms of the electron $g$-factor.

The accelerated frame is a natural choice of frame when discussing how the spin polarization effect is related to the Unruh effect effect. But beyond that, the description in the accelerated frame gives a conceptually simple way to study the quantum effects of electrons in a storage rings. Under the ideal conditions considered here, the electron is described as a harmonic oscillator coupled to a two-level system with transition probabilities determined by correlation functions of the radiation field. It may be of interest to examine this approach further, beyond the simplest approximation used here.

## Acknowledgement

I thank Geoff Canright and my wife Eva for several useful comments on the manuscript.

## References

1. W.G. Unruh: Phys. Rev. D **14**, 870 (1976)
2. P.C.W. Davies: J. Phys. A **8**, 609 (1975)
3. S. Hawking: Nature **248**, 30 (1974); Commun. Math. Phys. **43**, 199 (1975)
4. J.S. Bell, J.M. Leinaas: Nucl. Phys. B **212**, 131 (1983)
5. J.S. Bell, R.J. Hughes, J.M. Leinaas: Z. Phys. C **28**, 75 (1985)
6. J.S. Bell, J.M. Leinaas: Nucl. Phys. B **284**, 488 (1987)
7. J.D. Jackson: Rev. Mod. Phys. **48**, 417 (1976)
8. J.M. Leinaas: 'Quantum effects in accelerator physics'. In: *Quantum Reflections*, ed. by J. Ellis, D. Amati (Cambridge Univ. Press, Cambridge 2000)
9. A.A. Sokolov, I.M. Ternov: Dokl. Akad. Nauk. SSR **153**, 1963 (1052) [translation Sov. Phys. Dokl. **8**, 1203 (1964)]
10. V.N. Baier: Usp. Fiz. Nauk. **105**, 441 (1971) [translation Sov. Phys.-Usp. **14**, 695 (1972)]
11. Ya.S. Derbenev, A.M. Kondratenko: Zh. Eksp. Teor. Fiz. **64**, 1918 (1973) [translation Sov. Phys.-JETP **37**, 968 (1973)]
12. W.G. Unruh: 'Acceleration radiation for orbiting electrons'. In: *Quantum Aspects of Beam Physics*, ed. by P. Chen (World Scientific, Singapore 1999)
13. J.R. Letaw: Phys. Rev. D **823**, 1709 (1981)

# 29 Bell's Spaceships and Special Relativity

Franco Selleri

Equivalent transformations (ET) of space and time between inertial systems are obtained from two assumptions: (1) The two-way velocity of light is $c$ in all inertial systems and in all directions; (2) Clock retardation takes place with the usual square-root factor for clocks moving with respect to a certain isotropical reference frame. The ET contain a free coefficient, $e_1$, reflecting the well-known synchronisation arbitrariness. The Lorentz transformations are recovered for a particular value of $e_1$. Many experiments are insensitive to the choice of $e_1$: Michelson type, aberration, occultations of Jupiter's satellites, and radar ranging of planets. Accelerations break the equivalence in the set of ET. The example of two equally accelerating spaceships – originally discussed by John Bell for different purposes [1] – is used to show that absolute simultaneity gives the best description of physical reality.

## 29.1 The Generalised Transformations

The one-way velocity of light has never been measured. The obstacle is reflected in the conceptual basis of the theory of special relativity (TSR), particularly in the question of clock synchronisation discussed by *Reichenbach* [2], *Jammer* [3], *Mansouri* and *Sexl* [4] and *Will* [5]. Following the opinion often expressed by *Poincaré* [6], and sometimes also by *Einstein* [7], all one-way velocities have been considered devoid of physical interest. This is not a pleasant situation, because one-way velocities seem to be rather natural properties, after all, given that light and objects go from one point to another in well-defined ways in nature.

Following the line of thought of *Mansouri* and *Sexl* [4], recently a set of theories has been defined [8] differing from one another in clock synchronisation only. The set includes the TSR as a particular case. They are briefly reviewed below. A priori one could hope to find empirical differences between these theories, but we will show that the following experiments, fundamental for the TSR, are explained equally well by all theories of the set: Michelson-type measurements, aberration of star light, occultations of Jupiter's satellites, and radar ranging of planets.

Given the inertial frames $S_0$ and $S$, one can choose two systems of Cartesian orthogonal coordinates and assume the following:

(i) space is homogeneous and isotropic, and time is homogeneous;
(ii) in $S_0$ the velocity of light has the same value, $c$, in all directions, so that clocks can be sychronised in $S_0$ and velocities relative to $S_0$ can be measured;
(iii) the observer in $S$ sees his origin ($x = y = z = 0$) coincident with that of $S_0$ at $t = 0$, and the observer in $S_0$ sees his origin ($x_0 = y_0 = z_0 = 0$) coincident with that of $S$ at $t_0 = 0$;
(iv) the origin of $S$ is seen from $S_0$ to move with a velocity, $v$, parallel to the $+x_0$ axis, according to the equation $x_0 = vt_0$;
(v) the planes $(x_0, y_0)$ and $(x, y)$ superimpose at all times.

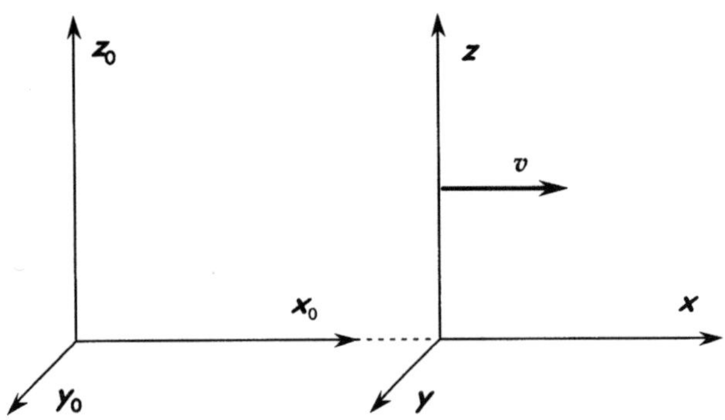


**Fig. 29.1.** Choice of Cartesian orthogonal axes in the inertial systems $S_0$ ("stationary") and $S$ ("moving")

By (iii)–(v) planes $(x_0, z_0)$ and $(x, z)$ must also superimpose at all times, while planes $(y_0, z_0)$ and $(y, z)$ must superimpose at $t_0 = 0$ only. Concerning (ii), note that this assumption is unobjectionable from the standpoint either of the TSR or of any plausible hypothesis involving a privileged frame; for the TSR it holds for all inertial systems, and in the other cases it is assumed to hold at least for the privileged system. In [8] it was shown that the previous conditions reduce the transformation laws from $S_0$ to $S$ to the form

$$\begin{cases} x = f_1(x_0 - vt_0) \, , \\ y = g_2 y_0 \, , \\ z = g_2 z_0 \, , \\ t = e_1 x_0 + e_4 t_0 \, , \end{cases} \tag{29.1}$$

where the factors $f_1, g_2, e_1, e_4$ can depend on the velocity $v$ of $S$ measured in $S_0$. The inverse one-way velocity of light $c_1(\theta)$ relative to $S$, in a direction forming an angle $\theta$ with respect to the velocity $\boldsymbol{v}$ ($x$-axis) is given by (ii) and (29.1). It is

$$\frac{1}{c_1(\theta)} = \left(\frac{e_1}{f_1} + \frac{R\beta}{cf_1(1-\beta^2)}\right)\cos\theta + \frac{R}{c}\left(\frac{\cos^2\theta}{f_1^2(1-\beta^2)^2} + \frac{\sin^2\theta}{g_2^2(1-\beta^2)}\right)^{\frac{1}{2}}, \tag{29.2}$$

where $\beta = v/c$ and

$$R = e_1 v + e_4 \,. \tag{29.3}$$

Elementary considerations show that the inverse two-way velocity of light, $c_2(\theta)$, from a point, $A$, to another point, $B$, and back to $A$ after reflection in $B$ is the average between the inverses of $c_1(\theta)$ and $c_1(\theta+\pi)$. A look at (29.2) then shows that

$$\frac{1}{c_2(\theta)} = \frac{R}{c}\left(\frac{\cos^2\theta}{f_1^2(1-\beta^2)^2} + \frac{\sin^2\theta}{g_2^2(1-\beta^2)}\right)^{\frac{1}{2}}. \tag{29.4}$$

Two well-established empirical facts can be used to restrict the generality of the transformations (29.1):

- The constancy of the two-way velocity of light,

$$c_2(\theta) = c\,, \tag{29.5}$$

  agrees with precise measurements and holds for all $\theta$ [9].
- Clock retardation with respect to $S_0$. A clock in the origin of $S$ satisfies $x_0 = vt_0$, and the last equation in (29.1) implies, for the time $t$ it marks, $t = (e_4 + e_1 v)t_0$. Experiments show [10] that this must become

$$t = \sqrt{1-\beta^2}\,t_0\,. \tag{29.6}$$

Equations (29.5) and (29.6) can be satisfied only if

$$f_1 = \frac{1}{\sqrt{1-\beta^2}}\,, \quad g_2 = 1\,, \quad e_4 = \sqrt{1-\beta^2} - e_1 v\,, \tag{29.7}$$

so that in place of (29.1) one has

$$\begin{cases} x = \frac{x_0 - vt_0}{\sqrt{1-\beta^2}}\,, \\ y = y_0\,, \\ z = z_0\,, \\ t = \sqrt{1-\beta^2}\,t_0 + e_1(x_0 - vt_0)\,. \end{cases} \tag{29.8}$$

The only remaining unknown factor is $e_1$. This is a conventional term and is called the "clock synchronisation factor". Length contraction of rods moving with respect to $S_0$ by the usual factor $\sqrt{1-\beta^2}$ (independently of $e_1$) is also a consequence of (29.8). The inverse velocity of light compatible with (29.8) can be obtained from (29.2) and (29.7) and is

$$\frac{1}{c_1(\theta)} = \frac{1+\Gamma\cos\theta}{c}, \tag{29.9}$$

with

$$\Gamma = \beta + ce_1\sqrt{1-\beta^2}. \tag{29.10}$$

The transformations (29.8) represent the complete set of theories "equivalent" to the TSR: if $e_1$ is varied, different elements of this set are obtained, which, according to the conventionality thesis of *Reichenbach* [1, 2], should be equivalent for the explanation of experimental results. The Lorentz transformation is recovered as a particular case with $e_1 = -\beta/c\sqrt{1-\beta^2}$, and therefore $\Gamma = 0$ and $c_1(\theta) = c$. Different values of $e_1$ are obtained from different clock-synchronisation conventions. In all cases but that of TSR such values imply the existence of a privileged frame.

## 29.2 Michelson-Type Experiments

A laboratory is at rest in the inertial system $S$ moving with absolute velocity $\boldsymbol{v}$, and in it an interferometric experiment is performed with a beam of light split into two parts by a semitransparent mirror placed at point $P$ (Fig. 29.2). The first part propagates along $P - A_1 - A_2 \dots A_m - Q$, where a reflecting mirror is placed at every intermediate point, the second part along the similar path $P - B_1 - B_2 \dots B_n - Q$. Finally the two parts superimpose at Q, where they interfere, Q being an arbitrary point of an interference figure. On the first path we define the vectors $\boldsymbol{\ell}_{ai}$ (with moduli $\ell_{ai}$), $i = 1, 2, \dots, m+1$, coincident with the rectilinear segments described by light and all oriented from $P$ toward $Q$; on the second path we similarly define the vectors $\boldsymbol{\ell}_{bj}$ (with moduli $\ell_{bj}$), $j = 1, 2, \dots, n+1$.

The interference at $Q$ is determined by the time delay, $\Delta T$, between the two rays. In the TSR light propagates in all directions with the same speed, $c$, also with respect to $S$ and one has

$$\Delta t = T_B - T_A = \frac{L_B - L_A}{c}, \tag{29.11}$$

where

$$L_A = \sum_{i=1}^{m+1} \ell_{ai}, \quad L_B = \sum_{j=1}^{n+1} \ell_{bj}. \tag{29.12}$$

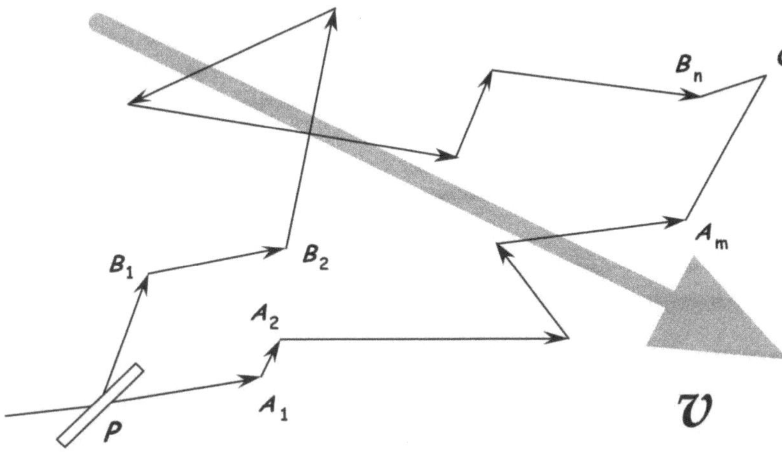


**Fig. 29.2.** A semitransparent mirror $P$ generates two coherent beams of light which follow different paths until they meet and interfere at $Q$. The *large grey arrow* represents the laboratory absolute velocity

Next we calculate $\Delta T$ from the equivalent transformations, according to which the inverse velocity of light relative to $S$ is given by (29.9). One has

$$\Delta T = \sum_{j=1}^{n+1} \frac{\ell_{bj}}{c_1(\theta_{bj})} - \sum_{i=1}^{m+1} \frac{\ell_{ai}}{c_1(\theta_{ai})}, \tag{29.13}$$

where $\theta_{ai}(\theta_{bj})$ is the angle between $\boldsymbol{\ell}_{ai}$ and $\boldsymbol{v}$ ($\boldsymbol{\ell}_{bj}$ and $\boldsymbol{v}$). By inserting (29.9) into (29.13), we find

$$\begin{aligned} \Delta T &= \frac{L_B - L_A}{c} + \frac{\Gamma}{c} \sum_{j=1}^{n+1} \ell_{bj} \cos\theta_{bj} - \frac{\Gamma}{c} \sum_{i=1}^{m+1} \ell_{ai} \cos\theta_{ai} \\ &= \frac{L_B - L_A}{c} + \frac{\Gamma}{c} \left( \sum_{j=1}^{n+1} \boldsymbol{\ell}_{bj} - \sum_{i=1}^{m+1} \boldsymbol{\ell}_{ai} \right) \frac{\boldsymbol{v}}{v} . \end{aligned} \tag{29.14}$$

The last term vanishes because the two terms in parenthesis separately equal the vector joining $P$ and $Q$. Thus (29.11) and (29.14) are the same. Therefore $\Gamma$ (containing $e_1$) disappears from the result, and all theories based on the equivalent transformations lead to the same predictions for interferometric experiments of the Michelson type (Michelson–Morley, Kennedy–Thorndike, Q. Majorana, etc.).

## 29.3 Jupiter's Satellites Occultations

Let a satellite of Jupiter (e.g., Io) be in any state of motion on the $x_0$-axis of the inertial system $S_0$ of Sect. 29.2 (Fig. 29.3). Io sends a light signal

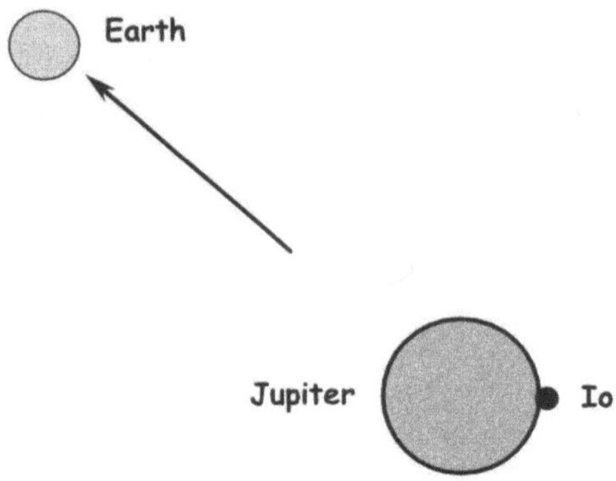


**Fig. 29.3.** The light coming from one of Jupiter's satellites (e.g., Io) is suddenly interrupted when it enters the shadow cone of the planet

(occultation) at the $S_0$ time $T_0$ when it is at the point $x_{0\mathrm{I}}$. The equation of motion of the signal in $S_0$ is

$$x_0 = x_{0\mathrm{I}} + c(t_0 - T_0) \ . \tag{29.15}$$

The Earth moves with constant speed, $v$, and constitutes the inertial system $S$. Its equation of motion as seen from $S_0$ is

$$x_{0\mathrm{E}} = vt_0 \ . \tag{29.16}$$

The signal reaches the Earth at time $t_{0r}$, which must satisfy

$$vt_{0r} = x_{0\mathrm{I}} + c(t_{0r} - T_0) \ ; \tag{29.17}$$

therefore

$$t_{0r} = \frac{cT_0 - x_{0\mathrm{I}}}{c - v} \ . \tag{29.18}$$

The Earth position at the time of reception is

$$x_{0r} = vt_{0r} \ . \tag{29.19}$$

We wish to find the time $t_r$ marked by the clock on Earth when the signal is received. From $S_0$ to $S$ the transformations (29.1) apply. Because of (29.19) one must have

$$t_r = (e_1 v + e_4)t_{0r} \ ; \tag{29.20}$$

thuse, using (29.18) and the last equation of (29.7),

$$t_r = \sqrt{1 - v^2/c^2}\, \frac{cT_0 - x_{0\mathrm{I}}}{c - v} \ . \tag{29.21}$$

On the right-hand side $v, c, x_{0\mathrm{I}}, T_0$ are measured in $S_0$ and are exactly the same in all equivalent theories independent of clock synchronisation in $S$ and of $e_1$. Therefore all the equivalent theories predict the same occultation time will be observed on the Earth.

## 29.4 Aberration

The following notation will be used:

- $\boldsymbol{v}$: Absolute velocity of Earth (variable during the year).
- $\boldsymbol{c}$: Vector velocity of a light ray with respect to $S_0$. Given isotropy its modulus has the constant value $c$.
- $\boldsymbol{c}_r$: Velocity of the same light ray relative to the Earth.

We start by reviewing the classical treatment. With reference to Fig. 29.4 one has obviously $\boldsymbol{c}_r = \boldsymbol{c} - \boldsymbol{v}$, which is the same as

$$-c_r \cos\theta = -c\cos\theta_0 - v\ , \qquad -c_r \sin\theta = -c\sin\theta_0\ ; \tag{29.22}$$

therefore

$$\tan\theta = \frac{c\sin\theta_0}{c\cos\theta_0 + v}\ . \tag{29.23}$$

We now come to the general solution of the problem. The transformations (29.1) from $S_0$ to the Earth frame, $S$, in two dimensions and in differential form are as follows:

$$\begin{cases} \mathrm{d}x_{\mathrm{E}} = f_1(\mathrm{d}x_0 - v\,\mathrm{d}t_0)\ , \\ \mathrm{d}y_{\mathrm{E}} = g_2\,\mathrm{d}y_0\ , \\ \mathrm{d}t_{\mathrm{E}} = e_1\,\mathrm{d}x_0 + e_4\,\mathrm{d}t_0\ , \end{cases} \tag{29.24}$$

where the index E denotes quantities calculated with respect to the inertial frame in which the Earth is instantaneously at rest.

Consider an object of any nature and define its velocity with respect to $S$ and $S_0$:

$$u_{\mathrm{E}x} = \frac{\mathrm{d}x_{\mathrm{E}}}{\mathrm{d}t_{\mathrm{E}}}\ , \quad u_{\mathrm{E}y} = \frac{\mathrm{d}y_{\mathrm{E}}}{\mathrm{d}t_{\mathrm{E}}}\ , \quad u_{0x} = \frac{\mathrm{d}x_0}{\mathrm{d}t_0}\ , \quad u_{0y} = \frac{\mathrm{d}y_0}{\mathrm{d}t_0}\ . \tag{29.25}$$

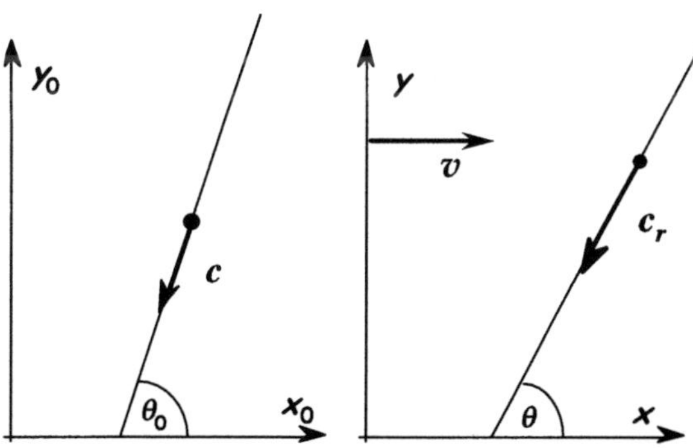


**Fig. 29.4.** If a light ray coming from a star has the inclination $\theta_0$ relative to the fundamental inertial system $S_0$, the same ray has a different inclination, $\theta$, relative to the Earth rest frame

Dividing side by side the first two equations of (29.24) by the third one gives

$$u_{\mathrm{E}x} = \frac{f_1(u_{0x} - v)}{e_1 u_{0x} + e_4} , \quad u_{\mathrm{E}y} = \frac{g_2 u_{0x}}{e_1 u_{0x} + e_4} . \tag{29.26}$$

Referring to the propagation of a light pulse and writing

$$\begin{aligned} u_{\mathrm{E}x} &= -c_r \cos\theta , \quad u_{\mathrm{E}y} = -c_r \sin\theta , \\ u_{0x} &= -c\cos\theta_0 , \quad u_{0y} = -c\sin\theta_0 , \end{aligned} \tag{29.27}$$

we obtain

$$c_r \cos\theta = \frac{f_1(c\cos\theta_0 + v)}{e_1 u_{0x} + e_4} , \quad c_r \sin\theta = \frac{g_2 c \sin\theta_0}{e_1 u_{0x} + e_4} , \tag{29.28}$$

which is the generalization of (29.22). Dividing side by side the two equations of (29.28) we find

$$\tan\theta = \sqrt{1-\beta^2} \frac{\sin\theta_0}{\cos\theta_0 + \beta} , \tag{29.29}$$

the square root being a consequence of (29.7). We thus see that the difference between (29.29) and (29.23) is only due to higher-order terms in $\beta$. Clearly all consequences of the classical theory, in particular the well-known (approximate) aberration formula, hold also in the present general approach based on the transformations (29.8), and (29.29) agrees with experimental evidence [11].

The most important point is however the following: All terms on the right-hand side of (29.29) $(v, c, \theta_0)$ are measured in $S_0$ and are independent of clock synchronisation in $S$. They are the same in all equivalent theories, and the angle $\theta$ perceived on Earth (and its variations) are thus predicted to be exactly the same in all such theories [12].

## 29.5 Radar Ranging of Planets

Let the equations of motion, written in $S_0$, of Earth, of Venus, and of a radar signal sent from Earth toward Venus (Fig. 29.5) respectively be

$$x_{01} = vt_0 , \quad x_{02} = v_2 t_0 + d_0 , \quad x_0 = ct_0 . \tag{29.30}$$

The time, $t_{0R}$, at which the reflection of the radar pulse on Venus takes place must then satisfy

$$ct_{0R} = v_2 t_{0R} + d_0 \quad \Rightarrow \quad t_{0R} = \frac{d_0}{c - v_2} . \tag{29.31}$$

During the return journey of the pulse from Venus to Earth the latter still obeys the first equation of (29.30), while the pulse must satisfy

$$x_0 = x_{0R} - c(t_0 - t_{0R}) . \tag{29.32}$$

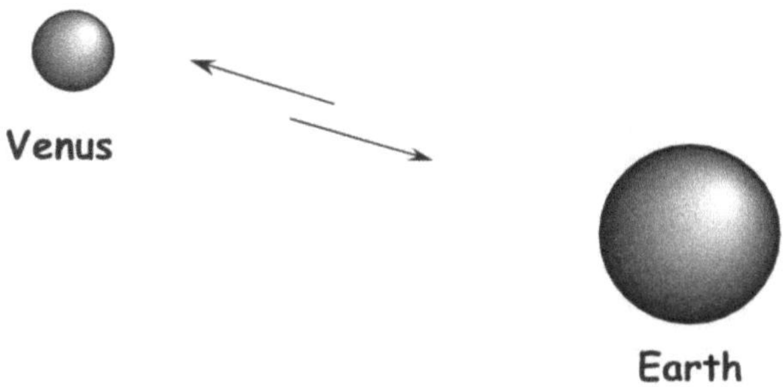


**Fig. 29.5.** A radar signal sent from Earth toward Venus requires some 10 – 20 min for its two-way trip. During this time the Earth displacement due to orbital motion is about 30 000 km

In (29.32) we introduced

$$x_{0R} = ct_{0R} = \frac{d_0 c}{c - v_2} , \tag{29.33}$$

the position occupied jointly by Venus and the pulse at time $t_{0R}$. Therefore the arrival time, $t_{0A}$, of the pulse on Earth satisfies

$$vt_{0A} = x_{0R} - c(t_{0A} - t_{0R}) . \tag{29.34}$$

By using (29.31) and (29.33), (29.34) gives

$$t_{0A} = \frac{2d_0 c}{(c - v_2)(c + v)} . \tag{29.35}$$

From $S_0$ to $S$ (Earth rest frame) the transformations (29.1) apply, and for the pulse arrival time measured on Earth one has

$$t_A = e_1 x_{0A} + e_4 t_{0A} = (e_1 v + e_4) t_{0A} , \tag{29.36}$$

where $x_{0A} = vt_{0A}$ is the Earth position at the time, $t_{0A}$, at which the radar signal returns. From (29.35) and the third equation of (29.7) it follows that

$$t_A = \sqrt{1 - v^2/c^2} \frac{2d_0 c}{(c - v_2)(c + v)} . \tag{29.37}$$

As one can see, on the right-hand side all the quantities $c, d_0, v, v_2$ are measured in $S_0$ and therefore have values which are exactly the same in all equivalent theories, independent of clock synchronisation in $S$ and thus of $e_1$. Once more all equivalent theories predict that exactly the same arrival time of the radar pulse will be observed on the Earth [13].

## 29.6 Proofs of Absolute Simultaneity

Our results constitute a natural development of the research carried out by *Mansouri* and *Sexl* [4], *Tyapkin* [14] and *Croca* and *Selleri* [15]. The authors of [4] worked with a "test theory of special relativity" [equivalent to our transformations (29.1)] in an approximate way, while we took into account the well-established experimental facts expressed by (29.5) and (29.6) and made rigorous calculations starting from the transformations (29.8). The case for a very large empirical equivalence of all transformations of the type (29.8), independent of the parameter $e_1$, is strengthened by the discussed results. That the equivalence is not complete will be shown next by demonstrating that the hypothetical indifference of the physical reality to clock synchronisation exists only insofar as accelerations are neglected. Accelerations modify the conceptual situation to the point that the choice of a well-defined clock synchronisation becomes obligatory [16].

Consider the inertial reference system $S_0$, and assume, as before, that it is isotropic: the one-way velocity of light relative to $S_0$ has the usual value $c$ in all directions. Assume that clocks in $S_0$ have been synchronised with the Einstein method, by using light signals. Two identical spaceships, $A$ and $B$, are initially at rest on the $x_0$-axis of $S_0$ at a distance, $d_0$, from one another. Their internal clocks are synchronised with those of $S_0$. At time $t_0 = 0$ the spaceships start accelerating in the $+x_0$ direction; they do so in exactly the same way and have the same velocity $v(t_0)$ at every time $t_0$ of $S_0$, until at time $t_0 = \bar{t}_0$ they reach a pre-established velocity, $v = v(\bar{t}_0)$, parallel to $+x_0$. For $t \geq \bar{t}_0$ the spaceships are at rest in a different inertial system, $S$ (which they concretely determine), in motion with velocity $v$ with respect to $S_0$. Seen from $S_0$, $A$ and $B$ maintain the distance $d_0$ from one another during the whole trip (Fig. 29.6).

We can now easily show that the relationship between the coordinates of $S_0$ and $S$ is given by the inertial transformations (and not by the Lorentz transformations), because the delay between the times marked by clocks on board $A$ and $B$ and those in $S_0$ does not depend on position.

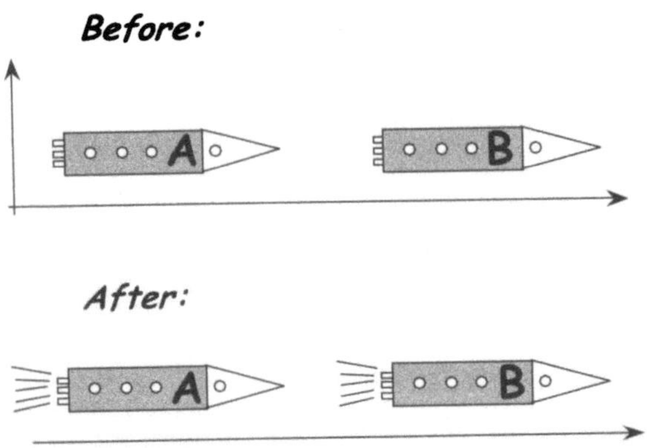


**Fig. 29.6.** Two identical spaceships, $A$ and $B$ are initially at rest on the $x_0$-axis of the inertial system $S_0$. After having accelerated in exactly the same way, $A$ and $B$ are at rest in a different inertial system, $S$, which they concretely constitute

The proof is based on a trivial remark: since $A$ and $B$ have at every time exactly the same velocity, their clocks accumulate exactly the same delay with respect to clocks at rest in $S_0$. Therefore two events simultaneous in $S_0$, taking place at different points of space near which $A$ and $B$ are passing, must be simultaneous also for the travelers in $A$ and $B$, and thus also in the rest system of the spaceships in $S$. It is thus clear that we are facing a situation of absolute simultaneity which cannot be accounted for if the Lorentz transformations are applied, but which is a basic feature of the "inertial" transformations, defined to be those given by (29.8) with $e_1 = 0$.

Not only does the absolute simultaneity arise spontaneously in $S$, but it provides the only reasonable description of the physical reality. To see this point, let us suppose that in $A$ and $B$ there are two passengers ($P_A$ and $P_B$, respectively) who are homozygous twins. Naturally nothing can stop them from resynchronising their clocks once the acceleration has ceased and the ships are at rest in $S$. However, when they do so, they discover they have different biological ages at the same time (resynchronised) in $S$, even if they traveled with exactly the same velocity, as stipulated above. Everything is regular, instead, if they do not implement any asymmetrical modification of the times shown by their clocks.

In fact we have already concluded that clocks in $A$ and $B$ are retarded in the same way, and that the transformations $S - S_0$ must be the inertial ones. Also the aging of the twins must have been the same, since at every time before, during and after the acceleration they were in identical physical conditions. Therefore the twins have the same age when the times shown by their clocks are the same if they have been synchronised in $S_0$ before departure and never modified after. Naturally $P_A$ and $P_B$ can inform one another of their biological ages (e.g., via telefax) by exchanging pictures in which the times they were taken are marked: the twin receiving a picture can check in his archives that at the time shown on his brother's picture he had exactly the same look, and therefore the same age.

Naturally the twins $P_A$ and $P_B$ can use a different clock synchronization, if they wish, e.g. Einstein's synchronization, implementing the validity of the Lorentz transformation between $S$ and $S_0$. To do so they must send a light signal, e.g. from $A$ to $B$, and they must reset at least one clock. We can even suppose that each twin has two clocks and keeps the first one set on the naturally produced time, while he regulates the second one to show the Lorentz time. More exactly we assume the following:

- $P_A$ has a first clock, $T_A$, marking natural time, $t_A$;
- $P_A$ has a second clock, $\hat{T}_A$, marking Einstein time, $\hat{t}_A$;
- $P_B$ has a first clock, $T_B$, marking natural time, $t_B$;
- $P_B$ has a second clock, $\hat{T}_B$, marking Einstein time, $\hat{t}_B$.

After a certain initial time interval during which $t_A = \hat{t}_A = t_B = \hat{t}_B$, only $T_B$ is resynchronised, in the following way: At a certain pre-established time a light signal is sent from $A$ to $B$. The convention that the one-way velocity of light in $S$ is $c$ forces the observer in $B$ to rotate the hands of his clock $\hat{T}_B$ in such a way that the time necessary for the signal to cover the distance, $d$, from $A$ to $B$ is measured to be $d/c$.

Clearly, after resynchronising, at a given time $t_0$ of $S_0$, one will have $t_B = t_{A'}$, but $\hat{t}_B \neq \hat{t}_{A'}$. The simultaneity of $\hat{T}_A$ and $\hat{T}_B$ is now different from that of $T_A$ and $T_B$! If $P_A$ and $P_B$ exchange pictures of themselves in which also the times marked by the clocks $T$ and $\hat{T}$ are shown, they discover they have the same age at time $t$, but different ages at time $\hat{t}$. This provides a strong argument in favour of the inertial transformations, because not all natural "clocks" can be synchronised: irreversible processes exist, such as the ageing of $P_A$ and $P_B$! Nature itself favours the inertial transformations from $S_0$ to $S$ for describing the time of inertial systems concretely produced.

Other phenomena also exist which lead to exactly the same conclusion, $e_1 = 0$, that is, to absolute simultaneity. To these we will here devote only a few lines for reasons of space, in spite of their great conceptual importance:

- The propagation of light along the rim of a circular rotating disk requires, in any reasonable theory, a velocity of light locally identical (at any point, $P$, on the rim) to the velocity of light relative to the inertial system instantaneously endowed with the same velocity as $P$. Unfortunately this condition is not satisfied in the standard relativistic approach. Given that all the possible points $P$ are physically symmetrical, one needs only to measure the velocity of light for a complete tour around the disk, with the advantage that one does not need to go through synchronisation procedures. It has been shown that to correctly calculate on the disk the fundamental time delay between light pulses propagating in opposite directions around the disk, one must use the velocity of light given by the inertial transformations, those with $e_1 = 0$ [17]. The same result allows one to explain the Sagnac effect, which up to now received only unsatisfactory theoretical treatments [18]. In this way a clear improvement has been made over the relativistic theories which contain a discontinuity between the rim of large slowly rotating disks and the locally comoving inertial systems.
- There are now numerous convincing experimental proofs that light can sometimes propagate with a group velocity larger than its usual value $c$ [19]. In all theories with $e_1 \neq 0$ (in particular in the TSR) this implies the existence of horrible causal paradoxes, for example the one, that Recami calls the "Tolman paradox" [20], which we review next. Consider two new space travelers Francis ($F$) and Gina ($G$) and suppose that $F$ travels in space away from $G$ with a large subluminal velocity. Supposing that they can exchange superluminal signals (SLS), at some time $F$, having had a local cancellation in his hard disk, sends to $G$ a SLS asking the following question: "Tell me the nineth word in the line 82 at page 1176 of volume XIV of the British Encyclopaedia." $G$ looks in the indicated place carefully but quickly and answers $F$ with another SLS: "The word you need is:

'hauberk'." Well, the TSR predicts that $G$'s answer will reach $F$ *before* he sends away his question. Having satisfied his curiosity, $F$ can now decide not to send the question. But then, how could $G$ answer exactly such a complicated question, if she did not receive the message from $F$? We are obviously facing a logical impossibility. The TSR does not give an answer; it can only declare that the SLS are impossible. Only the theory of the inertial transformations, with its choice $e_1 = 0$, can provide a logical space for the existence of SLS [21].

The previous considerations show clearly our probable point of arrival, a new theory in which clock retardation is not a relative effect anymore, but depends only on the velocity with respect to a privileged system. This eliminates all paradoxes of relativism. For example, absolute simultaneity allows one to get rid of the "block universe" paradox discussed by *Popper* [22]: with the absolute simultaneity all inertial observers have the same present and therefore the same reality. Finally the existence of a vacuum endowed with physical properties (ether) becomes fully acceptable. Naturally the new theory should be checked experimentally, but in a sense its predictions have already been verified, at least in the case of the Sagnac effect [17].

## 29.7 The Inertial Transformations

We have shown that the condition $e_1 = 0$ has to be adopted. This gives rise to the following transformation of space and time:

$$\begin{cases} x = \dfrac{x_0 - \beta c t_0}{R(\beta)} , \\ y = y_0 , \quad z = z_0 , \\ t = R(\beta) t_0 , \end{cases} \tag{29.38}$$

where $\beta = v/c$. As already stressed by Mansouri and Sexl, such transformations would have been the logical consequence of the development along the lines of thought of Lorentz–Larmor–Poincaré: they are the very relations one would write down if one had to formulate a theory in which rods shrink and clocks are slow by the usual factor when moving with respect to the ether. That the actual development went along different lines was due to the fact that "local time" was introduced at the early stage in considering the covariance of the Maxwell equations. The one-way speed of light predicted by (29.38) can easily be found by taking $e_1 = 0$ in (29.10):

$$\frac{1}{c(\theta)} = \frac{1 + \beta \cos\theta}{c} . \tag{29.39}$$

The transformation (29.38) can be inverted and gives

$$\begin{cases} x_0 = R(\beta)\left(x + \dfrac{\beta c}{R^2(\beta)}t\right), \\ y_0 = y, \quad z_0 = z, \\ t_0 = \dfrac{1}{R(\beta)}t. \end{cases} \tag{29.40}$$

Note the formal difference between (29.38) and (29.40). The latter implies, for example, that the origin of $S_0$ ($x_0 = y_0 = z_0 = 0$) is described in $S$ by $y = z = 0$ and by

$$x = -\frac{\beta c}{1-\beta^2}t. \tag{29.41}$$

This origin is thus seen to move with speed $\beta c/(1-\beta^2)$, which can exceed $c$, but cannot be superluminal. In fact a light pulse seen from $S$ to propagate in the same direction as $S_0$ has $\theta = \pi$ and thus [using (29.39)] has speed $c(\pi) = c/(1-\beta)$, which can easily be checked to satisfy

$$\frac{c}{1-\beta} \geq \frac{c\beta}{1-\beta^2}.$$

It is clear from (29.41) that the velocity of $S_0$ relative to $S$ is not equal and opposite to that of $S$ relative to $S_0$. Such reciprocity holds in the TSR, where it is a consequence of the particular synchronisation used, but cannot be expected to hold more generally.

Consider now a third inertial system, $S'$, moving with velocity $\beta' c$ and its transformation from $S_0$, which of course is given by (29.38) with $\beta'$ replacing $\beta$. By eliminating the $S_0$ variables, one can obtain the transformation between the two moving systems $S$ and $S'$:

$$\begin{cases} x' = \dfrac{R(\beta)}{R(\beta')}\left(x - \dfrac{\beta'-\beta}{R^2(\beta)}ct\right), \\ y' = y, \quad z' = z, \\ t' = \dfrac{R(\beta')}{R(\beta)}t. \end{cases} \tag{29.42}$$

We proposed to name (29.38), (29.40) and (29.42) "inertial transformations". In its most general form (29.42), the inertial transformation depends on two absolute velocities ($v$ and $v'$). When one of them is zero, either $S$ or $S'$ coincides with the privileged system $S_0$ and (29.42) becomes either (29.38) or (29.40).

By studying the multiplication properties of the inertial transformations, it has been possible to show that they do not form a group. There are no problems with the existence of the identical and inverse transformations, and also the associative law can be satisfied, but it is not always possible to write a meaningful product of two inertial transformations, due to the presence of two absolute velocities, $v$ and $v'$, in the transformation. If $\Omega(v, v')$ denotes (29.42), it is easy to understand that the product $\Omega(v, v')\Omega(v'', v''')$ is not an inertial transformation if $v'' \neq v'$.

A property implied by (29.38), (29.40) and (29.42) is *absolute simultaneity*: two events taking place at different points in $S$ but at the same time $t$ are judged to be simultaneous also in $S'$ (and vice versa). The existence of absolute simultaneity does not imply that time is absolute: on the contrary, the $\beta$-dependent factor in the transformation of time gives rise to time-dilatation phenomena similar to those of TSR. A clock at rest in $S$ is seen from $S_0$ to run slower, but a clock at rest in $S_0$ is seen from $S$ to run *faster*, so that both observers agree that motion relative to $S_0$ slows clocks. The difference with respect to TSR exists because in a meaningful comparison of rates a clock, $T_0$, at rest in $S_0$ must be compared with clocks at rest in different points of $S$, and the result is therefore dependent on the "convention" adopted for synchronising the latter clocks.

Absolute length contraction can also be deduced from (29.38), (29.40) and (29.42): All observers agree that motion relative to $S_0$ leads to contraction. The discrepancy with the TSR is due again to the different "conventions" concerning clock synchronisation: the length of a moving rod can only be obtained by marking the *simultaneous* positions of its end points, and therefore it depends on the very definition of simultaneity of distant events.

## References

1. J.S. Bell: 'How to teach special relativity'. In: J.S. Bell: *Speakable and Unspeakable in Quantum Mechanics* (Cambridge Univ. Press, Cambridge 1987) pp. 67–80
2. H. Reichenbach: *The Philosophy of Space and Time* (Dover, New York 1930)
3. M. Jammer: 'Some fundamental problems in the special theory of relativity'. In: *Problems in the Foundations of Physics*, ed. by G. Toraldo di Francia (North Holland, Amsterdam 1979) p. 202
4. R. Mansouri, R. Sexl: Gen. Relat. Gravit. **8**, 497, 515, 809 (1977)
5. C.M. Will: Phys. Rev. D **45**, 403 (1992)
6. H. Poincaré: Rev. Metaphys. Morale **6**, 1 (1898)
7. A. Einstein: *Relativity, The Special, The General Theory* (Methuen, London 1920), see statement at p. 18
8. F. Selleri: Found. Phys. **26**, 641 (1996); Found. Phys. Lett. **9**, 43 (1996)
9. Presently the two way velocity of light in vacuum is known to be: $c = (299\,792\,458.8 \pm 0.2)$ m/s. See: P.T. Woods et al.: Appl. Opt. **17**, 1048 (1978); D.A. Jennings et al.: J. Res. Natl. Bur. Stand. **92**, 11 (1987)
10. J. Bailey et al.: Nature **268**, 301 (1977)
11. For example, see: T.E. Phipps Jr.: Am. J. Phys. **57**, 549 (1989)
12. F. Goy: Found. Phys. Lett. **9**, 165 (1996)
13. S.J. Ostro: Rev. Mod. Phys. **65**, 1235 (1993)
14. A.A. Tyapkin: Nuovo Cimento Lett. **7**, 760 (1973)

15. J.R. Croca, F. Selleri: Nuovo Cimento B **114**, 447 (1999)
16. F. Selleri: 'Noninvariant one-way velocity of light and locally equivalent reference frames'. In: *New Developments on Fundamental Problems in Quantum Physics*, ed. by M. Ferrero et al. (Kluwer, Dordrecht 1997) pp. 381–394
17. F. Selleri: Found. Phys. Lett. **10**, 73 (1997); F. Selleri: 'On a Physical and Mathematical Discontinuity in Relativity Theory'. In: *Open Questions in Relativistic Physics*, ed. by F. Selleri (Apeiron, Montreal 1998) p. 69
18. M. Langevin: Comptes Rendus **173**, 831 (1921); Comptes Rendus **200**, 49 (1935); E. Post: Rev. Mod. Phys. **39**, 475 (1967)
19. G. Nimtz, A. Haibel: Ann. Phys. (Leipzig) **9**, 1 (2000)
20. E. Recami: Int. J. Mod. Phys. A **15**, 2793 (2000)
21. F. Selleri: in preparation (2002)
22. K. Popper: *Unended Quest. An Intellectual Autobiography* (Fontana Collins, Glasgow 1978)

# 30 John Bell and the Ten Challenges of Subnuclear Physics

Antonino Zichichi

John Stewart Bell in Erice

"*There is no new Physics without new projects*"

The contribution of John Bell to the furtherance of experimental physics and our physics technologies is reviewed in order to acknowledge the value of a great theoretical physicist in supporting new trends in physics outside the mainstream. If John Bell was with us, he would be a strong supporter of the ten challenges.

## 30.1 Introduction

The next decade will be a decisive one for subnuclear physics. It is necessary to think in a critical and constructive way in order to be sure that its future will be granted.

The last decade has been devoted to the search for the first example of superparticles (be it the neutralino or the gravitino) and for the lightest Higgs. Neither has been found. On the other hand, no one is able to "predict"

A picture – during the official celebration ceremony – of the bronze plaque which reproduces the original document signed at CERN. The first signature is that of John Bell. The Minister of Internal Affairs, Dr Enzo Bianco, uncovered the bronze plaque. John Bell played a vital role in establishing the existence of the Ettore Majorana Centre in Erice. Not only did he give a series of impressive lectures at the first "Erice School" in 1963, he was also a co-founder of the Ettore Majorana Foundation and Centre for Scientific Culture (EMFCSC), as can be seen from the bronze plaque, which reproduces the constitutive act signed by J.S. Bell, P.M.S. Blackett, I.I. Rabi, V.F. Weisskopf and A. Zichichi at CERN on 8 May 1962

the energy threshold for the two challenges in our field: spontaneous symmetry breaking (SSB) and supersymmetry (SUSY). These two thresholds, on a qualitative ground, should be nearly degenerate with the Fermi scale and therefore be potentially "round the corner".

Nevertheless, the point I want to emphasize is that it would be a serious mistake to focus all our attention on SSB and SUSY. There are in fact other challenges which are the real proof of the vitality of subnuclear physics.

I had the privilege for thirty years of having my office opposite that of John Bell. My heavy engagements in the experimental areas were periodically interrupted by intervals of time spent in the office; these periods – spent a few metres from John's office – were among the most remarkable of my scientific life. I would like to acknowledge the great patience of Mary Bell when she came into John's office to go home and we were heavily engaged in discussing hot topics; she was always smiling and never showed signs of impatience.

## 30.2 The Remarkable Value of John Bell's Support to My Physics

I owe a lot to John, as you will see. He encouraged all my experimental "tricks" and valued these as much as the theoretical "tricks". His support was of vital importance in many of my experiments. For example, in the search for a new heavy lepton with its own neutrino. These were times when very distinguished theoretical physicists at CERN were stating: "*Why should Nature be so stupid as to have two neutrinos if she can do all that is needed with only one?*" Think: even the second neutrino was considered unnecessary. The third, just nonsense. Nevertheless, if a third lepton with a mass as low as 1 GeV and its own leptonic number was there, its existence would have escaped detection. In fact, the abundance of the muon was an incredible coincidence, i.e., $m_{\mu} < m_{\pi}$. Why it is so easy to have muons and why a third lepton of just 1 GeV mass could exist and be undetected were arguments which were out of the mainstream at the time. The mainstream was the sequence of discoveries of new mesons and baryons. Why care about leptons? The standard wisdom was that to understand Nature, it was necessary to study the dozens and dozens of new particles, even if their theoretical understanding was far from being deep and complete.

Of course, there was the Gell–Mann and Ne'eman SU(3) global symmetry, but the mathematical formalism which was supposed to describe the interaction of these particles was missing. The only available mathematical structure was in fact the relativistic quantum field theory (RQFT). But the basic ingredients of an RQFT were the "source of a field" and the "field quantum". When Yukawa formulated his theory of the nuclear forces, the source of the nuclear field was the nucleon and the field quantum, the pion. The proliferation of baryons and mesons provided too many "sources" and too many "quanta". How were the most elementary and the best to be selected? This is why a new mathematical formalism came in, the S-matrix, with the "nuclear democracy" advocated by G. Chew. These new theoretical developments were the negation of RQFT.

The present status of subnuclear physics reminds me of the time when – some four decades ago – with John we were talking about the dominating bubble-chamber technology and the continuous flux of discoveries of mesons and baryons. Would our understanding of the Laws of Nature increase from this "flux"? And would the bubble-chamber technology open up new frontiers in physics?

It was in 1963, at the first Erice School when John Bell in a discussion session emphasized his conviction that "*There is no new physics without new projects*". And new projects are generated by new "tricks" in technology. These "tricks" must not be abstract but aimed at solving specific problems in physics.

Why invent the experimental "trick" to implement [1] high-precision magnetic fields of polynomial shape? Why build a very intense beam of negative

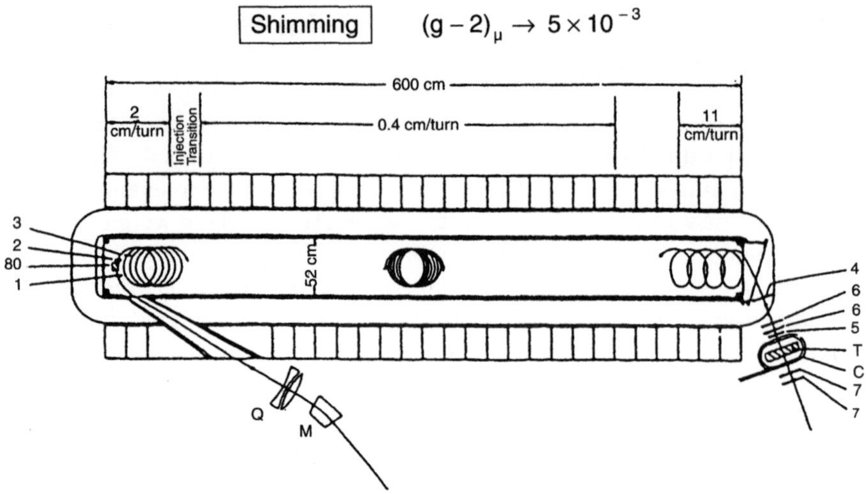


**Fig. 30.1.** General plan of the 6-m magnet. *M*: bending magnet; *Q*: pair of quadrupoles; 1, Be, 2, 3: injection assembly consisting of Be-moderator and counters 1, 2, 3; *T*: methylene-iodide target; counters 66′, 77′: "backward" and "forward" electron telescopes. A stored and ejected muon is registered as a coincidence $4, 6, 66'\,\bar{7}$, gated by a 1, 2, 3 and by an either forward or backward electron signal

particles if, for the bubble chamber, a few negative particles were already enough? Why invent the "trick" of the "preshower" in order to improve, by two orders of magnitude, the selection of lepton pairs in a large background of hadrons? Why determine with great accuracy how a muon penetrates through many metres of an absorber in order to improve its selection in a large background of hadrons? Why invent a "trick" in order to detect neutrons with high efficiency and high precision of the detection point? Why build a neutron spectrometer in order to measure – without using the masses – the "pseudoscalar" and the "vector" mixing angles in the mesonic "nonets"?

In terms of physics, the above-mentioned experimental "tricks" were aimed at obtaining results that I will briefly review in honour of John Bell.

1. The "trick" invented is now known as the "shimming" technology and allowed us to construct the longest magnet with high-precision magnetic fields of polynomial form in order to capture, store and eject muons after thousands of turns, thus allowing the measurement of the deviation of the muon magnetic moment from its original direction and check if this deviation was as expected by quantum electrodynamics (QED). The purpose was to check [2] if the second charged lepton, called a "muon", had the same virtual phenomena as predicted by QED or if some extra effects were present, thus opening (we were hoping) the way to understand why its mass was 200 times that of the electron.

The general plan of the 6-m magnet with all devices needed to perform the experiment is shown in Fig. 30.1.

**Photo 30.1.** The 6-m "flat magnet" where a sequence of high-precision magnetic fields has been implemented using the "shimming technology"

Photo 30.1 shows the magnet, which, despite its very large dimensions (the largest in the world), had six high-precision magnetic fields: one for the injection region, two for the transition, one for the storage, another one for the transition and the last one for the ejection magnetic field.

The results shown in Fig. 30.2 represent the proof that, within five parts in a thousand, the "radiative effects" in the magnetic moment of the second charged lepton were as predicted by QED.

Having proved that – within five parts in a million – the muon magnetic moment had nothing else but pure QED effects, the next step was "What about its weak charge?" In fact accurate measurements of the weak charge were lacking. Two attempts were affected by a serious problem: the rate dependence of the muon lifetime. Having available an excellent muon beam, built by us and investigated in all its details (as, for example, its polarization structure, i.e. within the momentum range of the muon beam the polarizations were different), we decided that it was worth making a high-precision measurement of the weak coupling. Once again the support of John was decisive in convincing the various committees that my request for beam time was "theoretically" fully justified and strongly supported.

And this is how we could perform [3] the most accurate, rate-independent, measurement of the muon weak charge; it turns out that this is the clearest way to determine the Fermi coupling. All the others involving hadrons are not "pure". The results are shown in Fig. 30.3.

*Proceedings of the 1962 International Conference on High Energy Physics, pp. 476–480*

## (g-2) AND ITS CONSEQUENCES

G. Charpak, F. J. M. Farley, R. L. Garwin, T. Muller, J. C. Sens and A. Zichichi
CERN, Genève

*(presented by A. Zichichi)*

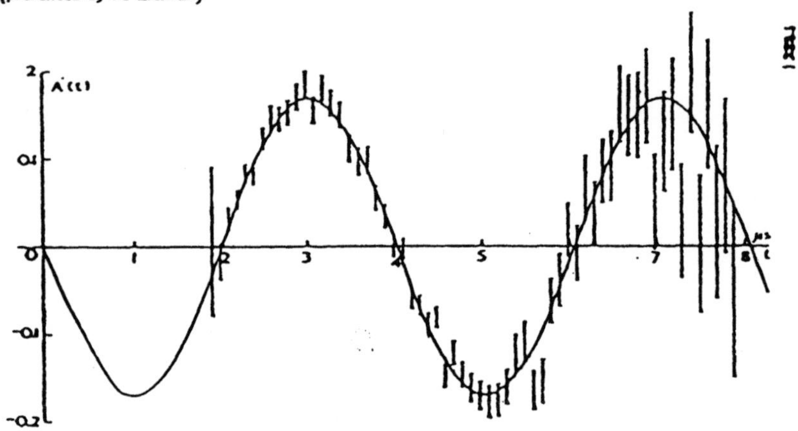


Fig. 5 Experimental data of the (g-2) experiment. Observed electron decay asymmetry $A(t)$ as a function of storage time. The curve represents the best fit of the data.

RESULTS OF G - 2 EXPERIMENT

| Experimental | Theoretical |
|---|---|
| | $\frac{g-2}{2} = \frac{\alpha}{2\pi} + 0.75\left(\frac{\alpha}{\pi}\right)^2$ |
| $\frac{g-2}{2} = 0.001162 \pm$ | $= 0.001161 + 0.000004$ |
| 0.000005 | $= 0.001165$ |

Muon mass $= (206.768 \pm 0.003) m_e$

Charge of muon $= (1.00000 \pm 0.00005) e$ 18 December 1961.

Charge of $\nu_\mu$ $= (0.00000 \pm 0.00005) e$ .

**Fig. 30.2.** The first high-precision measurement of QED radiative effects outside the electron and photon world

2. The proof that the antideuteron, $\overline{\mathrm{d}}$, the first example of nuclear antimatter, exists. The construction of the very intense beam of negative particles [4] was aimed at establishing if the antiprotons and the antineutrons

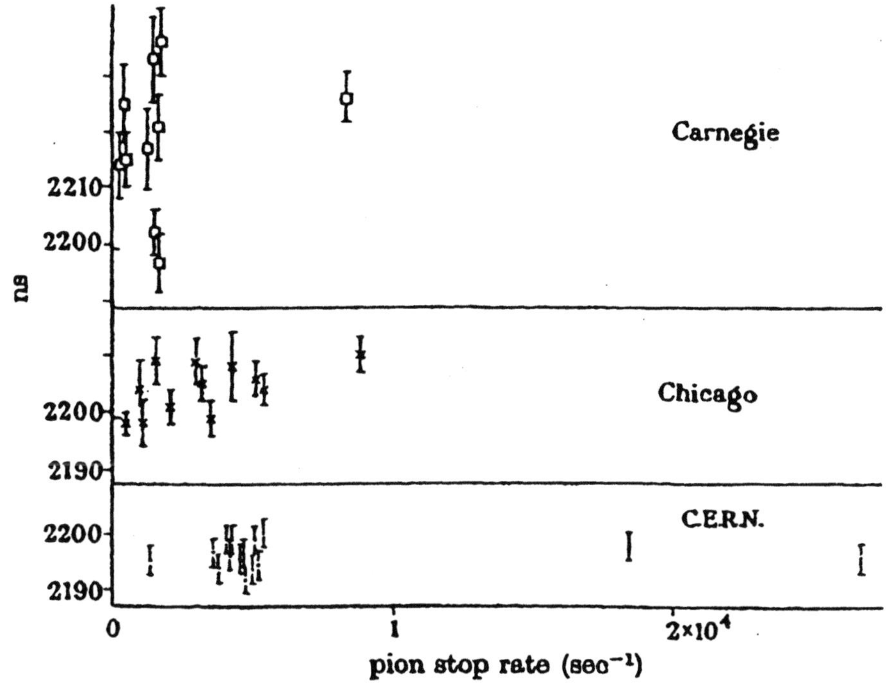


$$\tau_\mu^{CERN} = (2{\cdot}198 \pm 1)\ \text{n sec}$$

Combining all runs and making a least squares fit with sloping background as explained above, we find $\tau = 2198 \pm 1$ ns. I would like to emphasize, however, that this is a preliminary number and the measurements are still in progress.

This result is smaller than the accepted figure but is not really inconsistent with the data of Lundy *et al.*[1], and of Reiter[2] (shown in Fig. 1), and with a result of Eckhause ($2202 \pm 4$ ns)[3]. We believe that our figure is more correct because of the elimination of the systematic effects mentioned above. ✱

The new value brings the muon lifetime further away from the value calculated from the $O^{14}$ lifetime with conserved vector current theory. Because of the complexity of the nuclear $\beta$ decay we must regard the muon lifetime as the best measure of the weak interaction coupling constant. $(G_\mu^F)$ This is why it appears worth while to measure the lifetime as carefully as possible.

**Fig. 30.3.** These results, obtained at CERN, are the first high-precision measurements of the weak charge. The diagram shows that the experimental results on $\tau_\mu$ obtained in Chicago and Carnegie were affected by a rate-dependent systematic effect which invalidates the data. The CERN result is the first without this trouble were interacting as the protons and the neutrons do. If this was the case, the antideuteron,

$$\overline{\mathrm{d}} \equiv (\overline{\mathrm{p}}\,\overline{\mathrm{n}}) \,,$$

had to exist.

Let me recall that the discovery of CP breaking in 1964 by Christenson, Cronin, Fitch and Turlay, the triumph of the S-matrix and the lack of experimental evidence at the $10^{-7}$ level had raised serious doubts about the validity of RQFT, since the existence of antimatter is granted by the CPT theorem and the CPT theorem is a consequence of RQFT.

I recall a meeting in John's office with Heisenberg. Both were strong supporters of my proposal to search for the existence of the antideuteron at the lowest possible value of the production rate.

The antideuteron, $\overline{\mathrm{d}}$, was searched for and found not to be there at the level of one $\overline{\mathrm{d}}$ per ten million $\pi^-$ produced:

$$10^{-7} = \frac{\text{Number of } \overline{\mathrm{d}}}{\text{Number of } \pi^-} \,.$$

In order to reach the detection level

$$10^{-8} = \frac{\text{Number of } \overline{\mathrm{d}}}{\text{Number of } \pi^-}$$

it was necessary to build the most intense negative beam "partially separated" (i.e., of no value for the bubble-chamber technology) and to invent an electronic technical "trick" in order to reach a resolution in time-of-flight (TOF) at the level of

$$\pm 70\,\mathrm{ps} \quad (\mathrm{ps} = 10^{-12}\,\mathrm{s}) \,.$$

We were able [5] to detect the existence of $\overline{\mathrm{d}}$ at the $10^{-8}$ level. This was in 1965 and gave "confidence to the search for a field theoretical basis for the strong reactions," as stated by Luciano Maiani on the occasion of the 30th anniversary of the discovery of nuclear antimatter [6].

The details of the beam and of the sophisticated experimental set-up able to measure TOF with unprecedented accuracy ($\pm 70$ ps) are shown in Fig. 30.4. The same electronic "trick" was later used for high-precision tracking in plastic scintillators [7].

3. Here the aim was the search for the third lepton, with its own leptonic number. If its mass was in the GeV range, it could be detected only if we were able to separate an acoplanar (e$\mu$) pair from the very high level of hadronic background. In order to open the way to the search for the third lepton, it was necessary to prove that the (e$\mu$) acoplanar pairs can be detected in a very high level of hadronic background. The experimental set-up needed the "preshower" "trick" [8] for the electron detection and the well-calibrated

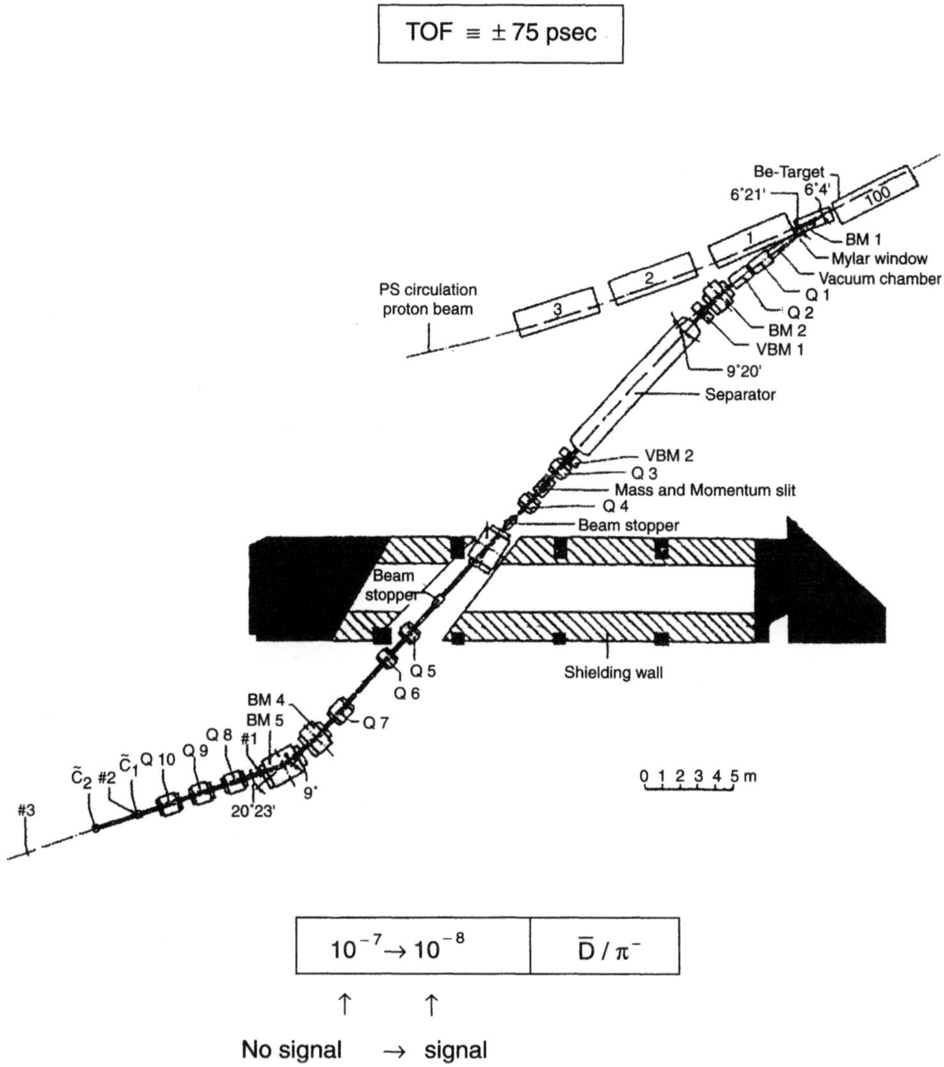


**Fig. 30.4.** The most intense negative beam ever built, with the details of the mass spectrometer, to search for *antideuterons* [5] and the high-precision TOF system

muon filters [9] for the muon detection. A drawing of the project is shown in Fig. 30.5.

The experimental set-up was able to detect, simultaneously, $(e^+e^-)$, $(\mu^+\mu^-)$ and $(\mu^\pm e^\mp)$ pairs. When we planned the search for a heavy lepton $(HL^\pm)$, no one knew that the proton has a strong "time-like" electromagnetic structure, $F_p^{em}(q^2)$. At our value of $q^2$, this structure produced a depression of a factor 500 in the expected electromagnetic cross-section for the production

**PRESHOWER $\pi/e = 5 \times 10^{-4}$**

**MUON PUNCH-THROUGH**

**To simultaneously detect $\mu^{\pm}e^{\mp}$ final states in $\bar{p}p$ annihilation the first experiment to search for the THIRD LEPTON (HL $\equiv \tau$)**

$(HL^{\pm}) \equiv \tau^{\pm}$ & $F_p^{em}(q^2)_{\text{time-like}}$

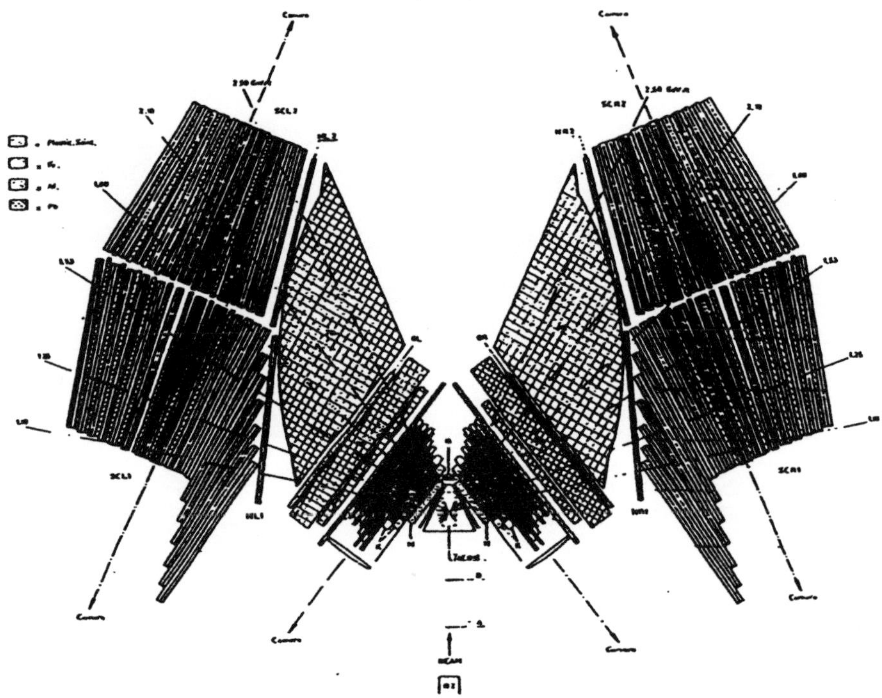

**Fig. 30.5.** The experimental set-up implemented to search for the third lepton (HL $\equiv \tau$) and for a time-like electromagnetic structure of the proton $[F_p^{em}(q^2)_{\text{time-like}}]$. The set-up was able to detect simultaneously the $(e^+e^-)$, $(\mu^+\mu^-)$ and $(e^{\pm}\mu^{\mp})$ final states of the $(\bar{p}p)$ annihilation

mechanism of the third lepton $HL^{\pm}$,

$$p\bar{p} = (HL)^+ + (HL)^- . \tag{30.1}$$

With the experimental set-up illustrated in Fig. 30.5, we succeeded in establishing the existence of a strong electromagnetic form factor of the proton in the time-like region [10], and we concluded that the production process for $(\mathrm{HL})^{\pm}$ had to be the electron–positron annihilation channel:

$$e^+e^- = (\mathrm{HL})^+ + (\mathrm{HL})^- \ .$$

Thanks to the support of John Bell and Viky Weisskopf, we were allowed to move to Frascati [11], where the Bruno Touschek $(e^+e^-)$ collider, called ADONE, was being implemented. The results obtained with ADONE are shown in Fig. 30.6. With the ADONE energy we could establish that the $(\mathrm{HL})^{\pm}$ mass had to be higher than 1 GeV [12].

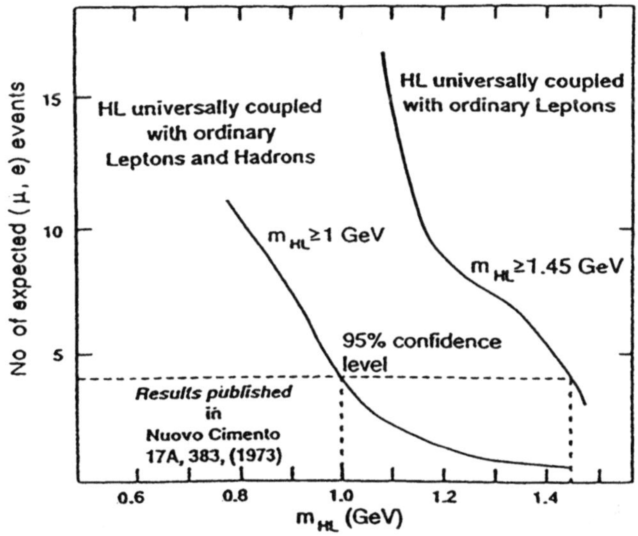


**Fig. 30.6.** The expected number of $(e^{\pm}\mu^{\mp})$ pairs versus $m_{\mathrm{HL}}$, i.e. the heavy lepton mass, for two types of universal weak couplings of the heavy lepton

In Fig. 30.7 there is a reproduction of the first results obtained at SLAC in 1975 [more than 10 years after we had proved at CERN the validity of the $(e\mu)$ method and of the associated technology]. The origin of the third family of leptons has its roots at CERN and Frascati [13].

As we will see later (Fig. 30.35), had John Bell been the Director of the Frascati Laboratory, the $(\mathrm{HL})^{\pm}$, now called $\tau^{\pm}$, would have been discovered there, at least five years before it was SLAC. The present status of the flavour space is shown in Fig. 30.8. The dates when the third lepton started to be investigated at CERN (1960), and at Frascati (1967) and the date of its discovery at SLAC (1975) are shown.

4. The direct measurement (i.e., without using the mass formulae) of the very different mixings in pseudoscalar and vector mesons, the respective angles being $\theta_{\mathrm{PS}} \cong 10°$ and $\theta_{\mathrm{V}} \cong 50°$. These results were obtained using the

STATUS OF SUBNUCLEAR PHYSICS

A. Zichichi
CERN, Geneva, Switzerland

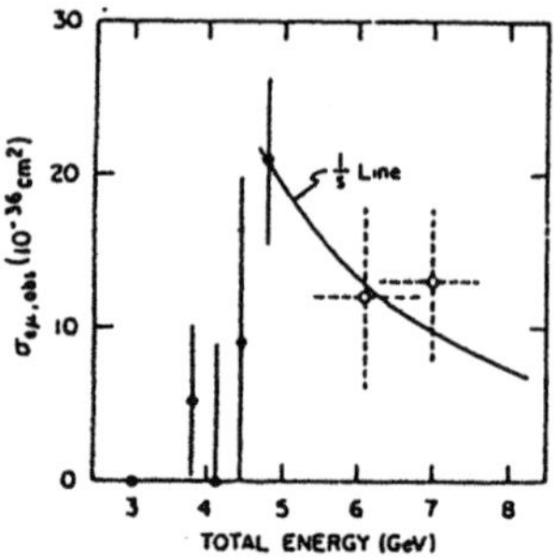


◀ Fig. 9 The observed cross-section for events of the type e and μ, with no other particles in the SLAC-LBL magnetic detector. The two high-energy measurements (dashed lines) are preliminary. These data have not been corrected for momentum and angle cuts and for the geometry of the detector. This correction can be a factor of 2 to 10, depending on the origin of the events.

4.4 Observation of ($e^{\pm}\mu^{\mp}$) pairs above 4 GeV

Preliminary results of a new effect have been reported by Dr. Feldman[4]. At 4.8 GeV the SLAC Group has observed the production of electron and muon pairs, i.e. the reaction

$$e^{+} + e^{-} \to e^{\pm} + \mu^{\mp} + \text{anything (provided it gives no detectable particles)} \quad (4.1)$$

This implies that the observed events (4.1) are not accompanied by other tracks. This selection is done in order to have good e and μ rejection against background from multihadron production. The interesting aspect of these preliminary results is the fact that the number of like pairs, $e^{-}\mu^{-}$, $e^{+}\mu^{+}$, is consistent with zero (background level $\sim$ 1 event), while the unlike pairs, $e^{\mp}\mu^{\pm}$, are 21 events with 4 to 6 expected as background. The authors exclude that the effect could be accounted for by systematic effects, and the search is going on. It should be recalled that (eμ) events of like and unlike charges have been thoroughly looked for at Frascati by the BCF Group[6] and that no signal had been detected up to 3 GeV total centre-of-mass energy. The BCF search was motivated by the high-mass leptons. No charm or gauge theories were available at that time, and it is not excluded that old heavy leptons are at the origin of the (eμ) events discovered at SLAC. According to Dr. Feldman's report it seems that the threshold for this new phenomenon is at about 4 GeV, as indicated in Fig. 9.

**Fig. 30.7.** Reproduction of page 1325 of my closing lecture at the 1975-EPS Conference in Palermo

electromagnetic decays of the mesons, thanks to the high rejection power of ($e^{+}e^{-}$) pairs and the neutron missing mass spectrometer. The "trick" needed was [14] the construction of a counter (plastic scintillator) able to detect the neutron-impinging-point in real time. This was accomplished by having at both ends of the counter two phototubes. The difference in arrival time of the light emitted by the neutron (interacting in the plastic material and producing a proton) was the measurement of the emission angle of the neutron, thus giving the mass of the produced meson. This "trick" was needed to solve the problems connected with the construction of the plastic scintillator counters

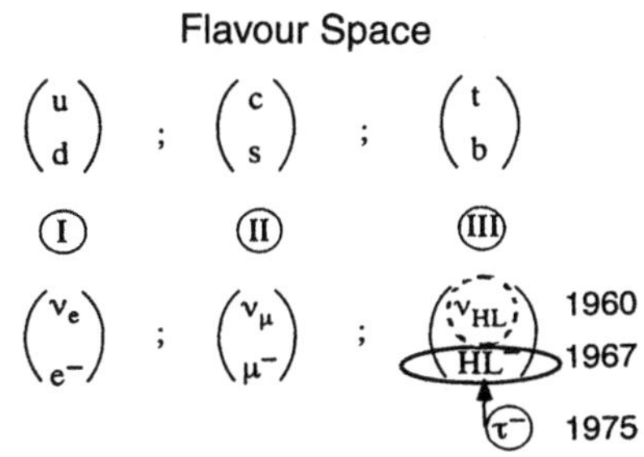


**Fig. 30.8.** The present status of flavour states classified in terms of three families, each one being composed of a quark pair plus a lepton pair. The experimental search for the lepton pair of the third family started at CERN in the early 1960s and moved to Frascati, where the $(e^+e^-)$ ADONE collider began to operate in 1967

(never built before with the needed properties), with their calibration, with the timing of all counters within ±100 ps and with the selection of $(e^+e^-)$ pairs. In this selection we were already "masters in the world": it was our "preshower" "trick" once again. The experimental set-up, which was called a "non-bubble-chamber" (NBC) instrument, is shown in Fig. 30.9.

This NBC instrument allowed us to measure many properties of the mesonic states which could not be determined with the bubble-chamber technology. The most important one, i.e. the direct measurement of $\theta_{\rm PS}$ and $\theta_{\rm V}$, is shown in Fig. 30.10.

It is probably interesting to remark that in the experimental set-up shown in Fig. 30.9 there is also a system of alternate layers (a "sandwich") of lead sheets plus a plastic scintillator in order to give a signal for the total energy deposited. This was in fact the first example of what is now known as a "calorimeter". A different view of the same set-up which clearly shows the "calorimetry" structure is in Fig. 30.11.

With this instrument we could measure many properties of pseudoscalar and vector mesons. One example is the two-gammas-decay channel of the meson – called at that time $X^0$, since no one knew in which multiplet it was supposed to be. A proposal was to put it in the tensor octet.

We found that this mysterious $X^0$ meson had a $2\gamma$ decay mode,

$$X^0 \to \gamma\gamma ,$$

and thus it could not be a tensor. It could be a pseudoscalar. This is how it became what is now known as the ninth member of the pseudoscalar nonet: the $\eta'$. I only let you imagine the many discussions with John on all these problems. He was also very happy that this set-up was called NBC (non-bubble-chamber). In fact it produced many dozens of papers on mesonic physics [15] which – as mentioned before – could not be obtained with the dominating bubble-chamber technology.

5. The proof that in violent (pp) collisions, the protons do not break into quarks, despite "scaling" discovered at SLAC in DIS. How can it be

Neutron Missing-Mass Spectrometer

TOF | ± 100 psec

Preshower: $\pi/e = 5 \times 10^{-4}$

Calorimetry

$\theta_{PS} \neq \theta_V$

**Fig. 30.9.** The "preshower" technology implemented in the CERN experimental set-up for the study of the rare decay modes of vector mesons (1963) in order to measure in a direct way – without the use of the mass formula – the pseudoscalar, $\theta_{\mathrm{PS}}$, and vector, $\theta_{\mathrm{V}}$, mesons mixing angles. This is the first example of what is now "standard" in experimental subnuclear physics: very large acceptance detectors. The apparatus was called a "non-bubble-chamber" (NBC) instrument. With this NBC set-up many properties of mesons were measured

that at SLAC they discovered "scaling" and there are no quarks around? My group had already been looking for quarks in cosmic rays. I was at the Vienna Conference when Jerry Friedman reported the discovery of "scaling"

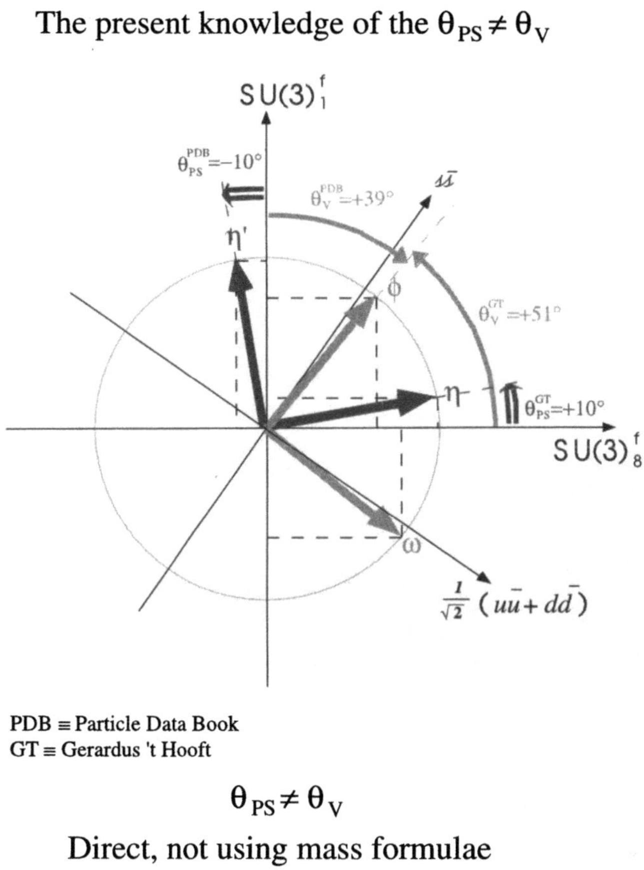


**Fig. 30.10.** Direct measurement of $\theta_{PS}$ and $\theta_V$ using the NBC instrument

in DIS, using electrons and protons. What about DIS using protons against protons? A very large magnet, called the Split-Field-Magnet, was available at the Intersecting Storage Ring (ISR), the new CERN collider. Another "trick" was needed [16] in order to put long plastic scintillator counters inside the magnet. For quarks with charge $\frac{1}{3}e$ the pulse height expected was nine times less than a minimum ionizing particle. Here the "trick" was simple. We needed a scintillator with a "gradient" in its scintillation power and a long absorption length. No one in industry was able to give us what we wanted. This is why we had no other way out but to build the counters ourselves. Once again the support of John was decisive in overcoming all the difficulties naturally arising when you want to do something which is not "mainstream". The experiment, which was quite complex in its implementation, is shown in Fig. 30.12.

**Fig. 30.11.** A view of the non-bubble-chamber (NBC) set-up. The fellow pictured is Julien Berbiers, my most devoted technician

Fortunately the protons obey confinement, and therefore, despite scaling in DIS, the quarks cannot come out as free particles. If protons would not obey confinement, at least up to a reasonable energy level, we could not exist. The results of the experiment [17] were the best upper limits on quarks.

A consequence of our engagement with the Split-Field-Magnet turned out to be very interesting. And this brings me to the next point. Here it was not a question of upper limits. Here we had the opposite situation. Lots of data could easily be obtained, but apparently of no value. The reason was what Gribov defined the "hidden side of QCD".

6. The "hidden side of QCD" was the total diversity in the multihadronic final states as if they had the memory of the interacting particle pairs: ($\pi$p), (pp), (kp), ($\bar{\text{p}}$p), ($e^+e^-$), ($\nu$p), etc. In fact each pair of interacting particles was associated with different multihadronic final states. As stated by Gribov [18], there was a sort of gentleman's agreement not to speak about data in conflict with the new-born theory describing the interaction between quarks and gluons. If the source of hadronization was the same, it was difficult

$\left(\frac{dE}{dx}\right)$ with $\lambda_{abs}$ = 25 meters and TOF ± 100 psec

DIS in (pp): no quarks

**Fig. 30.12.** The Split-Field-Magnet interaction region at the CERN ISR. Clearly visible in the foreground are the large-area scintillation counters for TOF. The $\mathrm{d}E/\mathrm{d}x$ counters for quarks are inside the Split-Field-Magnet

to explain why there was no sign of a common feature in the multihadronic states produced. Using the Split-Field-Magnet and a very simple multitrack device, with a minimum-bias trigger, we could find out that the multihadronic final states produced in (pp) interactions at the ISR had universality features, provided that a new quantity was introduced in their analysis. How can it be that no one before could see this effect? The answer is very simple. It was necessary to analyze thousands and thousands of multihadronic final states and have an exact knowledge of the new quantity needed: the effective energy.

The (pp) collisions at the ISR were with a fixed nominal energy of 62 GeV. But each interaction could produce a final state with effective energy ranging from a few GeV up to the maximum value of 62 GeV, when the two protons had to be at rest in the CM system (nearly coincident with the lab system, thanks to the collider mode of operation proper for a machine such as the ISR: the first proton collider in the world).

The selection of the effective energy to associate with each multihadronic final state was done in real time, event by event, detecting the "leading" proton of the event itself. The introduction of the new "trick", the effective energy, $E^{\mathrm{HAD}}$ [19], allowed us to establish the existence of universality features in the multihadronic final states, no matter what the nature of the interaction (strong, electromagnetic or weak) was. The results obtained at the ISR were ignored by the Theory Division at CERN but attracted the interest of a leader in quantum chromodynamics (QCD), Vladimir Gribov. Here is what he said [20]: "*Each pair of interacting particles, when producing systems consisting of many hadronic particles, had its own final state. No-one knew how to settle this flagrant contradiction. When I read the paper 'Evidence of the same multiparticle production mechanism in (pp) collisions as in ($e^+e^-$) annihilation', I realized that something very interesting had been found. In fact the introduction of the 'Effective Energy' in the analysis of (pp) collisions at the CERN-ISR gave a totally unexpected result.*" Once again this is the proof of how much John was of help in encouraging new trends in experimental physics. In fact, soon after *G. 't Hooft* stated [20]: "*Theoreticians were unable to prescribe what experimentalists had to look for to establish the universal nature of these final interactions. The experimental results were discouraging: scattering experiments yielded different final states for each pair of interacting particles. So it happened that these aspects of QCD had to wait until experimentalists themselves came with the right idea. The showers come with what is now called an 'Effective Energy' and, in terms of this quantity, universality could be established.*"

It was in Erice, during the first year of the Subnuclear Physics School – as mentioned above – that John made his famous statement emphazing that, for the new projects, of vital importance are what John liked to call experimental "tricks". It was in Erice in 1964 that Dick Feynman gave another strong push in the correct direction. Let me recall his statement: "*Experimental 'tricks' are (at least) as important as theoretical 'tricks'. So in both theory and experiment, to improve the physics of the future, I would like to see a littlc more pride in the work.*"

I have summarized in Table 30.1 the list of experimental "tricks" strongly supported by John. These represent the roots of another project of mine, the LAA Project, as we will see in the next section.

**Table 30.1.** Summary of the roots of the LAA Project

| Experimental "tricks" | Results |
|---|---|
| • High-precision magnetic fields, $\frac{\Delta B}{B} = \pm 10^{-4}$ | Fig. 30.1 and 30.2 |
| • TOF $= \pm 70\,\mathrm{ps}$ | Fig. 30.3 and 30.4 |
| • Preshower $= \pm 5 \times 10^{-4} \equiv \pi/e$ | Fig. 30.5, 30.6 and 30.7 |
| • Muon punch-through | Fig. 30.5, 30.6 and 30.7 |
| • Calorimetry | Fig. 30.9, 30.10 and 30.11 |
| • Neutron missing-mass spectrometer ($\pm 100\,\mathrm{ps}$) | Fig. 30.9, 30.10 and 30.11 |
| • High-precision $\frac{\mathrm{d}E}{\mathrm{d}x}$ with $\lambda_{\mathrm{abs}} = 25\,\mathrm{m}$ | Fig. 30.12 |

## 30.3 The LAA Project in a Few Words

The LAA Project is the analytic continuation of my "tricks", strongly supported by John Bell and Dick Feynman. In fact, my physics interests made me very much concerned about the future of subnuclear physics. This concern lies at the origin of my activity devoted to the implementation of new projects. The LAA Project is for subnuclear technologies able to allow us to make experiments with the next generation of colliders, where the energy and the luminosities need to be taken care of in an original set of ways. The LAA Project [21] came at the right moment at CERN, when all efforts were being made to build LEP and no one could have cared less about the future of sub-

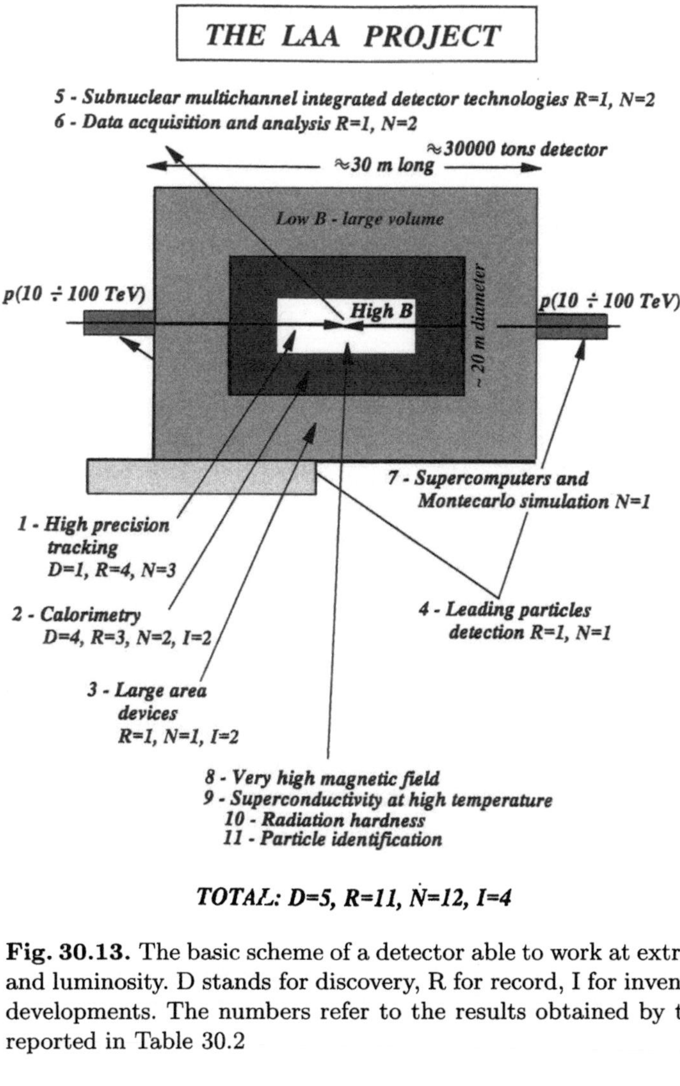


**Fig. 30.13.** The basic scheme of a detector able to work at extreme values of energy and luminosity. D stands for discovery, R for record, I for inventions, and N for new developments. The numbers refer to the results obtained by the LAA Project, as reported in Table 30.2

**Table 30.2.** Main achievements of the LAA Project in terms of discoveries (D), records (R), new developments (N), and inventions (I)

| | D | R | N | I |
|---|---|---|---|---|
| 1. High-precision tracking | | | | |
| a) Gaseous detectors | | 2 | 1 | |
| b) Scintillating fibres | 1 | 2 | 1 | |
| c) Microstrip GaAs | | | 1 | |
| 2. Calorimetry | | | | |
| a) High-precision EM | 2 | | 1 | |
| b) compact EM + hadronic | 1 | 2 | | 2 |
| c) "Perfect" calorimetry | 1 | 1 | 1 | |
| 3. Large-area devices | | | | |
| a) Construction | | | | 2 |
| b) Alignment | | 1 | 1 | |
| 4. Leading particle detection | | 1 | 1 | |
| 5. SMIDT | | | | |
| a) Microelectronics | | | 2 | |
| b) New, radiation-resistant, technologies | | 1 | | |
| 6. Data aquisition and analysis | | | | |
| a) Real-time data aquisition | | | 1 | |
| b) FASTBUS RISC Computer | | 1 | | |
| c) Fine-grained parallel processor | | | 1 | |
| 7. Supercomputers and montecarlo simulations | | | 1 | |
| Total | 5 | 11 | 12 | 4 |

nuclear technologies. Since this represents the future of subnuclear physics, let me again pay a tribute to John Bell, without whose support the LAA Project would not have overcome the innumerable number of difficulties it faced. These were taken care of by a group of devoted fellows, who helped me implement the project. Let me mention the most important one: Horst Wenninger. The summary of the results achieved are shown in Fig. 30.13 and Table 30.2.

In Fig. 30.13 there is an ideal "$4\pi$ detector" whose purpose is to be the guide for all R&D work in those technologies needed to allow the full exploitation of the extreme energies and luminosities of the ELN supercollider (see Sect. 30.6). The seven items indicated in Table 30.2 are those of the 11 components of the LAA Project where discoveries, records, new developments and inventions have been obtained [22]. The numbers refer to the main results achieved.

## 30.4 Facilities and the Basic Steps

Having outlined the consequences of John's statement for the future, let me summarize in Table 30.3 the results obtained in subnuclear physics together with the facilities needed.

**Table 30.3.** Facilities and the basic steps needed to obtain the results we have in subnuclear physics

| Facilities | |
|---|---|
| Past and present: | BNL (COSMOTRON, AGS, RHIC) – CERN (SC, LEAR, PS, ISR, SPS, S$\bar{\text{p}}$pS, LEP, LEP II) – DUBNA (LHE), NOVOSIBIRSK (VEPP-4M), SERPUKHOV (UNK) – INFN (ADONE, DA$\Phi$NE) – INFN-GRAN SASSO – DESY (DORIS, PETRA, HERA) – SLAC (SPEAR, PEP, SLC, PEP II-B-Factory) – CORNELL (CESR) – FNAL (TEVATRON) – SUPER-KAMIOKANDE – KEK (PS, TRISTAN, B-Factory, KEK-B) |
| Future: | (10%) ELN $\equiv$ LHC $\rightarrow$ TESLA $\rightarrow$ ELN |

**Basic Steps**

Present: the standard model Future: the open problem

① RGEs $[\alpha_i\,(i \equiv 1,2,3);\ m_j\,(j \equiv \mathrm{q,l,G,H})]:\ f(k^2)$
- GUT ($\alpha_{\mathrm{GUT}} \cong 1/24$) and GAP ($10^{16} - 10^{18}$)GeV
- SUSY (to stabilize $m_{\mathrm{F}}/m_{\mathrm{P}} \equiv 10^{-17}$)
- RQST (to quantize gravity)

② Gauge principle (hidden dimensions)
— How a fundamental force is generated: SU(3), SU(2), U(1)

③ The physics of imaginary masses: SSB
— The imaginary mass in SU(2) $\times$ U(1) produces masses ($m_{\mathrm{W}^\pm}, m_{\mathrm{Z}^0}, m_{\mathrm{q}}, m_{\mathrm{l}}$), including $m_\gamma = 0$
— The imaginary mass in SU(5) $\Rightarrow$ SU(3)$\times$SU(2)$\times$U(1) or in any – not containing U(1) – higher symmetry group $\Rightarrow$ SU(3)$\times$SU(2)$\times$U(1) produces monopoles
— The imaginary mass in SU(3)$_{\mathrm{c}}$ generates confinement

④ Flavour mixings and CP $\neq$, T $\neq$
— No need for it, but it is there

⑤ Anomalies and instantons
— Basic feature of all non-abelian forces

| Note: | | | | | |
|---|---|---|---|---|---|
| q | $\equiv$ | quark and squark | $m_{\mathrm{F}}$ | $\equiv$ | Fermi mass scale |
| l | $\equiv$ | lepton and slepton | $m_{\mathrm{P}}$ | $\equiv$ | Planck mass scale |
| G | $\equiv$ | Gauge boson and gaugino | k | $\equiv$ | quadrimomentum |
| H | $\equiv$ | Higgs and Shiggs | C | $\equiv$ | Charge conjugation |
| RGEs | $\equiv$ | Renormalization group equations | P | $\equiv$ | Parity |
| GUT | $\equiv$ | Grand unified theory | T | $\equiv$ | Time reversal |
| SUSY | $\equiv$ | Supersymmetry | $\neq$ | $\equiv$ | Breakdown of symmetry operators |
| RQST | $\equiv$ | Relativistic quantum string theory | | | |
| SSB | $\equiv$ | Spontaneous symmetry breaking | | | |

In fact it is the existence of the facilities listed in Table 30.3 which has allowed us to reach the basic steps outlined. This is another proof of the validity of John's statement for progress in our field. Let me briefly review this progress.

It is based on the following five points:

- The renormalization group equations (RGEs) which tell us that all quantities change with the energy scale, thus giving rise to the possibility of a Theoretical grand unification (GUT), to Supersymmetry (SUSY) and to the relativistic quantum string theory, which should, hopefully, one day become the theory of everything.
- The gauge principle which tells us that a fundamental force of Nature is generated by requiring the local (point by point in space–time) invariance of the Lagrangian for symmetry operations in complex hidden dimensions which can be, in number, one [U(1)], two [SU(2)] or three [SU(3)].
- The way to produce physical masses without creating theoretical catastrophes, i.e. the introduction of imaginary masses into the Lagrangian. This produces spontaneous symmetry breakdown (SSB).
- This SSB mechanism is not at work in the breakdown of CP and T symmetries. Moreover these non-SSB mechanisms appear to be associated with another mystery: the mixing of flavour states.
- Anomalies and instantons are basic features of the non-Abelian fundamental Laws of Nature.

I will now sum up all these basic points, together with other known facts of experimental origin but which are theoretically open, in order to see where our understanding of the world around us is, another point so dear to John. All we are doing, he liked to say, is trying to understand our world. Let us see where we stand.

We have understood that the origin of all fundamental forces of Nature is the same: the gauge principle. We have discovered that the number of families of the fundamental building blocks of matter (quarks and leptons) is three. When viewed using the Fermi forces, which provide the fuel for all the stars and control their rate of burning, quarks and leptons appear mixed. All basic quantities vary with energy. Not only real but imaginary masses exist. The existence of imaginary masses has an important effect on our description of the vacuum. In turn, the vacuum determines the masses and the interaction properties of the real particles. Neutrinos appear to not be massless.

We have very good reasons to believe that all gauge couplings converge towards the same value at an energy level two orders of magnitude below the maximum conceivable energy, the Planck scale. Here very intense new theoretical developments are taking place.

In order to explain how it is possible to keep separated two energy scales, the Fermi and the Planck scales, which are seventeen orders of magnitude apart, Nature should have a symmetrical structure which puts fermions and bosons on equal grounds.

The point-like description of our world must be abandoned in favour of a non-point-like (string or membrane) elementary basic entity. Our four-dimensional space–time is the result of an expansion of hidden dimensions

– both of bosonic (10 + 1) and fermionic (32) nature – whose number could be as high as 43: this is the superspace.

There is no other field of human intellectual activity where the past, present and future are so rich in fundamental discoveries and conceptual breakthroughs.

These basic developments would not have been possible if the facilities listed at the top of Table 30.3 were not available with all their "tricks" and innovative technologies.

## 30.5 The Ten Challenges

Without this set of basic developments, the ten subnuclear challenges (Table 30.4) could not exist. On each challenge I will present a few comments concerning its relevance for the future, and also in connection with its implementation.

**Table 30.4.** The ten subnuclear challenges

| |
|---|
| ① The physics of imaginary masses: SSB |
| ② CP $\neq$ , T $\neq$, CPT $\neq$, matter–antimatter symmetry |
| ③ Supersymmetry – SUSY |
| ④ Non-perturbative QCD |
| ⑤ Anomalies and instantons |
| ⑥ Flavour mixing in the quark sector |
| ⑦ Flavour mixing in the leptonic sector |
| ⑧ The problem of the missing mass in the universe |
| ⑨ The problem of the hierarchy |
| ⑩ Physics at the Planck scale: the gap and the number of expanded dimensions |

### 30.5.1 The Physics of Imaginary Masses (SSB)

The imaginary masses play a central role in physics. This is illustrated in Table 30.5, where we include superfluidity and superconductivity as examples of effects which are produced as if an imaginary mass was there, even if actually it is not there. An example are the Cooper pairs which enter into the Lagrangian as if a particle with twice the electron charge and negative mass was there. Superconductivity exists, but the particle with double-negative electric charge and imaginary mass is not there.

In SU(2)×U(1) the imaginary mass provides the very heavy $Z^0$ and $W^{\pm}$ masses and a massless photon; but we do not know how many Higgs particles really exist. For example, if we want an electric dipole moment for the neutron, an Isovector Higgs should exist, in addition to the others expected to exist. Finally, we should not forget that the Higgs mass in the standard model could be as heavy as 2 TeV. This is the reason why we have performed a series of Montecarlo simulations using the (μμμμ) mode as a signature for

**Table 30.5.** Imaginary mass and (B–F) symmetry

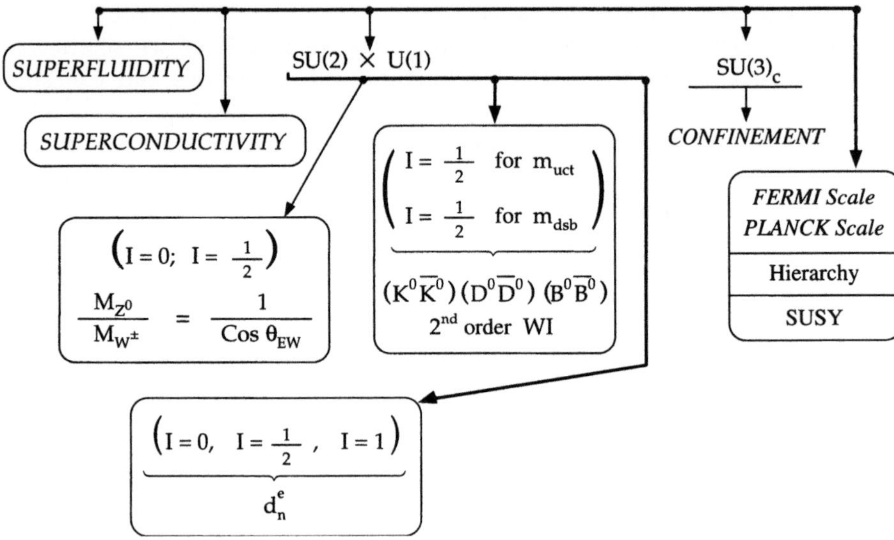


the Higgs. We have focused our simulations on heavy Higgs ($m_H > 2M_Z$) and searched at three energies: $(\sqrt{s})_{pp} = 16, 40$, and 200 TeV. The results are presented in Fig. 30.14 for $(\sqrt{s})_{pp} = 16$ and 40 TeV and $m_H = 1$ TeV. In Fig. 30.15 we show the results for $(\sqrt{s})_{pp} = 200$ TeV. By comparing the

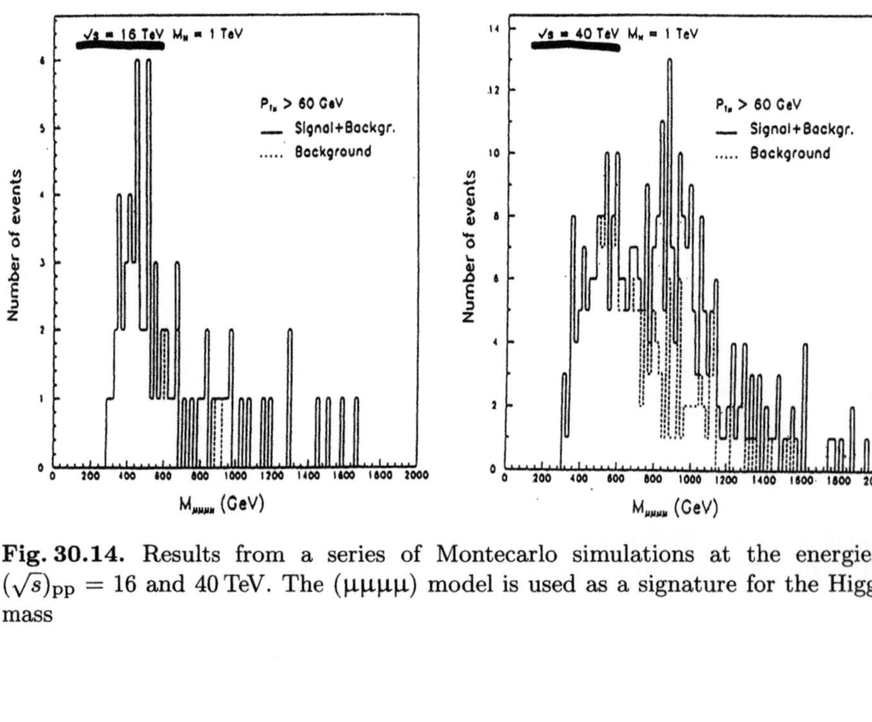


**Fig. 30.14.** Results from a series of Montecarlo simulations at the energies, $(\sqrt{s})_{pp} = 16$ and 40 TeV. The $(\mu\mu\mu\mu)$ model is used as a signature for the Higgs mass

signal/background the conclusion is straightforward. If a heavy Higgs is there, there is no question that the best energy is $(\sqrt{s})_{pp} = 200$ TeV, and this means that the ELN Project provided the best facilities [23].

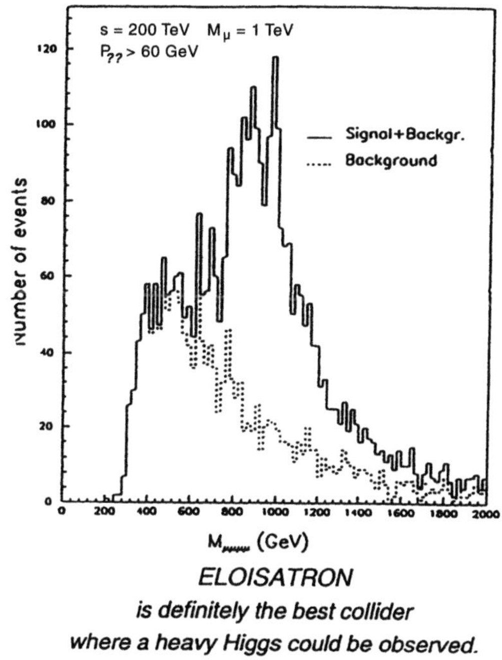


**Fig. 30.15.** As in Fig. 30.14, but for $(\sqrt{s})_{pp} = 200$ TeV

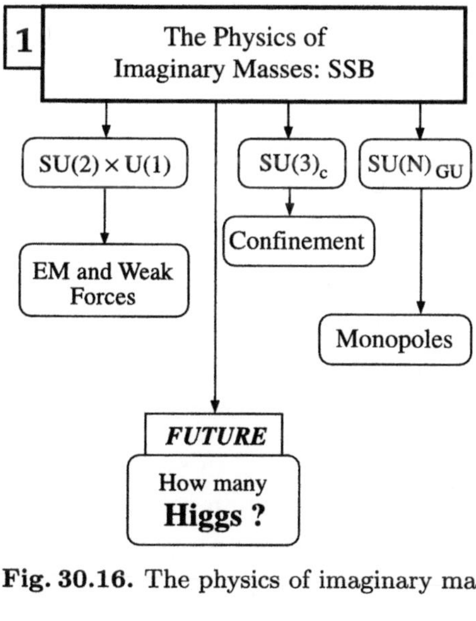


**Fig. 30.16.** The physics of imaginary masses (SSB): a summary

A synthesis of the physics of imaginary masses (SSB) is shown in Fig. 30.16, where "confinement" and "monopoles" are recalled as effects generated by the existence of imaginary masses.

### 30.5.2 Matter–Antimatter Symmetry

The main point here is to clarify a basic misunderstanding and to point out the two new frontiers.

The basic misunderstanding was discussed last year at the International Workshop on "Neutrino Telescopes" in Venice [24]. The experimental result on the mass difference between long- and short-lived $\mathrm{K}^0$-mesons,

$$\Delta m_{\mathrm{K_L K_S}} ,$$

is not a proof of matter–antimatter asymmetry. To be more precise, the fact that $\Delta m_{\mathrm{K_L K_S}} \neq 0$ is neither a proof of "meson–antimeson" mass asymmetry, nor a proof of "quark–antiquark" mass asymmetry, nor a proof of "confinement" mass asymmetry, nor a proof of "nuclear-binding" mass asymmetry.

There are in fact 3 types of masses:

- intrinsic (basic fermions),
- confinement (QCD),
- binding (nuclear and electromagnetic).

The first new frontier is that there are no experimental measurements on nuclear binding. If a large asymmetry were detected in nuclear physics, for example, between the deuteron and the antideuteron masses, $\Delta m(\mathrm{d\bar{d}})$, at the level of a percent, this result would be perfectly consistent with the value of $\Delta m_{\mathrm{K_L K_S}}$.

Our present theoretical understanding says that the existence of nuclear antimatter cannot be predicted on the basis of fundamental theoretical reasons. Experiments to improve our knowledge of the validity of CPT in the nuclear forces, i.e. in the QCD-colour-neutral effects, must be given high priority. It is, in fact, not true that, since

$$\frac{\Delta m_{\mathrm{K_L K_S}}}{m_{\mathrm{K_L K_S}}} \approx 10^{-14} ,$$

the value of $\Delta m_{\mathrm{K_L K_S}}$ is a proof for matter–antimatter asymmetry at the $10^{-14}$ level of accuracy. Even if $\Delta m_{\mathrm{K_L K_S}}$ was zero, the above-mentioned mass difference, $\Delta m(\mathrm{d\bar{d}})$, could be very large. A point to recall is that no one is able to make the transition from QCD to nuclear physics.

The second new frontier is the search for Antimatter in Space.

A synthesis of matter–antimatter symmetry is given in Fig. 30.17.

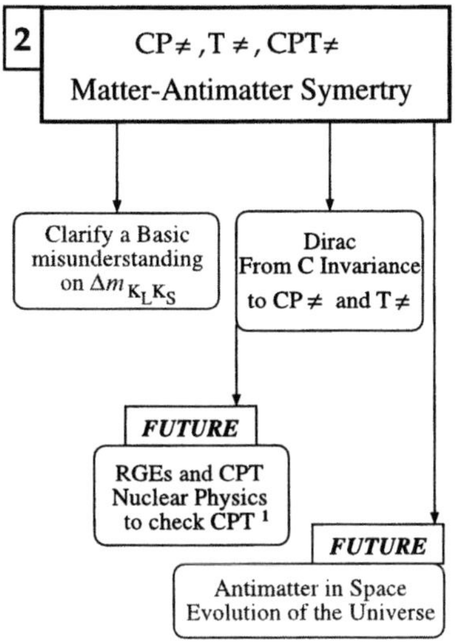


**Fig. 30.17.** Matter–antimatter symmetry: a summary

### 30.5.3 Supersymmetry

First, there are two points to be clarified; then I wish to point out what the future has in store.

The first point is in regard to an awareness of the incorrect claims on the SUSY threshold [25]. We do not know how to generate SUSY breaking. All attempts so far have been based on ad hoc assumptions which determine the SUSY breaking scale. This ad-hoc procedure is the only one we know; it makes use of the convergence of the three gauge couplings $(\alpha_1, \alpha_2, \alpha_3)$ using the renormalization group equations (RGEs). Here the only interesting news, not to be forgotten, is the EGM effect [25].

EGM stands for "evolution of gaugino masses". We have proved that the inclusion of this evolution in the RGEs has a very interesting effect. It lowers the SUSY threshold by nearly three orders of magnitude, more exactly, by a factor of 700.

Inclusion of the EGM effect allows the most accurate evolutions of the gauge couplings $(\alpha_1, \alpha_2, \alpha_3)$, as shown in Fig. 30.18, where the evolution of the Universe from the Planck scale down to the present day is reported in the upper part. These evolutions are consistent with the SUSY breaking scale being degenerate with the Fermi scale. If this was the case, the discovery of the first supersymmetric particle should soon be quantified.

For the immediate future it should be pointed out that the search for the first supersymmetric effect should not be biased in favour of the neutralino, $\chi_1^0$, the lightest superparticle. The reason is that it could very well be the gravitino, $\widetilde{G}$, the particle with the lightest mass. This is the reason why we devoted a lot of attention with L3 at LEP to search in $(e^+e^-)$ final state

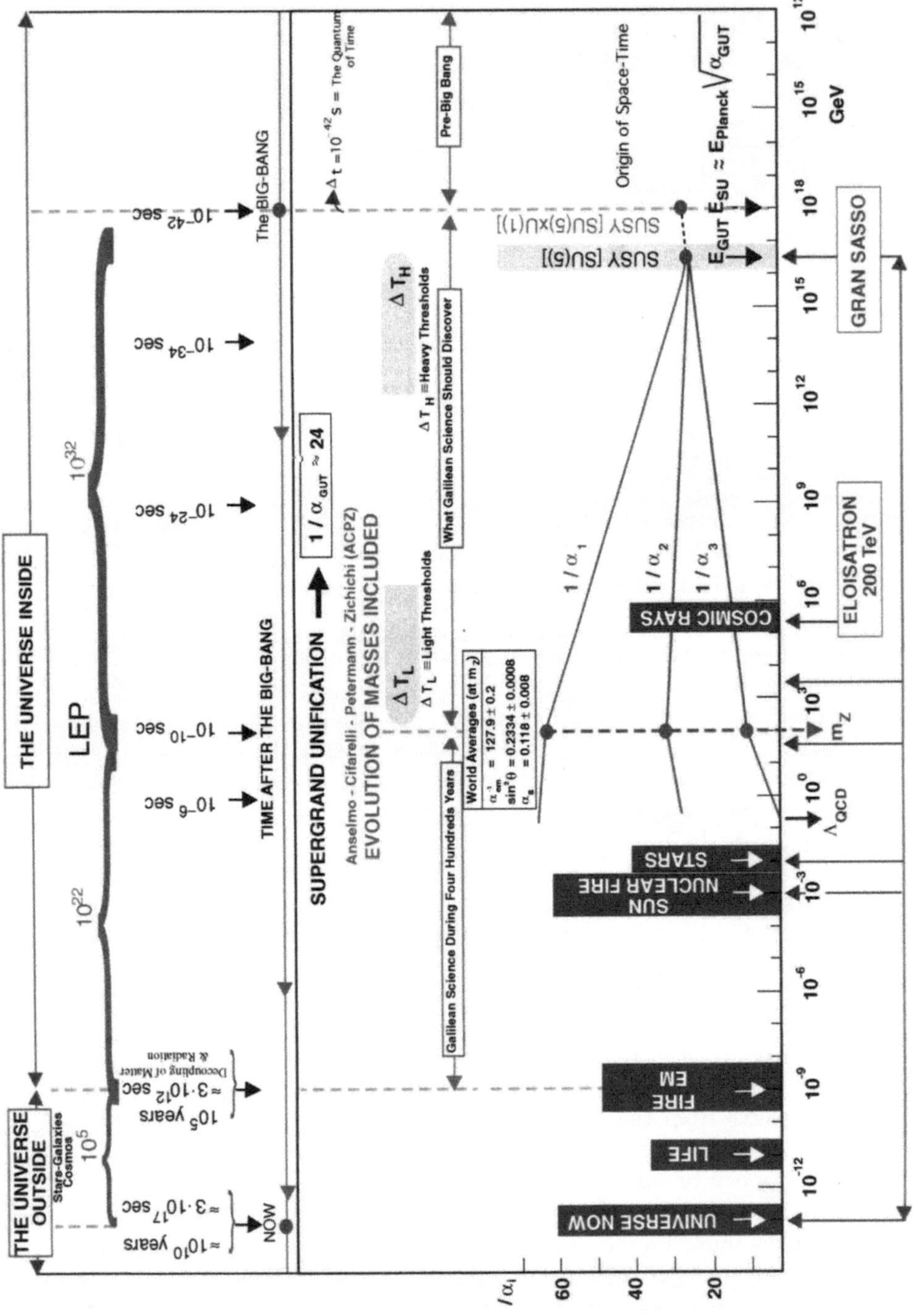


**Fig. 30.18.** Evolution of the universe (*upper part*) and the gauge couplings $(\alpha_1, \alpha_2, \alpha_3)$ (*lower part*). The EGM (evolution of gaugino masses) effect is included

events with only one photon and missing energy. For many years the search concentrated only on the two photon final states plus missing energy.

Let me briefly illustrate this point with the corresponding Feynman diagrams:

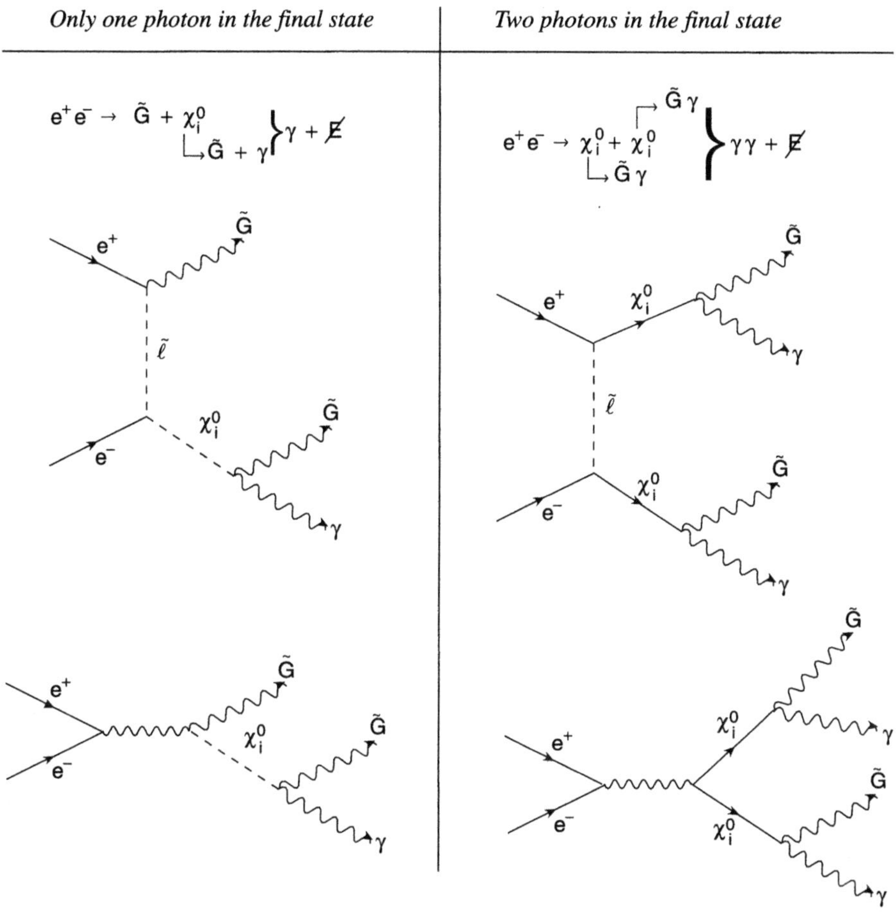


Let us recall that, so far, only upper limits exist on a SUSY signal. A synthesis of all I have said on supersymmetry is given in Fig. 30.19.

### 30.5.4 Non-perturbative QCD

One of the most interesting non-perturbative QCD effects, in addition to confinement, is the existence of the "effective energy". This has been described in a special volume edited by *Lipatov* [20]. It is remarkable that the multi hadronic final states produced (no matter whether the interaction is strong,

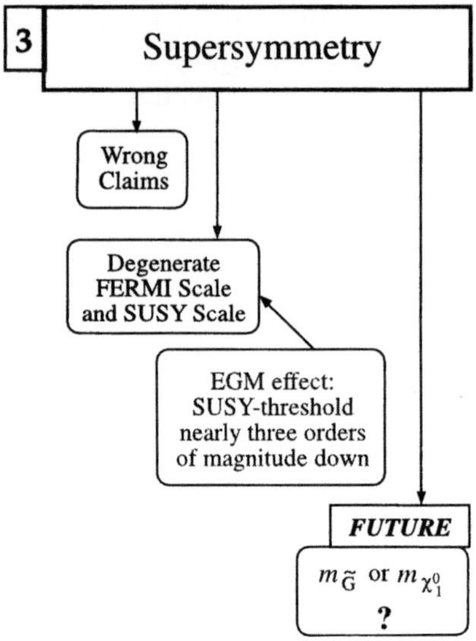


**Fig. 30.19.** Supersummetry: a summary

electromagnetic or weak) at different nominal energies show universality features (i.e. the same $\mathrm{d}E/\mathrm{d}x, < n_{\mathrm{ch}} >$, etc.) if the new variable "effective energy" is introduced in the analysis.

An example of the relevance of the effective energy is shown in Figs. 30.20 and 30.21, where the fractional energy distributions of multihadronic final states produced in (pp) and ($e^+e^-$) interactions are shown to be in excellent agreement [26], despite the very different nominal (pp) energy.

The challenge for the future is the study of the effective energy in the quark–gluon plasma with ALICE. We will thus study the physics of the coloured QCD world, where the proton should be very light, since there is no confinement mass in play, and $\pi$ should be heavy, since there is no spontaneous chiral symmetry breaking in action.

A synthesis of all I have said on the non-perturbative QCD is given in Fig. 30.22.

### 30.5.5 Anomalies and Instantons

Anomalies in physics have played an important role, starting with the celebrated ABJ anomaly, which explains why $\pi^0$ decays so fast in ($\gamma\gamma$), requiring the QCD colours to be three. In quantum fluid dynamics (QFD) the anomaly destroys renormalization, because the axial currents are coupled to the gauge fields. But, the anomaly is zero if there is an equal number of quarks and leptons in each family. This is how the existence of the top quark could be predicted.

When gravitational forces come into play, our theoretical friends tell us that, in order to have anomaly-free conditions, the choice must be SU(3)$\times$

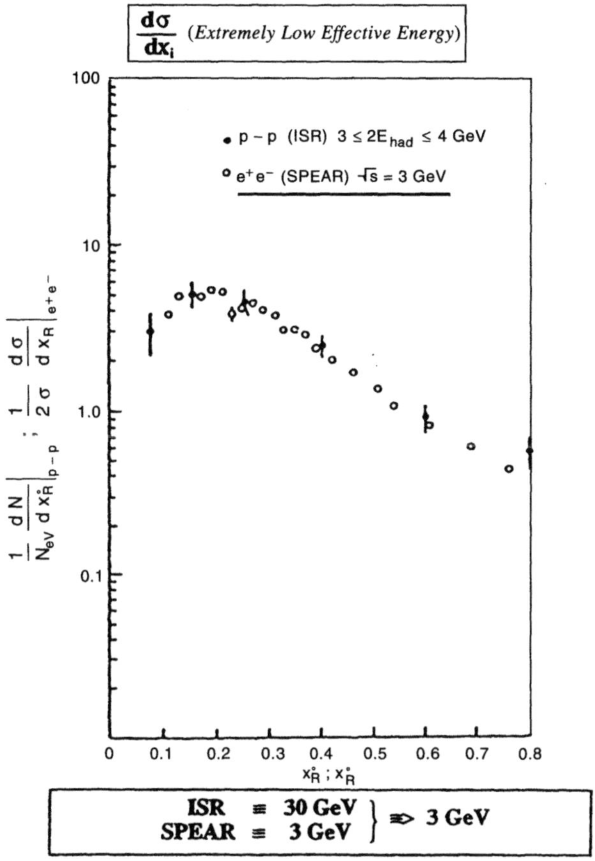


**Fig. 30.20.** Fractional energy distributions of multihadronic final states produced in (pp) and ($e^+e^-$) interactions with extremely low effective energy

SU(2)×U(1). In other words, we could finally understand why the standard model has the SU(3)×SU(2)×U(1) structure.

Let me emphasize an aspect of the ABJ anomaly which I consider very interesting: it has produced an impressive set of consequences not only in physics but also in mathematics. In fact algebraic topology is a very rich subject. The most interesting one is probably the existence of "topological" solutions in 1–4 dimensions, thus producing domain walls, vortices, magnetic monopoles and instantons.

Instantons typically have the effect of explicitly breaking U(1) symmetries. There are two kinds of Instantons, one for QCD and one for QFD.

In QCD they provide the "non-spontaneous", i.e. direct, breakdown of "chirality" invariance. This explains why the $\pi^0$ mass is non-zero, why the $\eta$ mass is heavy and why $\eta'$ is so much heavier than the $\eta$. Instantons allow us to understand the reason why the meson mixings are so different in the

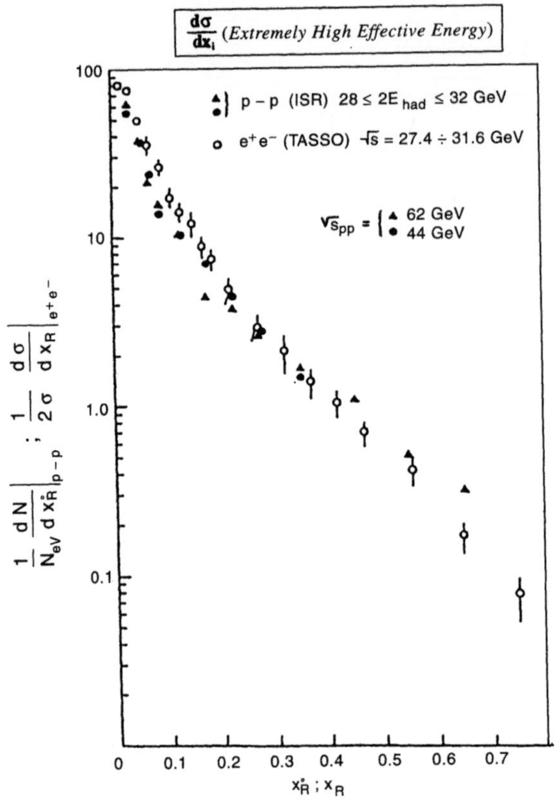


**Fig. 30.21.** As in Fig. 30.20, but with extremely high effective energy

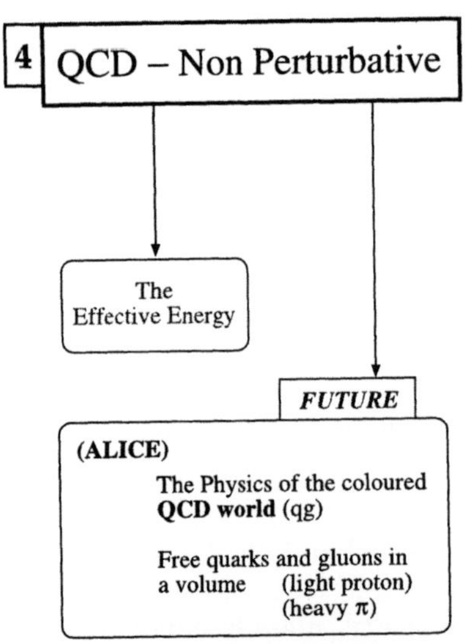


**Fig. 30.22.** Non-perturbative QCD: a summary

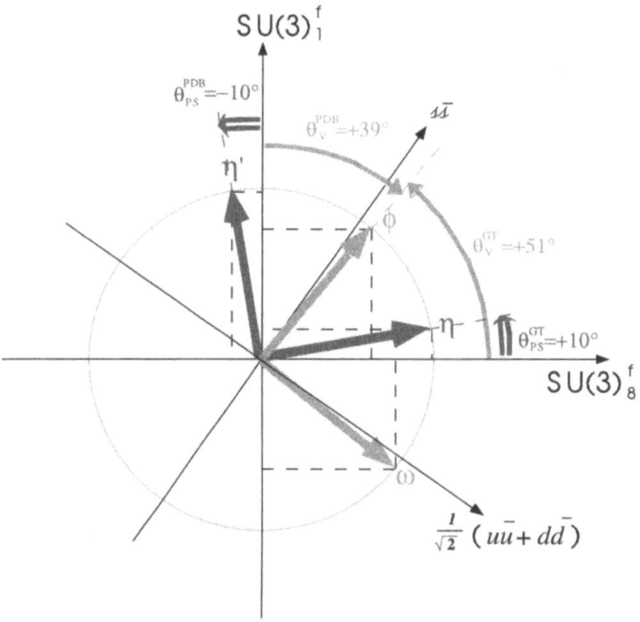


PDB ≡ Particle Data Book
GT ≡ Gerardus 't Hooft

**Fig. 30.23.** Present status of $\theta_{\mathrm{PS}}$ and $\theta_{\mathrm{V}}$

pseudoscalar (PS) and vector (V) cases. These two mixings are measured by the well-known angles $\theta_{\mathrm{PS}}$ and $\theta_{\mathrm{V}}$, which are far from being equal, $\theta_{\mathrm{PS}} \neq \theta_{\mathrm{V}}$. In Fig. 30.23 we show the present status of these angles, which is the same as the one known to us a few decades ago (as reported in Fig. 30.10).

The only basic novelty is the work of 't Hooft with instantons. It is the physics of the instantons which allows us to understand why $\theta_{\mathrm{PS}} \neq \theta_{\mathrm{V}}$.

It is interesting to point out that the effects produced by instantons can be understood in terms of the properties of the Dirac sea. In fact, the vacuum made of fermions has fermionic properties.

The problem of great interest for the future is to understand whether similar non-perturbative interactions also contribute significantly to the total cross-section in QCD. And this brings me to the physics of a project with which I am involved, ALICE. In this experiment we will search for new physics and the signature will be:

- cross-section much larger than that expected between two heavy nuclei $\sigma$ (quark–gluon plasma) $\gg \sigma$ (nuclei $\times$ nuclei);
- the existence of pseudoscalar quantities, which would prove the existence of the quark–gluon plasma, as suggested by *Lee* and *'t Hooft* [27].

Another interesting problem for the future is the following: In $\mathrm{SU}(2)_{\mathrm{L}}$ the effect of instantons is linked to the fact that the non-Abelian gauge

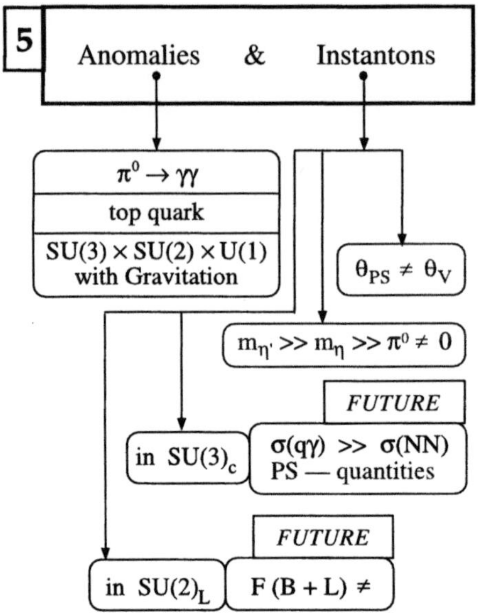


**Fig. 30.24.** Anomalies and instantons: a summary

force, QFD, acts only on left-handed states, and instantons generate baryon-number-non-conservation, which is another U(1) breaking. The violation of the fermion number, $F \neq$ ($F = B + L$; where $B$ is the baryonic and $L$ is the leptonic number), conservation could not take place in an Abelian gauge theory. The first condition for $(B+L) \neq$ is that the gauge force is non-Abelian, because the vacuum structure is non-trivial: it has many states characterized by an index which is an integral number ($n = 0, 1, 2\dots$). Thus transitions can take place between different vacuum states with ($n = 0, 1, 2, \dots$). The second condition is that the gauge force distinguishes left from right: it has to be chiral, since in this case there is an ABJ anomaly which is proportional to the number of fermions (quarks and leptons). In $\mathrm{SU(2)_L}$ the right components of the fermions (quarks and leptons) are inert, only the left components are active. The third condition is that the transition between the vacuum states (for example, from $n = 0$ to $n = 1$) takes place not via the tunnelling rate (which is typically of the order of $10^{-80}$) but via the "thermal fluctuation" rate, which could be very large. Needless to say that we hope the F $\neq$ condition will be active in the next LHC heavy ion collisions with ALICE. A synthesis is given in Fig. 30.24.

### 30.5.6 Flavour Mixing in the Quark Sector

Neither flavour mixings nor the violation of the symmetry operators CP and T have theoretical reasons to exist. QFD, in addition to acting on each flavour separately (flavour non-changing neutral currents), acts on mixed-flavour states (flavour changing charged currents). In the quark sector, the weak gauge coupling, $g_2(\alpha_2 = g_2^2/4\pi)$, is given, not only to a single quark flavour, but also to a mixture.

The KM matrix introduces a privilege in QFD:

- either the up type quarks are pure and the down type are mixed
- or vice versa.

It is difficult to believe that this is really the case. Our present knowledge of QFD can be formulated [28] in a perfectly symmetric way between $\psi_{\mathrm{up}}^{\mathrm{mixed}}$ and $\psi_{\mathrm{down}}^{\mathrm{mixed}}$ states, where

$$\psi_{\mathrm{up}}^{\mathrm{mixed}} \equiv \begin{pmatrix} \psi_{\mathrm{u}} \\ \psi_{\mathrm{c}} \\ \psi_{\mathrm{t}} \end{pmatrix} \quad \text{and} \quad \psi_{\mathrm{down}}^{\mathrm{mixed}} \equiv \begin{pmatrix} \psi_{\mathrm{d}} \\ \psi_{\mathrm{s}} \\ \psi_{\mathrm{b}} \end{pmatrix} .$$

QFD induced transitions between quark mixed flavour states.

The KM matrix corresponds to the product of these two mixtures. The point to be emphasized is that the two mixtures

$$\psi_{\mathrm{up}}^{\mathrm{mixed}} \quad \text{and} \quad \psi_{\mathrm{down}}^{\mathrm{mixed}}$$

must be different. In fact, the reason why no mixtures can be seen in the neutral currents is because, in this case, the two mixtures are necessarily the same and cancel out in the $(\bar{\psi}\psi)$ structure, which is needed for the standard structure of all currents.

An interesting development in this field would be to disentangle this QFD feature. Suppose that the two mixtures $\psi_{\mathrm{up}}^{\mathrm{mixed}}$ and $\psi_{\mathrm{down}}^{\mathrm{mixed}}$ are the real structure of the QFD currents. In this case it would be possible that the two phase angles

$$\mathrm{e}^{\mathrm{i}\delta_{\mathrm{up}}} \quad \text{and} \quad \mathrm{e}^{\mathrm{i}\delta_{\mathrm{down}}}$$

would be nearly the same and therefore in the $\bar{\psi}\psi$ structure they would appear as

$$\delta_{\mathrm{up}} - \delta_{\mathrm{down}} = \delta_{\mathrm{KM}} ,$$

where $\delta_{\mathrm{KM}}$ is the effective value appearing in the KM matrix. The reason why it is small would be the consequence of the two mixtures "up" and "down" being different but with nearly the same "phases". Let me conclude this "challenge" by recalling that the violations of the symmetry operators CP and T are not induced by the SSB mechanism. If this was the case the ratio $\varepsilon'/\varepsilon$ could not be so small. Once again this needs to be understood by our theoretical friends.

A synthesis of all I have said on flavour mixing in the quark sector is given in Fig. 30.25.

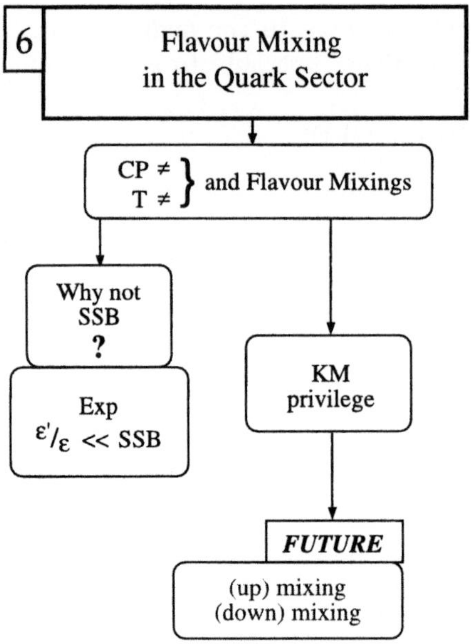


**Fig. 30.25.** Flavour mixing in the quark sector: a summary

### 30.5.7 Flavour Mixing in the Leptonic Sector

No one knows the origin of these mixtures, nor if these mixtures (strongly present in the quark sector) are responsible for the global-symmetry break-

**Table 30.6.** Nature of flavour mixing in the leptonic sector as proposed by H. Fritsch

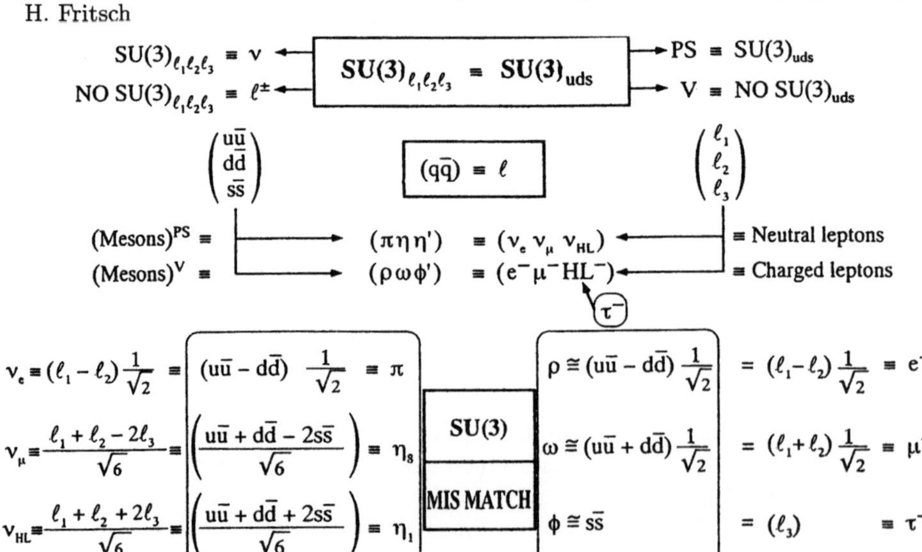

down of CP and T. All we can do is measure these effects. In the quark sector CP and T violations have been known to exist since 1964. To establish their existence in the leptonic sector is a challenge for the future. The nature of the flavour mixing in the leptonic sector is another interesting challenge. In this respect I like the proposal made by Harald Fritzsch. This is illustrated in Table 30.6.

All experimental data agree with this proposal, which puts on the same ground the flavour mixing in the pseudoscalar mesons with the leptonic flavour mixing in the neutrinos, and the flavour mixing in the vector mesons with the leptonic flavour mixing in the charged leptons. This analogy is illustrated in Fig. 30.26.

Problems for the future are the following: To establish whether the Majorana neutrinos exist and to understand the neutrino mass spectrum. Furthermore to find whether the existence of a neutrino mass is a signal coming from the Grand Unification Energy scale.

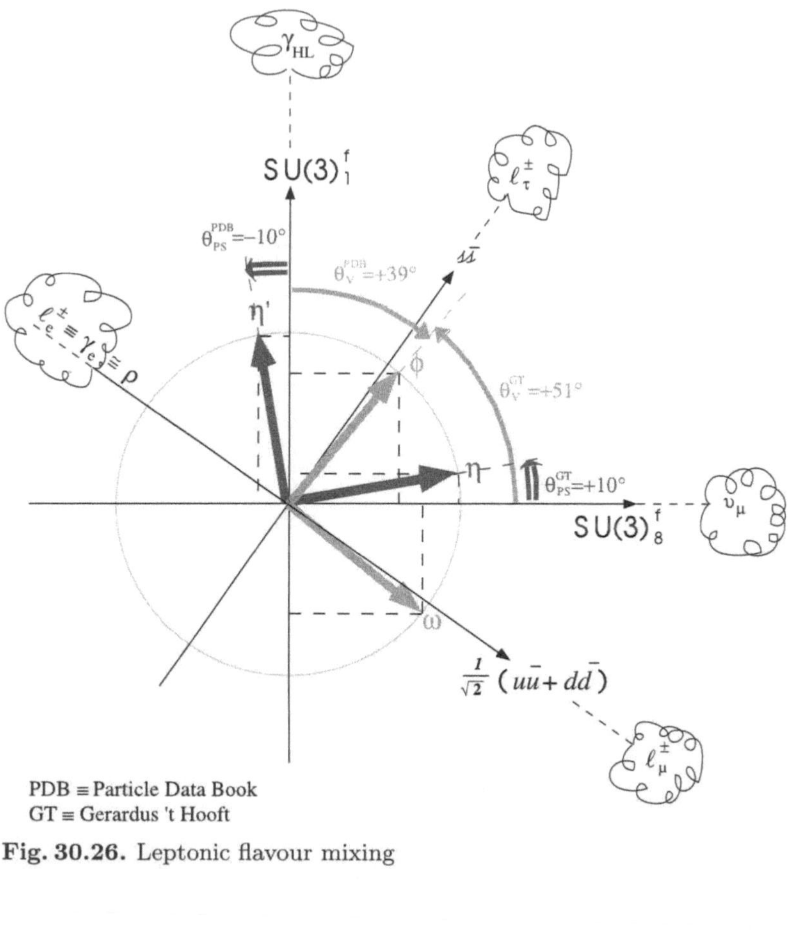


**Fig. 30.26.** Leptonic flavour mixing

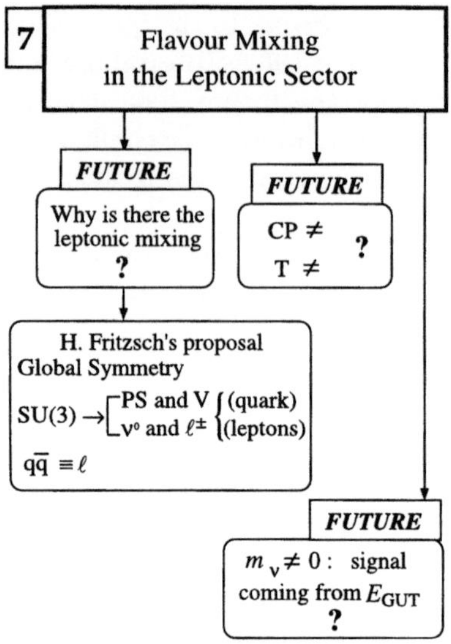


**Fig. 30.27.** Flavour mixing in the leptonic sector: a summary

A synthesis of all I have said on flavour mixing in leptonic sector is given in Fig. 30.27.

### 30.5.8 The Problem of the Missing Mass in the Universe

There is a range of missing-mass candidates. This goes from the axion with a mass as low as $10^{-5}$ eV to very massive objects from $10^2$ to $10^6$ solar masses. On the other hand we should not forget that if a new form of matter, which is only gravitationally coupled with our matter, exists, the problem is to invent a detector extremely sensitive to the "dark matter". We have been thinking in this regard and even tried to build such a detector, but I do not want to speak about this since we have no results so far.

In Table 30.7 is shown the present status of our knowledge in the field, including the known modification of Newton's law and a new attempt to study if it is possible to get gravitational confinement. Suppose that the non-Abelian nature of a fundamental force would generate "asymptotic freedom" and "confinement". We know that asymptotic freedom is not possible with the present theory of gravitation. We are working on the "confinement". If this was there, one of the main reasons why "dark matter" is considered necessary would disappear.

A synthesis of all I have said on the missing mass in the universe is given in Fig. 30.28.

**Table 30.7.** Current knowledge with regard to missing-mass candidates

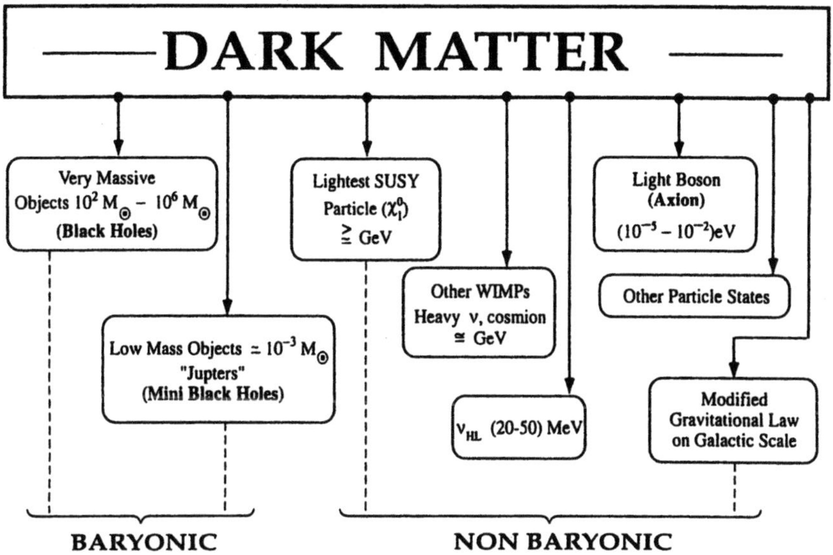


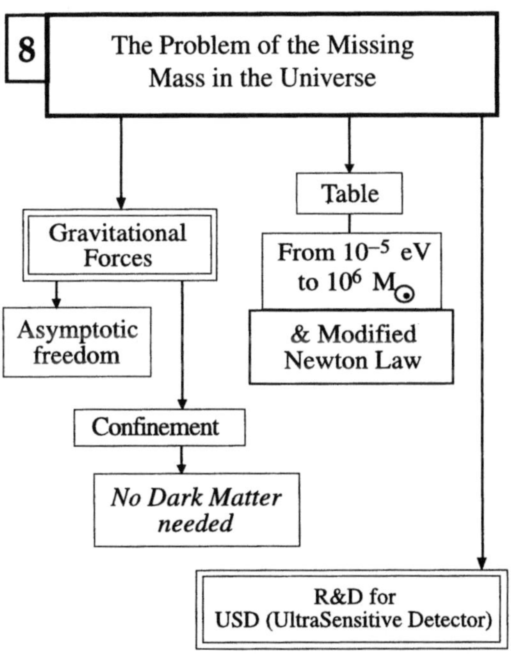


**Fig. 30.28.** The problem of the missing mass in the Universe: a summary

### 30.5.9 The Problem of Hierarchy

To understand this problem is a great challenge for our theoretical friends. In fact it is the existence of imaginary masses bound to scalar particles which brings the Fermi scale up to the Planck scale, i.e. the grand unification scale. Here a crucial set of experiments for the future is that on proton decay: i.e. to reach the most sensitive level of detection, including the various predicted (and unpredicted) decay channels.

Another set of basic issues is to be understood:

- Why the Universe we live in is so big.
- What we can learn in subnuclear physics from the observations in cosmology.
- What we can learn in cosmology from the discoveries in subnuclear physics.
- How the baryon-dominated Universe is born. Where is it created? In the domain-walls?
- Are we really sure that we understand the 16 orders of magnitude which separate the B-asymmetric Universe from the B-symmetric one?

A synthesis of all I have said on the problem of hierarchy is given in Fig. 30.29.

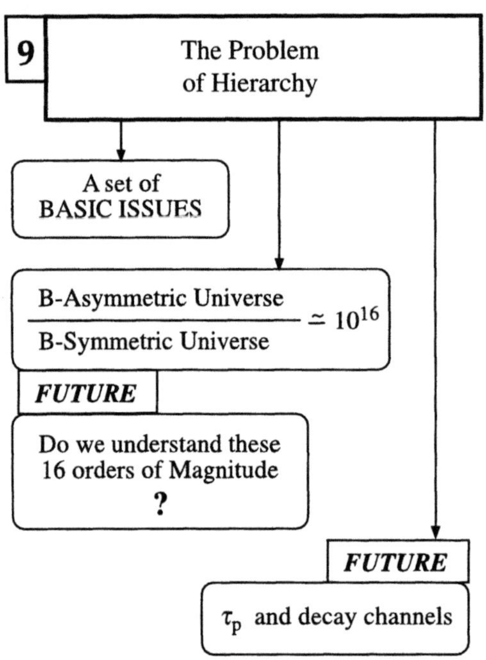


**Fig. 30.29.** The problem of hierarchy: a summary

### 30.5.10 The Physics at the Planck Scale: The Gap and the Number of Expanded Dimensions

There are two basic issues here. One is whether there is a gap [29] between the energy where the gauge couplings $(\alpha_1, \alpha_2, \alpha_3)$ converge and meet at the same point $(E_{\mathrm{GUT}})$ and the string unification scale $(E_{\mathrm{SU}})$. The other is whether we are missing a great opportunity.

Let me first offer a few words on the gap. A detailed study has been performed by us [29] in order to establish the confidence level of the convergence of the gauge couplings. The results are shown in Fig. 30.30, where the following is proved: If we take the world average values (WA) of the gauge couplings and push to the values indicated all the mass ratios involved in the transition from the unification symmetry group SU(N) to SU(3)×SU(2)×U(1), the result is that $E_{\mathrm{GUT}}$ can reach a level as high as $10^{18}$ GeV and the gap disappears. However, if the mass ratios are all $\approx 1$ and the WA values are taken without any deviation, the gap is there.

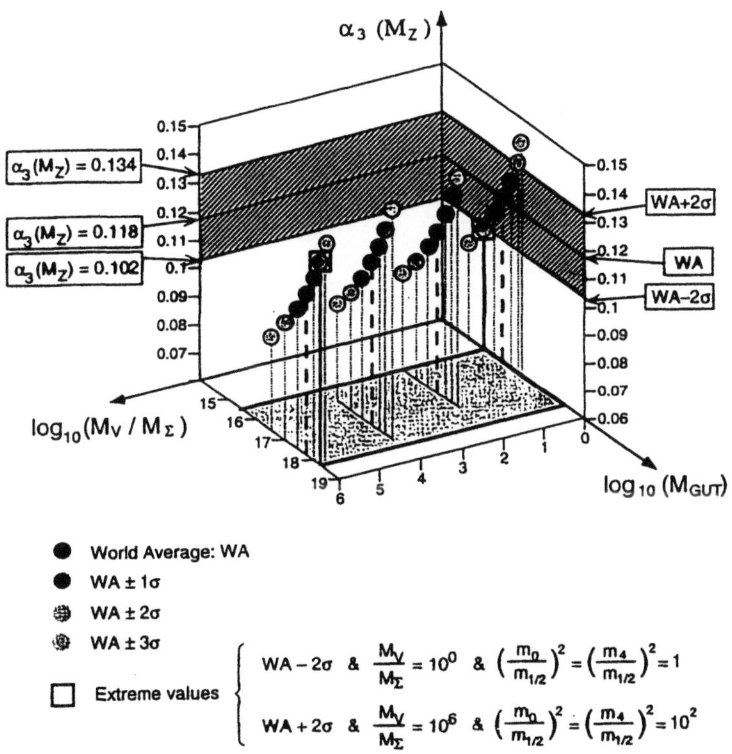


**Fig. 30.30.** Results of a study to establish the confidence level of the convergence of the gauge couplings

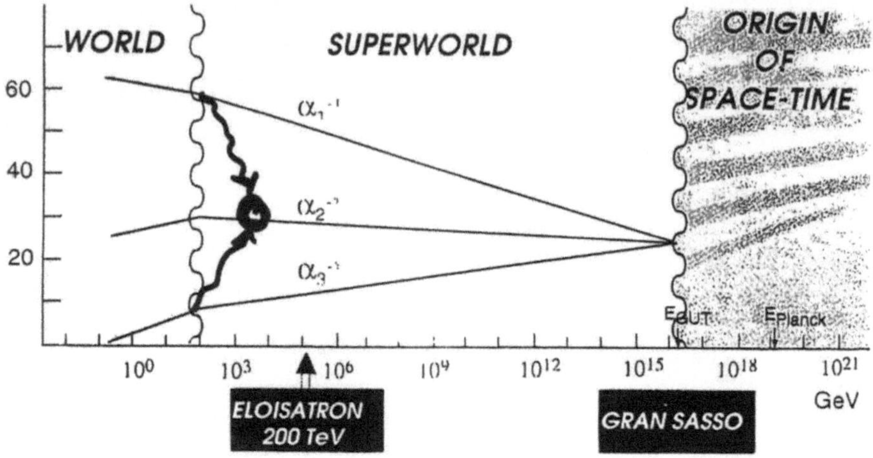


**Fig. 30.31.** The energy scale

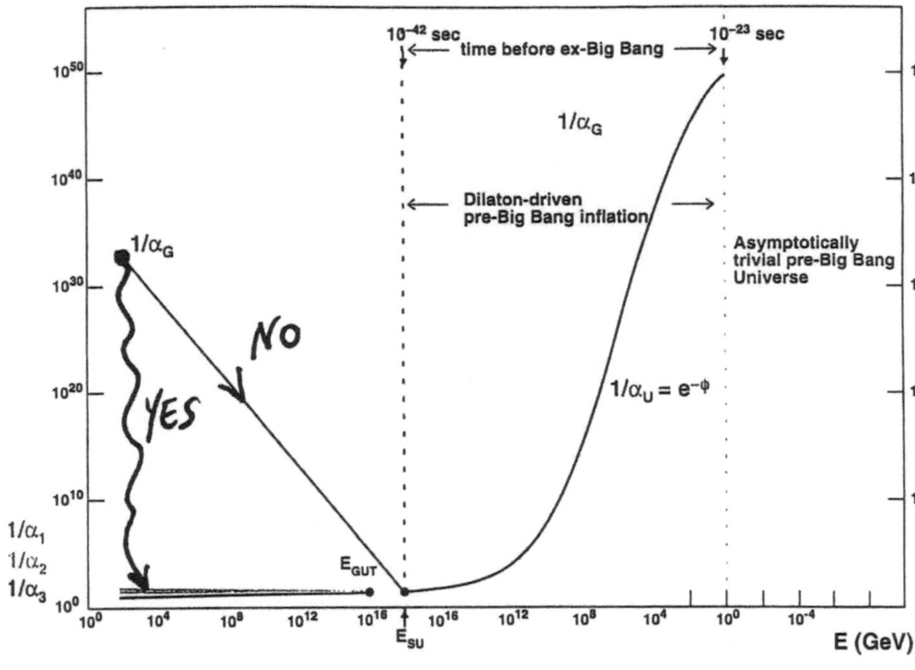


**Fig. 30.32.** The time scale

If the gap is there, the best candidate to fill it is SU(5)×U(1). In this case there would be no magnetic monopoles. It should be pointed out that SU(5)×U(1) is the simplest string-derivable supergravity model.

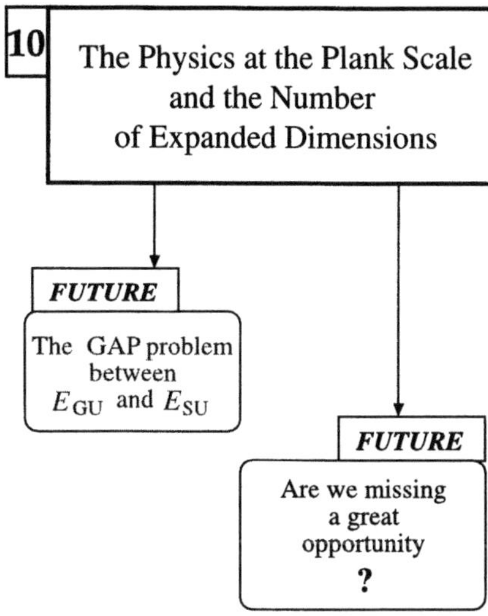


**Fig. 30.33.** Physics at the Planck scale and the number of expanded dimensions: a summary

In all we have said so far, the Planck scale is our dream for the unification of all fundamental forces of Nature. And now I turn to the second basic issue: have we been missing a formidable opportunity?

Our four-dimensional space–time $(3 + 1)$ is the result of an expansion of hidden dimensions, of both bosonic $(10+1)$ and fermionic (32) nature, whose number could be as high as 43: this is the superspace.

Is it possible that an extra dimension can manifest itself in a range of distance which is very large compared with the Planck range ($10^{-33}$ cm), thus producing effects on the TeV scale? If this were the case, the Planck scale would come down by 15 orders of magnitude. This is illustrated in Figs. 30.31 and 30.32 [30].

A synthesis of all I have said on physics at the Planck scale and the number of expanded dimensions is given in Fig. 30.33.

## 30.6 The ELN Project in a Few Words

The ten challenges need a new generation of supercolliders. Their asymptotic limit is the ELN Project mentioned on a few occasions, and especially in Sect. 30.5.1, with reference to a heavy Higgs. It is interesting to recall that the problem of having the facilities able to make the ten challenges not appear like dreams has confronted us since 1979. It is probably appropriate to illustrate in a few words the ELN Project and strategy.

The INFN Eloisatron (ELN) Project started in Italy, in 1979, with a series of basic steps going from the construction of the superconducting cyclotron (CS), in Milan, to the construction of 50% of the superconducting dipole magnets for the 800 GeV proton ring of HERA, in Hamburg. An intensive R&D Project to study new technologies for subnuclear physics, briefly mentioned in Sect. 30.3, has also been implemented at CERN through the LAA Project, which is part of the ELN strategy.

Since the very beginning, the conception of this project was based on a new way of tackling the problem of big projects, based on the followings points:

- do not start to build a new laboratory if the work needed can be carried out within existing structures;
- involve the industry in the implementation of the project, which also means that the R&D work should be done in the laboratories of the industries in close collaboration with our labs.

So far these basic ELN steps have been the merging point for an effective collaboration between various universities, institutions and industries.

In order to be sure that subnuclear physics will stay in the forefront of scientific research, we need right now, not in a decade, to study the multi hundred TeV physics in terms of theoretical goals, machine problems and new detector technologies.

This is why we have been active for many years with the ELN Project whose Working Group structure is as follows:

(*i*) theoretical and phenomenological studies in subnuclear physics: hadron interactions at extreme energies, spanning from presently or shortly (LHC at CERN) accessible values up to the multi hundred TeV regime;
(*ii*) theoretical studies on the machine: maximum reachable energy ($100 + 100\,\text{TeV} \Rightarrow 500 + 500\,\text{TeV}$) and luminosity ($10^{33}\,\text{cm}^{-2}\,\text{s}^{-1} \Rightarrow 10^{36}\,\text{cm}^{-2}\text{s}^{-1}$) levels for a future hadron supercollider;
(*iii*) R&D on a new generation of powerful superconducting magnets;
(*iv*) R&D on new detectors capable to operate at extremely high energy and luminosity (i.e. continuation of the LAA).

As far as point (*ii*) is concerned, a conceptual design of ELN has already been performed by *Kjell Johnsen* and collaborators [23]. This study shows that there are no conceptual difficulties in the realisation of a $100 + 100$ TeV proton collider able to work with a luminosity of the order of $10^{33}\,\text{cm}^{-2}\,\text{s}^{-1}$. The goal of the extreme energy and luminosity has also been theoretically well fixed, as shown in Fig. 30.34.

Concerning point (*iii*), the role of innovative superconducting materials, at low and high $T_c$, is particularly relevant. And the working group is bringing together experts in different fields, namely accelerator, subnuclear and solid-state physics, in the very spirit of the ELN Project.

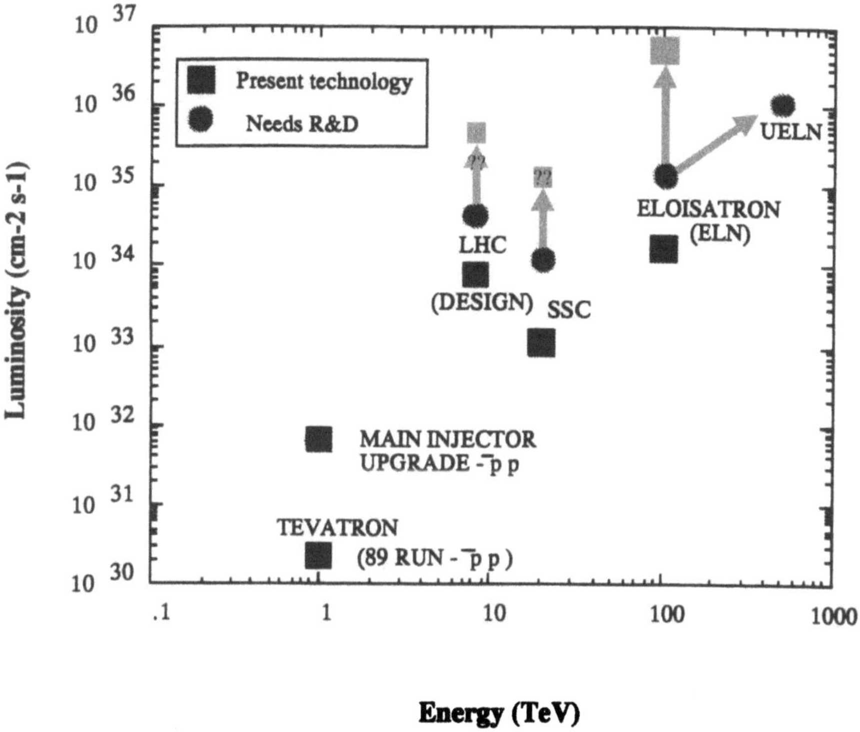


**Fig. 30.34.** The ultimate (UELN) energy and luminosity of ELN, compared with all other colliders in the world

## 30.7 Conclusions

The point I want to emphasize is that we should not follow the "mainstream". We should not over-emphasize the importance of SSB and SUSY. In fact we should be prepared to face the situation where their existence could not be established in the next energy step. If higher energy is needed, we must be able to continue working, having in mind the great vitality of our field.

This vitality is illustrated by the ten challenges of Sect. 30.5. I have no doubt that, taking into account the experimental (missing) results of the last (nearly two) decades and the theoretical (lack of) understanding of the energy threshold for new physics, there is one and only one wise conclusion today. Our physics needs not a small, but the biggest possible jump in energy, and it is therefore very risky to wait and see the results of LHC (< 10% ELN), whose energy could still be too low for the physics beyond the standard model to be discovered.

We should not be afraid of possible criticisms. When John was granting his support to my ideas in inventing new experimental "tricks" and in proposing new searches outside the "mainstream" of the time, life was not easy. Let me recall two famous statements against myself.

The first was: "Zichichi cerca farfalle" (Zichichi is looking for butterflies). The butterflies were: the J/Ψ and the third lepton. All this is illustrated in Fig. 30.35, where the horizontal axis is the energy of the colliding ($e^+e^-$) pair. As stated in Sect. 30.2, and as you can easily see, if John had been the Director of the Frascati Laboratory, the J/Ψ and the third lepton would have been discovered there, at least five years [13] before SLAC.

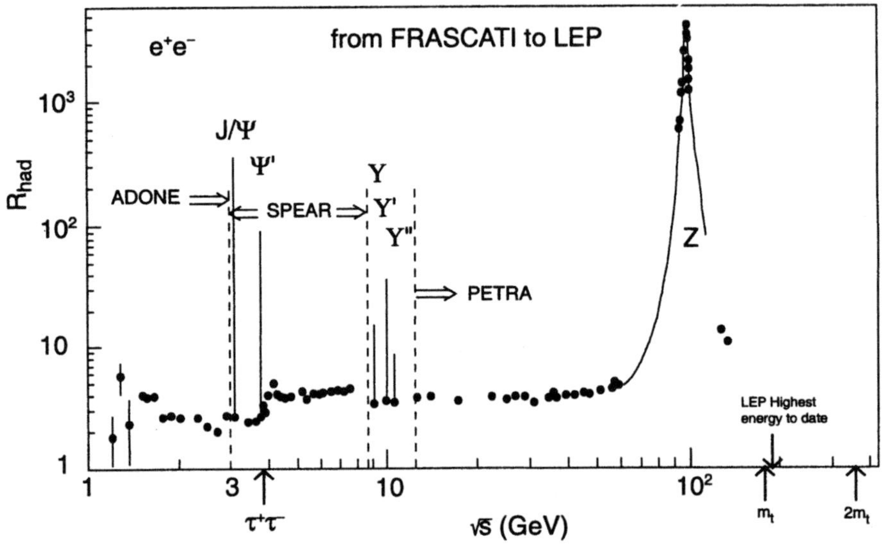


**Fig. 30.35.** J/Ψ and the third lepton

The other well-known statement was: "Zichichi is dreaming". Strongly supported by John's statement "*There is no new Physics without new projects,*" I was able to go on promoting new projects, and, in fact, the Gran Sasso Project is now the largest underground laboratory in the world, the superconducting proton-ring at HERA exists, and the 5-m (instead of 3-m) diameter of the LEP tunnel (whose length is 27 km, not 13 km as wanted by others during the crucial years of the approval for the LEP Project) exists and represents the basic reason for the LHC magnets to be installed in the same LEP tunnel. The "dreams" sometimes become real facts.

These past achievements in project realization are mentioned in order to corroborate my optimism and enthusiasm in encouraging new actions and new ideas for the future of subnuclear physics in Europe and in the world.

The ELN Project at present is a "dream", even if, thanks to its existence, many results have been obtained in our real life.

We should never forget that "*There is no new Physics without new projects.*" Let me pay tribute once again to his extraordinary wisdom: John Bell was a theorist who greatly valued projects and experiments. The best gift I have received from John is his dedication of his book "Speakable and unspeakable in quantum mechanics", as shown in Fig. 30.36.

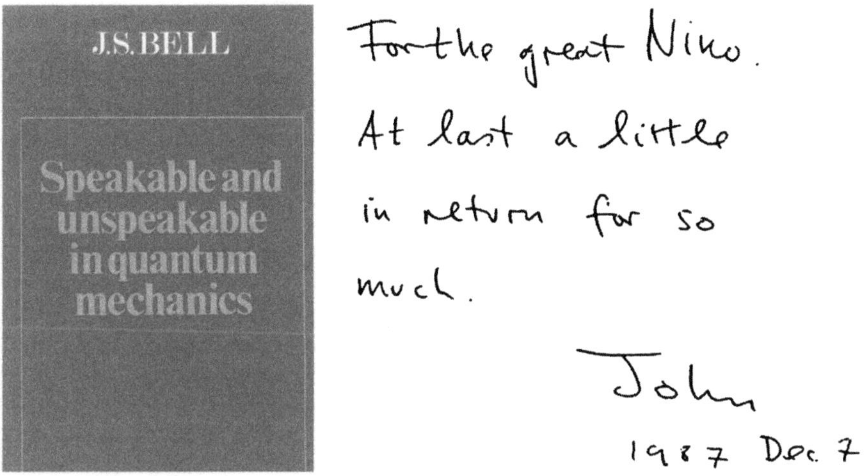


**Fig. 30.36.** Dedication made by John Bell in his book "Speakable and unspeakable in quantum mechanics"

## References

1. G. Charpak, L.M. Lederman, J.C. Sens, A. Zichichi: Nuovo Cimento **17**, 288 (1960)
2. G. Charpak et al.: '$(g-2)$ and its consequences'. In: *Proc. HEP Conference, Geneva, 4–11 July 1962* (CERN, Geneva 1962) p. 476; G. Charpak et al.: Nuovo Cimento **37**, 1241 (1965)
3. S. Gorodetzky, T. Muller, M. Port, A. Zichichi: Phys. Lett. **2**, 133 (1962); F.J. Farley et al.: 'A measurement of the $\mu^+$ lifetime'. In: *Proc. HEP Conference, Geneva, 4–11 July 1962* (CERN, Geneva 1962) p. 415
4. G. Brautti et al.: Nuovo Cimento **38**, 1861 (1965); G. Brautti et al.: 'A high-intensity enriched beam of kaons and antiprotons'. In: *Proc. HEP Conference, Dubna, 5–15 August 1964*, Vol. II (Atomizdat, Moscow 1966) p. 556
5. T. Massam et al.: Nuovo Cimento **39**, 10 (1965)
6. L. Maiani, R.A. Ricci (eds.): *The Discovery of Nuclear Antimatter, Proceedings of the Symposium to celebrate the 30th Anniversary of the Discovery* (Italian Physical Society, Bologna 1995) Vol. 53
7. D. Bollini et al.: Nucl. Instrum. Methods **81**, 56 (1970)
8. T. Massam et al.: 'A Telescope to Identify Electrons in the Presence of Pion Background'. In: *CERN Report 63-25* (CERN, Geneva 1963); D. Bollini et al.: Rev. Phys. Appl. **4**, 108 (1969); M. Basile, et al.: Nucl. Instrum. Methods **101**, 433 (1972)

9. A. Buhler et al.: 'Range Measurements for Muons in the GeV region'. In: *CERN Report 64-31* (CERN, Geneva 1964); A. Buhler at al.: Nuovo Cimento **35**, 759 (1965)
10. M. Conversi et al.: Nuovo Cimento **40**, 690 (1965)
11. M. Bernardini et al.: *A Proposal to Search for Leptonic Quarks and Heavy Leptons Produced by ADONE,* INFN/AE-67/3 (1967)
12. V. Alles-Borelli et al.: Nuovo Cimento **4**, 1156 (1970); M. Bernardini et al.: Nuovo Cimento A **17**, 383 (1973); A. Zichichi: R. Nuovo Cimento **4**, 498 (1974)
13. C.S. Wu, T.D. Lee, N. Cabibbo, V.F. Weisskopf, S.C.C. Ting, C. Villi, M. Conversi, A. Petermann, B.H. Wiik, G. Wolf, 'The Origin Of The Third Family'. In: *World Scientific Ser. 20th Cent. Phys.* ***20***, ed. by O. Barnabei, L. Maiani, R.A. Ricci, F. Roversi Monaco (World Scientific, Singapore 1998)
14. D. Bollini et al.: Nuovo Cimento A **61**, 125 (1969); D. Bollini et al.: Rev. Phys. Appl. **4**, 301 (1969)
15. D. Bollini et al.: 'Study of electromagnetic decays at CERN'. In: *Proc. Int. Seminar on Vector Mesons and Electromagnetic Interactions, Dubna, 23–26 September 1969* (JINR, Dubna 1969) p. 387; A. Zichichi: 'The basic SU(3) mixing: $\omega_8 \leftrightarrow \omega_1$'. In: *Evolution of Particle Physics* (Academic, New York 1970) p. 299; A. Zichichi: Ann. Phys. **66**, 405 (1971); M. Basile et al.: 'A review of the results obtained by the Bologna-CERN multipurpose NBC set-up'. In: *Proc. 3rd Int. Conf. Experimental Meson Spectroscopy, Philadelphia, USA, 28–29 April 1972* (AIP, New York 1972) p. 147
16. P.E. Kristensen, A. Zichichi: 'Project for a Laboratory to Produce Plastic Scintillators with Standard and Special Properties', CERN Internal Note, 20 October 1972 (CERN, Geneva 1972); S.A. Gabriele et al.: Phys. Lett. B **42**, 504 (1972); S.A. Gabriele et al.: Nucl. Instrum. Methods **108**, 431 (1973); A. Zichichi: 'Developments in plastic scintillators'. In: *Proc. Int. Conf. Instrumentation for High-Energy Physics, Frascati, 8–12 May 1973* (CNEN, Frascati 1973) p. 565; S.A. Gabriele et al.: Nucl. Instrum. Methods **113**, 465 (1973)
17. A. Zichichi: 'A summary of all quarks experiments'. In: *Proc. Int. Conf. Elementary Particles, Heidelberg, 1967* (North-Holland, Amsterdam 1968) p. 268; T. Massam, A. Zichichi: 'Quark search at the ISR', presented at: ISR Users' Meeting, CERN, Geneva, 10–11 June, 1968; M. Basile et al.: 'Is the absence of quarks as well established as is generally believed?'. In: *Proc. EPS European Conference on Particle Physics, Budapest, 4–9 July 1977* (KFKI, Budapest 1977) p. 335; M. Basile et al.: Lett. Nuovo Cimento **18**, 529 (1977); M. Basile et al.: Nuovo Cimento A **40**, 41 (1977); M. Basile et al.: Nuovo Cimento A **45**, 281 (1978); M. Basile et al.: Lett. Nuovo Cimento **29**, 251 (1980)
18. V.N. Gribov: In: *Effective Theories and Fundamental Interactions, Proc. Subnuclear Physics School 34, Erice 1996* (World Scientific, Singapore 1997) p. 500
19. M. Basile et al.: Phys. Lett. B **92**, 367 (1980); A. Zichichi: 'Multiparticle hadronic systems produced in high-energy (pp) interactions, and comparison with ($e^+e^-$)'. In: *Proc. EPS Int. Conf. High-Energy Physics, Lisbon, Portugal, 9–15 July 1981* (EPS, Geneva 1982) p. 1133; M. Basile et al.: 'What we can learn from high-energy, soft (pp) interactions'. In: *Proc. Subnuclear Physics School (Erice, 1981): The Unity of the Fundamental Interactions* (Plenum, New York 1983) p. 695
20. V.N. Gribov, G. 't Hooft, G. Veneziano, V.F. Weisskopf: *The Creation of Quantum Chromodynamics and the Effective Energy*, ed. by N.L. Lipatov, Academy

of Sciences and University of Bologna, INFN, SIF, World Scientific Ser. 20th Cent. Phys. **25** (World Scientific, Singapore 2000)
21. A. Zichichi et al.: 'The LAA project', Report N° 2, CERN/LAA, 25 June 1987 (CERN, Geneva 1987)
22. D. Acosta et al.: R. Nuovo Cimento **13**, 10 (1990); A. Zichichi et al.: 'The main achievements of the LAA project'. In: *Proc. Subnuclear Physics School (Erice, 1990): Physics up to* 200 TeV (Plenum, New York 1991) p. 327; A. Zichichi et al.: 'The Main Achievements of the LAA project', Report N° 7, CERN/LAA/91-1, 1 March 1991 (CERN, Geneva 1991)
23. M. Basile et al.: ELOISATRON (The European LOng Intersecting Storage Accelerator), INFN/AE-83/7, June 1983; INFN/AE-84/2, January 1984; Presented at the Galileo Galilei and Alfred B. Nobel Celebrations *Science for Peace*, San Remo and Rome, Italy, 1–11 May 1983; A. Zichichi: Nuovo Cimento **21**, 12 (1998); A. Zichichi, 'Why 200 TeV, Opening Lecture'. In: *Proc. of the INFN ELOISATRON Project on New Technologies for Supercolliders, Erice, 1990* (Plenum, New York 1991) p. 1
24. A. Zichichi: 'Antimatter'. In: *Proc. 8th Int. Workshop on Neutrino Telescopes, Venezia, 23–26 February 1999*, ed. by M. Baldo Ceolin, (University of Padova 1999) p. 287
25. F. Anselmo, L. Cifarelli, A. Petermann, A. Zichichi: Nuovo Cimento A **105**, 1179 (1992)
26. A. Zichichi: 'Universality properties in non-perturbative QCD'. In: *Proc. Subnuclear Physics School (Erice 1985): Old and New Forces of Nature* (Plenum, New York 1988) p. 117
27. T.D. Lee, G. 't Hooft: private communication
28. A. Zichichi: *Lecture Notes on Subnuclear Physics* (Bologna University, Bologna 1983)
29. F. Anselmo, L. Cifarelli, A. Zichichi: Nuovo Cimento A **105**, 1335 (1992)
30. A. Zichichi: *Subnuclear Physics – The First Fifty Years: Highlights from Erice to ELN*, ed. by O. Barnabei, P. Pupillo, F. Roversi Monaco, Academy of Sciences and University of Bologna, World Scientific Ser. 20th Cent. Phys. **24** (World Scientific, Singapore 2000)

# Subject Index


4 o'clock tea, 35

aberration formula, 420
absolute simultaneity, 427
action-at-a-distance, 67f., 84, 192, 202
Aharonov–Bohm effects, 369
analyzer quality, 114
anomalous
– anomalous coincidences, 93f.
– anomalous commutators, 378
– anomalous symmetry breaking, 377
anomaly, 458
– axial anomaly, 379
antideuteron, 436
APS policy, 72
asymmetry term, 171ff.
atheist, 5
atom based experiments, 109ff.
atomic radiative cascades, 137, 147
automorphism, 263

BCH inequality, 105ff.
BCHSH inequalities, 128
beables, 43, 58, 89f., 290, 311
beam attenuation, 360
beauty system, 180
Bell inequalities, 104
– Bell inequalities in massive systems, 164ff.
Bell's theorem, VIII, 51, 54, 61ff., 76, 120, 129, 155, 244, 259
Bell's Whisky, 47
Bell–Jackiw–Adler anomaly, VII
Bell-basis, 245
Bell-state measurement, 233, 248
Bertlmann's socks, 43
blackbody radiation, 343
Bohm's model, 52
Boulder experiment, 110
Broglie–Bohm model, 25

calorimeter, 441
Cauchy–Schwarz inequality, 94
CH inequality, 84ff.
changeable, 311
charge conjugation, 167
charge operator
– charge operator in the soliton sector, 388, 393
– charge operator in the vacuum sector, 387
– partial charge operator, 393
– partial charge operator with fuzzy edges, 394
Chern–Simons terms, 380
CHSH inequality, 81f., 104, 166, 176f., 214
classical physics, 133
Clebsch parameterization, 381
clock synchronisation, 427
clock synchronisation factor, 415
coherence
– spatial coherence, 337
– spectral coherence, 337
coherence function, 353
coincidence rates, 106
collapse of wave-function, 15
collisions with gas molecules, 345
complementarity, 242
completeness, 66ff., 209
consistency, 28
contextual hidden variables models, 54
continuous spontaneous localization model (CSL), 300f.
Copenhagen interpretation, 15, 63

correlation
– composite correlation function, 409
– correlation coefficient, 122, 133
– correlation function of the (rest frame) electric field, 409
– correlation function of the magnetic field, 404
– perfect correlation, 78
$CP$ eigenstates, 169
$CP$ parameter $\varepsilon$, 169, 178
CPLEAR experiment, 169f.

dark matter, 466
decay processes, 291
decoherence parameter $\zeta$, 172, 180
decoherence theory, 340
degenerate ground states, 388
density matrix, 274
Descartes, René, 26f.
detection efficiency, 115, 156
determinism, 83, 131
deterministic system, 310f.
deterministic theory
– entanglement, 314
– quantum coherence, 314
– quantum interference, 314
Diósi's scheme, 320
diatomic molecule, 111
Diffie–Hellman key exchange protocol, 211f.
diffraction
– diffraction at a grating, 339
– single-slit diffraction, 337
Dirac equation, 386
double-slit experiment, 242, 333
duality, 30
– duality principle, 361
– global duality, 32
– local duality, 32, 33f.
dynamical reduction program (DRP), 287

$E_8$ lattice, 264
eavesdropper, 215, 251
eavesdropping analysis, 216
effective
– energy, 445, 457
– temperature, 409
Einstein locality, 160, 227
Einstein's causality, 132
Einstein's synchronization, 423
electrons
– electrons in a storage ring, 407
– linear acceleration of electrons, 403
element of reality, 83, 124, 186, 209, 243
ELN Project, 471f.
Emeleus, Karl George,. 10
energy conservation, 94
– point-wise, 94
– statistical average, 94
entangled state, 121, 134
entanglement, 68, 149
– key distribution, 213
– purification, 218
– swapping, 249
Einstein–Podolsky–Rosen (EPR), 17, 67, 103f., 209
EPR–Bohm Gedankenexperiment, 121
equivalence classes of states, 312
equivalent transformations (ET), 413
evolution of gaugino masses, 455
Ewald, Peter Paul, 12
– Ewald sphere, 12
expectation value, 391
experiments
– four-channel experiments, 88
– two-channel experiments, 88

fair-sampling assumption, 142
Faraday, Michael, 18ff.
Fizeau phase shift, 356
flavour mixing, 462, 465
four-particle-state, 284
four-photon entanglement, 230
fractional eigenvalues of the Number operator, 388
fractional quantum Hall systems, 398
Fuchs' counterexample, 275
fullerenes, 336
Furry's hypothesis, 172

gauge couplings, 469
general covariance, 323
geometrical effects, 368
Ghirardi–Rimini–Weber (GRW)
– jumps, 297
– model, 292
– theory, 287

Gleason's theorem, 53, 259, 268
gravitational phase shift, 347, 355
gravitational self-energy, 324
Greenberger–Horne–Zeilinger (GHZ)
– contradiction, 228
– multi-photon entanglement, 229
– setup, 221
– states, 246
– theorem, 100, 226
– three-particle entanglements, 100
– three-photon entanglement, 233
– three-photon state, 226
ground state, 312
– ground state in the soliton sector, 388, 393
– ground state in the vacuum sector, 386

Hadamard transformation, 275
Hamiltonian density, 313
heavy lepton, 437
Heisenberg cut, 16
Heisenberg Uncertainty Principle, 15, 52
hidden variable theory, 17, 75f.
hidden variables, 103f., 124f.
hierarchy, 468
Higgs particles, 451
homomorphism, 261

imaginary masses, 451
inertial transformations, 422, 425f., 426
information, 187f., 245, 272
– loss, 312
instantons, 459
interactions
– electric interactions, 347
– magnetic interactions, 346
– quasi-static interactions, 346
internal clock, 348
Intersecting Storage Ring (ISR), 443
ion trap, 159

Josephson effect, 359
Jupiter, 417

Kerr, Lesley, 11
key distribution problem, 211
Killing vector, 321
kinetic helicity, 380
kink solution, 385
Klein configurations, 264f.
knowledge, 272, 279
Kobayashi–Maskawa (KM) matrix, 463
Kochen–Specker (KS) theorem, 259
Kocher–Commins experiment, 80
Kogut–Susskind lattice regularisation, 390

LAA Project, 447
lattice chain, 396
local realism, 227
local realistic theory, 84, 234
locality, 26, 209, 259
– assumption, 131
locally deterministic approaches, 309
loophole
– communication loophole, 108
– detection efficiency loophole, 108
– detector-efficiency loophole, 86
– locality loophole, 95, 115f.
– spatial correlation loophole, 107
Lorentz invariance, 296

Mössbauer crystal, 324
macro-objectification problem, 289
magic moments, 35ff.
– ratio of moments, 38
Maharishi Effect, 56ff.
Majorana neutrinos, 465
many-world interpretation, 189
mass eigenstates, 169
matter–antimatter symmetry, 454
maximum violation, 129
measurement, 14ff., 187, 197, 290
measurement problem, 201
Mermin's inequality, 234
Michelson-type experiments, 416
mind-body duality, 196
missing mass, 466
multihadronic final states, 445
multiparticle states, 283
muon
– muon magnetic moment, 432
– muon weak charge, 433

von Neumann measurement, 191
von Neumann projection postulate, 189

von Neumann's proof, 66, 258
von Neumann's theorem, 53
neutral kaons, 167
neutral-kaon system, 163
neutron
– interferometer, 282, 351
– interferometry, 241
– optics, 281
– properties, 354
no-enhancement assumption, 86
non-bubble-chamber (NBC), 441
non-contextuality, 259
non-local interactions, 76
non-perturbative QCD effects, 457
noncommutative geometry, 58f.
nonlocality, 22, 26, 149
nonrelativistic moment, 37
nonseparability, 22
normalization, 87f.

objective local theory, 84
objective reduction of the quantum state (OR), 326
observable, 268
one time pad, 210
one-channel polarizer, 136
ontological states, 311
ontologically interpretable, 24

parametric down conversion, 146
parents, 7
Paris experiments, 109
parity operator, 167
partial boolean algebra, 261
particle-like character, 92f.
Peierls' condition, 277
Peierls' criterion, 274
Peres–Mermin proof, 266
phase shift, 352
phase-echo method, 366
pilot wave model, 202
pilot-wave theories, 75
Planck scale, 309
plastic scintillator, 440
Plato, 194
pointer states, 191
polarization, 410
polyacetylene, 389
postselection procedures, 362
potential of Bell and Bertlmann, 42
pre-potentials, 381
predeterminism, 315
predictable measurement, 190
probability
– à priori probability distribution, 64f.
– classical probability, 63
projection postulate, 269
public key systems, 211
purity, 113

quack, 72
quanglement, 327
quantum cryptography, 200, 251
– polarization-entangled quantum cryptography, 252
quantum dense coding, 247
quantum fluctuations, 308
quantum fluid dynamics (QFD), 458, 462f.
quantum gravity, 319
quantum jumps, 193
quantum mechanics
with spontaneous localizations (QMSL), 292
quantum privacy amplification, 218
quantum teleportation, 248, 327
quantum-state reconstruction, 368
quarkonium, 29
quasispin eigenstates, 169
qubits, 209

Rabi resonance flipper, 358
radiation theory
– neoclassical radiation theory, 90f.
– semi-classical radiation theory, 86, 89
rationality, 27
Rayleigh scattering, 344
realism, 26, 28, 61, 65
reality
– empirical reality, 25
– Mind-Independent Reality, 24
reduced density matrix, 341
reduction of the state vector, 123
reduction principle, 190
relativistic quantum field theory, 431
relativistic reduction mechanisms, 303
relativity, 204
– general relativity, 58, 307

– special relativity, 58, 84, 307
RSA encryption system, 211f.
Russel, Bertrand, 194

Sagnac effect, 347, 424
Sagnac term, 355
scaling, 441
Schopenhauer, Arthur, 193
Schrödinger cat-like state, 364
Schrödinger–Furry hypothesis, 68, 74, 79, 91
second quantization, 307
selection rules, 112
self-contradictory, 285
self-interference, 354
shimming technology, 432
Sloane, Richard, 10
soccer player, 333
Sokolov–Ternov effect, 405
soliton sector, 385
Solvay Conference, 63
spaceships, 422
spatial correlation, 113
spatial separation, 160f.
spin polarization, 405
spin superposition, 357f.
spin-singlet entangled state, 74
spin-singlet state, 164
Split-Field-Magnet, 443
spontaneous symmetry breakdown (SSB), 450
squeezing, 365
standard religion, 69f.
state, 268
stationary world line, 406
stochastic formalism, 300
strangeness number, 164
string theories, 308
student at Queen's, 10
superluminal signals, 424
superposition, 288
superspace, 471
supersymmetry (SUSY), 450
– breaking, 455
supplementary parameters, 124f.
switching, 161
switching of polarization
– periodic switching of polarization, 143
– random switching of polarization, 148
symmetry, 268
– $4\pi$ symmetry, 357

technician at Queen's, 8f.
teleportation, 149
temperature formula for a black hole, 404
theory of everything, 195
theory of special relativity (TSR), 413
thinking, 23
time variable, 310
topological effects, 368
topological phases, 369
transversality, 378
trigger mechanism, 293
twins, 423
twistor theory, 330
two-channel polarizers, 141
two-particle interferometry, 100
two-photon selective excitation, 138

unitary time evolution, 175, 176
Unruh effect, 401, 411
Unruh temperature, 402

vacuum sectors, 385
vegetarianism, 8
vegetarians, 4
velocity of light
– one-way velocity of light, 414
– two-way velocity of light, 415
Venus, 420
vibrational transitions, 342

W-states, 246
wavefunction, 296
Wigner functions, 367f.
Wigner's theorem, 268
Wigner–Weisskopf approximation, 168
Wigner-type inequality, 178
Wittgenstein, 196
Witting polytope, 264f.

Zeno-type Bell inequalities proof, 222
zero-modes, 392

Location: http://www.springer.de/phys/

***You are one* click *away***
***from a* world of physics *information!***

***Come and visit* Springer's**

# Physics Online Library

### Books

- Search the Springer website catalogue
- Subscribe to our free alerting service for new books
- Look through the book series profiles

**You want to order?** **Email to: orders@springer.de**

### Journals

- Get abstracts, ToC´s free of charge to everyone
- Use our powerful search engine LINK Search
- Subscribe to our free alerting service LINK *Alert*
- Read full-text articles (available only to subscribers of the paper version of a journal)

**You want to subscribe?** **Email to: subscriptions@springer.de**

### Electronic Media

- Get more information on our software and CD-ROMs

**You have a question on an electronic product?** **Email to: helpdesk-em@springer.de**

**Bookmark now:**

**http://www.springer.de/phys/**

Springer

**Springer · Customer Service**
**Haberstr. 7 · D-69126 Heidelberg, Germany**
**Tel: +49 6221 345 200 · Fax: +49 6221 300 186**
d&p · 6437a/MNT/SF · Gha.

MIX
Papier aus verantwortungsvollen Quellen
Paper from responsible sources
FSC® C105338

If you have any concerns about our products,
you can contact us on
**ProductSafety@springernature.com**

In case Publisher is established outside the EU,
the EU authorized representative is:
**Springer Nature Customer Service Center GmbH**
**Europaplatz 3, 69115 Heidelberg, Germany**

Printed by Libri Plureos GmbH
in Hamburg, Germany